PERMISSION TO RECOVER

BY
CHEYNE CURRY

Bossy Pants Books

PERMISSION TO RECOVER

NOTE: If you purchased this book without a cover, you should be aware that it is stolen property. It was reported as "unsold and destroyed" to the publisher, and neither the author nor the publisher has received any payment for this "stripped book."

This is a work of fiction. All characters, locales and events are either products of the author's imagination or are used fictitiously.

PERMISSION TO RECOVER

Copyright © 2024 by Cheyne Curry
www.cheynecurry.com

All rights reserved. No part of this book may be reproduced in any manner whatsoever without written permission from the publisher, save for brief quotations used in critical articles or reviews.

Front and Back Cover design by AnnMcMan

A Bossy Pants Book
Published by Bossy Pants Books
Columbus, Ohio 43229
bossypantsbooks@gmail.com

ISBN-13: 978-1-945124-05-1

First Edition, March, 2024 (copyrighted with Library of Congress in 1987)

Printed in the United States of America and in the United Kingdom

ACKNOWLEDGMENTS

Editor Nat Burns and contributing editors Renae Hunt and Day Peterson. They did their jobs, there were just certain things I was too stubborn to give into.

Karen Badger and Barbara Sawyer from Badger Bliss Books for being fabulous (and formatting)

The one and only Ann McMan and Treehouse Studio, Famous Author (AMFA), for her cover design.

I know I'm forgetting people (because it's taken so damned long to get this book out) so here's a pre-emptive apology in case one of those people is you.

DEDICATIONS

My lovely, understanding, tolerant and former Army MP wife, Brenda.

Renae, who, since we started working on this book, had her only daughter, Claire, graduate junior high school, high school, university and now has two grandchildren from her.

Heidi, who started it all.

My Army buddies Gail Trabucco, Cindy MacDonald, Denise Carlson and Dennis Dale

And all the women who served before, during and after the OSUT experiment.

For Paladin

AUTHOR'S NOTES:

One Station Unit Training was real. The basic training courses and Military Police School drilling and instruction were real for that time. The women's struggles and triumphs were real. The story is and the characters are fiction.

This is not an Uber or converted fan fiction. It was originally copyrighted with the Library of Congress in 1987. It was first submitted online in 2009 and has been adjusted and refined over the past several years.

I know there is a lot of narrative in this novel. I also know that there are times I switch into omniscient voice. Don't blame the editors. They tried. If I'd written all that out to accommodate Show, Don't Tell, the book would have been 1,000 pages or more and I would have had to label the paperback, The Hernia Edition. Even though this is mostly Dale's story, it cannot be told from just one POV. I've been majorly stubborn about this project, meaning there won't be novel norms in this book. Dates are not spelled out because you don't do that in the military. Acronyms are everyday military jargon. I try to explain them when I can. Some may have to be looked up. With that in mind, I hope you enjoy this story and that female vets find something in the story that clicks with their training experiences.

ALSO WRITTEN BY CHEYNE CURRY AND OFFERED BY BOSSY PANTS BOOKS:

Clandestine

The Tropic of Hunter

Renegade

The End - Book One of The Sanctuary Series (co-written by Roselle Graskey and Cheyne Curry)

The Resistance - Book Two of The Sanctuary Series (co-written by Roselle Graskey and Cheyne Curry)

CHEYNE CURRY

PART ONE

PERMISSION TO RECOVER

CHEYNE CURRY

PRELUDE

The two uniformed figures moved swiftly up the hill toward the enemy bunker, silently giving one another hand signals. The ominous sounds of combat surrounded the duo as they closed in on the small, camouflaged and reinforced shelter.

They knew that, down in the village, soldiers were overtaking the enemy ground. A young captain had been ordered to lead a group of twenty to surround two buildings until they were precisely positioned to storm the inside. They, no doubt, awaited their captain's signal.

On the hill, there were a total of five bunkers and each one was in the process of being stormed by at least two soldiers. The attacking soldiers gave visual signs as they utilized their three-to-five second rush to cover. They had almost made it all the way to the bunker when an enemy soldier stuck his head above the barricade to scout the area just in time to see Private Oakes dive down for concealment. Landing next to Private Bradshaw, as gunfire besieged them, Oakes muttered quietly. "He saw us. Guess we need to do it by the book now."

"Yeah, we might as well go for it. They've probably got us on closed circuit TV somewhere." *Bradshaw rolled, hiding behind a tree. She readied herself, getting in position to move. She nodded at Oakes.*

"Ready!" *Oakes yelled.*

"Cover me!" *Bradshaw yelled back.*

"You're covered!"

"Moving!" *Bradshaw leapt up, ran forward then hit the ground again in five seconds. The closeness of the bunker evident, Bradshaw looked back at Oakes.* "Bunker go!"

"Cover me!"

"Shifting fire!"

"Moving!" *Oakes rose and rapidly moved through the*

trees, to the hill fortress, then leaned against the side wall.

"You lost, ladies, we saw you fifteen feet ago," the enemy GI said.

Bradshaw joined Oakes on the hill.

"Sacrificial lambs, boys," Oakes said with a smile. "Your objective was to shoot us. And you were too late." She tucked a stray, dark hair out of view inside her helmet.

"Bullshit. We did shoot. Fifteen feet ago."

"That's exactly my point." Oakes gestured behind them. "Revell and Schwartz had their rifles pointed down your necks at twenty feet."

The two men turned to see Revell and Schwartz standing behind them, M16s aimed right between their eyes.

"So how come you didn't shoot?" the GI asked.

"We said bang," Revell said.

"Bang? You said bang?" He smirked at his fellow GI. "Big deal. What, are you afraid to fire the big bad gun?"

Bradshaw fired off one of her blank rounds at a sandbag. The sound made the two men jump and look back at Bradshaw as if she were a lunatic. Then they looked at the olive-green sandbag, which was now singed black, smoke curling up from the charred canvas.

"I didn't know you were into pain," Bradshaw commented, amused.

The male GIs exchanged looks and broke into laughter. "Lieutenant Wilder's gonna shit," one of them said, shaking his head in disbelief.

Schwartz, who had since relaxed her grip on the M16, offered them one of her cap grenades. "You could always blow yourselves up and save yourselves a worse death later."

Dale Oakes awoke when Bojangles, her cat, jumped on her bed. She had been reliving the Army again, this time an early training exercise between her unit and a male basic training company.

Once the cat stopped kneading and curled up on the pillow next to her, Dale rolled onto her stomach, sighed and stared at the wall. Within minutes, she had drifted back to sleep.

The vehicle sped around the corner of the unpaved road, tires grinding against the gravel. Headlights highlighted dust and dirt loosened by the chassis that fishtailed in its attempt to quickly get where it was going.

The sedan screeched to a halt at the most secluded part of the rifle range. The back door flew open and a bloodied body hit the ground, motionless.

The door slammed shut and wheels spun on the dry, soft soil. The rear tires slid laterally enough to alter the position of the car. As the victim's foot was crushed under the tire, everything went black.

Dale bolted upright. She was drenched with sweat and could not catch her breath. The panic and terror that now surged through her was just as real as it had been in her dream. It took her a moment to realize that she was at home and safe in her own bed.

She threw back the covers and sat on the edge. She held her stomach tightly and waited until the waves of nausea passed.

After several minutes, Dale rose and shuffled to the kitchen. She wondered if she would ever stop having this recurring nightmare and laughed ruefully. Maybe if it hadn't actually happened. Maybe if she could recall all of the details.

Maybe she was better off with the amnesia. She breathed deeply. Sleep would elude her for the rest of the night.

CHAPTER ONE

October 1977 Vermont

It had taken no time at all for Dale Oakes to adjust to civilian life. The Department of the Army had sent her home on convalescent leave to await medical discharge. The fact that it was the United States Army's time and money just made it all the easier. And, if the Army tried to contact her for one reason or another, she had left specific instructions with whoever answered her phone to pass on the message that she had dropped dead. She had even promised not to haunt them so there was no chance they could attempt to repossess her soul for more active duty.

Her military superiors, however, knew exactly how to get a response from her. Send Lieutenant Colonel Anne Bishaye. Dale knew whenever Bishaye appeared anywhere near her, it was due to something Dale had done, hadn't done, or was going to have to do. Dale had spent many a time sliding down drainage pipes and climbing out windows just to avoid confrontations with the colonel.

She didn't dislike the colonel. Quite the contrary, she held an enormous amount of respect for her. Anne could probably give orders for Dale to fry in hell and have them carried out. There was also that little thing about Dale's dormant infatuation with her former commanding officer.

Dale suspected something was brewing when she got the message to meet LTC Bishaye at the Rutland Airport at 1400 hours. She considered ignoring the call but Dale knew the Rutland Airport was no place to abandon a stranger, in fact, just being *at* that airport was like being stranded.

The main thing that convinced Dale to meet Anne was the reassuring thought that she had done nothing lately to earn the colonel's wrath. And that her forthcoming discharge prohibited her from going back to work.

    Colonel Bishaye was easily recognizable as she stepped off her husband's privately owned jet. Dale automatically smiled at the sight of this striking woman approaching her in her Class-A uniform. She had hoped her former company commander had lost some of her appeal since the last time they had seen one another. No such luck. Dale's pulse sped up as the colonel got closer. Several deep breaths brought her heartbeat under control.
    Anne, at forty-five years of age, was a tall and extremely fit woman, whose healthy complexion always seemed tanned, regardless of her location or the time of year. Her short, burgundy-colored hair framed chiseled features and dark blue eyes. She was, sometimes to her disadvantage, naturally and breathtakingly beautiful. As Anne approached the fence, Dale felt those blue eyes studying her.
    At five feet, five inches tall, Dale was four inches shorter than Anne and she was acutely aware of that height difference as the colonel now stood in front of her.
    "So, like, am I supposed to salute you or what?" Dale was the first to speak. She swept her long, dark hair off her neck with her hand.
    Anne's smile broadened into a contagious laugh that made Dale's heart flutter. "Why bother? Military courtesy was never a strong point of yours before. Don't go getting formal on me now, Dale, I don't think I could handle it."
    "Well, then, you'll forgive me if I'm just a little bit careful until I find out why you're here. If you had arrived in civilian clothes, you might have gotten away with making me think you were in the area and just dropped in to say hi. Instead, you're strutting around like a decorated pine tree and that tells me to proceed with extreme caution."
    They silently appraised one another, Anne's eyes revealing nothing. Dale didn't know if that was unfortunate or not.
    "I'm surprised you showed up at all. This is one of the few times I haven't had to go looking for you. Or are you trying your hand at diplomacy."
    "Why should I be diplomatic? I'm with you," Dale said, as

they walked toward her car.

Anne chuckled and slipped an arm around Dale's shoulder and pulled her close, briefly executing a binding hold as they walked. The contact made Dale shiver. "You're impossible. You haven't changed."

Releasing Dale, Anne fell a half step behind her. "You're walking well, Dale. I'm very pleased. How's your foot?"

Turning, Dale responded. "Why do I feel that question is like a hangman asking me on the strength of my neck?" The position of the sun in her eyes made it impossible see Anne's expression.

"You're talking as though you don't trust me."

"I trust you. I just don't trust me especially when you're around. You do strange things to my head. You can talk me into doing insane things because you make them sound plausible. You could make the Spanish inquisition sound like a trip to Disneyland, expenses paid."

"I didn't know I was that good."

"Yes, you did," Dale said, smiling. They stopped walking when they reached Dale's old, dusty, years-old, Datsun 240Z "What are we doing now? Are you hungry?"

"I'm a little hungry," Anne admitted.

Dale pointed to a restaurant at the top of the drive that led to the airport. "We can get something there, if you'd like." She glanced back at the jet. "Where's Jack?"

"He's got a ton of phone calls to make at the terminal while we're here," she said, "He won't be joining us."

"How long are you staying?"

Anne looked at her watch. "About an hour or so. I've got to be back at McCullough by twenty hundred hours. That's eight o'clock in case you've lost your military bearings already."

Dale smiled but her suspicions were confirmed. Anne didn't take a day off to go sky hopping from Alabama to Vermont just because she missed her young colleague and had a sudden desire to reconnect with her. While that would have been nice, Dale knew it wasn't true.

After they were seated at a booth inside the small country restaurant and their orders were taken, Anne seemed to relax.

Dale, however, was reluctant to let her guard down.

"So...how is your foot?" Anne asked again.

"You already know. You have too many spies— I'm sorry, I mean *connections* —not to know."

"I want to hear it from you." They locked stares. Anne broke the intense gaze when the glass of iced tea she ordered was brought to the table. She waited for the man to leave. "Why did you stop working out with that physical therapist I hired for you?"

Dale studied her. "You know, I hate to put this thought into your head but I actually think you've gotten better looking in your old age, if that's possible. And if I didn't like you so much, I really wouldn't like you much."

Dale looked down at the table and played with her folded napkin. One word that was rarely used to describe Dale was shy unless she was under the concentrated examination of Anne Bishaye. Although it remained unspoken, it seemed both women were aware of the effect that Anne had on Dale.

"Old age, huh?" Anne said, suppressing a smile. She dipped her fingers into her water glass and flicked them at Dale. "Stop trying to change the subject. Answer me."

"He was a dick. I worked it out on my own."

"Dr. Solberg is one of the best in the state of Vermont, if not the entire VA system. You're supposed to be working out with him every day," Anne reprimanded softly.

"Why?" Dale was irritated. "He's a retired Army doctor. He gave me ultrasound, made me do some circular exercises, put me in a whirlpool, then made me do push-ups against a wall, not to mention the fifteen times across the floor with the tippy-toe-to-heel bullshit. I did that exact same thing when I had tendonitis and it didn't help that, either. I don't need him. It's a lot better. I'm fine. I'm not even limping anymore," she protested. Then, without missing a beat, she leaned back in her chair. "Did I just tighten my own noose?"

Anne laughed, took a sip of tea and paused, as though considering. "Is your foot fine enough for you to go back to work?"

"I knew it." Dale made a gesture like she was hanging herself.

"It's important, Dale, I need you for a job." Anne's tone was gentle yet firm.

Dale sat up, crossing her arms. "I don't do jobs anymore. I'm being discharged, remember?"

"You're not out yet, Lieutenant Oakes," Anne reminded her. The colonel must have seen the instant recognition of betrayal in Dale's expression and looked down at the table.

"Are you pulling rank on me, Colonel Bishaye? Because if you are, you'll forgive me if I tell you exactly where to stick your oak leaves."

"Forgive you? Need I remind you of everything I've already forgiven you for? Or do you have selective amnesia about your younger years in my company?" There was a knowing smirk rising on Anne's lips.

"Hey—" Dale began but stopped when Anne held up her hand in a halting motion.

"I know, I know. Your calamities usually happened after a night of drinking all the guys in second platoon under the table in the name of female GIs everywhere and, obviously, I am the *only* one who doesn't understand the significance of that."

"Damn. I wish you were the one with selective amnesia."

Dale suddenly remembered who she was dealing with. Anne was not a total stranger, nor one of the new drinking buddies she had acquired since her arrival home. Those whom she could dazzle with her war stories or intimidate with her incomparable experiences. This was the woman who started it all, the one who overlooked Dale's mischievous side, saw Dale's potential as a good undercover police officer and gave her a push in the right direction.

This was the woman who ignored the military's fraternization rule between officers and enlisted personnel and became Dale's friend at a time when she needed one the most, sparking an unrequited crush that Dale went to great lengths to hide. This was the woman who encouraged her to complete her college education and the one who came to her rescue by hiding her away in Officer's Candidate School when Dale's life was being threatened.

She was the person Dale saw when she awoke at the hospital after what the military deemed an unfortunate mishap, resulting in Dale's impending discharge. This was a woman who knew and understood Dale better than anyone, a woman Dale honestly idolized, all sexual confusion aside. Suddenly Dale struggled with her impulse to draw Anne into an argument.

As usual, Anne clearly saw through her and tried to diffuse the tension with a disarming smile. "You want to piss me off, don't you? You want to yell and scream and provoke me into yelling and screaming back."

Dale said nothing.

"Don't waste your energy. You know it won't work. Dale, look, I don't have much time and I have something important to talk to you about."

"If it involves the Army, save your breath because I'm not interested."

"Actually, you don't have much of a choice," Anne said, her smile disappearing.

"Yeah, huh, I'm coming to that conclusion," Dale said quietly.

"Hey," Anne said, softly. "Look at me." She waited until Dale's eyes met hers. "Do you honestly think I would involve you in anything that would get you hurt again? Do you?"

Dale broke the gaze and looked away from the table, shaking her head. "I guess not. I don't know. I don't know what to think anymore."

"If this was something bigger, I wouldn't have the authority to ask you to do it. But as it just affects my immediate area, they have given me the choice of several Criminal Investigation Division agents. I chose you."

"But I'm not available. I'm medically unfit." Dale said, a feeling of dread washing across her. "Aren't I?"

Anne sat back and waited while the owner of the restaurant placed her salad order before her. When he left, she shook her head at Dale. "No, not anymore, CID has been keeping close tabs on you. They have reports, not confirmed by any doctors, but reports that you've recuperated. You're rehabilitated. Which makes you available until the Department of the Army comes through with your discharge," Anne glanced at Dale. "Which could be months."

Dale leaned forward. "That's not fair!"

PERMISSION TO RECOVER

"Uncle Sam usually isn't fair, dear. You should be used to that by now."

"But this isn't Uncle Sam, *dear*, this is you. What's going on here?" Dale's question was greeted with silence. "Don't do this to me, Anne. I want to get out. I have to get out. I've had it."

"Look, Dale," Anne began, calmly. "If you don't agree to this job, they'll be calling you back for something else. This way we can work together. I can keep an eye on things. If you're working on this case, at least I know the job will get done. I trust you, Dale, and I'll keep you safe. You've proved yourself on much more dangerous assignments. Besides, I've already taken some of the pressure off. You'll be working with a partner." Anne was obviously waiting for a reaction and Dale did not disappoint her.

Eyeing Anne dubiously, Dale looked around the restaurant with exaggerated vigilance. She leaned in even closer and spoke barely above a whisper. "I don't know about you but I feel as though I should bend over and prepare for an attack of the Green Weenie." Dale tried to diffuse her growing anger by joking. She wasn't prepared to deal with Anne possibly being a sellout.

Anne smiled at the familiar expression. "Oh, it's just a small attack, you'll get over it."

Dale ignored Anne's flippant response. "Just what is this job?"

Anne seemed visibly relieved by Dale's interest. "You'll love it. It's cut out for you."

"Really? Well, if it's cut out just for me, how come I wouldn't be working alone?

"I'll tell you. We've been having trouble with one of our training companies at McCullough. For the past three training cycles, several of our most respected drill sergeants have been accused of sexual misconduct. We think they have been set up. Alpha-10 —"

"Alpha-10? That's my old training unit. I didn't know you'd been given command of 10th battalion."

"Yes, I know," Anne replied. "Try calling me sometime. That way you won't be surprised as much."

"The phone rings both ways, you know."

"Not at your end. You rarely answer your phone,

remember?"

"And with good reason." Dale finished her coffee. "Okay. What's going on?"

"10th battalion has always had some of the best MP and law enforcement training companies on post, you know that, and their drill sergeants have been some of the finest. Sure, you get a bad one every now and then, you can't avoid it, but —"

Dale rolled her eyes and said, "Please. Spare me the commercial. This is me you're talking to."

Anne nodded. "All right. Women have been pressing charges against at least two drill sergeants every cycle. I was disappointed the first time it happened, disgusted the second and very suspicious by the third. It's now the opinion of battalion headquarters and the provost marshal that somebody's got a grudge against the unit and is setting these men up. After these kinds of charges are brought up against them these men might as well kiss their careers goodbye not to mention what it does to their personal lives."

"So, you would want me and this other agent to go through AIT as new recruits to find out what's going on, right?"

"Basic and AIT," Anne corrected.

"Basic?"

"We have this new program called OSUT. It stands for One Station Unit Training. It starts with basic training and goes right through to the end of Advanced Individual Training."

Dale moaned. "What about coming in as an insert after basic?"

"Dale, you couldn't possibly get to know these people as well as you'd need to if you came in after basic training. Not only that, since we're training males and females together for the first time, it's essential you get in there from the beginning. We're expecting some of this bullshit to start the last few weeks of basic training when the females gain a little more freedom."

"Wait, wait, what? You're going to train men and women together, side by side? No separation through basic?"

"The only separation will be billeting. The company will be integrated and platoons will be arranged by alphabetical order."

Dale felt her competitive nature awaken.

Anne obviously noticed it, too. "Oh, now you're

interested?"

She hated that Bishaye could read her so well. She attempted to act indifferent. "It still doesn't sound to me as if it couldn't be handled by one agent, preferably the other one."

"The drills know we put spies in some cycles, they just don't know who or when. But if they make one of you there's still the other one."

"What appropriate phrasing under the circumstances."

"I'm not trying to be clever. I'm serious. We had two plants in there last cycle, a man and a woman, and they didn't catch anything."

"Well," Dale said and laughed. "They're luckier than most trainees."

"Dale..." Anne's tone warned.

"Maybe there isn't anything to catch."

"Yes, there is," Anne insisted. "The incidents slipped by us because we didn't know what we were looking for. We didn't suspect a set up then."

"We? You have fleas?"

She ignored Dale's sarcasm. "Me, Alpha-10's training officer and Raymond Sedakis, the regional CID commander. The post commander also agrees with us."

"I'd need a hell of a lot more evidence than coincidence to make me sit up and take any notice."

Anne looked across the table at Dale and, said in a teasing tone, "I'm surprised you didn't sit up and take any notice at the mention of Sedakis' name. He sure remembered yours."

"Sedakis...Sedakis..." Dale concentrated. "That does sound familiar. Do I know him?"

"Oh, you've met him. Colonel Sedakis used to work in cooperation with the provost marshal's office at Fort Ord. One night when he was coming through main gate, he was late for an appointment and he didn't slow down enough for your liking. You somehow managed to let his left headlight run into your nightstick and then you cited him for speeding and having a headlight out."

"Oh, yeah, him," Dale nodded. "Now I remember. That was Sedakis?"

"Yes and, at the time, that was his brand-new Mercedes,

too."

"Well, he should have slowed down like everyone else has to. Just because he's a high-ranking officer doesn't give him special privileges and just because he's high up in CID doesn't mean he can break post rules. He should be setting an example for the rest of us, not making himself an exception. You know, that's what used to irk me about that place, you were burned if you didn't do your job and burned if you did it to the wrong people. So, old lead foot remembered me, huh?"

"He certainly did and not too fondly, either. But I sold him on all your good qualities and reminded him of how STRAC you used to look, especially on main gate."

"Skilled, Tough, Ready Around the Clock, that was me, all right," Dale said and rolled her eyes. "Besides, how would you know? You used to go so goddamn fast through main gate yourself, I'm surprised you saw me at all."

"I don't recall that," Anne said and grinned.

"Of course not."

"I told Sedakis I would find out if you were physically able to handle this. I know what the reports say. I want to know what you say. I want an honest answer from you. Is your foot really up to it?"

"Would you believe me if I said no?" Dale paused and shook her head. "Honestly don't know if I could get through something as physically stressful as basic training again." She looked pointedly at Anne. "Would this positively be the last thing I had to do?"

"Let me put it this way. According to my sources, your discharge should be coming through right around the time your AIT ends."

Dale closed her eyes, frustrated. "How convenient." She pinched the bridge of her nose and focused back on Anne.

"And then you'll be free. I promise."

"My, but the shit's getting deep in here."

"You think I'm lying to you?"

Dale took a deep breath to stop herself from saying something that crossed the line. "Boy, this last promotion really must have gone to your head. You've gone real military on me."

"It's all politics at this rank."

"It's politics at every rank. It's just you never got involved before. You always seemed to be able to rise above it."

"Things are different now." Anne sat back and looked at Dale. "I understand your hostility —"

"The hell you do!" Dale snapped.

"Okay," Anne's voice was soothing, "I'm giving you the benefit of the doubt. You were almost killed, and I will always feel partially responsible for that because I started the whole thing by not kicking your ass out of the Army when it came time for your first reenlistment. But I know you. And I know how you work. I'd like to have you work for me one more time before the government releases you. Humor me, Dale. Don't let me think that oddly brilliant mind of yours is going to waste by just sitting around, feeling sorry for yourself and collecting a pension."

Dale sat in thoughtful silence. She didn't want to part company with her newfound freedom but she *was* bored. Not only that, the woman telling her that she didn't really have a choice was also the woman Dale would do absolutely anything for.

Not only that, Dale knew that she wasn't kidding anyone to pretend she wasn't itching to get back in the game. She desperately wanted to play Army again but the fear of her getting too deeply involved, like before, almost outweighed her adventurous nature. At any rate, up until that point, Dale figured her only future entanglements dealing with undercover assignments would be by reading about them or going to the movies. Anne's mission did have its disadvantages, but the bait she was dangling was just too inviting to ignore.

"Why did you even come and ask me? You could have just sent me one of those official government letters telling me where I had to be and when?" Dale looked at Anne for a long moment. "I'll do it because obviously I have to but if anything happens to my foot after I've practically got it back to normal, I will sue Uncle Sam for every red cent he has."

"Oh, come on, Dale. Sue Uncle Sam? You're being ridiculous."

"Am I? This is my life you're fucking around with here." Dale no longer hid her resentment. She thought about her discharge so close, yet so far. "I knew it was too good to be true.

Some discharge. You guys are terrific at making promises and not so terrific at keeping them. When did you get so loyal, anyway? You're supposed to be my friend."

"I am your friend! I'm doing you a favor, believe me."

"So I see," Dale said.

"You think sending you one of those official letters would have made me more of a friend? That would have been a lot easier for me."

Silently fuming, Dale wouldn't look at Anne. She tapped her fingers heavily on the table in an unconscious rhythm before she spoke. "When do I have to be there?"

"The cycle doesn't start until the fifth of December but you'll be going in with the first group of females on twenty-two November."

"That's only a month away!"

"Actually, I'd like you to fly down next Monday and meet with the training officer, Lieutenant Henning. She can give you more details on what's been happening. Lieutenant Henning is responsible for requesting two special agents being placed in the next cycle. Basically, it was her idea."

"Remind me to thank her," Dale responded, still not looking at Anne.

"The company commander, Captain Colton, has been on leave and he'll be informed when he returns. Lieutenant Henning is a good person and I think after you've met her, you'll feel better about this assignment."

"I bet Lizzie Borden's nanny was told that, too."

"Oh, Dale, stop it, for Christ's sake," Anne finally gave in to her exasperation. "There aren't too many people who are hand selected to do these jobs unless their reputation is excellent and that goes double for a woman. Someone like you, who still has so much to offer is too valuable for us to lose. I wish there was something I could do to keep you in permanently. We could really use you."

"Yeah, well, I've been used by the Army enough, thank you," Dale said.

Anne studied Dale for a moment "You know, you've done more in six years of service than most men do in thirty," she said. "You should be proud of your accomplishments."

"I can't take pride in stupidity. I'm lucky to still be breathing much less walking and you know it. The next phase for me if I stayed in would, no doubt, be death. I've stepped on too many people's toes."

Anne nodded in agreement. "Don't expect me to argue with you on that. There's too much truth in that statement and it frightens me, Dale. This is why it would be beneficial for you to finish out your obligation on my watch."

She glanced up at the wall clock. "I need to get going or I'm never going to get back to Alabama in time." She looked at her half-eaten salad then signaled for the check.

"Dale," Anne said with a sigh as she reached out and grasping Dale's wrist. "Don't be angry with me. My goal today was not to come here to trap you or make your life worse."

Dale tried hard to ignore the heat that blazed through her body just from Anne's touch. She cleared her throat before she spoke and tried to collect her thoughts. "I know. I know. You wouldn't be pressuring me if someone weren't pressuring you. But, goddamn it, I just feel like I'm never going to get out of the clutches of Uncle Sam."

"You will," Anne said in a gentle, reassuring voice. "You will."

They strolled down to the terminal entrance, through the gate and walked toward the direction of the small plane. Jack Bishaye was pointing to his watch and waving his wife toward him, signaling for her to hurry.

"I think Jack would like you to haul your ass, although I can't understand why. It's not like there's so much traffic he won't be able to get a clearance," Dale said, dryly.

Anne looked around. "Yes, we are a little isolated, aren't we?"

"This is populated. You sure can tell you're a flatlander."

"Flatlander? That kills me. When you leave this Green Hill State and come back into the real world, you'll remember what actual mountains look like, especially when you have to climb one of them at about your fourth or fifth week of basic training."

"You just continue to make this whole assignment so appealing." Dale said. "So, I meet with Henning on Monday. I guess I'll meet the other agent then?"

"You won't. I had wanted you both there on Monday but we couldn't clear her temporary duty assignment that soon, so she won't be available to meet with Henning until Friday. Plus, it's better this way. I'd rather not take the chance of having you two seen together before training starts. Averill, Alabama is only so big, remember?"

"Oh," Dale said, disappointed. "I would have liked to have met her beforehand, you know, to get to know her, how she works. What if we hate one another? It's not easy working with someone you hate."

"You're both down there to do a job and your only concern should be to get that job done. Besides, I've met her and I think you two will get along just fine."

"What's her name?"

"Walker. She's a second louie just like you." As they reached the jet, Anne grinned and put her hand on Dale's shoulder. "It's going to be good knowing you're around again, giving my ulcer a reason to act up." She leaned in giving Dale a half-hug.

Turning into the contact, Dale seldom allowed herself to be touched without reservation but she reveled in the sensation of the nearness of Anne Bishaye. *If only.* No, she had to stop thinking that. Anne was married and Dale had only ever had boyfriends. And neither of them were lesbians. All she had were fantasies which were supposed to be healthy and normal. *Heh,* Dale thought, *define normal.*

Jack Bishaye descended the plane's stairway, bent and gave Dale a kiss on the cheek, bringing her out of her short-lived daydream.

"How are you, kiddo?" He spoke in a warm, Texas drawl that, regardless of his mood, always sounded friendly. She wondered how jovial he'd be if he was aware of the crush Dale had on his wife. "You look pretty good, considering."

Anne had always told her if Jack didn't pick on her, it meant he didn't like her.

"You don't look so bad yourself," Dale said, truthfully. "Considering."

She studied the two people before her. Jack and Anne were

a striking couple. Dale had always referred to them as Barbie and Ken.

Jack was one of the most attractive, personable men she'd ever met. He was formerly a flight instructor for an international airline but he now taught at a private school. He was an easy-going gentleman of forty-five, tall, muscular and tanned. He had a full head of thick brown hair, light green eyes, a wide enthusiastic smile and a very alluring quality about him. Anne and Jack had been married for fifteen years and Dale figured they had a pretty active sex life. The thought of the colonel and her husband in bed together automatically segued into an almost whimsical incarnation of Dale and Anne rolling around between the sheets and it made Dale blush suddenly, an event, which so rarely happened, both Jack and his wife noticed.

"You okay there, Lieutenant?" Jack asked.

"Yeah, fine, couldn't be better," Dale answered, silently admonishing herself. She had to rein in these surreptitious feelings toward her superior officer, especially now that they would be working together again.

"Dale, I hate to be rude but, Anne, we have got to move," Jack was now addressing his wife. "We're late."

He turned back to Dale. "Take care of yourself, kiddo, and come see us in Alabama."

"I don't think I have much of a choice. I'll have to look at her mug whether I want to or not. You," she said, smiling at Jack. "I will visit freely." Jack playfully punched her shoulder and disappeared back inside the plane.

Dale turned to Anne and said, strictly for her benefit. "I wouldn't kick his sandals out from under my caravan, I'll tell you that."

"You can't kick with two broken legs," Anne countered, good-naturedly.

"Yeah, you and what battalion?" Dale deflected.

Anne laughed as she climbed two steps and turned around to face Dale. "Look, I've really got to go. It was great seeing you again. I wish it could have been under different circumstances but I'm glad I had an excuse." She regarded Dale seriously. "I've really missed you. More than you'll ever know or I want to admit. I need you near me, Dale, I can't lose you again."

The two women locked meaningful stares and Dale's throat went dry. There was something different in Anne's expression, in her voice, that Dale could not quite identify. Was she misinterpreting the sentiment? Was Anne implying what Dale thought she was or was it just wishful thinking on her part? Feeling as though her heart had stopped beating momentarily, Dale's capricious musings were once again interrupted by the impatient voice of Jack Bishaye.

"Anne, let's go! Finish your goodbyes and get in here, we've got to leave. Sorry, Dale." Jack once again returned to the cockpit.

Yeah, me too, Dale thought, as Anne shrugged, waved, and stepped into the plane, pulling the hatch closed behind her.

Dale watched as the small jet taxied down the runway and lifted off, vanishing into a multitude of clouds. "What the fuck just happened?" Dale mumbled to herself, perplexed, as she made her way back to the parking lot toward her car.

CHAPTER TWO

October 1977 Texas

"My last goddamned working day before I go on an unsolicited and unwanted temporary duty assignment, and I have to pull MP duty officer? Fuck!" Lieutenant Shannon Walker muttered her indignation to an empty room. "TDY, my ass!"

She remembered the phone call repeatedly and became more agitated at every recollection. "Lieutenant Walker? Captain Rosenberg is ill and Captain Alvarez cannot be reached. Being that you were available and haven't cleared your gear yet, we have scheduled you for duty officer this evening. Can you comply?"

No, I cannot comply, I have an inflamed hangnail that is killing me and would definitely prevent me from performing my duties in a normal military manner. Can I comply? Are they joking? she thought angrily.

If it hadn't been a request directly from the MP battalion commander himself, she might have put up a bigger fight. Walker cussed vehemently at the couch, the wall and anything else that had the misfortune of being in the same room with her, as she broke starch and donned her MP gear.

She continued to swear passionately every time she had to shift gears on her short drive to the station. She hoped she'd get the annoyance out of her system by the time she pulled into the parking lot and locked up her red and black Trans Am. She took a deep breath and reached for the doors that said *MPs ONLY*.

After getting settled into duty, she discovered that no one had bothered to restock the coffee and, by this time of the evening, both main and branch commissaries were closed. Fighting off a caffeine fit, she distracted herself by lighting up three cigarettes and smoking them all at once. She had promised herself that she would, without a doubt this time, give up smoking by 2400 hours and, suddenly, at approximately 2130 hours, she found two forgotten packs of cigarettes in one of the pockets of her fatigue jacket. In accordance with the vow to herself, she either had to smoke them all by midnight or give them all away, in which case, she would rather have given away her first-born male child, if she'd had one. So, she figured she'd better smoke them all or die in the process and the way she was feeling after sitting inside that station averaging twelve cigarettes an hour, dying wouldn't have been a bad idea.

When the desk sergeant informed her that the green of her fatigues was clashing violently with the shade of green on her face, she decided to get out into the nippy night air and make the rounds. Her duty driver picked her up in the watch commander's car and they began their patrol.

The monotony finally broke when a call came in. It was a code two run, meaning use lights but no siren and get there cautiously but as soon as possible. 'Cautiously' was apparently a word her duty driver did not understand and once they arrived on scene, she wondered how they ever got there in one piece. Walker couldn't scold him, though, because, as an enlisted MP, she remembered going code three (lights and siren) just to go to chow, especially on graveyard shift.

A young MP approached her, saluted, and waited for her to return the courtesy, which she did as quickly as possible to get it done.

"Ma'am, we've got a problem here," the GI told her, nervously. His nametag read Prauss and the chevron and rocker rank pinned to his collar of his fatigue shirt told her he was a private first class.

She glanced at the two military police sedans, painted olive drab green and white, with their light-bars flashing, and then at two men in handcuffs standing near the patrol cars. The female with them also sported restraints.

Lieutenant Walker strolled to the MP who greeted her at the scene. "What have we got?"

"We pulled them over for deuce but when we got the driver out of the vehicle for a field sobriety test, the female and the other dude came out, too. Swinging."

Deuce was a slang term for drunk driving. Walker observed the casual street attire of the detainees and their late sixties model Chevy. It had been modified into a low-rider complete with furry dice dangling on the rearview mirror. It sported no visible military decal. Walker stopped and turned to Prauss. "They're civilians?"

"Yes, Ma'am."

"What are they doing on post?"

"As much as Aguilar can get out of them, which is next to nothing, they were just taking a drive."

"They never should have been allowed on without a pass or specific purpose. Which gate did they come in?"

"Tonio gate, Ma'am."

"Who's working that gate?"

"McCarthy."

"I want to see him at the end of shift. You'll give him that message, won't you, Private?"

"Yes, Ma'am."

"Outstanding," Lieutenant Walker responded. Prauss' compliance really wasn't outstanding. It was just a standard military response she used when she wanted to be encouraging but not go too far. She then looked back at the three civilians. "Why is that woman looking at me like a psychopath?"

"Because I think she is. She's causing the most trouble. She claims to be an active member of the *Nuestra Familia*."

Active member? An inactive member was no doubt a dead member, Walker thought.

"I doubt she really is," Walker said. "If she was, the last way you'd find out is by her telling you. She's trying to scare you."

Upon closer scrutiny, the lieutenant noted that the woman didn't need to say anything menacing as her appearance said enough. She guessed the female to be in her mid-to-late twenties, but unlike her two male companions, she did not look to be of Hispanic descent. She appeared to be the leader.

The woman stood defiantly, more than a few inches taller than Walker and outweighing her by a good fifty pounds, which looked to be all muscle. Her face might have been considered pretty if her eye makeup wasn't so heavy. Her clothing was all black leather adorned by chains and her demeanor matched her wardrobe.

"Well, whoever or whatever she is, she was really raising some hell. That's why we put her in cuffs, Ma'am."

"Call a female out here to search her and let's get them back to the station."

"You're the only female on the road tonight, Ma'am. The two women we have, one's TDY, in the middle of traveling and the other one's on leave."

Of course. Naturally. She hated performing searches, especially strip searches. She wished they would issue some kind of hand protection for that. Gratefully, Walker only had to conduct a routine, clothed search this time.

Walker sized up this hostile Amazon she was about to frisk for weapons, and decided, for her own safety, to leave the shackles on.

Walker nodded toward the two men. "Have you searched those two?"

"Yes, Ma'am. They've also been read their rights in English and in Spanish."

"Outstanding." Walker forced the woman to lean forward against the squad car at which point the woman proceeded to scream what Walker assumed to be obscenities in Spanish, but the lieutenant continued to search, unperturbed. She started to emit a sigh of relief until she felt a long, hard object on the woman's calf. She lifted the woman's leather pant leg, where she discovered an ice pick, secured by masking tape.

"Jesus," Walker said. She removed the object, leaving the majority of the tape on the woman's leg. She called her duty driver to her and handed him the weapon. "Get them out of here. I'll meet you back at the station," she instructed Prauss. "And don't forget to radio in your starting mileage."

Walker shook a little bit when she got back to the car, a combination of absence of nicotine, lack of caffeine and the *what might have been* scenario if she hadn't found the weapon. She called in to dispatch and cleared them from the scene and then informed her driver that if he drove back in the same manner he drove to the site, she would not hesitate to use the confiscated ice pick on him.

The trio was inside the rear of the station being processed by the back of the MP desk, with three police officers guarding them, when Walker arrived. All restraints had been removed and the female had seated herself on a bench with her two buddies in chairs, flanking her. She had not lost that cold, hard glare that had given Walker an unsettling feeling back when she first encountered the woman at the traffic stop.

"Can I have a cigarette?" the female asked.

"No, you can't," Walker answered, signing her incident report.

"I wasn't asking you, *puta,* I was asking my boyfriend."

"You still can't have one," Walker stated firmly, noticing out of the corner of her eye that the guys guarding the prisoners fidgeted. Clearly they sensed trouble.

"Why can't I have a cigarette?" she persisted.

"Are you asking me or your boyfriend?"

The woman's eyes narrowed as she glowered at Walker. "Why can't I have a cigarette?"

"Because people under apprehension at this station are not allowed to smoke, drink coffee, eat cake or eat shit." Walker stood and tossed the report on the desk.

"Puta," the woman spit out.

"Yeah, I heard you the first time," Walker told her.

The female suspect stood. The callousness in her eyes had grown worse. The tense atmosphere suddenly ratcheted up a few notches as Walker stood before her.

"Miss Villard, please sit back down," Walker directed the prisoner, forcing politeness, remaining calm.

"Are you asking me or telling me?"

"I'm asking you. Don't make me have to tell you."

Rose Villard stood there obstinately, arms folded across her ample chest. The look on her face just dared Walker to try something, anything. Walker's right leg started to tense involuntarily, as if her nervous system perceived calamity but she knew it was too late to back down. Villard shouted something to her boyfriend in Spanish, causing the three suspects to look at Walker and laugh.

Walker tilted her head in question.

"You want to know what I told them?" Villard smirked. "I told them only a woman who wanted to be a man would do this kind of a job so she could prove what kind of a man she really is."

"Well, if that's true, what does that make a woman who wants to prove she's tougher than me?"

"You go take a flying fuck!" Villard roared.

Walker addressed the desk sergeant but never took her eyes off Villard. "Hey, Sergeant, isn't there a regulation against airborne copulation on this post?"

As the desk crew laughed, Walker casually took a step closer to Villard, who read that as an invitation to rumble and took a swing at the lieutenant. It's not that she didn't expect it, Villard just moved faster than Walker thought she would, but since Villard had telegraphed the punch, Walker was able to avoid it. With split second timing, Walker put her right hand on Villard's left shoulder and brought her left arm under Villard's chin and sat her down by force.

Returning a cold, hard look, Walker peripherally observed everyone in the MP station stop dead in their tracks, completely astounded by the swift action. Villard was quite startled herself, taking a moment to regain her momentum. As soon as Walker hinted at turning back toward the desk, Villard rose up, grabbed a chair out from under one of her companions, dumping him on the floor. She heaved the object at Walker.

Walker ducked as two MPs grabbed the two male detainees to make sure they didn't get involved. The third MP, who was supposed to be guarding Villard, just stood there, looking as if he'd been struck by lightning.

Villard charged and caught Walker directly in the chest, knocking her against the MP desk. Trying to get her breath back, Walker realized that even being shoved by this woman seemed to have the same impact as being kicked by a horse.

Spinning, Walker swept Villard's feet out from under her, and knocked the aggressive woman to the floor, face first. Walker straddled Villard's hips in an effort to subdue her long enough to restrain her but Villard bucked like a bronco with a burr under its saddle, making the undertaking very difficult until MPs appeared from everywhere to assist. They held Villard down while Walker slapped metal cuffs on her wrists and flexicuffs on her ankles. Walker composed herself and decided to guard her prisoner from behind the duty officer's desk.

Taking a deep breath Walker sat and propped her feet up, then watched as Villard started smashing her chin on the floor, repeatedly shouting. "I'm going to tell them you beat me up! I'm going to show them my bruises and tell them you did it!"

"Make sure you get the name right, then," Walker told her. "It's Walker, Shannon B., O-1, second lieutenant."

"Fuck you, bitch!"

"Sorry. You're just not my type and, besides, I'd probably get frostbite."

"I'm going to kill you, *pendeja!* My family is going to get you, *coño y madre!* Don't set your motherfucking ass in town, 'cause I'll have it blown away!"

Walker's face lit up. "Blown away, huh? Kinky," she commented, almost blandly, which provoked Villard even more.

"I'm going to kill you! Do you hear me? You're going to die!"

Walker leaned forward and held out the gold bar on her collar. "Could you speak a little louder into my butterbar, please? I want to make sure they're getting all this in the back."

"You fucking white whore!"

"Sweetheart, I've got news for you. According to your civilian police records, the only Latin in you is by injection," Walker smirked.

The back and forth continued for the rest of the shift and beyond until the three suspects were ordered released to off-post authorities. As civilians, the trio would probably get off with a slap on the wrist but Rose Villard did end up being charged with carrying a concealed weapon, communicating a threat on a military reservation, trespassing and one count of battery. Her boyfriend was cited with trespassing and driving under the influence and the third member of the group was just charged with trespassing. Walker doubted she would ever see them in a courtroom, military or civilian, however.

After Villard and her companions were hustled out the door and escorted off post, Private McCarthy was disciplined for allowing the trio on the reservation in the first place. Following McCarthy's ass chewing, Walker breathed a little easier, knowing she had proven herself once again and subsequently became incensed that she had to continue to prove herself at all. Then she remembered she'd be away from it all within a few hours.

She looked at the clock. 0230 hours. Outstanding. Walker contemplated her last half-filled pack of cigarettes. It had been crushed in the scuffle but took out a broken one anyway. She removed the bent filter, stuck it in her mouth, lit it and went back to work. Three and a half more hours to go and suddenly she felt exceedingly sick to her stomach.

October 1977 Vermont

The tall, good-looking man jogged alongside Dale. He had dark blond hair which he wore not too short but conservatively nonetheless, hazel eyes, a mustache and a look Dale described to everyone as latently animalistic. His name was Keith, he was British and he co-owned one of the town's more popular hangouts called CK's Tavern. He and Dale had met at a softball game just before she had gone into the Army and there was an intense attraction to one another. But it took nearly a year for anything to develop because distance and complications kept them apart.

The initial chemistry between them was undeniable and the relationship had been physical from the beginning, only later evolving into something remotely resembling love. In reality, even Dale hadn't realized she was using him to assuage her guilt about not really being sexually attracted to men at all. She wasn't quite ready to give up on guys and definitely not ready to admit it, especially for fear of military retribution. Keith was comfortable and a good excuse not to have to get involved with anyone else.

"God, you're running well," he commented.

If nothing else, she loved to hear him talk. Even after five years of a strained, mostly absentee relationship, she never tired of listening to his accent.

"I think so, too. It hurts less and less each time."

They strolled in silence, cooling down.

"I still can't believe you're leaving again. You just barely got back," he said finally.

"Keith, let's not get into this again, okay? It's only fifteen weeks or so, maybe less. We made it through a separation of almost two years once, so three months should be no problem."

"But that defines our entire time together. Separation. I thought it would be different now. It makes me mad. What happens if you die this time?"

"This isn't that dangerous," she advised him, wiping beads of sweat off her forehead with the back of her hand. At least not the assignment. Her being around Anne Bishaye again was another matter.

"Neither was your last mission. Supposedly."

"I'm just going down there to try to find out what's going on with these drill sergeants. If I die, it will be because I'm killing myself going through another basic training and AIT, that's all. Then, when I'm done, I'm going to politely excuse myself from the Army."

"I've just gotten used to having you around again. I don't want you to leave, Dale."

"I have no doubt you'll find good company while I'm away. You always do." Dale said, smiling wryly at him.

Keith rolled his eyes and let out an annoyed breath. "I knew you were going to bring this up. You just said yourself that sometimes you're away for years at a time. What do you want from me? And I never flaunted anything in front of you."

"That's very true. You are very discreet. It's just your gossipy little drinking buddies who feel it's their undying duty to inform me you took Mary Jo Shmoe to Atlantic City or Dena Douchebag to Maine."

"So, what you're telling me is that you can go out on me but I can't go out on you then, eh?" Keith stopped to face her, placing his hands on his hips.

"Going out is one thing. To me, going out with someone is going to a movie or to dinner or…or bowling. To me, going out with someone because you want companionship and the someone you really want to be with you can't be with at that particular time is a lot different than spending a week or two in another state with women whose reputations would make the Happy Hooker blush. Do you get what I'm saying here?"

"And you're being a hypocrite. You've slept with other men since our relationship started."

"If you expect me to deny that, I won't. But even when I've done it here in town, it's been where no one can throw it up in your face and with the understanding that nothing else would come of it."

Keith was gaping at her. "You've slept with other guys...here, in Rutland?" He could not hide his astonishment. "I knew you'd had a few flings in places where you were stationed but I am bloody amazed that you've done it right under my nose!"

"Not other guys. Other guy. Just one. And that's not my point, Keith..."

"Just what is your bloody point, Dale? Huh?" He was agitated.

"Okay. I'm considered your girlfriend and you won't even take me to West Rutland, let alone Atlantic City or Maine. I'm good enough to fuck six ways from Sunday but not good enough to openly claim. I've been back here how many months, Keith? Everything you and I have done together with the exception of running or drinking at your bar has been inside and usually involves a bed. I am not something to be ashamed of and I'm tired of being hidden."

"Then stop hiding."

"What? What does that mean?" Dale asked, defensively. *He couldn't know...could he?*

"What do you think it means?" he countered, just as defensively.

"Oh, no. No, no, no. You said it, you explain it."

Keith heaved a sigh and ran his hand through his hair. "You don't think I figured it out? About you and Anne Bishaye? Do you honestly think I'm that stupid?"

It took Dale a moment to find her voice. She had to make sure, when she responded, that she didn't sound too culpable or too blasé. She was certainly not ready to personally confront this issue yet, how was she supposed to meet it head on with Keith? And if he had figured it out, there was not a slim chance in hell that Anne hadn't, especially after the colonel's parting remarks.

"I can honestly tell you that there is nothing other than a professional relationship going on between Colonel Bishaye and me," she said calmly.

"If that's true, which I highly doubt, then it's only a matter of time."

Exasperated but, more than that, curious, Dale looked up at the sky then back at her boyfriend. "Why are you saying this, Keith? Because once again I have to choose my job instead of you? You knew going into this relationship that my work required me to be away. Is your ego so fragile that rather than just accepting that, you have to accuse me of being in love with someone else?" It was a low blow but Dale was trying to deflect the allegation.

"In love? Who said anything about being in love?" His comeback hit even farther below the belt.

"You implied it, okay?" she argued. "We always go through this, Keith. Every time I'm here and then have to go back. Why make it so personal this time?"

"Because you were supposed to be discharged. That was supposed to be final. At last, I was going to have you full time, all to myself, and then that fucking Anne Bishaye snaps her fingers, as usual, and nothing else matters except pleasing her!"

"I am still under government contract. I have to do it. They could get very nasty with me and force me to go up before another physical evaluation board, which I'm sure, seeing the condition I am in now, would consider me at least fit for reclassification or make me resign. Which means I would owe them money because I didn't fulfill my contract. I was told if I play along, I will get my discharge when this is through. Fifteen weeks and I'm done with it forever."

"Yeah, as told by Anne Bishaye," he threw in.

"Because I am going to be working for her!" Dale argued, infuriated. "What is it that you'd like me to do? Tell the Army to go fuck themselves and just accept the consequences? I would think you would rather see me as Anne Bishaye's bitch than someone's at Leavenworth." Under the circumstances, that probably wasn't the best visual to put in his head, she thought, immediately. Dale closed her eyes, took a deep breath and looked at him again, composing herself. "Fifteen weeks. I promise."

"Don't! Don't you dare make a promise you can't or won't keep! You won't be back in fifteen weeks. You'll get this crap back in your system again and that's all the reason you'll need. Cut you and you'll bleed olive drab green. You will never commit to me because you're already committed to the goddamn Army...and your precious colonel."

"Commit? Who said anything about commitment? I just said it would be nice to be recognized as your girlfriend, nice to know that you're proud to be my boyfriend. Then you go and create a diversion about Colonel Bishaye and me to take the focus off the real issue. God, Keith, that's just pathetic. Maybe it's good all this came up. We'll save a lot of time not wasting it on each other."

"Wait a minute! You are not going to turn this around and make this about me. But you know what, Dale? You're right. It is good all this came up. Nice to know you can give me up so easily." Keith began stomping away, then stopped and turned back toward his now ex-girlfriend. "Fuck you, Dale! Fuck you and your colonel and your Uncle Sam!"

Dale kept her temper in check, watched him walk away, get into his car and drive off. For the second time in two days, Dale scratched her head, wondering what the hell just happened. Hurricane Anne had swept into town again, rapidly and skillfully destroying anything that threatened to get in the path of her mesmeric control of Dale. Tears filled Dale's eyes as she watched Keith's jeep drive out of her field of vision. It wasn't that she mourned the sudden loss of the relationship, she was more disturbed by the fact that she really didn't care.

CHAPTER THREE

October 1977 Alabama

It was a seven-hour ordeal from Rutland and then by plane from Albany, New York to Averill, Alabama, via stops in Philadelphia and Atlanta. Dale tried to pretend she was asleep most of the way so that the miscreants who always seemed to be assigned seats next to her would leave her alone.

She inclined her head toward the small window of the passenger jet and let the humming of the aircraft lull her into a fitful sleep.

Dale had enlisted right out of high school. She had left for boot camp the day she passed her Armed Forces Entrance Examination test, or AFEES, knowing that if she had too much time to think about it, she might change her mind. Dale wasn't all that patriotic. She had just been bored. She would go nowhere in her small town, with or without a higher education and she had desired to be away from the watchful eye of nosey relatives.

In 1971, Dale had hopped off the bus at Fort McCullough, a post widely known for its Women's Army Corps or WAC training and its occupancy in the middle of Alabama. She had breezed through the all-female basic training. When it had come time to decide what her occupational specialty would be, however, the decision had been taken out of her hands. The powers that be had sent her across post to a new barracks for advanced individual training, or AIT, in the US Army Military Law Enforcement School, where she had trained side by side with male students. The WAC indoctrination wasn't exactly easy but neither did it prepare her for when she went on to that next phase. It was like basic once again except everything was male oriented and a lot tougher.

After school, Dale was assigned to her first permanent duty

station, a military police company at Fort Ord. By day, Dale performed her MP responsibilities and at night she was allowed to attend college on post, permission granted by her company commander, Captain Anne Bishaye.

With six months experience as a patrol officer and with Captain Bishaye's recommendation, she had requested to go to Military Police Investigator School. Dale, as the only female MP in Bishaye's company, had caught Bishaye's eye early on. Especially as she did her job without complaints or special treatment and could give guys back as good as she got. As immature as Dale could be at times, when she was on duty, her work was exemplary. Dale was sent before the MPI board where she passed oral and written exams and then was returned to McCullough for additional training.

At Fort Ord, the undercover police units needed females and even though Dale was not a commissioned or warrant officer, the Criminal Investigation Division, or CID, authorized her to work with them. Bishaye had repeatedly emphasized Dale's ability to fall into manure and come up smelling like flowers, and when Dale returned from school, they sent Dale out on her first assignment alone.

A job that was supposed to be a simple observe and report exercise, involving the post commander's wife, turned into a complicated extortion racket. The next mission had her working at a neighboring county's recruiting station as a civilian clerk, to keep an eye on a young staff sergeant who was reportedly stealing supplies. The next case led to several high-ranking enlisted personnel lying about enlistment quotas to collect thousands of dollars illegally from the government.

Reports of her success started to circulate through the Army grapevine and before long, other CID units across the country requested her services.

Dale's resume included stopping a drug supply and demand operation through a reputed airborne division in the south, infiltrating a military white supremacy group in Kansas and catching a serial rapist who preyed on Army wives left alone while their husbands were in the field for months-long training. Her achievements continued to build but so did open threats against her life.

PERMISSION TO RECOVER

For example, if CID had not gotten her out of Kansas the night of that bust, they most likely never would have found her alive, perhaps never even found a body. In fact, just thinking of Kansas made the little hairs on the back of Dale's neck stand up.

She was told there was a contract out on her life in the state of Texas pertaining to the time she interfered with a slew of non-commissioned officers and local civilian authorities who were about to sell a helicopter filled with expensive, new-age technology weapons to foreign agents. Her presence seemed to be a joyful welcome until she brought half the MP battalion, along with the FBI, CIA and ATF, down on the group the night they intended to carry out the crime. She led the bust, catching the offenders with the arsenal, a damaging amount of paperwork and other incriminating evidence. She was commended for her achievement on that one but she knew she had jumped the gun and gone in too soon. The mastermind and the pilot who was to fly the helicopter were still anonymous and at large. Yet, had she waited any longer, the plans and any evidence would have been destroyed. The Army applauded her choice. Dale regretted it.

To let events cool down, she was sent back to Anne at Fort Ord, who arranged for a reluctant Dale to go to OCS and to re-enlist indefinitely. She breezed through the OCS basic training and a ten-week period of instruction and then was assigned to Fort Jackson in South Carolina as a training officer. Dale passionately resented the duty because it took her out of the action, even though she knew, for her own safety, she had to keep a low profile and try to stay out of trouble.

Trouble found her anyway.

While at Jackson, Dale exited the officer's club one evening and was on her way to her car when she was approached from behind and jabbed in the back by the barrel of a .357. She was ordered to get into a silver sedan with Florida license plates, where she was driven into the wilderness and pistol whipped into unconsciousness by two men she did not recognize. They threw her out of the car, leaving her for dead, and the right back tire ran across Dale's left foot, crushing it, as they drove away. She was found early the next morning by a platoon of basic trainees marching with their drill sergeant to one of the rifle ranges.

Her bruises healed in time but she was told she would never be able to walk again without a pronounced limp. Bishaye, who had flown in to see her, got her the best physical therapist the Army could provide to come work with Dale. With the help of the therapist and Dale's own determination, she managed to narrow it down to a slight, painful hobble. By then, the physical evaluation board had decided to award Dale a medical discharge, even though she had quite a bit of time left to serve. They sent her home to Vermont to recuperate on her own and wait for the final papers. After months of determined self-rehabilitation, she had gotten most of her athletic ability back. She could run pretty well, softball was effortless, tennis was slightly strenuous and long-distance jogging still gave her some problems but she had eliminated the limp and that had to count for something.

It was now clear that she had done herself no favors.

It was after dark when her taxi entered through the main gate, Dale perked up as they drove by the prefatory sign that read: *Welcome to Fort McCullough, Alabama, Home of the United States Army Law Enforcement School, Training Center and Fort McCullough*

Dale was hit with a sudden wave of nostalgia as she recalled the first time she was driven onto the installation and saw that greeting. She was thrumming with the excitement of starting her new journey and plagued by dread because of the rumors of how horrible this decision could have been. She knew this route well as the cab then swung by Raburn Hall, by the post theater and down toward the main exchange. It was almost like coming home and yet, it wasn't.

"Building 1801 South, right?" the driver drawled.

"Yes, sir, that's right," Dale replied, quietly. Few people were about as the taxi pulled up to the sidewalk of Alpha Company, tenth battalion.

"Four twenty-five," the driver told her, and Dale handed him a five-dollar bill. She grabbed her overnight bag and stepped out of the vehicle.

Dale stopped after the cab pulled away and she took a deep breath. It was autumn and the temperatures were mild but damp, as Alabama was known to be heading into winter. She smelled two distinct odors that instantly brought her back to her training days— wood smoke and gunpowder. *Some things never change*, she thought as her lips curled into a smirk.

She looked around as she walked to the Orderly Room, surprised but grateful not to see any trainees milling about. Without knocking, she opened the door and walked inside and was immediately glared at by the charge of quarters, or CQ, and the CQ runner.

Dale was attired in faded, patched jeans, a navy-blue pullover with a hood and torn Adidas sneakers that looked like they may have been white at one time. Her hair hadn't been combed since she left Albany and she knew it gave her a somewhat wild appearance. She addressed the young man in fatigues sitting behind the first desk.

"Hi. Lieutenant Henning wouldn't be around by any chance, would she?" Dale asked. She was tired from traveling and hoped this wouldn't take long.

"Who should I say is asking?" he sneered, looking her up and down.

"I'm just a friend of hers. If she's here, I need to see her, please." *You pompous little shit,* Dale thought. She was not in the mood for attitude at least not someone else's.

He continued to eye her suspiciously as he stood and walked toward a closed door. He knocked and stuck his head inside. "Ma'am? There's a lady out here who says she's a friend of yours."

Second Lieutenant Karen D. Henning opened the door and stared at Dale. She looked up at the CQ, a bit confused, then looked back at Dale. "I'm Lieutenant Henning. Can I help you?" she asked, in a thick Texas accent.

Dale briefly studied the company training officer. Lieutenant Henning was a short, pretty woman with large light brown eyes.

"Lieutenant Henning, I think you're expecting me," Dale stated, handing Henning her military identification and badge.

Henning examined the small document and nodded at Dale. She reached and shook Dale's hand with the grip of someone used to doing a lot of physical training. "Nice to meet you. Please, come in," she gestured into the first sergeant's office.

Dale followed Henning into the office.

"I wasn't expecting you until tomorrow," Henning said, closing the door.

"The colonel told me to be here tonight." Dale glanced at her watch. "Do you always work this late?" she asked.

"Sometimes I do but tonight it's because I'm reviewing all your paperwork and getting it ready." Henning unabashedly inspected her. "I hope you don't take this wrong, but you weren't at all what I was expecting."

Dale smiled. "I can imagine. The way Bishaye most likely described me, you were probably expecting Quasimodo."

Henning returned Dale's smile. "Close." She studied Dale for a moment longer. "How's your foot?"

"You know about that?"

"Yes. Colonel Bishaye and I had a long talk. She seems to have a lot of faith in you."

"Yeah, I know. Maybe too much," Dale said, vaguely, as she checked out the office.

"Really? She speaks very highly of you. Are you sure you're up for this?"

"I wouldn't be here if I weren't."

"Somehow I doubt that. Colonel Bishaye says you're pretty hardcore."

"I may be hardcore in the colonel's opinion but I'm not an idiot. This foot has got to last me the rest of my life, I'm not about to take a chance on ruining it."

"No, I suppose you wouldn't." A knowing smile crossed Henning's face. "She also said you're pretty good at making up excuses. She said since you've been in the Army, your father has died eight times."

"Yeah, well, he was a pretty sick man," Dale told her, dryly. "And sometimes the colonel talks too much."

Henning opened the file in her hand. "How much did Colonel Bishaye tell you?"

"She skimmed the surface of what's been going on around here and that it was your idea to call in some snitches." Dale drew a deep breath. "I have to be upfront with you, Lieutenant Henning, I'm not all that thrilled about being here. I think that we might be wasting our time. But Colonel Bishaye is a friend of mine and she believes something is out of line. She's seldom wrong and I respect her opinion. So, what happens now?"

"Yes, I was warned about your reluctance to be here. Hopefully, I can change your mind. You will be coming in on the 22nd of this month with about eight other females. You'll be in the first set. Lieutenant Walker will be in the last set. The rest will arrive the four days in between. You'll be a private, E-nothing, so will Walker and your hometown will be the same since according to today's readouts, no one else will be entering the company from there. Here," she handed Dale the folder. "This is your personal and medical file."

Dale glanced at it. "It seems pretty accurate up to what I've been doing for the past six years. It won't take long to memorize this other information."

Henning watched Dale study the file. "Look, Lieutenant Oakes, I've been here since January of last year and I've seen some exceptional, dedicated men get burned. I've felt helpless. So, this set up is important to me.

"All I'm trying to say is what if these men really did prompt the fraternization?" Dale suggested, receiving a sharp look from Henning. "I've been in your shoes. And I've seen drill sergeants at work. Some of these guys treat female trainees as if trading sexual favors is part of the program."

"Not in my company," Henning responded, crisply.

"Okay, maybe not all of them but there is a possibility that at least one of them wanted to fool around, right?"

"Of course, there's a possibility. But only once, since I've been here anyway, did we have an incident with a drill sergeant who was clearly on the make. And he wasn't even one of the ones accused. These men have had spotless records. I *know* these men. They wouldn't let something like this happen to them."

"How do you know?" Dale countered quickly. "They're human. They have weaknesses. It did happen to them, which means that, as hard as it is to believe, they had a flaw in their systems somewhere. My job is to find out where and how, and why. What about the female drills? Haven't they been bothered?"

"Not to my knowledge. Why do you ask?"

"Six years ago, when I was in training, there were a few

incidents with female drill sergeants and trainees. Female trainees."

"In Alpha-10?"

"No. Incidents over in WacVille." Dale read the investigative report on the events and copies of the blotter entries. She looked up at Henning halfway through the paperwork. "What the military needs is to build the perfect soldier who would never become disobedient, damaged or amorous and then they should clone them."

"For sure." She leaned back against the filing cabinet. "Unfortunately, even the most experienced soldiers aren't immune."

"They're probably the worst because they know the boundaries of regulation and they get bored and test them anyway." Dale nodded toward the door, changing the subject. "Is the CQ out there going to be around when I get here as a trainee?"

"No. He's waiting on orders now. I'll inform the colonel that he and the CQ runner have seen you so I'm sure she will arrange to have them both out of here very shortly."

"How many females do you have upstairs right now?"

"Twelve, I think."

"All waiting on orders?"

Henning nodded.

"And I believe Bishaye told me there was a female spy in the last cycle?"

"Yes, she's still here. Her name is Linda Boehner and she's an E-5. The CO and I never would have known who she was if this situation hadn't come up again. She had to come forward and give MPI everything she knew. The drill sergeants still think she's a trainee."

"They didn't suspect her at all?"

"No. Well, if they did, I never heard about it. The people they did suspect are already gone."

"Could you call Sergeant Boehner down?"

"You bet. Hold on a second." Henning went outside to the intercom system, which sat on the first sergeant's desk. It was nicknamed the bitch box and through this device, the drill sergeants bitched at the trainees and, when they thought the drill sergeants could no longer hear them, the trainees returned the courtesy. The box connected with a speaker up in the ceiling of all the bays. Supposedly when the little red light on the amplifier was flashing, it meant someone in the vicinity of the box was either talking or listening. When the little red bulb in the ceiling was dark, it indicated communications were shut off. However, Dale knew from using one herself that the intercom could be activated without the red light working, therefore, the trainees could be eavesdropped on. It was a very sneaky way the cadre used to find out information that the new recruits would rather have kept among their friends or to themselves.

Within a minute, Henning returned. "She'll be down shortly."

Dale's feet were propped up on the desk and she was again studying the MPI reports. "These last two females sure seemed to know what they were doing. They got these guys right where they wanted them. I mean, damn, look at this stuff. One has no alibi, the other was in a car with one of the girls and each case has witnesses who've signed sworn statements that they and heard these drill sergeants openly flirting with the two victims. They've got their bases covered."

"That's what I mean. Maybe with you two looking out for it, we can break this thing wide open."

Dale noticed the hope in Henning's voice. She seemed convinced it was some kind of conspiracy. Dale had learned to never take anything at face value. She'd been disappointed too many times by believing in something the same optimistic way Karen Henning was doing now. "Am I correct in assuming that Private Stuart and Private Willensky and their witnesses are gone?"

"Yes, battalion arranged to get them out of here as soon as possible."

Dale looked disgusted. "I wonder why battalion did that. I really needed to talk to those women."

"Private Stuart, of course, was discharged."

Jesus, Dale thought, *Anne really didn't tell me anything*. She was about to ask why Stuart was discharged when there was a knock on the door.

"Excuse me, Ma'am, Boehner is here."

A woman in civilian clothes, about the same size as Dale, entered the office. She closed the door behind her, turned and stood before both women. Dressed in sweatpants and a long-sleeved T-shirt with *Tenth Battalion* emblazoned across the front, Linda Boehner, stood, somewhat defiant. She was a plain woman with a hardened expression that belied her twenty-four years. "You wanted to see me, Ma'am?"

"As you were, Sergeant Boehner," Henning commanded, waiting for the woman to relax, something the sergeant didn't seem to do completely. "I'd like you to meet Lieutenant Oakes."

"Sergeant," Dale acknowledged, shaking Boehner's hand.

"Ma'am." She regarded Dale with hesitancy.

Dale smiled. "She's your Ma'am," she indicated Henning. "You can call me Dale. Sit down, please."

Boehner sat down in a chair in front of the desk. "What can I do for you, Ma'am?"

"Dale." Dale corrected, annoyed with the snitch already. There was something in this woman's demeanor that sandpapered Dale immediately, something in her bearing she couldn't quite put her finger on.

"Dale." Boehner said it not so much uneasily as suspiciously.

"Look, Sergeant Boehner, we both know military courtesy ordinarily forbids my being this informal with you, which is bullshit, but I don't make the rules, okay? You're going to be seeing me again and when you do, you'll either be calling me Dale or Oakes or something you might not want to say in front of Lieutenant Henning, so I need you to get out of the habit of calling me ma'am now." Dale tossed the closed file onto the desk. "Tell me everything you know about this drill sergeant business."

"You mean the fraternization thing? Now?" Boehner looked at her watch.

"That's why I'm here. I'm one of your replacements as snitch. So, I need you to tell me anything that might help us, things the women did, things they said, anything that might have made you think these drill sergeants were about to be set up."

Boehner sat back and looked at her. "I told all this to MPI."

"Tell me. I wasn't there and I want to know what *didn't* go into those reports."

Boehner sighed, apparently annoyed, not only at the questions but also by Dale's presence. "I didn't know them super well. I never hung around them that much. One was a real flirt and the other was very pretty but an unlikely candidate for the charge," Boehner said and then snickered.

"What is it?" Dale asked.

"Well, Carolyn Stuart was always bragging that she wanted to make it with a drill sergeant if it was the last thing she did here. And it was."

Dale's head snapped to Henning. "If she was always bragging and everyone heard her, how come you couldn't call the charge on that?"

"Because," Henning said. "Carolyn Stuart admitted she was a lesbian, Dale. She was bragging about wanting to make it with a female drill sergeant. That's why her accusations seemed so believable. Why would a lesbian go after a male drill sergeant?"

Dale sat silently for a few minutes, blinking back this new information. "Is that why she was discharged? Because she admitted she was gay?"

"Yes. Some of the other females had been grumbling about her for some time, not that she openly tried anything with anybody. She was just so verbal about it. It made some of the younger women nervous. When we questioned her about these admissions mid-cycle, she denied it all. However, after the fraternization incident, she gave us a written confession that was backed up by several other women in her platoon and she was dismissed," Henning offered.

"Honorable trainee discharge?" Dale asked. Henning nodded. Dale shook her head. "I don't know. Something is weird here. Most of the lesbians I know, especially in the military, almost have this secret code of honor to keep their mouths shut about their sexuality. They travel in their own little circles and leave everybody else alone. It's too dangerous not to. "She looked at Boehner and then at Henning. "These incidents happened within a week of one another, didn't they?"

"The first one was three weeks apart. The cycle before this one was longer than that. Week two and week six, I think," Henning volunteered.

"So, there's no set pattern," Dale sighed. "What about the other female?"

Boehner lit a cigarette. "Willensky? Drill Sergeant Halpin was always around her, although I never heard him say anything suggestive to her and then suddenly they're found together up on one of the range roads."

Dale thought for a moment. "According to the investigator's report the MP patrol assigned that night approached Halpin's vehicle from behind. It appeared that she and Halpin were just sitting there, talking, she on one side of the car, he on the other. As soon as the MPs turned their spotlight on, she was on top of him. She told the investigators that Halpin brought her up there to have sex with her. The way this report was written, it sounds to me like they definitely think it was a set up."

"They do," Henning confirmed. "But she cried attempted rape, insisted he was about to force himself on her. She even had his skin under her fingernails in an apparent effort to fight him off. They tried to break her by telling her what they had witnessed but she was so adamant, they had no other choice but to take action against Halpin. Before the incident, he'd had a couple of beers at the NCO club. He claimed he was on his way home when he stopped by the main Post Exchange and ran into a very distraught Willensky."

"Could he be guilty?" Dale asked, simply.

"Sure. But I doubt it" Karen said, quietly.

"Then what was he doing on an isolated range road with her? He must know those roads are off-limits after the hours of darkness. Even as a trainee I knew that. He should have known better."

"He said he didn't know how to handle her and that she had threatened suicide if he didn't take her somewhere to talk," Henning said, exasperated. "I would probably have done the same thing, especially when she never gave him any time to think about it."

Dale leaned back. "I agree."

Henning and Boehner stared at one another, clearly wondering if Dale realized she had just contradicted herself. Henning seemed to relax, giving Dale a smile she might have reserved for those suspected of being one bullet short of a magazine.

Knowing what was going through both their minds, Dale chuckled. "I would have taken her somewhere to talk but not up to an isolated range road. That was just stupid if he is innocent. Look, all you have is his word against hers and the speculation of everybody who wants to cover his ass. Before I go condemning either one of these women, I want to make goddamned certain they're guilty. I am sure if it were either one of you, you would want me to do the same thing. Women tend to get the raw deal in situations like this. So, I need to look at every detail as objectively as I can. I don't want you, me or Uncle Sam to be wasting our time or the taxpayer's money." She looked at Boehner. "What do you remember?"

"Everything happened so fast that it surprised me, especially with the first trainee-drill sergeant incident," Boehner said.

Dale flipped through the paperwork to pinpoint the section Boehner referred to. The names jumped out at her. "You mean Stuart and Sergeant Franciosa?" she asked, for confirmation.

"Yes," Henning verified.

"But you weren't surprised with Willensky and Sergeant Halpin?" Dale asked.

"Yes, of course I was. I always had the deepest respect for Drill Sergeant Halpin. He was my platoon sergeant and I got to know him quite well," Boehner told her.

"Then you would say what he was accused of didn't sound to you like something he would do."

"That's right. For him to take her up on a range road alone in a car would have been way out of character."

"Did Willensky have a boyfriend in the company, or any male in particular that she hung around with?"

"No," Boehner answered. "She was popular with everybody. She spread herself out pretty evenly. There never seemed to be anyone special."

"Sexually popular or just plain friendly?"

"Just friendly. From what I know, anyway."

"What about Stuart? Anyone in particular with her?"

"No. Not that I knew about."

"Could Sergeant Franciosa have been one of those men who pursue lesbians because they feel it is their calling in life to convert them?" Dale challenged.

"No," Boehner said, firmly.

Dale rubbed her forehead. For some reason, Sergeant Boehner's condescending tone was beginning to greatly annoy her. It's not like she wasn't used to the attitude from NCOs who thought lower ranking officers were morons. "I sure hope this turns out the way you want it to, that these drill sergeants are innocent."

Dale sat up and straightened out the paperwork she was using for reference. "Sergeant Boehner, since you're going to be classified as a holdover for a few weeks, I would like you to study each one of the new females that come in carefully. Talk to as many as you can discreetly and if you start to notice any similarities at all to the two other females, report it to Lieutenant Henning so that she can bring it to Lieutenant Walker or me. We'll want to hear anything that either sounds out of place or even familiar to you." Dale looked back at Henning. "Anything you want to add to that?"

"No. I think you covered it all."

Linda Boehner stood up almost the same time as Dale. "I'll do what I can, Lieutenant Oakes, but I don't feel I should have to do your job for you and Lieutenant Walker."

Pushed to an edge by Boehner's attitude, Dale's voice suddenly became cool and professional. "I'm not asking you to, Sergeant. Boehner. I'm just asking for a little assistance on this. Do you have a problem with that?"

Boehner did not lose eye contact with her. "No, Ma'am."

"Fine. You may go now."

Boehner nodded. She looked at Henning, who nodded in agreement with the order. "Goodnight, Ma'am." Without acknowledging Dale again, she left the room.

"I don't think she likes me," Dale said.

"I don't think it's that so much as I think she's just upset and a little defensive because she failed at her task."

"What was her task?" Dale countered. "Nobody suspected a set up when she started, so her task was to be a normal cycle spy, right?"

A normal cycle spy's duty was usually only keeping an eye on the trainees generally and point out the usual troublemakers.

They weren't keeping a special eye out for any one thing. They were the eyes and ears of the inspector general, or IG, because they not only reported on trainees, they also reported on the cadre, which is why the cadre was normally not informed of their presence. Fraternization was one of the issues they looked for but the suspected set-ups were not the reason Boehner was there. "Obviously, she did that very well. So, what could she have failed at? I know the military expects you to be clairvoyant upon entry but very few people are. What's her real problem?"

Henning paced a little. "What's yours?" They looked directly at one another. "You talk a good game about being objective but you're quick to point out that you don't think these drill sergeants were set up."

"I never said that," Dale protested, a little taken aback. "Nor do I mean to imply that. I'm just trying to be fair. Besides, it's not what I think that matters. I need to investigate this case thoroughly and properly and in order to do that every angle has to be explored. Listen, Alpha-10 is special to me, too. I took AIT with this company six years ago. Bad things are going to come up, they always do during an investigation, which is why I need to know everything about this case so that there are no surprises. Or let's just say the fewer the better."

Henning hesitated and then shrugged. "You know this business better than I do."

"Well, I'm sure if Lieutenant Walker is worth her badge, she'll say pretty much the same thing." Dale gathered up her file and information. "Is this everything?"

"Yes."

"Thanks. I'll look at it all again tonight. I may have some other questions in the morning." Dale yawned. "Now I'll let you go home and get some sleep."

"Where are you going now?"

"I was going to stay at the Bishaye's but they are conveniently out of town. Not that I am accusing it of being anything other than a coincidence," Dale added, dryly. "So, it's either the post guest house or the Journey Inn in Averill. At this late hour, I'm sure it will end up being the guest house. Would it be too much trouble to get a lift? If you're finished here, that is."

Henning nodded. "I'm done. Listen...Dale..." she hesitated. "Why don't you stay with me? I've got a couch that makes into a very comfortable bed."

"I really don't want to put you out. I appreciate the offer, but —"

"So, you'd rather go to the post guest house, normally overrun with screaming, spoiled dependents where you can never get a decent night's sleep?"

Once again, Dale started to resist. There was a slight tension between them and Dale wasn't exactly comfortable around strangers but the thought of the usual racket of guest house dependents made her reconsider. "Sure. I'd love to if it's no trouble."

Dale moved behind Henning to exit the CQ office, realizing for the first time how short Henning really was. It was an observation that would have to wait for a later time to joke about if they ever reached that level of comfort with one another.

Dale looked around as they headed toward Henning's brown and beige Malibu. She looked around the company area one last time as a free woman, shaking her head, mumbling, "I must be fucking nuts."

Dale got into the car and Karen Henning drove them away.

Karen Henning's quarters were located beyond the other side of the post, which gave enough time for her to reveal quite a bit about herself. There were the usual *why'd you join the Army* questions and in her answers she exposed herself to be a woman with a very charming personality, who could be engagingly persuasive when she wanted to be. She appeared to be easy-going but Dale got the impression she could be one tough customer, too. She did possess a substantial amount of tenacity, a persistence that, if properly developed, could rival Anne Bishaye's.

There was something unambiguous about Henning. Probably it was that she lacked the usual vanity attached to new officers. Dale found that uncommon since most officers just out of OCS had an infallible and arrogant way about them. Most were reluctant to admit a fear of not knowing everything that was happening especially around their own company. Henning was not.

"I'm apprehensive about being relatively new and holding an authoritative rank, even if it's low man on the officer totem pole," Karen said. They had settled into chairs opposite one another. Her place was half of a small duplex in an area popular with GIs who wanted affordable, off-post housing. "I'm aware that the drill sergeants and other cadre compare my military knowledge as a second lieutenant to that of a basic trainee and it bothers me that they find it so amusing when I make mistakes."

Dale sat quietly while Karen unloaded her frustrations on someone she maybe saw as a kindred spirit.

Karen continued. "I fear that I'm trying so hard to fit in sometimes that I might be making a fool of myself, but I know I have the potential to become a good officer."

"Do you feel like sometimes you're resented for being a female who holds a higher rank and is younger than most members of your company?" Dale asked.

"Or just for being a female, period," Karen responded, the frustration in her voice clear.

"I understand. I encountered many of the same aggravations and prejudices when I was a training officer at Fort Jackson. I *did* know more about the Army when I became an officer and a majority of the cadre still treated me as if I were an exceptionally slow-witted trainee."

"How did you like that assignment?"

Dale smiled. "I didn't. The job at Jackson was my least favorite assignment. It felt, in more ways than one, like a step backward."

"Really?" Karen seemed surprised by that statement.

"I enjoyed the challenge of undercover work and was doing well in it but to be thrown back into a basic training environment seems almost parallel to punishment. My only advantage above a trainee, aside from rank, was a little more freedom and that's all."

Karen nodded and glanced at her watch. "Oh, I didn't realize how late it was getting. Let me put the water to boil for some tea and we can make up the couch."

While Karen was out in her kitchenette, pouring two cups of mint tea, Dale thumbed through some new military field manuals. She wasn't really concentrating on what she was doing. She was thinking that she found Karen Henning attractive and was tormenting herself as to why she was starting to frequently have these thoughts about women.

Karen returned to the living room, interrupting Dale's introspection and handed Dale her cup of tea.

"Thank you," Dale said. "Maybe now I can relax and stop being such a bear."

Karen smiled. "I understand, what with your discharge so close, you probably hate this. I don't blame you. If somebody threw another couple of months of training in my face, I don't think I could handle it."

"At least it'll be better, I won't be as intimidated as I was the first time."

Karen, who had changed out of her fatigues into old blue jeans and an oversized sweatshirt, sat in a chair, curling her legs underneath her. Dale sat down opposite her again and slowly sipped her tea.

"Do you remember your first night at McCullough?" Karen drawled.

"Oh, yeah." Dale thought back. "It was complete culture shock, different than anything I'd *ever* known. And I couldn't believe how nasty the drill sergeants were. It was frightening. I was just out of high school and somewhere along the way I picked up the notion that I knew everything. That thought sure didn't stick around for long. The Army taught me a lot, probably more than I needed to know, and a lot of reality hit me in the face between the time I stepped off that bus in WacVille and the time I stepped off that plane to my first permanent duty station. I grew up a lot." Dale paused to sip her tea. "Which is maybe why Bishaye is insisting I go back...so that I can grow up again."

"Well, just don't be too hard on my drill sergeants," Karen said and smiled. "They've been through enough as it is."

"Come on, drill sergeants need some lively trainees to keep them on their toes. Besides, I want to fu...play with their minds a little, the way they did with me and my fellow newbies in my first cycle."

Karen chuckled at her sudden verbal detour.

"Tell me," Karen began. "Out of curiosity, do you get a lot

of flak from civilians about being a woman in the Army? Do they usually try to label you as either a whore or a lesbian?"

"You get that, too, huh?" Dale smiled ruefully, not at all surprised by the question. "I just chalk it up to ignorance. I don't pay attention to it anymore but it used to bother me. Being from a small town and all, the thought of being categorized as either horrified me. Then, of course, I got to know the women I was in with and I realized that they were no different than I was regardless of their preferences."

Especially since those lines have become so personally blurred lately, Dale thought.

"I'm just starting to get to that place myself," Karen acknowledged.

"I soon discovered that being a female in the military is a fight all its own that is rarely made public. I think that civilians who really aren't familiar with today's Army think that every female who signs up becomes a grunt. The visual they like to put out there of a woman in full war gear ready to go into sole battle with the Middle East doesn't exactly help our image any. We're either portrayed that way or as if we're ready to take on entire battalions in the sack in record time. I just reached a point where I said, what difference does it make. People are going to believe what they want to anyway. We're dealing with stubborn, old-fashioned ideals that probably won't be changed in my lifetime, regardless of the strides women make, especially to the old fart, hardcore, career military men. Getting them to alter that opinion would be like asking the KKK to support the United Negro College Fund."

Nodding, Karen sipped on her tea, absorbing Dale's words. "What about relationships?"

"What about them?" Dale asked,

"I find it very difficult to maintain relationships with military men because their frame of mind is, well...let's just say, different. And, civilian men just don't understand *my* frame of mind. Plus, it is a nomadic life. The colonel told me that you have a boyfriend, someone you have been dating for about five or six years. How have you made that work?"

Sighing, Dale shook her head. "It doesn't. Not anymore. We broke up last week."

"Oh. I'm sorry."

"Don't be," Dale assured her. "It ran its course." Now it was Dale's turn to change the subject. "So how is the PT now? I remember that when I went through AIT here before they had a couple of exercises that defied human potential."

"The physical training is pretty mild or at least it has been. It's a lot of running, though. Think that'll bother your foot?"

"I hope not. If it does, I'll see what I can do about getting a profile."

"Think our wonderful Alabama winter weather will have any effect on it?"

"That much I am betting on."

Dale..." Karen hesitated. "I'm not sure I should broach this subject but I really want to know what happened to you. Colonel Bishaye touched on it briefly but never went into any detail. Would it bother you to talk about your incident?".

"No. I've talked about it to so many people already."

"Why so many people?"

"Oh, you know, investigators, military and civilian, like CID, FBI —"

"FBI?"

"Sure. And they were just the tip of the iceberg. You see, nobody knows who tried to kill me. And since I've been involved in so much covert stuff that involved both military and civilian personnel, the FBI, among other agencies, had to check it out."

"And there are still no leads?"

"If there are, no one is telling me about it."

"What was going through your mind when those men forced you into that car?"

"Just how much did she tell you, anyway?" Dale asked a little surprised.

"She was brief."

"Well, it scared the shit out of me. But what frightens me most is that I still don't remember a hell of a lot. It's just been recently that any of it has started to come back to me. I was totally blank for a long time. I do recall feeling unusually calm, like my time was up and I knew it and absolutely nothing I said or did would change it. When the barrel of that gun was stuck in my back, all I could think was, God, don't prolong it, just do it."

"Did any of them say anything to you to give you a clue as to who they were or why they were doing it?"

"Not that I can recall. I can vaguely picture that there were two men and a driver, that the car was a large sedan and that the license plate was from Florida. I didn't know I had pissed someone off in Florida."

"You were pretty lucky."

"Lucky?"

"At least you weren't killed," Karen said, gently. "You could have ended up maimed or crippled, but you didn't."

"And I'm able to come back into the Army, out in the open, ready to give whoever it was another shot at me. Hey, you only live once, right? Unless you're James Bond."

CHAPTER FOUR

November 1977 Alabama

Second Lieutenant Shannon B. Walker strolled into the downtown Averill, Alabama bar and stopped to let her eyes adjust to the dim, smoky room. Her light blonde hair reflected green, blue, red and yellow from the colored lights as she stood in the doorway. She surveyed the room quietly, taking a long drag from her cigarette and resting her free hand on her hip. It had been a while since she had traveled this turf and, although the place was exactly as she had remembered it, it was somehow different. But to the regulars, she was a stranger and because she was new, because she perceived herself as half-way decent looking and because she seemed in such control, every eye in the place was on her. She dropped her hand to her side and moved casually through the room to the bar, ignoring the attention she was getting.

Shannon made eye contact with the bartender. He wiped off a section of the counter directly in front of her, then leaned in closely and grinned. She counted at least two teeth missing before she wisely decided to stop studying his mouth. Shannon returned a very non-committal smile.

"And what would y'all like this fine evening?" he asked.

"I all would like a million dollars. But I'll settle for a brandy on the rocks, and easy on the rocks, please."

The bartender attempted a sly, sexy laugh that got caught in his throat and sounded more like he was clearing away an impure substance. "So, you're one of them brandy girls."

Shannon looked at him blankly. What did that *mean*? Then she had to remember the species she was dealing with— a horny, deep-woods, good ol' boy who was no doubt the missing link, or very closely related. "Yep, that's me. One of them notorious brandy girls." She waited patiently as he just stood there, leering at her. Finally, she looked at her watch. "Am I going to get it tonight?"

"If you want it, baby."

"I meant my drink."

"Sure, darlin,' comin' right up."

As he left to make her drink, Shannon glanced around the bar again. When she turned back, the bartender was there with her glass of brandy.

"How much?" she asked, digging into her pocket.

"It's on me."

"No, I insist," she said.

"No, honey, I insist."

Shannon backed down and shrugged. *Okay, asshole, if you really insist*, she thought.

He leaned in closely again and was starting to get on her nerves. All she wanted was a damn drink. Why couldn't men read signals that said *just because I walked in here by myself does not mean I don't intend to leave by myself or at least not with you*?

"What would you say if I asked you to meet me here after work tonight?" he asked her, giving her body an obvious once-over.

"Probably nothing." She picked up her glass. "I can't talk and laugh at the same time."

The bartender did not seem fazed. "That was cold, darlin.' Come on, baby, let me take you home tonight after I get off shift. I got somethin' special for you." He looked downward in the direction of his fly.

Shannon leaned across the bar, stared directly at his crotch, then blandly looked back up at him, smiling ever-so-sweetly. "Oh, goody. I'll be sure to bring a magnifying glass and some tweezers."

His smarmy smile was replaced by a scowl. "It's a buck twenty-five for the drink."

Shannon placed two dollars on the counter. "Keep the change. The way you try to sell yourself, you probably need it." Her smirk never faded and his never returned. She spun away from him, sipping her drink, looking for a female sitting alone, wearing a sweatshirt that had *Property of Fort McCullough* printed on it. Her eyes reached Karen Henning's and the lieutenant smiled.

Seeing the exchange, the bartender grumbled. "No wonder you didn't want me. Goddamn Army women."

Ignoring him, Shannon approached Henning's table. "Karen Henning, I hope?" Shannon suddenly thought if this wasn't the lady she was supposed to meet, she might have to get really creative to get out of it, without offending this woman. Anywhere near a military installation, that could have gotten complicated.

Still grinning, Karen held up her hand. "Relax, Lieutenant Walker. I'm Henning. I pegged you the minute you walked in the door. How was your trip?"

"Terrible. So, how are things at wonderful Fort McCowlick?"

"Does the term FUBAR ring a bell?"

"Ah. Fucked Up Beyond All Recognition. So, nothing has changed." Shannon sat down opposite Henning.

"Has Colonel Bishaye told you anything?"

"Other than to be here tonight and the basic stuff? No. She said you'd brief me."

"Okay, great. I really don't want to get into it here, though. You never know who's listening."

"Roger that," Shannon agreed, finishing her brandy. They both got to their feet. "Damn, you're short. Stand up."

Karen looked at her and rolled her eyes.

"What?"

"Gee, I've never heard that before," she said. Her tone of voice and expression indicated she'd heard it way too many times.

"I can't help that I'm so observant. I'm an MP," Shannon said, dryly.

"The night I met Lieutenant Oakes, she made that exact same comment to me."

"Lieutenant...Oakes? Is she the other agent working on this case?"

"Yes. I met with her on Monday. "

"It's a shame we both couldn't have met with you at the

same time. It would have been nice to spend some time with her beforehand so I could gauge how we might work together."

"She said the same thing."

Shannon associated the name with one from her past. "Just out of curiosity, what's her first name?"

"Dale."

Shannon whirled to face Henning squarely. "Dale? Dale Oakes?"

"Yes. Why?" Henning asked, leaving Shannon to drop their glasses off at the bar.

Shannon absorbed the news, then caught up with Henning.

"I knew a Dale Oakes long ago and far away. I never thought there could be two of them. The Dale Oakes I knew was enlisted, an E-3, when I last heard from her. She's most likely out by now. She wasn't very military oriented but then neither was I. What does this Dale Oakes look like?"

"Medium height, slender, athletic, long brown hair, brown eyes, attractive...distrustful. Of course, that's understandable with what she's been through."

"Are you serious? *Lieutenant* Dale Oakes?" They stopped in the middle of the parking lot. This was incomprehensible to Shannon. "Impossible. You'd have to know my Dale Oakes. There's no way. It couldn't be."

"It surely is. After I saw her, I got your paperwork and I read where you two had trained together. Why didn't she recognize your name?"

"Because the last time she heard from me, my last name was Bradshaw. I got married and divorced almost as quickly. Do me a favor. Don't say anything when you see her again. Colonel Bishaye obviously did this on purpose. I'd like to keep the surprise."

Henning wore a skeptical look. "I'm not so sure she's the type of woman I'd want to surprise."

"Oh, she ain't so tough," Shannon assured her, as they got into Henning's car and left.

November 1977 New Hampshire

Three weeks after Dale met with Henning in Alabama, it

was time for her to start her assignment. Dale peeked around the corner of the career counselor's office at the AFEES building in Manchester, New Hampshire. Dale was there to get sworn in again, starting the onset of her undercover assignment.

Dale saw a woman sitting at her desk diligently working and Dale smiled.

Sergeant Theresa Burke helped new recruits decide what they were going to do during their time committed to Uncle Sam. Based on their qualifications, desires, test scores, occupational availability and what the military needed, she aided them in selecting their Military Occupational Specialty or MOS.

She was also the same Theresa Burke who locked Dale out of the latrine at three o'clock in the morning after she had just come in from a night of drinking when they were both privates stationed at Fort Ord. Dale vividly remembered dancing in the corridor of the barracks while Burke made her promise to never drink and drive again. At that point, Dale would have promised to become a geranium if it had meant relief for her bladder.

"Hey," Dale entered the office, breaking the silence.

Theresa Burke looked up from her work and grinned, cheerfully. "Hey, slick! I've been waiting for you to get here. Getting close to zero hour, huh?" She stood up, rounded her desk and met Dale halfway as Dale came forward and gave her a hug.

"How are you doing, Tess?" Dale observed the short-haired, green eyed blonde at arm's length. "You look great. You know, I never knew you were down here until the recruiter in Rutland mentioned your name. I would have been here to party long before now."

"Oh, that's okay. I haven't been in the partying mood much," she said, as Dale sat down on the edge of her desk. Sergeant Burke sat down in her chair. "All I'm doing is marking time, really. I'm a two-digit midget. I'm getting out in ninety-two days."

Dale was stunned. "That can't be your decision. Are you being forced?"

"Not officially forced. But ever since they froze my rank and sent me out of Fort Bragg and here to this spot in the sticks, I've felt forced."

"Why did they freeze you at E-5 and why did you get sent

out of Bragg?"

"Oh, some bullshit came down after I dropped a dime on my company commander for sexual harassment."

"Did you press charges?"

"Yes," she answered, wearily. "I took it to the IG, got a JAG lawyer, the whole nine yards."

"And the charges stuck?" Dale asked, hopefully, regardless of how unrealistically. Knowing Burke the way she did, Dale was sure it had to have been bad for her to have gone as far as pressing charges.

"Get serious. They sent me here and they slapped him on the wrists. Then they reassigned him to the battalion across the street and promoted him to captain. Fuck up and move up."

"He commits the crime and you pay for it. Wonderful system we have here. Just lovely. So, you're getting out. That makes me very sad. If anyone was ever meant to be in the Army, it was you. Don't take that as an insult. What are you going to do?"

"I'm going to be a juvenile counselor."

"Well, you're certainly juvenile enough," Dale said, with a laugh. "Seriously though, I thought you didn't like dealing with kids."

"I don't but it pays really well, and it beats playing strip solitaire." She shook some hair away from her forehead, then picked up some papers and waved them at Dale. "I've got your contracts here." She set them back down. "Are you ready to let Uncle Sam exorcise the civilian life out of you again?"

"Shall I open a vein?"

"Not good enough." She became serious and pushed the paperwork toward Dale. "This is a Meddac form, an affidavit really, and a sworn statement that's been co-signed by the doctor about your foot. Read it before you sign it to make sure you are not waiving any of your disability rights. I read it and it looks on the level but you'd better look at it just in case." Tess Burke studied her as Dale perused the paperwork. "I got a letter from Cindy a couple of months ago," she said, referring to a mutual friend.

"Oh, geez, I've got to write to her. How is she?" Dale asked, as she signed the sworn statement and passed it back to Burke to witness.

"She's fine. She said you really got hurt."

"Yeah, I did. She was in my unit at Jackson when they found me." Dale changed the subject. "Bishaye called you and explained about this job, didn't she?"

"Yes, but she didn't go into any detail. It was good talking to her. Is she still as unbelievably gorgeous as she was at Ord?"

"Nah. She's a dog."

"Yeah? If that's the case, I'd gladly be her chew toy."

Dale laughed. "She's a dog and you're a fucking hound. Actually, she's even more gorgeous if you can believe that. Time definitely agrees with her."

"It wouldn't dare not to."

"Yeah, well, I wouldn't mind waking up one morning looking like Anne Bishaye."

"I wouldn't mind waking up one morning looking *at* Anne Bishaye," Burke said and grinned.

Dale thought, *hell, yeah, me too,* but instead, smiled back and answered. "At least you dream big, Tess."

Burke reached across and took the signed form as Dale started reading her contracts. "Even as close as you two have always been, I'm really surprised she picked you for this. According to Cindy, you were pretty mangled but you look worse than she said." She dodged just in time as Dale swatted at her. "Are you sure it wasn't a pissed off drill sergeant that nearly killed you? I mean, I *know* you..."

Dale laughed again, signing the contracts. "You're warped, you know that?" She handed back the pen.

"So, you're off to Fort McCullough, the happy hunting ground. Oh, that's right it's not a happy hunting ground for you. I keep forgetting how disgustingly straight you are."

Dale shifted, uncomfortably. Should she admit to someone she considered a close friend and confidant, her gradual realization that she may not be as heterosexual as originally thought? If she could talk to anyone about it, would Tess be the one to disclose these nagging feelings to? She would trust Sergeant Burke with her life, but she wasn't ready to trust her with this yet. Dale grinned at her. "Well, we all have our faults."

"And one of yours is you can be so close to someone like

Bishaye and not have any desire to take advantage of it. That just floors me," Burke shook her head.

"Want to trade jobs?"

"Gladly." Burke turned pensive. "I don't know. Maybe it's time to get out. It doesn't seem to be fun anymore. Example. Look what happened to me. Do you believe it? My attitude and they make me a career counselor? Remember how I used to say military intelligence was a contradiction of terms? Do you know that they were talking about giving me— me—the one who was infringed upon, an Article 91 for insubordinate conduct? Nothing has turned out the way I thought it would. Not only that, people seem to be playing for keeps now. With what happened to you and then I just found out that a girl, one I sent down four months ago, got murdered last week."

"Really? At McCullough?" Dale perked up.

"No. Here. She got kicked out of AIT."

Dale's curiosity piqued. "Why?"

"I'm not sure. It was a chapter five discharge, which could mean any number of things. She was in A-10. Isn't that where you're going?"

Dale's eyes darkened with suspicion. "Yes. What was her name?"

"Carolyn Stuart."

"Carolyn Stuart was murdered?" Dale's mouth went dry.

"Did you know her?" Burke asked, noticing the stunned look on Dale's face.

"No. But she's one of the reasons I'm going back to McCullough. What happened?" Dale couldn't locate Stuart when she had tried to look her up for questioning two weeks before. Nobody seemed to know where she was and, even if Dale had found her, rumor had it that Stuart wasn't talking. Now she *really* wasn't talking. She was dead. Why didn't anyone ever tell her anything?

"I really don't know what happened, the newspaper wasn't too clear about it. It was all very odd. We used to frequent some of the same bars and I ran into her a few times after her discharge and we talked but never about her being kicked out of the Army. She avoided the issue altogether. And she had definitely changed. She acted all jittery. She couldn't even look at me when we talked, which was unusual for Carolyn because she was always such a flirt. Eye contact was her thing. Instead, she looked everywhere else, especially at doorways and she was insistent about always sitting with her back up against a wall."

"Not a good sign."

"No. Then five days ago they found her body inside one of the condemned buildings in the south end. She was shot three times, twice in the back and once in the back of the head."

"Jesus," Dale commented, hit hard by the news. This complicated matters a great deal. "What kind of person was she?"

"She was immature. But she was fun. Her GT score was high enough so that when she requested MP, we gave it to her. I thought it was an odd choice for her and I told her so but she really seemed desperate to get in. She qualified so we sent her down for it."

"Why was it an odd choice?"

"Because I could never see her concentrating that hard or getting that serious. I liked Carolyn but she doesn't, didn't have a lot of boundaries."

"Then why was she accepted into the Army at all?"

"I guess we all thought basic and MP training would change her. She seemed determined and she was a bright girl. I was hoping she would realize her potential and straighten out. So to speak."

"You know, according to the people I spoke with at McCullough, she was really open about being a lesbian. Didn't you warn her about that before you sent her down there?"

"Yes, and I thought she understood. But, like I said, there was really something bizarre about the whole thing."

"Do they have any suspects? Any clues? A ballistics report? Anything?"

"I don't know about a ballistics report. The police around here wouldn't give that information out, anyway. And, according to today's newspaper, there is still no solid lead. I spoke to one of the reporters who told me that when they located Carolyn's lover, she said that sometimes a female would drive Carolyn home from her night art class but the lover never actually saw this woman. She told them that Carolyn was extremely upset on the evenings this mystery woman would drive her home. She said Carolyn never discussed the woman with her, in fact, her girlfriend said she never would have known this woman existed if she hadn't seen her drive away once. She thought Carolyn might have been having an affair and, if so, brings up at least one possible motive for murder."

"A reporter told you all that? Was that in the paper?"

"No. This came from a liaison reporter who works with us to put enlistment or promotion news in the community section."

"Interesting."

"Yeah," Burke agreed. "Regardless, paranoia wasn't normal behavior for her and that's all she seemed to display after her time at McCullough."

"So, no one has any idea who this mystery woman was? And what about the car? Did she get a good look at the car?"

"If they did, the police aren't giving out any info."

Dale checked her watch and stood up. "Do me a favor. Keep track of all this and send me the information. Try to get me a list of people in her art class, too. I don't think it's too wise for you to send me the newspaper clippings, though. If I get caught with them, my cover is as good as blown. Better yet, send everything to Karen Henning, the training officer. She's in on all this."

"Karen Henning?" Burke wrote it down. "Same address? A company, 10th battalion?"

"Yes. She's a second lieutenant. And try to get Anne Bishaye on the phone. She never mentioned any of this to me, so she might not know. I'd like to talk to her."

"Me, too," Burke said, waggling her eyebrows. "Do you have an autovon number for her?"

"No. Not with me. I have her home phone number, though."

"As much as I'd really love to have that, I would feel very awkward about calling her at home. I'll call the post locator and get her office number." Burke glanced at the wall clock. "You have to go get sworn in. In fact, you're late."

"I know. Don't go anywhere. I'll be right back." Dale hurried out of the office and down the hallway to the ceremony room, where an Air Force lieutenant was about to administer the enlistment oath.

"You're off to a fine start, young lady" the lieutenant said, scowling at her as she took her place in line in the back of the room.

"Sorry, Sir. I was in with the career counselor," Dale offered.

He grumbled something inaudible, then asked the eighteen assorted people in the room if anyone wanted to change their mind and leave and, if so, now was the time to cry uncle and make their move. Even though she and everyone else remained silent, Dale's mind suddenly screamed out that she didn't want to go. This so-called piece of cake assignment was getting more complicated by the minute. She didn't care for this case to begin with but now that murder was involved, she looked forward to it even less.

"Raise your right hand and repeat after me."

Eighteen right hands shot into the air.

"I...state your full name...do solemnly swear...that I will support and defend...the constitution of the United States...against all enemies, foreign and domestic...that I will bear true faith...and allegiance to the same...and that I will obey the orders...of the President of the United States...and the orders of the officers appointed over me...according to regulations...and the Uniform Code of Military Justice...so help me God."

November 1977 Maryland

Shannon Walker had learned the day before that one of the alleged victims in this case had been found dead.

She wished she'd had this information before she had tracked down and cornered two ex-drill instructors and three former trainees included in the questionable fraternization charges. The two drill sergeants were still incensed and confused and their stories echoed the investigation reports. The three females stuck to their initial accounts unless Shannon threw in a trick question. Then they played stupid or just didn't talk at all. Later that week, Shannon, who used the false surname of Robertson, decided to employ a different approach after hearing about Carolyn Stuart's death from Bishaye. She went back to see Andrea Willensky and tried to intimidate her into talking.

"You again?" Willensky observed, annoyed, as she closed

the door of the MPI office at Fort Mead, Maryland.

"Sit down, Private!" Shannon ordered.

"Look, Lieutenant Robertson, I told you —"

"And I don't believe you! You're in serious trouble, Private, *very* serious trouble. Who paid you to set up Drill Sergeant Halpin?"

"Nobody! I didn't set him up, damn it, he tried to rape me!"

"That's bullshit. Why don't you just stop playing games with me? Look, I'll give you one more chance. You tell me everything right now, and I mean *everything*, and when we pick this guy up, you'll be given consideration for a much lesser charge."

"I don't know what you're talking about, *Ma'am*. Hey, I want a JAG lawyer here or something. There are laws against this kind of harassment. Drill Sergeant Halpin was in the wrong! The man tried to rape me and I reported him for it and now I'm being tormented about it and by another woman, too. I would think, as a female, you'd understand it. It was a traumatic experience, so with all due respect, Lieutenant, get off my fucking back! I am trying hard to forget it, why won't you let me?" There were tears in her eyes.

Shannon leaned up against the wall and glared at her. She clapped her hands together slowly, deliberately. "Sensational performance. Who are you auditioning for? No one is here, dear, not even Allen Funt and his candid camera. Just me. No two-way mirrors, hidden microphones, nothing. So, if you're done acting, I'd like you to listen to me very carefully. Your AIT buddy, Stuart, the other party to this little snow job, didn't make it. She's dead. She was murdered. Shot three times, twice in the back and the third shot almost took her head clean off."

Shannon watched Willensky's eyes close and tighten on that one. "I don't know if you really realized what you got yourself into but these people mean business. So, go ahead...sit there and play innocent all you want but let me tell you something, Andrea, they know we're onto them," she lied. "And they're scared. We're just waiting for them to make another mistake. Their first one was having Carolyn Stuart murdered. The second one might be having you killed. If we're lucky enough to catch them before that happens, you'll be busted right along with them and the other women under nice little charges like conspiracy and withholding evidence. If you are as smart as you're supposed to be, you'll talk before somebody else gets hurt, somebody like you."

Shannon let her words sink in. "So, I hope whatever you were promised or whatever amount you were paid, it was worth it and whoever you're protecting appreciates it."

Shannon also hoped that she was right, that Willensky hadn't been assaulted by Halpin because if this turned out *not* to be some kind of set up, she would never forgive herself for being so cavalier to Private Willensky's pain. She knew rape in the military was no joke and never something to take lightly, no matter how much the male brass tried to sweep the crime under Uncle Sam's rug. With the evidence she had however, regardless of how circumstantial, her instinct told her she was not wrong.

After approximately a minute of thought, Willensky stood up. Her expression and attitude had definitely changed. Her gaze was level, fixed on Shannon and her voice was cool and even. "I have nothing more to say to you except you're wrong. I'm sorry for what happened to Stuart. I don't know what she was into but it didn't include me," Willensky responded. "May I go now, Ma'am?"

Well...she tried. This perpetrator, whoever he or she was, knew how to choose accomplices. They might have made excellent prisoners of war. Even under duress, they remained calm. Although Shannon had yet to resort to torture, it was beginning to be a thought.

"Okay, Willensky," Shannon said with a sigh, somewhat defeated. "I still don't believe you and I sincerely fear for your safety. I also urge you to think about everything I have told you today. If you do change your mind, get in touch with Specialist Lancer here at this office. He'll know how to reach me."

Shannon watched her leave the room and wished she had been gifted with the ability to read minds.

November 1977, New Hampshire

It was unfortunate that Anne had not been available to discuss this new development before Dale took off but she knew

PERMISSION TO RECOVER

Tess would pass the information along to Bishaye as soon as she could get her on the phone. That was if she wasn't already in possession of this news and, if she was, why in hell hadn't she passed it along to Dale? Of course, she could have and Dale may not have been around to answer her phone.

This definitely changed things, Dale reflected, as the jet taxied down the runway of the small Manchester airport. Carolyn Stuart's murder changed everything. Too many related events were happening for it all to be coincidental. Somebody was quite a bit more invested in the cause than even Anne Bishaye had originally suspected.

But it was November twenty second, doomsday. She was on the first leg of the journey back to Alabama to get right in the middle of it where she was going to be practically under lock, key and guard. She and Lieutenant Walker were going to try and figure it out from the inside now. With the murder of Carolyn Stuart this case had become far more serious, it was now deadly.

CHAPTER FIVE

November 1977 Georgia

The United Service Organization room in the Atlanta Municipal Airport was located on the lower level of the main terminal and when Dale saw it again, the past flashed before her eyes so suddenly and vividly, it sent a momentary wave of panic through her. She paused at the doorway, composed herself and made a quick, visual search of the room.

Most USO rooms were the same. They usually included couches, chairs, shelves of not-so-current reading material, and some assortment of game tables—pool, card, chess, backgammon, a raised shuffleboard, foosball, etc. They also offered soda and candy machines and a television set which, ironically, always seemed to be showing something military. Currently, everyone was being entertained by a showing of *GI Blues*.

Dale found an empty seat toward the back of the room among the large assemblage of recruits. For the time being, she chose to sit and observe. There was a lot of getting acquainted, nervous laughter and paranoia going on.

And then there was Helen Zerby, a little chatterbox who had enlisted for a communications slot and had taken her oath in New Hampshire with Dale. Helen had talked non-stop from New England to Georgia and Dale got hoarse just listening to her. Helen had already made some new friends so Dale didn't have to worry about that distraction for a while.

Sooner than Dale had expected, however, her concentration was broken and her interest was drawn back in Helen's direction when a rather athletic, masculine-looking female loudly expressed her opinion about the Elvis Presley movie that had just started on TV. Her remarks were heavily sexual in content and immediate attention was cast upon her. The various conversations resumed after the initial curiosity wore off and the woman's foul mouth gradually blended in with the rest of the noise.

The familiar sounds and even the smells of the crowded USO room started to close in on Dale. An abrupt rush of claustrophobia sent Dale outside of the terminal for a breath of fresh air. She leaned against a wall as she inhaled deeply several times and wondered again for the hundredth time what she was getting herself into.

When a uniformed employee walked by her. Dale asked him how long it would be before the bus got there.

"Well, they said it would be here in two hours but that was four hours ago."

"Thanks, I think," Dale said. Not even at McCullough yet and the hurry up and wait game had already begun.

No more than ten minutes later, a young man carrying a track bag came outside to the loading area.

"They just announced the bus," he told her.

Dale nodded.

He looked at her. "Can I ask you something?"

"Sure."

"What are you down here for? What field are you going into?"

"Military police."

"I just heard that the guys and girls will be training together for basic."

"Yeah. I heard that we'll all be together straight through until the end of law enforcement school."

"Shit," he said, unable to hide his disappointment. "What a joke."

His reaction didn't really surprise Dale but it did frustrate her. She shrugged at him. "Time to come out of the dark ages, it's almost 1978. Things change. Besides," she added with a smile. "Maybe you'll learn something."

Dale could tell by his expression that he was about to respond in a negative manner but never got the chance because the capacity of the USO room came spilling outside to join them and he got swallowed up in the crowd.

She took her time going back inside and getting her suitcase out of the locker she had stored it in. When she returned outside, the bus was being loaded. Her body automatically followed commands to board, but her subconscious kept sending her very negative feelings about even leaving the airport.

Once she was settled in a seat and the bus was underway, she evaluated her assignment with a sense of foreboding. The case had changed completely the moment she'd learned of Carolyn Stuart's murder. Hell, assassination, because that's how it looked to Dale with two in the back and one kill shot to the head. But then maybe it was just a coincidence and had nothing to do with any of this. She heard laughter in the back of the bus and the loud, strident voice of that one female again but Dale really couldn't make out what she was saying nor did she much care. If this woman were going into MP training, Dale's first instinct would be to eliminate her as a suspect because of her blatancy. On the other hand, Stuart was brazen, too, but then common sense should make a drill sergeant steer clear of a trainee with similar characteristics. She needed to get more background on everyone. But who could she trust for that background intel, who was reliable? Definitely Bishaye and Burke but what about Henning, or Boehner? Or even Lieutenant Walker for that matter?

"This is ridiculous," she mumbled. She needed to take a deep breath and focus.

Dale stretched out as much as she could without disturbing the dozing recruit next to her and tried to eavesdrop on bits and pieces of conversation. The discussion going on directly behind her had caught her attention. She put her ear to the space that divided the two seats.

"So, what are you going to do now?" a warm, light soprano voice asked. Dale placed the accent to be from the upper Midwest.

"I'm getting out as soon as I get there. I ain't puttin' up with any of their bullshit. I'm getting out," answered a hard-edged, reedy voice, her accent definitely inner city.

"Well then why didn't you get out at AFEES before you even took the oath? They give you the chance," the first woman asked.

"What could I do? My parents were standing right there and they wanted me in. Then later when I asked the career counselor he told me they'd handle it at McCullough. But he told me not to worry, there shouldn't be any problem."

Dale shook her head sympathetically. She knew that old trick and the young female recruit was not getting out that easy. One of the biggest games the military played was called pass the buck. She knew of many a confused soldier who ran into unnecessary difficulty due to a supposedly qualified individual's response of *I can't do anything for you, your next station will have to take care of it*, which would only compound the GI's problems. By the time the soldier got to the next station and got anyone else to listen, it was usually too late to help him. It was this kind of impersonal behavior that led to bad attitude charges and AWOL offenses.

The conversation behind her dwindled into silence and Dale stared out the window at the dark night.

November 1977 Alabama

"Everybody off the bus, let's go, we've got a lot to do!" It wasn't a drill sergeant shouting those words and that surprised Dale. Usually, as soon as the bus pulled to a stop, and sometimes before, at the reception station, a drill instructor stomped onto the bus and scared the living feces out of everyone. But to Dale's observation, this guy seemed to be just a normal, everyday, annoyed and tired staff sergeant who wanted to hurry this whole mess up so he could go home and go to bed.

The new recruits dragged their luggage off the bus and hurried into the building where they were lined up in the hallway and awaited their next command.

"Okay, listen up. There are forms laid out on top of the individual desks inside the lecture hall." The staff sergeant gestured with his head toward the double doors behind him. "After I divide you up into groups I'm going to send you inside and I want you to be seated by group. Don't worry about your suitcases. Just push them up against the wall and leave them here, they will be guarded. Men, when I call off your name, please give me a copy of your orders and go inside."

The sergeant alphabetically yelled names and when he reached the end of the list on his clipboard, he looked up. "Is there any male whose name I did not call?"

There was no reply, so he continued. "All right, you women are going to be divided into two groups, Delta-2's and Alpha-10's. The Delta-2's are regular WACs and the Alpha-10's are MPs. Deltas, after I call your name, give me a copy of your orders and go inside where the specialist will show you where to sit. Alphas, the same will go for you." He proceeded to call off more names until the crowd had shrunk to just seven women. "You must be the MP recruits," he said, almost managing a smile as the seven females silently appraised one another.

"Hey, Sarge." It was the loudmouth from the USO room at the airport. "Are we the only females training with all those men?" There was an anticipating smile lingering on her lips.

Dale waited for the non-commissioned officer to react with hostility at being referred to by a nickname only Beetle Bailey was stupid enough to use. Instead, he sighed. "Number one, my name is not Sarge, it's *Sergeant* Pulaski, and number two, I'm sorry to disappoint you but you are just the first group of several to come in. Please answer when I call your name. Almstead, Alexis."

"Here."

He collected a copy of her orders as she went inside. "Caffrey, Michelle."

"That's me," the husky, booming voice of the blustery female acknowledged and followed the same procedure as the recruit before her.

"Ferrence, Debbie." There was no answer. "Ferrence, Debbie?" He looked around, scowled and then scribbled something on his notes. "Kirk, Jascelle."

"Here and I have to talk to somebody about a problem, can —" Dale identified her voice as the one coming from the seat behind her on the bus, the one who wanted out.

"All problems will be discussed at your designated unit, there is nothing we can do about it here except hold everybody up and it's late enough as it is, don't you think?" He and the very young-looking black woman stared at one another until she just shrugged helplessly and went inside.

"Kramer, Brigitte." Again, there was no response. He focused on the four females left. "None of you is Brigitte Kramer?" Dale exchanged glances with the other women. If one of them was, no one was admitting it.

"Damned computers," he mumbled. "Wave of the future, my ass." He scrawled something on his clipboard, then smiled. "Well...there are only four names left so we should come out even here. Kotski, Laurel."

"Here," a very tall woman with the face and body of a model, spoke up.

"Oakes, Dale."

"Here." Dale fought the instinct to respond with the proper *here, Sergeant* but since he had yet to instruct anyone on military courtesy, she certainly did not want to sound any more knowledgeable than the rest. Dale followed Kotski inside, then to where the other women in their Alpha group were seated in a section way up in the back, away from the men and the Delta women. Dale took the seat next to Kotski.

"We must have cooties or something," Michelle Caffrey sneered, as the last two recruits, Marilyn Segore and Renee Troice joined their group.

Pulaski, the clearly sleep-deprived NCO hopped up on the stage and explained how to fill out the tax forms, insurance forms and allotment information. It seemed to take an eternity before everyone was finished. When they were finally through, he dove into his required speech, supposedly to put everyone more at ease about what was probably going through their minds at that point in the process.

"If he knew what was going through my mind," Caffrey commented. "I'd probably be arrested."

Dale looked down the row at her then briefly studied the rest of the women in her group. If they were nervous, they hid it well. The only one who showed any signs of agitation was the Kirk girl who, at a closer examination, looked even younger than the first glance.

The sergeant was now telling them about the notorious amnesty box, which, in reality, was a trash barrel in the passageway behind the stage and informed them that this was their last chance to get rid of their contraband, such as books, magazines, candy, gum, weapons, drugs, marijuana, basically anything that kept them occupied and happy as a civilian. Whether a person had anything to dispose of or not, he or she still had to walk through the corridor before they were allowed into a classroom that provided bag lunches on a metal table in the back.

He then instructed them to go get whatever contraband they had and to meet him downstairs. Dale had picked up a pack of gum at the airport, which she tossed into the bin as she walked by. Behind her she could hear the mostly muted complaints from a few of the recruits about having to toss good stuff.

Downstairs, in the classroom with the bag lunches, the women sat in groups or by themselves to put their heads down on their desks to nap. Dale took a seat near Helen Zerby and found herself sitting beside Alexis Almstead and in front of Michelle Caffrey.

"What's up?" Dale asked Helen, who had finally stopped gabbing and looked like she was ready to collapse.

Helen rubbed her eyes with the heels of both hands. "I think I made a mistake. I wish I were home with my boyfriend."

"Yeah, I'm horny as hell, too," Caffrey stated, as she opened up one of the bag lunches. It consisted of a stale peanut butter and jelly sandwich and a hard-boiled egg. "Look at this! It's green!" She held the now peeled egg between two fingers, its over-cooked yolk showed through the white as green.

"They're just trying to get you used to the color, Caffrey," Almstead said and laughed.

"I can't eat this shit," she protested and stuffed everything back into the bag, crushing it into a ball. "I'd rather be eating something else," she said, eyeballing an oblivious Kotski with a lewd smile and some lecherous eyebrow waggling.

Dale blinked at that statement. Evidently gender wasn't an issue when it came to fulfilling Caffrey's sexual needs.

"Oh, Christ, is sex all you can think about?" Kirk muttered. She was sitting behind Caffrey. Her head was down on her desk, resting on her folded arms.

"Is there anything else?" Caffrey turned around, slowly.

"Besides, it's better than thinking up ways to cause trouble."

Kirk raised her head and looked at Caffrey, amused. She broke into a grin. "Caffrey, I'm not going to be here long enough to cause trouble. Especially with you."

Caffrey mirrored her expression. "That's good, Kirk, 'cause I don't want to fight with you, either." She turned to face Helen Zerby again and then said loud enough for everyone to hear. "I happen to like sex. And the more it hurts, the better I like it."

Dale rolled her eyes and wondered why there always had to be a Caffrey in every group of new recruits.

"Hey, I'm just being honest," Caffrey continued. "And I don't particularly care if I offend anyone. I'm the type of girl who says what's on her mind."

"If you really said what was on your mind, you'd be speechless," Kirk said into her desk.

Caffrey ignored her and concentrated on Zerby.

"You didn't offend me," Helen told her, but the startled look in her eyes contradicted that statement.

"Hey, Caffrey, why don't you tone it down a little, huh?" Almstead suggested, soothingly.

"Yeah, Caffrey, most of us don't give two shits if you like it between two sheets or two slices of bread. Some of us are trying to sleep," Kirk growled, her face buried in her arms.

"You wimps," Caffrey said, scoffing. She tried to make a hook shot from her seat to the wastebasket in the corner with her balled up paper bag, but she missed. She stood up and walked across to throw it away.

"Do you two know each other?" Dale asked Almstead.

"Not really. We're both from Wisconsin, though. So is Kotski. We all met in the USO room. We were the first girls to hit the USO room, so we started to talk," Almstead explained, as Caffrey returned to her seat. "What was your name again?" she asked Dale.

"Dale Oakes."

Almstead, the young tomboyish blonde, introduced herself and shook Dale's hand. "My friends call me Lex."

"Sexy Lexy," Caffrey intoned, soulfully.

Almstead rolled her eyes and looked away, shaking her head.

"I'm Michelle Caffrey," Caffrey told Dale. "And my friends call me a lot of things. Some of them are even nice. But my nickname is Mitch." She openly ogled Dale as though she were next on the menu.

Good God, she's obnoxious, Dale thought.

"Mitch the bitch," Kirk muttered, barely audible.

Caffrey swung her body around, threateningly. "If you know what's good for you, you'll knock it off, Kirk the jerk."

Kirk lifted her head up not at all fazed by Caffrey's attitude. "Kirk the jerk?" she repeated, almost incredulously. "Oooh, floats like a bee, stings like a butterfly. See me shaking in my high tops?" Kirk was trying not to laugh. "Where are you from, Caffrey?"

"Lacrosse, Wisconsin," she answered, defensively.

"Well, listen, Miss Cheesehead 1977, I'm from the Dee-troit ghet-to. Do *not* fuck with me."

"Then don't call me names."

"No problem, Butch," Kirk taunted, putting her head back down.

"It's Mitch," Caffrey said, sharply.

"Sure it is," Kirk burrowed her face in the crook of her arm.

"Fuck you."

"In your dreams."

Dale was actually beginning to enjoy the exchange. Caffrey, the one who obviously thought she was, and pretended to be, the bully, talked the talk but wouldn't walk the walk. Kirk, the younger and smaller of the two, refused to back down to her and verbally handled her like a pro.

"And you guys are our future MPs?" a Delta female spoke up, breaking the tension. "Great."

"Hey," Almstead jumped in, almost protectively. "Don't let that toughness or nickname fool you. She gave every last man who walked through that USO room door a one-to-ten rating. Most of them rated an eleven. Including the janitor."

"Jesus," Dale commented. "He was four days younger than God. Not very discriminating, are you?"

She could almost be likable, Dale thought, if she wasn't just so damned determined to be the center of attention. As Dale stood up to walk outside the classroom, she looked at Jascelle

Kirk again and wondered what her story was. She was on her way to the water fountain when she saw Laurel Kotski in front of her.

"Is it me or is that room too hot?" Kotski asked.

Dale recognized her voice as the one sitting next to Kirk on the bus.

"It's you," Dale responded, with a straight face.

The tall woman stopped and studied her, not quite knowing whether she was serious or not until Dale cracked a smile. "It probably isn't the classroom as much as it is the tension. Let's face it, we're all strangers thrown together in an even stranger environment. It's tiring, it's lonely and for the next few weeks it will probably get worse before it gets better."

"Is this supposed to be a pep talk?" Kotski looked at her, skeptically, and took a sip of water. "I mean, I understand it won't get easier."

"They say after a while it does." *Especially if you've been through it a couple of times before*, Dale thought, as she bent down to take a drink. "I wish that stupid bus would get here. This waiting is for the birds."

"I know." Kotski lingered as Dale took another drink. "Didn't they say we could go outside if we wanted a cigarette?"

"I'm sure I heard someone say that."

"Good. I need a cigarette."

"I don't smoke but I sure could use some fresh air." Dale said, as they walked outside and Kotski immediately lit up.

"What's going on with Kirk? Do you know?" Dale asked, after they had introduced themselves,

"Oh, that. All I know is what she told me on the bus, which is that she is only seventeen and her parents made her sign up. The recruiter and her father were good friends and they forced her to enlist. I guess she got kicked out of high school and her folks said it was the last straw."

It was the type of story Dale had heard many times.

"She said there are thirteen kids in her family, she's the youngest and her parents just didn't want her around anymore, especially after getting expelled. She said they gave her a choice of enlisting or juvenile hall. Or something like that. Anyway, she said they altered her test scores and everything."

"Does she have proof of that?" Dale asked, reminding herself to stop talking like an investigator.

"I don't know. I doubt it. She's determined not to stay here, though. Do you think she'll get out?"

"I don't know," Dale lied.

If nothing else, she expected Kirk's immediate future to be interesting, if not exasperating. She knew the system well enough to know that they didn't give up on anybody without a fight. She knew that the military couldn't easily accept someone who was intelligent enough to realize early on that he had simply made a mistake and that he and Uncle Sam would not be compatible. However, these circumstances were different. Kirk was coming forward before the training even started with what Dale believed to be a valid reason for being released from her contract - an illegality called fraudulent enlistment. There were a few kinks in the story but then Dale was hearing it secondhand, too. If Kirk was telling the truth and was there against her will, it was more than likely she had no evidence to prove it. Recruiters who used outright deception to enlist a civilian were usually clever enough to cover their tracks. Having worked months on that type of case, Dale knew from personal experience just how tricky those certain individuals could be.

Dale had forgotten how easily one got sucked into other people's personal problems while in basic training and somehow knew, instinctively, she was going to get herself tangled up in this one. She silently promised herself to remain neutral this time because if she didn't and wasn't careful, it would only be a matter of time before her true identity was revealed and she really didn't care to face an irate Anne Bishaye. She would rather face a firing squad. She was sure it was psychosomatic but suddenly her foot started to ache.

Just after midnight, the bus finally arrived. It was not like the plush coach that had dropped them off, however. This poor excuse for transportation looked like a dilapidated school bus that had been painted an unattractive shade of green - olive drab, to be exact. The women loaded up and took a five-minute ride from the newer steel reinforced concrete constructed reception center to the older white with green trim wooden firetraps used as housing for the trainees who were to become members of the Women's Army Corps.

This section of Fort McCullough was nicknamed WacVille because it catered almost exclusively to the females with their own dispensary, post exchange, or PX, churches, recreation areas, classrooms and PT sections, not to mention the battalions that were made up of just women alone. At that particular time, McCullough was one of the three Army basic training posts in the United States where the women greatly outnumbered the men.

The bus, groaning as the driver shifted it into a lower gear, coughed its way into the parking lot of 2nd battalion, where only for a minute, the atmosphere seemed almost transcendental. The abstract mood was shattered all too quickly when the vehicle jerked to a complete halt, the door squeaked open with a bang and this mean looking man in a Smokey the Bear hat materialized from nowhere and stood next to the driver.

"All you no good, lazy excuses for females who have been assigned to Delta company, second battalion, you have just thirty seconds to get your lazy, fat asses off this bus and twenty-nine of them are already gone!" He bellowed this out in one thunderous breath.

There was mass confusion as all the Delta women tried to exit the bus at the same time. Once outside, the women who, by that hour, barely knew their own names, much less anyone else's, were ordered to line up in alphabetical order.

"What? You ain't finished yet?" It had only been five seconds since the command. *"My grandma's slow but she's old! This will not do, ladies! Get down and knock me out some push-ups! Hit it!"*

Twenty-three females reacted as if they had just entered the "Twilight Zone."

Inside the bus, the silence was deafening. Even Caffrey was too stunned to comment as they watched the women on the field.

Out of the corner of her eye, Dale spotted two women, both armed with suitcases, fleeing the barracks toward the bus, almost as if they were prisoners running for their lives.

"Are you the two Alpha females?" the voice boomed at them. Two heads bobbed up and down. They boarded the bus,

barely touching a step, ran to the back of the vehicle and sank down into a seat.

"What is going on out there?" Renee Troice, the only one who dared to speak, asked them.

"I don't want to talk about it," the girl with the short, blonde hair and glasses answered, as though dazed. "I keep hoping Scotty will beam me up."

"You guys wouldn't be...uh...Kramer and Ferris, would you?" Almstead wondered.

"Ferrence," the other blonde woman corrected. "That's us. Why? Oh my God, don't tell me we're in trouble already."

"No, it's just that they called your names off back at that reception place and you weren't there."

"We got here early this afternoon, so they told us to process in with the Delta group and then they just brought us here and told us to wait for the next bus. I keep thinking I'm going to wake up and find myself home in my nice, soft bed and this will all have been a very bad dream. But I've dozed off and woke up three times already and I'm still here," Ferrence moaned.

"Yeah," Kramer agreed, gazing off into the distance. "I was always under the impression that you had to die before you went to hell."

"It's that bad?" Marilyn Segore, a stocky redhead, asked, her voice laced with panic.

"And this isn't even the beginning," someone else said.

Ferrence shook her head. "I have never heard anyone shout an entire speech in *italics* before tonight."

"Italics can't be spoken," Troice said.

"Did you *hear* him?" Ferrence asked, gesturing out the window.

The bus started up and moved away from the group of females who were trying unsuccessfully to maintain some sort of orderly fashion to the commanded exercise. If it didn't seem like such a forewarning to the rest of them, the women on the bus would have gotten a good laugh out of it. Dale remembered back to the day she got down and knocked out her first set of push-ups for Uncle Sam. Her arms shook for days.

The Alpha women dreaded the short journey's end and rode to their destination in speechless shock. No one seemed to have trouble staying awake now and Dale tried very hard to suppress a smile.

For those who had never known the species, they were about to have a close encounter with an alien being that the Army fondly referred to as the drill sergeant. A drill sergeant was, questionably at times, a human male or female who had spent weeks at a special school of instruction, being humiliated by teachers, who sometimes held a lower rank and showed them, somewhat gleefully, what they had been doing incorrectly throughout their military life. The instructors were stuck with the thankless task of breaking bad habits and retraining these usually career-oriented GIs by making them repetitiously practice what they would eventually preach. Hours, days and weeks were spent screaming cadences and executing to perfection movements with such precision and unison that it would have rivaled a Broadway chorus line.

Yet, somehow, even those who asked for the privilege of being a drill sergeant, felt rather degraded and demoted by being put back into a situation where they were subjected to severe inspections and disciplinary action. The resentment silently built because most of the military personnel undergoing training held the pay grade of E-4 and above and were, in most cases, already in a position to conduct such inspections and administer harsh discipline. Unable to completely cope with once again being treated like a brainless wonder but having to conceal the hostility in order to graduate, by the time they were assigned to a training unit, these superior soldiers were ripe for revenge. This feeling closely reflected what both Dale and, she was sure, Lieutenant Walker would soon feel.

The broken-down military coach rasped and ground its way into another parking lot, nearly stalling out on the slight hill that wound around to a steel and concrete quad structure called the new barracks.

The bus ceased to move, and the women sat tensely, not breathing, staring apprehensively toward the front of the coach, ready for a repeat of what they saw at Delta Company. When the vehicle's doors slammed open, the sound nearly made everyone jump but no one boarded the bus.

"This is it, ladies," the PFC behind the wheel announced.

"The end of the line. Everybody gets out here."

Slowly, each woman stepped off the bus, almost disappointed by this hospitality. They had prepared themselves for a few indignities but being ignored was not one of them.

"I'll go tell the CQ that you're here," said the private as he hopped off the bus behind the last female, who happened to be Kotski. He walked backward, staring appreciatively at the tall brunette most of the way across the patio to the CQ office. When he returned, he was with another young man in an identical uniform.

The soldier next to the bus driver spoke. "Are you sure you're supposed to be here tonight? Is that what your orders say? Twenty-two November?" He was wearing a black felt armband with the white felt letters *A CO CQ* sewn to it.

"Yes," Caffrey answered, seriously. "All of us couldn't have made the same mistake." It was the first chaste sentence she had uttered all evening.

"That's true," the CQ agreed. "One of you maybe but not nine of you." He sighed, disgusted, then looked at the bus driver. "Naughton, you stay with them, and I'll go get the staff duty NCO."

He ran through the open area, across the south patio and disappeared around the corner of the building. He was gone only a moment when a drill sergeant, flanked by the CQ, walked directly to the CQ office.

Seconds later, the PFC dashed down to the confused group of women. "Get your gear and follow me."

They picked up their luggage and walked behind him to the Orderly Room door.

"Put your suitcases down and listen up," he told them. "Line up in alphabetical order here to the left side of the door and get a copy of your orders ready. Now, Drill Sergeant Boyle will be out in a second to talk to you. If he asks or tells you anything, answer with yes, Drill Sergeant, no, Drill Sergeant. Whatever you say to him, make sure you add Drill Sergeant to the end of it."

The women nodded at him and again exchanged names with one another, so that they were in proper order. As the CQ raced to another destination, the Orderly Room door opened and there *it* stood. He looked menacing and the dimly lit patio did nothing to improve his appearance. He strolled back and forth, his fists resting on his belt, reminding Dale of Darth Vader in a pickle suit, as he carefully scrutinized each of them. He then addressed them all as if they were deaf.

"Obviously there have been some crossed wires here! Training isn't even supposed to start until 2 December, so what you're all doing here now, I don't know! We'll have to find out in the morning, won't we?"

It was more of a statement than a question, so the women continued to listen and quake.

He lowered his voice but was no less ominous. "What I want you to do is this. After I go back inside the orderly room, which is this door here," he pointed to the CQ office. "I want you to come in one by one, give me your full name, rank, which you should know by now, and social security number. I will then assign you a meal card, so that you can all eat tomorrow. *I know you're all anxious to start your daily regimen of bread and water!"*

He picked Brigitte Kramer out of the crowd and stuck his face right in hers. *"Do I scare you?"*

"No," Kramer gasped, sounding frightened.

The CQ, now accompanied by a young woman in fatigues, approached the drill sergeant. The young woman's nametag read Lanigan, and she had no rank on her cap and collar. "Lanigan was fireguard, Drill Sergeant," the CQ told him.

"You stay outside and keep these ladies in order," Boyle instructed Lanigan, in a normal tone of voice. "Send them in individually in about a minute."

"Yes, Drill Sergeant," Lanigan said, as Boyle disappeared inside the office.

There was a brief silence, then Kramer spoke up, timidly. "Why are they so mean?"

"They castrate them," Dale deadpanned. "I hear it's a requirement."

Kramer leaned forward and stared at her in wide-eyed gullibility. Ferrence and Segore regarded Dale as if they were ready to believe her, too.

"They're not so bad," Lanigan said, with a smirk. "He's just putting on a show. They all will until you get used to them."

"Do they have to shout like that?" Ferrence asked.

"I told you, he's just trying to make an impression," Lanigan said.

"Well, he can stop already. I'm impressed," Kotski stated.

Lanigan looked inside the office. "He's ready for you now," she pointed to Almstead. "You can go in."

Alexis Almstead shook her short, blonde hair out of her eyes, looked back at the rest of her group then bravely went inside. Less than a minute later, she came back outside, still intact. She was holding a small, rectangular, laminated white card that had a serial number printed on it. Instead of the soldier's name and unit in the space provided were the two letters RA, standing for Regular Army, stamped in red, and at the bottom, it was pre-signed by the company commander. This was her meal card. It was the first military document a trainee was issued, and the last thing turned in before he or she left their basic training post.

When Dale's turn came, she entered the Orderly Room quietly and walked to the far desk where Drill Sergeant Boyle was sitting. She thought back to the night she met Henning there and took a quick glance up to the company roster to see if CQ's name was listed from that night. It was not. Neither was the name of the previous male cycle spy, Specialist Eastman. Linda Boehner's name was still there, though, and that didn't particularly cheer Dale up. She could sense Boehner didn't like her but in all fairness she wasn't crazy about Boehner, either. She was relieved that Boehner wasn't going to be working with her on this case but then, not having met Lieutenant Walker yet, for all she knew, the rapport between them might even be worse.

"Name."

"Oakes, Dale, Drill Sergeant," she said, handing him a copy of her orders.

"Is that O-A-K-E-S?" He asked, checking it with the spelling on the sheet of paper she had just given him.

"Yes, Drill Sergeant."

He wrote it down and looked up at her. "Any relation to Robert Oakes? He's an E-7 at Fort Sam."

She responded as though he were speaking a foreign

language. "No, I don't believe so, Drill Sergeant."

"Service number?"

Dale knew her service number was also her social security number. She rattled it off and he handed her a meal card then instructed her to go outside and wait. After she joined the others, she looked at the plastic-coated piece of construction paper. Her number was 6427 and it was signed by Rory D. Colton, CPT, MPC, Commanding, someone she was very eager to see, since he was the Alpha-10 company commander.

Drill Sergeant Boyle dismissed Lanigan when he came outside, then led the recruits out of the company area and around the corner of the building to the tenth battalion headquarters, where he was on twenty-four-hour duty. Dale momentarily wished Anne Bishaye was running around her office just so she could permeate her with dirty looks. Boyle brought them inside the headquarters orderly room, which was nothing more than an outer office, where he told his runner to call Sergeant Carey, who was in charge of supply. The private behind the desk picked up the phone and dialed.

"While we're waiting for Sergeant Carey to wake up so he can open the supply room, are there any questions?" Boyle asked, almost civilly.

"Yes, sir —"

"Sir means officer," he snapped at Kirk. "I am not an officer. I work for a living."

Dale rolled her eyes at his originality. If she heard that statement one more time from someone enlisted, she was going to retch.

"What was your question?" The drill sergeant looked at Kirk.

"How do I get out of this prison?" Kirk asked.

The women laughed, nervously.

Boyle wasn't amused. In fact, his unpleasant expression turned even more disagreeable. He glared at Kirk long and hard enough to bore holes through her and the other women immediately shut up.

"I don't play favorites," he said. "I'll drop you as fast as any man." He looked at the others coldly, individually. "Or any one of you who thinks this is a joke." His gaze fell back on Kirk. "I didn't make you raise your right hand, young lady."

"No, my parents did."

"That's something you should have corrected at AFEES before you left."

"They told me that you would handle it at this end."

"I can't help you, young lady. I'm not even in your company. You'll have to take it up with your company commander."

After they left the HQ orderly room, Boyle led them back outside and down to another company's patio. The women caught their first glimpse of Sergeant Carey as they rounded the corner and paraded into Delta-10's company area at the bottom of the hill on the opposite side of the quad. Sergeant Terian Carey was half-asleep and as Boyle approached with his nine shadows, Carey belted his baby blue terrycloth robe around himself tightly.

Carey seemed to be somewhat more pleasant than Boyle, if the expression on his face were any indication, even after he had been dragged out of a deep sleep in the middle of the night. He eyed the ladies carefully as he unlocked the supply room doors and with the help of Delta-10's CQ runner, he started to issue bedding.

"Listen up, ladies," Carey announced, as they stood around the supply room. "I want you to get this right so that we don't have to stay up any longer than necessary. These drill sergeants have no mercy."

He winked at Kotski, who just seemed to stand out, then flashed a grin to Boyle who didn't acknowledge him. During his explanation of the linen issue to the new trainees, he cracked a few jokes to ease the nervousness, but the women always looked to Boyle for a reaction. When he didn't display any, they all kept straight faces. None of them wanted to be dropped – not that they knew what Boyle even meant by that.

Each woman signed for and carried back to the Alpha company area two wool US Army blankets, two sheets, one pillow and one pillowcase and were then sent up to the second floor. There, Private Lanigan took them inside the barracks.

"It's called a bay. Don't ask me why," she told them as she

showed them an empty row of bunks.

Dale tossed her linen on a bunk and went downstairs to retrieve her suitcase. When she was back upstairs, she looked at the women in bed on the left side of the bay. The bay was an open room, large enough to house at least fifty occupants in single beds. It allowed for no privacy, no real personal space, no individualism. The four uniform rows of bunks were marked off with vertical elements, in this case, standing lockers at the head or foot of each bed, that divided the huge room into two aisles. Dale counted seven lumps in their bunks, so she estimated there were eight female holdovers, including Lanigan, but then it was dark going toward the back so there could have been more bodies up farther. At that point, however, she was too tired to care, and she knew she would find out within a few hours when they would more than likely be rudely awakened by five o'clock at the latest.

As she threw her bedding together on her bunk carelessly, she heard her eight comrades bombarding Lanigan with questions.

Already knowing most of the questions and, more importantly, all of the answers, Dale didn't feel like participating in the quiz, so she settled down and dozed off in no time. Even when more female recruits came in an hour later, she only vaguely heard the commotion. She knew in the months to come she was not going to be getting much sleep, so she was going to take advantage of every available opportunity to close her eyes and nothing, except possibly an act of God, was going to rob her of it.

CHAPTER SIX

Zero five hundred hours. The silence was broken by a male voice coming out of the speaker in the ceiling. "Wake up, ladies! It's time to get your lazy asses out of bed! You have to be downstairs to fall-in in twenty minutes. Is everybody up?" the deep voice asked in a sickeningly sweet tone.

"Yes, Drill Sergeant," someone answered. The voice sounded like Linda Boehner's.

The small, red, blinking light on the speaker went off and, in between moans, groans and mild swearing, everybody crawled out of their bunks, some with little more than two hours sleep behind them.

"The goddamn sun ain't even up yet, why should I have to be?" someone yelled.

"What does fall in mean? What do we have to fall into?" another voice said.

"Why aren't I home in bed like a normal person?" Ferrence wailed.

"I didn't make you raise your right hand, young lady," Kirk recited, as she walked by Ferrence's bunk toward the bathroom.

"It was a moment of insanity, I should be pardoned at least. Where's Gerald Ford when you need him?" Ferrence countered, as her eyes fought to stay open against the assaulting brightness of the florescent ceiling lights.

Most of the new recruits said nothing as they made their way to the latrine. Any unnecessary conversation would just take up time they didn't have. They needed to shower, dress, clean up, make their bunks and be downstairs in less than fifteen minutes.

The drill sergeant they had been greeted by the night before was not the same drill sergeant who was now standing before them, holding a sheet of paper. This NCO's name was Robbins and, though he was strict, he appeared to be much better natured than Boyle.

He called off all their names and made sure they had all been issued a meal card. Then he put them in groups of two and marched them down a flight of ten steps to the mess hall for morning chow. The modern mess hall was located in the middle of the four companies–Alpha, Bravo, Charlie, Delta– that made up the 10th battalion and it had a reputation for being one of the better eating facilities on post.

The group waited in line inside the building where they were to wolf down mostly all their meals for the next five or six weeks. Individually, they signed their name, rank, company and meal card number on a mess hall roster, grabbed a heavy, plastic tray and pointed out what they wanted for breakfast. Most of it was nice greasy, starchy food, a perfect morning booster for anyone not trying to stay healthy or thin. However, most of the women seemed too nervous to consume anything other than a cup of coffee and their own fingernails.

Cautiously, Dale looked around at the tables, and with much relief, recognized no one except Linda Boehner. She then scanned the room, picking out the eight new females who had come in a few hours ago and also really took a good look at the eight she had arrived with. The new trainees would have been easy to spot, even if they hadn't been the only females in the mess hall in civilian clothes.

There were four Alpha women sitting at a table in front of Dale and she noticed them trying to converse discreetly, but they were caught by a patrolling female drill sergeant named Bradbury. The only difference between a male and a female drill sergeant, other than sex, was uniform. They dressed in identical attire, but the women's fatigue shirts were worn outside the pants instead of being tucked in. Also, instead of the brown ranger hat the men wore, the females sported an off-white Australian bush hat.

"*Inhale it and get out!*" she thundered. Bending at the table, she continued. "If you're finished, get out! If you're not, eat and shut up! This is not a gossip area! You're in the Army now, ladies, there are no special rules for you!" She pointed to one of the trays. *"You will finish everything on that plate! Take what you want but eat what you take!"*

The four women looked embarrassed at being made examples of and they became instantly aware of Drill Sergeant Bradbury from that point on. Dale knew that as time in basic training progressed, the trainees would learn to alert on certain voices and be able to read certain tones so that they would be pretty sure as to how much or how little they could get away with.

After chow, all the women returned upstairs to square away their areas and then they were ordered back downstairs to wait. A lot of the women were getting better acquainted with one another and some were catching holdovers before they were assigned details or went to class to ask them more questions.

Dale strolled to one of the picnic tables that sat on the north patio, where Linda Boehner was surrounded by three females. Kramer had asked her if the combat training was hard.

"No," Boehner said. "It's all attitude. If you say you can't do it, you won't. If you push yourself and say you believe you can, you will."

Dale passed the women and stopped. "Hey, Boner, I thought you weren't supposed to be talking to us?" Dale then walked away as Boehner attacked her with icy glares.

"It's Boehner," she spit out, saying *Bayner*.

"Well, it is spelled like boner," Kramer supplied.

"I should know how to pronounce my own name," Boehner snapped, still glowering at Dale.

Dale chuckled to herself, knowing she just gave Boehner a new basic training nickname. In fact, it was too easy to incorrectly read Boehner's nametag, so she was pretty sure the undercover sergeant had to battle being called Boner long before Dale started it. Dale then moved to where Jascelle Kirk was standing with Renee Troice and Marilyn Segore.

"Hey, Kirk, how's it going?" Dale asked.

"It's going, I'm not," Kirk answered.

"Have they let you talk to anybody yet?"

"No. The company commander is on leave, and they said I have to talk to the senior drill sergeant, somebody named Ritchie. But he won't be here until tonight or tomorrow. This whole thing is such a mess."

"Listen, hang on," Dale urged. "They'll get you out."

"How can you be so sure? I've never felt so trapped in my life."

Dale couldn't be so sure, so she decided not to commit herself to any further speculation. She smiled at the worried young woman. "Just stand your ground. If it'll help, I'm behind you."

"Yeah, me too," Segore agreed courageously.

"I really appreciate it, but you guys want to be here, I don't. It doesn't matter how much I get yelled at or get into trouble because their rules and regulations don't mean anything to me and, besides, I ain't staying around that long. But you two have to stay here after I'm gone, and it wouldn't be fair to fuck it up for either of you."

"Kirk, what is happening to you is not right," Dale said.

"I know," she sighed, distressed. "It's just a mistake, though. I don't want anyone else getting involved in it. For all I know, it might make things worse."

"Well, you could have a point there," Dale conceded. "But if you need someone to talk to, don't hesitate. I'm a good listener."

Kirk searched Dale's face for a hint of sarcasm or insincerity but found none. "Thanks, Oakes," she said, seriously.

Kotski came up on the conversation. "Anyone see that wench in the mess hall this morning yelling at us? Wasn't she something? She sounded like a fishwife."

Troice shrugged. "That probably isn't anything compared to what we will all be in for once we get in uniform."

"I heard that," Kirk said. "That's why this kid's going to be long gone when uniform time comes around."

Dale, Troice and Segore fixed their gazes on Kirk. None of them wanted to be in her shoes. She was clearly miserable, and Dale knew her fight hadn't even begun yet.

Four hours after they had been awakened, Drill Sergeant Robbins loaded them onto an Army bus similar to the vehicle that had dropped them off the night before, only this one was in a little better condition.

They were again on their way to the reception center, this time under the supervision of a female specialist fourth class named Harriman who, by Dale's careful observation was, no doubt, in training to be a shrew. The specialist, a noticeably unattractive creature, was not the friendliest person Dale or the others had met since they'd been there. To Dale, it seemed like she found her job of escorting new trainees to specific points on post about as enjoyable as cleaning up elephant waste and she made each female feel solely responsible for her being appointed to that particular detail. Once she got them to their destination, the poker-faced specialist couldn't deposit them or disappear fast enough.

"Just what I needed to see," a tall, slender, brunette recruit named Antoinetta Sherlock said, as they waited. "Someone who is enthusiastic about their job." Sherlock was one of the later overnight arrivals to Alpha-10.

"This is a little ridiculous," commented the redhead behind Sherlock, after forty-five minutes had crawled by. "What does the Army train us for? Standing in line?"

Ten minutes later, Dale and the rest were reunited with the Delta women and the young men who were eventually, after processing and orientation, going to be in Alpha-10. They were all seated in the lecture hall provided with instruction on how to fill out more forms on legal assistance, religious persuasion, postal service, service benefits, dependent allotments, life insurance, medical benefits, leave policies, pass policies, recreational facility information, correspondence, policies on civilian clothing, pay and allowances.

When lunchtime approached, Specialist Harriman boarded the new MP female recruits onto the bus and made them disembark at a WAC facility for lunch. She led them all inside and instructed them to sign in. Then she left them on their own to line up, go inside and eat whenever the young soldier assigned to count the number of prospective occupants told them there was room.

As they stood in the hallway, waiting, Dale watched the new Alpha women observe some actual uniformed female basic trainees maneuver through the chow line. The WACs stood in line in uniform rigidity until it was the next two people's turn in the queue to move up and they took a step in a coordinated drill, returning to their original position. This exercise was repeated until everyone was in the serving line. To the newbies, it was quite impressive.

"Showoffs," Diane Tierni, a petite woman with medium-length brown hair, noted.

"We don't have to do that, do we?" Ferrence asked.

"Not yet," the male headcount answered. "How come youse girls ain't in uniform yet?"

"Oh, we'll never be in uniform," another new Alpha recruit, Tracy Travis, drolly said, immediately baiting him. "Because we're training for plainclothes and undercover work."

He stared at her, believing it. "That must be the new experiment they've been tellin' us about, goin' on at the new barracks."

"That's the one," Travis continued. "So don't mess with us, okay? Because if we're provoked, we can't be responsible for our actions. We have been given the divine right to kill without question."

A few of the women started to snicker and then suddenly he looked as though he realized that Travis was trying to make a fool of him, which he clearly did not appreciate. His glare did not appear to ruffle Travis in the least, in fact, she was the one who pointed out to him that the chow line was now empty. Stiffly, he allowed them to go inside.

This mess hall wasn't quite as modern as the 10th battalion facility, Dale noticed. The food was neither as tasty nor as warm. Some of the chow in the MP mess hall wasn't recognizable but at least it was hot. The only plus at the WAC building was that they could at least converse with one another openly.

Dale observed a lot of drill sergeants running around, mostly female, but they paid little attention to the seventeen women in civilian clothes. Their main concern seemed to be their own platoons and that's where they kept their focus. Until one female drill sergeant caught Dale's eye and the undercover lieutenant couldn't stop herself from staring.

It wasn't that Dale knew her and feared blowing her cover, it was just that she was possibly the most stunning woman Dale had ever seen and that included Anne Bishaye. The uniform and hat did absolutely nothing complimentary to a female's appearance and if this drill sergeant was that gorgeous in that get-up, Dale was just mesmerized at what she must look like in civvies. She was slender and obviously physically fit. Dale guessed her to be, maybe, five foot nine or ten, late twenties to early thirties at the most. She had jet black hair, pulled back tightly into a bun, dark expressive eyes, a flawlessly smooth complexion, a skin tone that seemed naturally tanned and the most perfect, whitest teeth Dale had ever seen. Which she only got a good glimpse of when she realized the woman was now staring back at her.

Shit, shit shit! Dale thought, as the drill sergeant made a beeline toward her. What the hell was the matter with her? Wasn't her hammering lust for Anne Bishaye perplexing and chaotic enough? She was now moving on to openly leering at other women? Well, honestly, this woman was too riveting not to stare at...wasn't she? Dale looked around quickly. Obviously, no one else thought so or if they had, they were able to control it a hell of a lot better than the soon to be mortified Lieutenant Oakes. *Shit, shit, shit!*

Before Dale could react, the drill sergeant was standing directly in front of her. "Stand up," she commanded in a direct but restrained *don't even think about fucking with me* tone of voice. "Now."

All conversation stopped and all eyes were now on Dale and the drill sergeant, as Dale rose to her feet. She peripherally saw she was the center of attention, much to her dismay. Glancing quickly, she read the woman's nametag. Cassidy. Snapping her eyes back to the female's face, Dale thought she didn't look Irish...she looked more...Greek. With some Native American thrown in. She was even more breathtaking up close and personal. Way to go, Oakes, she reprimanded herself silently, way to keep yourself inconspicuous.

"Keep your eyes straight ahead, do not look at me," Staff Sergeant Cassidy's voice was firm. It was also low and husky with a hint of an accent. Texas, Dale guessed. All this did was add to her allure. "What's your name, Private?"

"Oakes, Drill Sergeant," Dale responded.

"What were you staring at, Private Oakes?" Cassidy's face

was now mere inches from Dale's.

One flustering phrase continued to cross Dale's mind—*extremely* fuckable. It was torture as suddenly Dale couldn't find any saliva with which to respond. Instead of this tactic intimidating her, it was turning her on. What the hell was happening to her?

"Well?" Cassidy waited for an answer.

Dale cleared her throat. "Nothing, Drill Sergeant."

Cassidy didn't move. "Nothing? You were staring right at me, Private, are you calling me nothing?"

Damn it! Dale knew better than to set herself up like that. All she really wanted to do at this point was gaze deeply into this woman's eyes and plead insanity. But she knew better than to look at her, not only because it would probably result in, at the very least, a dime in push-ups but also she didn't trust what she might convey with that visual connection. "No, Drill Sergeant. I didn't mean to stare at you. I was just...staring...at air...I didn't get a lot of sleep last night..."

"So, what you're saying is you were staring at air, and I got in your way."

Oh, that voice. "Yes, Drill Sergeant." Thank God, Cassidy wasn't a member of Alpha-10's cadre. She would be much too distracting.

Cassidy took a step back. Dale could have sworn the drill sergeant gave her a once-over, but it was probably just wishful thinking. "A suggestion, Private Oakes? Always be conscious of your surroundings. And always be prepared to defend your position. You may sit back down now."

Wow, Dale thought, *that was astonishing —useful advice given with apparent sincerity.*

"Yes, Drill Sergeant," Dale acknowledged and took her seat again as Cassidy walked away, immediately zeroing in on her own platoon of trainees. Well...it could have been much worse. Returning her attention to the other three women at her table who were still holding their breath, Dale looked at them. "What?"

"Nothing," Tracy Travis answered. "Other than we're zero for two. I wonder which one of us will piss off a drill sergeant for dinner."

After noon chow, the women were back at the reception center. They were lined up, single file, outside an office where they were to be blood-typed and immunized against tuberculosis, mononucleosis and rubella. That morning, they had each been given a yellow document that was folded into book form with their name, rank and social security number stamped on it. Dale looked at the international certificate of vaccination, noting the style and format hadn't changed much. They were all informed to hang onto that little register because throughout their military career, any shots they received would be recorded in this booklet and stamped, signed and/or initialed by the physician administering the inoculation.

"Look at this!" Kirk announced, displaying both arms to Dale. "I look like a junkie with bad aim."

While the Delta and Alpha women sat and waited for more instruction, Dale knew the men were taking their first step into the process of military cloning. They were getting their heads shaved. Personally, Dale liked seeing men with thick, neat, moderately long hair and the thought of seeing regulation haircuts, exclusively, for the next six weeks or so, depressed her almost as much as basic training again.

Most of the women were engaged in trivial conversation, with Dale nonchalantly trying to pick up anything sounding even remotely suspicious. There was too much being said at the same time, however, to take on any significant form.

That afternoon was mainly spent sitting around gabbing, laughing, and biding time, most voicing the question if this was how they were going to spend the rest of basic training. Dale heard some of the women say and agree that they were under the impression training began the second they stepped onto US Army territory. Others said had no idea what to expect. A few brought up the videotape that their recruiters had shown them, the one that had been documented on television's *60 Minutes*. It really didn't psychologically prepare them for much. Dale would have vociferously agreed with that sentiment, but she wasn't supposed to know any better, so she stayed quiet.

Later, when Dale and all the new Alpha women were

standing outside the reception center, waiting for the bus that was to take them back to the barracks, two platoons of WACs were marched by them. The MP recruits watched, soundlessly, as two drill instructors, one of each gender, appeared to Dale to decide to make an example out of their troops in front of the impressionable Alpha women. The uniformed WAC trainees were verbally abused, degraded and made to do push-ups, even though they were all equipped with what Dale guessed was their full, sixty-five-pound backpacks, web gear, steel helmets and rifles. The female soldiers were put in the front leaning rest position–body prone, arms locked straight out in front to support the body–their weapons resting on the backs of the hands. And there was definite hell to pay if the M16s touched the ground, as a drill sergeant slowly counted the cadence of the four-count push-up and the trainees responded with the repetitions.

Dale knew this illustration would give the women in civilian clothes a better look at what they had gotten themselves into. When the bus came, the seventeen Alpha women silently boarded, returning to the barracks in a catatonic state, some openly contemplating their sanity.

After chow, the mood still hadn't returned to normal. The gloomy atmosphere had spread to the fifteen new females who had arrived that day. Several women took their showers early and went to bed. Some talked to Boehner and other AIT graduates. Some read mail, some wrote letters, some took cigarette breaks but Dale noticed that no one was as jovial as they had been before the demonstration that had taken place that afternoon.

Especially not Kirk who had been called down to the Orderly Room to meet with the senior drill sergeant. From Kirk's description of Sergeant First Class Ritchie, who did nothing but belittle her, Dale was not looking forward to meeting him. He sounded like trouble and just exactly what they didn't need at this point in time.

CHAPTER SEVEN

The next day was Thanksgiving.

Dale was surprised when everyone was permitted to sleep until 0600 hours. Then at morning chow, surprisingly enough, they were even allowed to take their time and engage in conversation while they ate. After mess, they were all assigned company details such as cleaning the latrine, tidying up the barracks, sweeping and mopping the floors and patios, straightening up the laundry room and putting the CQ office in order.

When the details had been completed, the trainees were summoned back outside to learn about police call, which Dale knew had nothing whatsoever to do with law enforcement. The women were lined up, arm's length apart, at a specific point, usually the parking lot. They then walked forward, scanning the ground in front of them, picking up anything that wasn't indigenous, namely cigarette butts and litter.

"If it doesn't grow there, remove it!" Drill Sergeant Robbins yelled at them.

After fifteen minutes, everyone reassembled back on the second floor. The metal door swung open with a bang and a deep, male voice roared, "Man on the floor!"

"At ease!" one of the AIT females yelled.

"Carry on," the drill sergeant said before anyone had a chance to move.

He silently walked up one aisle and down the other. Dale discerned there was something unsettling about this man's demeanor, something tough and unyielding that made it obvious he was imposing, strict and quite professional before he even opened his mouth. He was a tall man, with a thick, brown mustache and that's all they were able to notice about him, because Dale knew no one would have the nerve to look at him long enough to see anything else.

Even though he had given the command of carry on, nobody dared to move or speak. By the hue of some of her fellow recruits' coloring, apparently some didn't even dare to breathe. He studied each bunk and personal area during his stroll, slowing down a few times, lingering long enough to scare certain women into thinking it was their turn to be singled out and chastised for something as devastating as a microscopic fuzz ball on the blanket.

"I am Drill Sergeant McCoy. Whose bunk is this?" he asked, pointing to a random bed.

Son-of-a-bitch, Dale thought, almost incredulously, *Am I sending out a scent or something?* "Mine, Drill Sergeant," she spoke up.

"Who are you, young lady?" His gaze of steel seemed to go right through her.

"Private Oakes, Drill Sergeant."

"Do you know how to make a military bed, Private Oakes?" His voice boomed.

"No, Drill Sergeant," Dale lied. The thought of folding another ninety-degree angled corner made her want to cry. She had done it so many times she could have willed it into precision.

"All right, ladies, I want you to gather around Private Oakes' bunk. I'm going to show you how to make a military bed and then you are all going to remake your bunks that way." His manner of speaking was grating and matter of fact. He talked in a loud monotone, dividing each word into deliberate syllables and described the procedure to them as if they were all mentally deficient.

Yet, despite that, there was something about him that Dale had instantly picked up on and she respected him for it. He did not introduce himself to the new recruits by promptly selecting anyone to personally degrade nor did he put them down as a group, such as telling them they were the worst looking bunch of females he had ever seen, which was a favorite ploy of most drill sergeants.

Being tyrannical was a quick way to establish power, to demonstrate who was in charge. But McCoy was different. He didn't appear to be into those kinds of head games, which pleased Dale to no end. He seemed to be the type of man who hung around just long enough to tend to the matters at hand and that was it. It appeared that Sergeant First Class Sam McCoy had better things to do with his time than to amuse himself by playing unnecessary mental sports with new recruits. They had enough on their minds as it was and clearly, for him, it was a waste of time, energy and emotion that could be channeled into other, more productive areas.

When he was finished, they all stripped their beds and redid them military style. McCoy went around to check and instead of reprimanding when someone hadn't done it exactly right, he pointed out what her mistake was and showed her how to correct it, telling her to do it again, until she got it right. His voice and temperament were still quite gruff, and everyone secretly hoped they didn't get him as a platoon sergeant. Everyone except Dale. She admired the no-nonsense type. They were less sneaky.

Drill Sergeant Ted Robbins was waiting for all thirty-two females down on the north patio after Sam McCoy had released them. Robbins separated them into two groups, then put them into ranks and patiently instructed them on drill and ceremony. Dale didn't want to jump to conclusions this early, but Drill Sergeant Robbins was going to have to be watched carefully. His unconventional good looks, charming personality and tolerant manner made him a perfect target for a setup, if there was to be one.

Dale's eyes persistently probed as unobtrusively as possible, gathering visual information to store in her memory banks for the future. She looked for insignificant, picayune little things right now, such as how the other women looked at Robbins and what she could read into those expressions. Also, which ones made it a point to establish some kind of contact and who pushed to be noticed by him.

Staff Sergeant Robbins explained about the two-part command, which most military drills had, the preparatory command and the command of execution. The preparatory command was the movement that was to be carried out. The command of execution was when to carry out that movement.

PERMISSION TO RECOVER

For example, in the command, About Face, the preparatory command was About and the command of execution was Face. He also told them about combined commands where the preparatory command and the command of execution were joined, such as the commands of Rest, At Ease and Attention.

He showed them Attention and Fall In first.

"When you assume the position of Attention or are told to Fall In, you should bring your heels together so that they are in line with your toes. Your toes should be pointed out, equally, forming a forty-five-degree angle. Your legs should be straight, not stiff. Do not lock your knees, you'll be able to hold the position longer. Your body will be straight, and your shoulders will be even. The weight of your body will be distributed equally on each foot. Keep your head erect and your eyes should look directly to the front. Your arms should hang straight down at your sides with the back of your hands facing outward, fingers curled and joined, your thumbs touching the seam of your trousers. This is what it should look like." He demonstrated. "Now, when I call you to Attention or tell you to Fall In, I want you to do what I did. Think you can handle that, ladies?"

"Yes, Drill Sergeant," the women answered in unison.

"I can't hear you!"

"Yes, Drill Sergeant!" The recruits said louder.

"That's better. Okay. Company! *Ah - tan - haun*!"

"What'd he say?" someone asked from the back.

Eventually, they all got to the position of Attention but, evidently, not fast enough for Drill Sergeant Robbins. Dale tried hard to bite back a smile, recalling how dumbfounded and uncoordinated she felt the first time she went through this instruction.

"What do you think this is, ladies? Bingo night at St. Peter's? You're slower than a gaggle of prostitutes walking into a paddy wagon. *Get your lead asses in gear, you're in the Army now*! When I call you to Attention, if you're not in the position in one second, you've taken too long. Let's try it again. Company...*Ah - tan - haun*!" Robbins commanded.

They did it five more times before he seemed even remotely satisfied with their speed. He also worked several other drills with them — At Ease, Stand At Ease, Parade Rest, Rest, Right

Face, Left Face and Fall Out.

"Now, we're going to cover a few simple rules. When a drill sergeant enters your area, the first person to see him or her will say immediately, if not sooner, *At Ease,* loud enough for everyone to hear and God help those who don't jump to that position. If you choose to ignore the command, you will be chosen to knock out a couple dozen push-ups. We are very fond of push-ups here in the Army. Some of you will find out just how much.

"You *will not* leave the position of at ease until the drill sergeant has told you to Carry On or has left your area. The same goes for when a drill sergeant enters the bay. If he is male, you will be forewarned by his yelling man on the floor as soon as he opens the door. The minute you hear that, somebody better call At Ease and you better drop what you're doing and be At Ease.

"Also, when you see that little red light flashing on the intercom in the ceiling, that generally means somebody down in the Orderly Room wants your undivided attention, so someone up in the bay better yell At Ease. And it better be in a tone of voice that everyone can hear.

"A word of warning about the female drill sergeants. They're luckier. They can sneak right in on you upstairs in the barracks...but the rules still don't change. The first one to see her still yells At Ease.

"One last little item of importance. We have a lieutenant running around here today. She is our training officer, and her name is Lieutenant Henning. You can recognize her by the little gold bar on her collar and cap. You're not in uniform so you don't have to worry about saluting just yet, however, if you see her and she talks to you or you talk to her, anything you say to her will be preceded or followed by the word ma'am. Female officers are addressed as ma'am and male officers as sir. Everybody follow me on that? Do you understand everything I have just told you?"

"Yes, Drill Sergeant," came the dazed, collective response.

"I can't hear you!"

"Yes, Drill Sergeant!"

"Any questions?" Robbins inquired.

"Drill Sergeant?"

"Identify yourself, young lady."

"Private Minty, Drill Sergeant."

"Yes, Private Minty?"

"Drill Sergeant, how do you respond when we see a drill sergeant in our area and there is no one else to say At Ease to? We don't scream it at ourselves, do we?"

"Good point, Private Minty. No, you do not. You just automatically assume the position of At Ease until he or she tells you to carry on or leaves your area. Any other questions? Remember — the only stupid question is the one that isn't asked."

Dale side-glanced at Katherine Minty. She was a tall eyeful from Oklahoma who gave the impression she had the brains to match her looks. Her questions didn't put her on Dale's suspicious list because she had asked frequent ones since her arrival, and they all brought up valid points instead of just making aimless conversation. Furthermore, Dale thought, if her intentions were directed personally at Robbins, she would have made her inquiries in as private a situation as she could manage.

On the other hand, after overhearing a conversation between Minty and a few other new recruits earlier, Dale decided that Minty did have a chip on her shoulder about being an Army brat. Minty claimed she had been brought up on military installations all around the world and had been around the Army all her life, therefore, she instinctively knew more about the Army's rules and regulations than the average recruit. That would also give her an advantage knowing what the drill sergeants would or wouldn't tolerate. Yet intuition told Dale that Minty was probably not one of the people they were looking for.

"No more questions? All right, ladies. *Ah-tan-haun!*"

They all snapped to Attention, some a little slow, but they looked pretty good in the long run. Dale once again reminded herself to keep her reflexes in slow motion. She needed to blend and look no better or worse than anyone else in the learning process.

"That's right. Eyes straight ahead. Don't look at me. Don't lock your knees, they'll buckle faster if you do. Okay. At Ease." They switched positions a lot more efficiently than Robbins obviously expected. He looked mildly surprised but did not comment on the group's fluidity. "Before I give you the command to Fall Out, I need a few volunteers to help Lieutenant Henning decorate the mess hall."

No one moved.

"I see someone has already warned you about volunteering for anything...which is redundant being that you're all here. I guess I'll have to select volunteers." He scanned his personnel roster.

"Almstead...Jaffe...Minty...Oakes...and Sherlock. Report to the mess hall directly after I dismiss the others. The rest of you are free until noon chow, which gives you about an hour and a half. You can hang around on either patio, the laundry room, the bay, or the concrete walkway in between the two patios. We will be setting up the nets and you're welcome, in fact, you're encouraged to play volleyball. The dayroom is off-limits to you. The AIT graduates may use it, but don't any of you get caught in there. Am I understood?"

"Yes, Drill Sergeant!"

"Good. One last thing. We have a motto here at A-10. We say Alpha-10, First and Best of the LE School, Sir! Because Alpha is the first of the four companies that make up 10th battalion and 10th battalion is the first of the three battalions that make up the Law Enforcement School. Now it takes on a new meaning because this experimental cycle you'll be going through is also a first. So that's our motto. Alpha-10, first and best of the LE School, Sir! Everybody say it."

"Alpha-10, first and best of the LE School, Sir!" the women chorused, awkwardly.

"Louder!"

"Alpha-10, first and best of the LE School, Sir!"

"Outstanding." He had them repeat it several times. "Whenever you are called together for attention as a group and given the command of At Ease or Parade Rest, I want you to come back with that motto as loud and as proud as you can and then snap to whichever position was commanded." He practiced with them a few more times.

Not bad, Dale thought, *but it definitely needed work.*

Robbins then called them to Attention and commanded

them to Fall Out. The group relaxed and moved away quietly except for the five women he had selected to help Henning. As those individuals headed down the steps toward the dining facility, Dale looked back and, much to her disappointment, saw no one seize the opportunity to monopolize Drill Sergeant Ted Robbins while he stood in the middle of the patio alone. Dale observed him watch all the women for several seconds, then shake his head, smiling, and walk back to the CQ office.

The selected women waited in the dining area of the mess hall, sitting in two booths between them.

"I don't know what I expected it to look like but it sure wasn't this," Dale said, trying to start a conversation.

"This is really nice," Minty said. "They never used to look like this. I mean, look around, this is more like a small restaurant with all these tables and booths. We're lucky, girls, we could have one like the one we ate in at the WAC side of the post."

"That wasn't so bad," Lex Almstead said and yawned.

"Speak for yourself," Dale commented, her encounter with Drill Sergeant Cassidy was still a fresh wound, for more reasons than just embarrassment.

"Look, a military mess hall is still a military mess hall no matter what they look like. The only difference is in a basic training mess hall, the alphabet soup only has four letters," Sherlock remarked. "Hey, Minty, how come you know so much about the Army?"

"My daddy is a sergeant major. I grew up around it," she beamed, grabbing a chance to boast.

"If you grew up all around it, then you know what it's like. What'd you do a dumb thing like enlist for?" Lesley Jaffe, a sweet-faced redhead asked.

"I am going to prove to my father that not just the men in our family can make it in the military. I'm going to move right up through the ranks like he did," Minty assured them.

No matter who it burns along the way, Dale thought. Well, fortunately for Minty, she was ambitious, and Dale secretly wished her the best of luck. There were two obstacles that were going to frequently cross her path to the top that her father didn't have to worry about, and that was harassment and discrimination all because of her gender. Those two barriers had been the deciding factor in many career-oriented women's premature resignation. Minty seemed to have that ruthless determination now but Dale wondered what her attitude would be three months from now.

"Well, you can have it," Dale said, tiredly. "We've been here —" she turned to Almstead. "How long have we been here?"

"A day and a half."

"That's all? It feels like a month. Anyway," Dale continued. "I've never wasted so much time in my life. The only thing that breaks the monotony is the Kirk thing."

"I hope she doesn't ruin it for the rest of us," Almstead said.

"If they get her out of here like she wants, she won't," Sherlock said, almost defensively.

"They never should have let her leave AFEES in Detroit," Jaffe said.

"But, other than that, I'm bored," Dale told them.

"You won't be, believe me," Minty drawled.

"I'm sure I'll regret saying that in two weeks but right now this whole thing is such a yawn. All we do is wait and sit around and stand around. Why aren't we more organized? Shouldn't we all be in uniform by now? Somebody said that next week will be spent processing in, too, and they also said that what we do in that amount of time could actually be accomplished in one day," Dale whined, hoping to sound like an impatient, inexperienced trainee. "This is ridiculous. I'd rather get into it as soon as possible instead of all this...whatever it is we're doing."

Even Minty nodded in agreement.

"Boy, I sure hear that," Sherlock growled. "The longer they draw this out, the worse my attitude is going to get. I wasn't too keen on this decision in the first place and I'm getting less keen on it by the hour."

"Hello, ladies." A new voice broke in and they all looked in its direction. Karen Henning withheld a smile at seeing Dale with the group and approached the two tables where the women were seated. "I'm Lieutenant Henning. Ready to get to work?" Her attitude was pleasant, but her voice wasn't without its authoritative edge.

"Yes, Ma'am," they answered, not all at the same time. They stood up just as randomly and followed her to a table that had a cardboard box on it.

"Come on, ladies, surely you can show me a little more enthusiasm than that." She laughed as she opened the box.

"Ma'am? I don't mean to be disrespectful, like I'm trying to pry or anything like that, but you don't look any older than the rest of us," Jaffe said. "Out of curiosity, can I ask how long you've been in?"

"Sure you can. I've been in...oh, let's see, it'll be a year in January. And I don't look any older than the rest of you because I'm not. Not really. I'm twenty-four. Now, let's get to work." She took out a handful of decorations and looked up at the five eight Minty on one side of her and the five ten Sherlock on the other side of her.

"Spotted any land on the horizon?" she kidded, dividing her handful between them. "You two can start at this end and work around. Strategically place them as best you can."

She reached inside the box and pulled out rolls of orange and brown crepe paper, masking tape and two pair of scissors and gave them to Jaffe and Almstead. "I don't think you can reach the corners by standing on the tables, so there are a couple of ladders in the kitchen. Just knock on that door there and tell them you're helping me decorate." She showed them where she wanted the streamers to go. "And you," she looked at Dale. "Come with me. We have to put paper tablecloths on every table, and I forgot to bring them down. They're in the supply room. Let's go."

"Yes, Ma'am." They walked upstairs toward Alpha Company's personal supply room, which was located on the south patio.

"You don't look happy at all, Private Oakes," Henning said, laughing, not being able to contain it any longer.

"Oh, you're wrong, Lieutenant Henning, I'm just enjoying the shit right out of this," Dale answered her, sarcastically.

"Anything yet?"

"No, not yet but I haven't been here that long and not everyone is here yet, including Walker. And, anyway, everyone is still too goddamned horrified to do anything obvious. So far, everyone is acting like a normal, fucked up civilian who has no idea what she has gotten herself into. This is all going to take a while to unfold. Unfortunately."

"How does it feel to really be back?"

"Let's put it this way...do you want to trade places?"

"Not on your life."

"That's exactly how it feels. Tough break about Stuart, huh?"

Henning shook her head, grimacing. "Horrible. I honestly hope it wasn't related to this case."

"Yeah, me too." But Dale really didn't hold out much hope of the murder just being a coincidence. She was pretty sure Henning didn't, either.

"What do you think of the drill sergeants so far?" They entered the stock room and retrieved the box containing the tablecloths. Dale waited until they got back outside to answer. They headed back toward the mess hall.

"I've only met two. Robbins and McCoy. It's just too early yet." She didn't want to go into her speculation of Robbins being a ladies' man. She could have been wrong about him, after all, it was only a first impression. "When does Colton get back?"

"As far as I know, he's already back and signed in. He just hasn't been around here."

Dale was about to bring up the Kirk situation when another young woman approximately their age, in civilian clothes, approached them from the front. "Hi, Karen, what's going on?"

"Hi, Connie, did you just get back?"

Dale waited, politely, and held the box of tablecloths as Henning and this lady chatted in a friendly manner. Dale later learned that Second Lieutenant Connie Clarke was 10th battalion's operations officer and worked under the command of Anne Bishaye, whom Dale would have gleefully strangled if she could have gotten her hands on her. Henning instructed Dale to take the box to the mess hall and wait for her while she continued to catch up with Lieutenant Clarke. She rejoined the five recruits moments later, where it was business as usual

Outside, it couldn't have been a more beautiful day. The temperature was an unusually warm seventy-six degrees. The last Thanksgiving Dale had spent at Fort McCullough wasn't quite as memorable. It had been a humid forty-nine degrees and it had chilled her straight to the bone. She had grown up learning how to fight the dry, bitter cold of the northeast and just wasn't prepared for the damp, unpredictable winters of the south.

The volleyball nets were set up in the open area between the patios but not everyone participated in the games. Some women remained upstairs in their depression to write letters or just get acquainted. Dale had talked Kirk into going downstairs and getting involved in the game and, as Dale had expected, it took Kirk's mind off her problems for a while, in fact, she almost enjoyed herself.

Just as the women were really beginning to get into the game, they were ordered to Fall In on the north patio. A majority of the women came to the position of Attention as best they remembered but Drill Sergeant Robbins, who had returned to the company area in his dress blue uniform, didn't push it.

The dining facility was handsomely decorated and offered the closest thing to a home cooked meal Dale knew the women would get while they were at McCullough. Again, as at breakfast, conversation was permitted and everyone talked freely. Dale was seated at a table with Laurel Kotski, Margaret Jane, MJ, Mroz and Diane Tierni, women who were all Dale's age. Kotski seemed a bit more settled now but the other two were still going through their second thought stage. They discussed the upcoming weeks as though they were a prison sentence.

"I feel like I've been forced to do time for a crime I didn't commit," Tierni commented.

"And we haven't even gotten into it yet," Kotski said.

"What's Henning like?" Mroz inquired, skeptically. "Hopefully not the temperament of the Red Queen from *Alice In Wonderland*."

Dale shrugged. "She seems okay. We couldn't really be sociable, of course, but she seems pretty on the level. Her disposition is nice enough. Time will tell, I guess."

"Minty said you two had to go to the stock room together. Didn't she even talk to you?" Tierni asked.

"Sure. Worthless conversation, questions like why'd you join and why'd you pick MP, that kind of stuff. We've been asked it a hundred times since we've been here and are likely to be asked it a hundred more."

"Did you remember to say ma'am after everything?" Kotski inquired.

"I said it when I thought of it, but she wasn't too pushy about it. I think she knows we're all confused right now. It didn't seem like that big of a deal."

"Bet it'll be a big deal soon," Tierni offered.

"No doubt," Dale agreed, knowingly.

The women wandered back upstairs to the Alpha company area on their own when they were finished eating and eventually picked up where they left off before dinner. The AIT graduates escaped constant questions by taking refuge in the dayroom. They clearly knew they were safe from pesky trainees there and also wouldn't risk getting snagged by a roving drill sergeant who had repeatedly warned them not to be telling unsuspecting recruits things they would be finding out for themselves sooner or later. Dale snickered, quietly, observing whenever a holdover spotted a confused looking female in civilian clothes heading in his or her direction, they ducked into the dayroom.

The dayroom was the military's answer to a relaxation area. Every company area usually came equipped with one. Most resembled the holding area at a local bus terminal. Several rows of chairs and couches faced one object, for example, a wall mounted television, or one another, however, there still never seemed to be enough seats to accommodate the entire company, if they desired to all congregate there at once. Dayrooms also usually came equipped with a foosball game and a round wooden table for playing cards and board games. The snack and soda machines were located outside on the patio, but the new Alphas hadn't earned any privileges yet, therefore, the vending machines and the dayroom were off-limits at this point and would remain so until those concessions were warranted by good behavior.

Later, after Dale had taken a shower and relaxed on her bed, she looked around at the cluster of young women who were going to be undergoing phenomenal changes within the next couple of months and she was curious about who was going to make it through and who wasn't. She also wondered if the set up women had been planted yet and, if so, who they were. This early in the game, it would have been foolish to try to pick out anyone in particular with abnormal or suspicious behavior. For the first couple of weeks, while adjusting to military life, *everyone's* behavior was abnormal and suspicious. And if Boehner had spotted or heard anything worth mentioning, she had yet to relay it to Dale or, to Dale's knowledge, Henning.

As the lights were switched off at nine thirty, or 2130 hours, per regulation, Dale punched and pushed on her coarse, cotton-ticking fabric pillow that was rough on the skin and took some getting used to, even with a thin linen case covering it. She tried to make the down, goose feather filling comfortable enough for her to relax her head at an angle that would be conducive to rest.

She listened to the women around her also try to adjust to their new sleeping arrangements on slim single mattresses on old, hard bedsprings set up on a metal frame. Some of the bunks were going to need to be oiled if she ever expected to attempt a decent night's sleep in the months to come.

Dale wondered again about Lieutenant Walker and hoped against hope that she wasn't the type of woman who was strictly hardcore military. Dale did want a good agent on the case with her, but she also prayed for someone with a sense of humor, and who knew how to raise a little hell but with enough discretion that she wouldn't draw unnecessary attention to either one of them. It was probably too much to ask for, Dale thought. Suddenly, for no reason, she pictured herself back home, driving up Route 4 toward Killington to party at one of the ski lodges. The vision only made her acutely aware of her definite loss of freedom and that very lonely feeling of emptiness and abandonment stayed with her until she drifted off to a restless sleep.

Bay details were posted the next day and the women were directed to the list after morning chow. Dale, along with Kirk, Kramer, a female from Pennsylvania named Melanie Mackey and a six-foot-tall Hawaiian named Kay Verno, was given the latrine and the showers to do. It was not her preferred duty but certainly not her least favorite detail. They split the job so that three of them cleaned the cubicles and sinks while Dale and Verno wiped down the showers and swept and mopped the floor.

The bathroom areas had not had a chance to get that grungy yet, so that was a relief. Dale recalled that because the latrines were cleaned every day by trainees, the stalls, commodes, sinks, fixtures and showers never really became unpleasantly odorous or unmanageable. The only time she was not thrilled with the detail was after the company spent time at bivouac crawling through the Alabama clay in the rain. It usually took three times as long to get the mire off the walls, floors and out of the grout in the tiles.

Dale refocused on scrubbing the shower drains and looked up at Kay Verno, who was the youngest of all the females there, other than Kirk. As Dale swept the floor, she just marveled at how Verno could reach the top tiles in the shower stalls without a boost and three times Verno turned around and caught Dale staring at her.

"What's the matter with you, Oakes? You keep looking at me funny. You're not a lezzie, are you?"

Nothing like putting it right out there. "No, no, it's not that. I'm sorry for staring, it's just...tell me, does anyone ever yell timber when you fall?"

Verno laughed. "No one has so far."

"I'm not making fun of your height, Verno, don't get me wrong, I wish I was that tall. I would have been the most popular girl on my basketball team. It probably would have been a bitch getting dates, though."

"Tell me about it," Verno responded with a glare.

Well, naturally it would be a sensitive subject, you twit, Dale admonished herself, and began asking questions about Hawaii. That bottled up frustration eventually had to be vented on someone and Dale did not want to be the target.

At the reception center later, the women lined up at a window outside the lecture hall and each got an advanced pay of one hundred dollars. The original seventeen women were driven to WacVille to get measured for their dress uniforms, fatigues, and combat boots. They were then told they would not receive their actual issue until the following Monday.

"Well, at least we'll finally be in uniform in a couple days," Jaffe said, while they were taking advantage of the smoke break, outside the building. Even though most of them, like Dale, didn't smoke, it was an excuse to get a few moments to themselves. "Maybe that will make a difference in our morale."

"Maybe not. Specialist Harriman doesn't look so motivated, and she's obviously been in a while," Minty said.

"I'll be glad when all this processing bullshit is through," Sherlock said. "Looking at Specialist Harriman's mustache every hour is making me lonesome for my boyfriend."

"Looking at Specialist Harriman's mustache is giving me nightmares," Minty said, joining in.

"Looking at Specialist Harriman's mustache reminds me that I need to shave my underarms," Jaffe said.

"Specialist Harriman is right behind you," Specialist Harriman announced to the mortified trio. "Get back inside! Smoke break is over!"

The three recruits, now united in matching shades of red, could not get back inside fast enough.

After noon chow, they were escorted back to the reception center to have their military identification cards made up. Dale did not look forward to that because that specific document was never flattering. It was usually a photograph that made everyone look like a member of San Quentin's inmate file and, unfortunately, unless a soldier lost it or was promoted to a rank above E-3, his military ID followed him around, haunting him for the entire period of his first enlistment.

The photographic equipment was obsolete, and the lights were so bright that one either had to squint to protect herself from going blind or her eyes grew twice as wide from the shock of what seemed like a nuclear blast. Hair was not allowed to be combed or touched up. After all, this was still a man's Army and most men by this time had no hair to comb or touch up, so the effects of the sixty mile an hour windstorm the Alpha women had just come inside from, never would have shown up on the men anyway. And make-up, of course, was supposedly forbidden.

"You're in the Army now, young lady, you don't have to impress anyone with your looks anymore," McCoy had told Sherlock when she had been asking around for lip gloss.

Consequently, the outcome of the pictures made the entire bunch look as if they had been tossed into a dungeon for a month.

A distinctly feminine, petite woman with classically structured facial features named Tanya Swinegar almost went into cardiac arrest when she saw her ID picture. "This doesn't look anything at all like me!" she gasped.

"Give it a few weeks, it will," the woman in charge of handing out the cards assured her.

"I don't know about the rest of you, but I look like I just stepped out of Rocky Horror," Tracy Travis, a woman with a heavy Boston accent, said. Dale noticed Travis appeared not sure whether to be pleased or disappointed.

When they returned to the Alpha company area, they were greeted by another non-commissioned officer. This NCO's name was Lenny Kathan, a staff sergeant, and though he was addressed as drill sergeant, he had yet to complete Drill Sergeant School. Instead of the ranger hat, he wore a helmet liner with his rank decaled on the front and, to his peers, what he wore was affectionately called a turtle shell and he was their trainee. But Dale knew he was just as dangerous as the real thing.

Kathan was a tall, dark, handsome man who had obviously gone through Dale's theory of the castration process prematurely. Every word that left his lips had a razor sharpness to them, warning the trainees that he was not a man to mess with.

His formidable quality eroded some, however, when he marched some of the females to the PX to pick up essentials and it was discovered that not only was he tone deaf, but he had no rhythm, either. He sang the cadences flatly and he threw the count off so that nobody's right foot hit the ground at the same time. At this point, it didn't mean much to any of the others but if it was one thing Dale *did* enjoy about basic training, it was marching to cadence and looking good when she did. Kathan drove her crazy and Dale prayed he wouldn't be there long enough to get assigned as an assistant platoon sergeant.

In the forty-five minutes the women were released to run loose at Fort McCullough's main post exchange, most of the thirty-two females lined up at the pay phone to put a call in to their parents, husband, boyfriend, girlfriend, or friends. They were restricted to buy only military requirements, soap holders, soap, shower shoes, toothbrush container, toothpaste, white only towels, shampoo, etc., and dire female necessities. No contraband such as candy, gum, munchies of any sort or civilian reading material was allowed, although cigarettes were begrudgingly permitted as long as the buyer limited herself to one carton. That was to last her until her next visit to the PX.

Dale strolled around the store with Kirk, who was falling into a deeper depression every day. Kirk tried to get to the phone twice, but it was just too crowded and, in no time, Kathan was calling them all back into formation.

That evening in the bay, an open argument broke out involving Kirk, Caffrey, Almstead and a new trainee, Shelley Creed, who had quickly become Lex's ally, pertaining to Kirk's dilemma.

"You know what, Kirk? You need to knock off the crybaby act," Caffrey told the young woman as she exited the latrine.

"Fuck you, Caffrey," Kirk responded, almost calmly. When Kirk tried to walk to her assigned bunk in the back of the room, she was intercepted by Almstead and Creed.

"No, fuck you, Kirk," Creed said, blocking her way. "You're going to spoil it for those of us who are really serious about being here."

"Right," Almstead chimed in. "You're going to have them all thinking we're like you."

"This isn't my fault!" Kirk said, defensively. "And I am not going to ruin my life by being forced to do three years of this mindless bullshit." She took a step forward, which put her in Caffrey's personal space. "Now get out of my way."

"Or what?" Caffrey placed her fists on her hips in defiance.

"Or I will move you out of my way," Kirk said, her voice a low growl.

"You'll have to move all three of us then," Creed told her.

Dale observed the situation with interest. She wanted to jump up and calm everyone down, but she was more curious to see how it all played out.

Sherlock then jumped up from her bunk and walked to the tense group, followed by Troice and Travis. "Why don't y'all just get off her case. It's the screwed-up system that's involving all of us, not Kirk. Stop blaming her. It's not like any of you are perfect."

Caffrey whirled and faced Sherlock, menacingly. "Really? You're one to talk. How many times have you and Minty been warned not to flirt with the AIT males?"

Minty looked up, obviously startled that she'd been drawn into this argument. "Leave me out of this, Caffrey, I have nothing to do with your issues with Kirk."

"Yes, you do. You and Sherlock are just as bad. You think you're special and that the rules don't apply to you," Almstead spoke up. "Caffrey's right. You've been told and told and tonight you were caught flirting with them again."

"You're just proving to them that the women can't be trusted," Creed said.

Dale cocked her head, focusing back on Minty and Sherlock. Creed had a point.

"Yeah and I don't know about the rest of you," Caffrey said, with a sneer. "I didn't enlist just so I could hunt men."

Minty clearly did not respond well to personal attacks. She rose up off her bunk and stalked to the group, where she stood next to Sherlock. "Yeah, Caffrey, we know. Me and half the other women up here are quite aware of what you joined the Army to hunt." Her eyes narrowed. "And if you come near any one of us, you will, without a doubt, be the first recipient of a blanket party."

"What the fuck is a blanket party?" Caffrey asked. Her expression showed her to be curious in both a good and bad way.

"You want to find out?" Minty challenged.

Dale hoped it wouldn't go that far. Basic training units were notorious for their blanket parties. They were given for various reasons but usually to troublemakers who lived in the barracks, the kind who repeatedly informed the cadre on the other trainees or constantly were responsible for more mental or physical agony than usual being thrust upon a platoon or an entire company. They were also given to individuals who could, but stubbornly would not, conform to the Army's way of life, which resulted, one way or another, in even more misery among barracks life.

The only blanket party Dale ever witnessed was during her first time in basic training. The beneficiary was a female who had diarrhea of the mouth and the others knew she was responsible for severe punishment being inflicted on them all by the drill sergeants. One night when this woman was asleep, ten females wrapped her in a wool US Army blanket, dragged her, kicking and screaming, into the showers, turned steaming hot water on her and then, with the blanket still securely around her, punched and kicked her senseless. It was a hard lesson, but it worked. Humiliation and fear of it happening again made her stay quiet for the rest of training.

The cool, professional voice of Quinn Brewer, so far the oldest woman there at twenty-nine, snapped Dale out of her recollection. "Hey! Knock it off, all of you. The last thing we needed right now, when training hasn't even started yet, is to be at odds with one another."

Everyone was quiet now and their focus was on Brewer. "This many women confined to such a small area together are bound to bring on problems, but if we're going to make it, we, at least, have to try to stick together," Brewer said, reasonably.

It didn't make any of them instant friends but the small amount of truth to her words did seem to pacify tempers for the time being.

Watching them from her bunk, Dale had not taken sides just as most of the other women had not. She knew that this cease-fire would not last long with such exact opposites living in the barracks and she also knew next time the blow up would be bigger and include more people. Quinn Brewer was going to be good to have around but her moderating would only go so far. The hostility was just going to have to work itself out.

Dale went to sleep troubled that evening. There was still at least one more group to come in and already the women weren't getting along. It wasn't unusual for antagonism to arise among females who were virtual strangers, thrown together and forced to cohabit for a considerable period of time. In Dale's experience, however, the opposition didn't ordinarily occur quite so soon in the cycle. Dale wondered what Boehner would have said or done about the incident, had she been there instead of off somewhere in Averill on a four-day pass, that lucky shit. This was perhaps the only time in the duration of their short acquaintance where Dale actually envied her.

Saturday night might as well have not bothered to come. After detail had been completed, the day ceased to exist for most of the women. They were restricted to the company area and once again forbidden to use the dayroom.

Sunday morning, another series of volleyball games broke some of the monotony but even that got old after a while. Growing restless, Dale returned to the bay to make sure her locker area was squared away. When that was completed, she pulled out some stationery to write a letter but couldn't think of who to write to or what to say.

She was about ready to roll off her bunk and put away her pen and paper when a shadow crossed her blanket. Glancing up, Dale saw Kirk who looked ready to climb the walls. "Hey, what's up?"

"Can we go downstairs and talk?"

"Sure. Just let me secure this stuff first."

Downstairs, they leaned against the picnic table closest to the laundry room. Dale had been observing Kirk with increasing concern, trying hard to stay outwardly neutral. She looked upon what was happening to the frightened young woman with almost as much dismay as Kirk herself.

"They're not going to let me out, Oakes, I know it," Kirk told her, as tears threatened the corner of her eyes. "They're just going to keep stringing me along, aren't they?"

"Don't ask me. I don't have the answer to that. If I knew what to tell you to make it any easier, you certainly would have heard it by now."

"Everyone upstairs hates my guts. They don't understand."

"Sherlock fights for you."

"She fights for me but she never talks to me about it. It's

like she'll jump on anything to sound righteous, but she doesn't know or even care what she's yelling about just as long as it gets her the attention she wants."

"Oh, I don't know about that. I don't think you're being quite fair. She doesn't have anything to gain by defending you, in fact, just the opposite. I'm sure there are a lot of women up there who are on your side but they don't know all the facts and they're a bit apprehensive about speaking up. This *is* still America but it's also the Army and freedom of speech has limitations now."

"How come you didn't come to my defense last night?"

"As I recall," Dale began, patiently. "You told me and two others to stay out of it because it might make things worse, remember?" Kirk nodded. "And, by the time I knew what was going on, the focus of the argument had shifted from you to Sherlock and Minty. Besides, do you really think I could have said or done anything to make a difference?"

"Brewer said something that made a difference."

"Well, Brewer was probably a diplomat in another life," Dale said and smiled.

Clearly, Kirk was not in the mood to smile back.

"Look, like I said, I don't have any answers or foolproof ideas to get you out of here but I'm sure if you continue to stand up for yourself, eventually they'll give in. I would think if you were going to fight them every step of the way, they would want you out of here as much as you want to get out. It's obvious to me they just want to see how much crap you can put up with. After a while, they'll get tired and let you out."

"If they don't drive me to suicide first." There was no trace of humor in Kirk's voice or expression.

"Then they'll have won, won't they?" Dale responded, sharply. "Just chalk up another casualty to old Uncle Sam, right? Who the hell cares, right?"

"What are you getting so pissed off about?" Kirk asked, defensively, at Dale's agitated tone. She looked at Dale for the first time since their conversation started.

"Because you sounded so serious about them driving you to suicide and you said it so calmly, like it would be no big deal."

"Well..." Kirk shrugged and looked off into the distance.

"Damn it!" Dale exclaimed, exasperated. She drew a deep breath and calmed down. "We shouldn't even be discussing this. You - listen to me —" She spoke in a tone of voice that prompted Kirk to meet her eyes. "You will get out of here and then you can go back to doing whatever it was you were doing before you got here. Suicide is a little extreme, don't you think? Death is forever, my friend. You're talking about something you cannot remedy later when this *temporary* situation goes away. I'm pretty sure you can put up with a few weeks of head games if it means your freedom in the end."

"I just got myself into something I had no business getting into. I shouldn't be going through this at all."

"That's very true. But you're here so you are just going to have to ride it through. Just hang on, okay? I'll give whatever support I can but remember, I'm just a lowly peon like you."

That made Kirk smile. Dale watched her light up a cigarette and Dale hoped that Kirk's remarks about suicide were just passing thoughts.

Just before she drifted off to sleep, Dale reflected on the Kirk situation and wondered if something similar had driven the person she was after to get revenge on the company. Desperation was an unpredictable emotion. Dale knew that firsthand.

Dale awoke to the sound of Tracy Travis shrieking out a Streisand song in the shower. When she was fully conscious, tiny horizontal rays of actual sunlight attacked her eyes and she knew, instinctively, it was well past 0500 hours or even 0700 hours. She checked her watch and moved toward the latrine to shower. She suddenly remembered that no new females had come in yesterday and wondered if anyone would arrive today. She glanced at her watch again and still couldn't believe it was going for nine. The blinds hadn't even been opened yet and most of the women were still in bed. Dale shook her head, a little bewildered and made a mental note to ask Anne why the first few days had been so slack. This was usually the time the cadre used their most inspiring Gestapo tactics.

CHAPTER EIGHT

Shannon awoke to the sound of her own head slamming against a desk after a specialist with a mustache named Harriman had knocked her supporting arm out from under her chin. It wasn't her fault that the prosaic little staff sergeant, who had bored her to sleep twice already since she arrived at the reception center an hour ago, had caused her to doze again. After Shannon and Harriman exchanged glacial glares, Shannon looked around at the ten other women she was grouped with, who would accompany her to Alpha-10 and it was obvious that with the exception of one or two, most of them would rather have been home in bed, also.

She had been seated next to a woman named Christine Wachsman, who was just going to be there for basic and law enforcement training and then would return to her home in Pennsylvania. She was one of a handful of females in the upcoming cycle either in the National Guard or army reserve and whose enlistment requirement would be one weekend a month and two weeks in the summer for three years, sometimes four. There was something about Wachsman that told Shannon they were going to get along well.

The last busload of females, Shannon's group, arrived in the Alpha company area almost exactly at chow time. The staff duty NCO, a drill sergeant from Charlie Company, assigned them all meal cards and sent them upstairs to find an empty bunk and locker. Shannon looked around for Dale but the only women on the second floor were the ones who had just returned from or skipped noon mess. She rushed to claim a bed near the rear exit door, put her suitcase away and hurried downstairs with Wachsman so they wouldn't be last in line.

Inside the mess hall, Dale took a seat opposite Deborah

Michaelson, a quiet blonde with an understated attractiveness, an almost startling appeal that radiated through despite the fact that she wore no make-up. She had the kind of smooth, clear complexion that made her envy of just about every female in the bay and sky-blue eyes that would surely send most of the male trainees, not to mention a few drill sergeants, into hot flashes just by making visual contact. Dale also guessed that a few female trainees would be eyeballing her, too, but she would not be one of them. Even if Dale did decide to quench a newly curious thirst, it would not be with a trainee, regardless of how appealing or willing. But she certainly was pleasant to look at.

The aloof Private Michaelson didn't appear to be aware that she possessed that kind of power, nor did she seem to care. Perhaps being the most attractive woman in the company, so far, did not, apparently, have any effect on her. Dale tagged Michaelson as a loner because she kept to herself, unless someone directly involved her in a conversation. If someone happened to make a complimentary reference to her appearance, she would smile shyly and politely deflect it.

Dale had yet to see Michaelson attach herself to any one person in the barracks or seek out anyone's friendship as Kirk had with Dale, or Creed had with Almstead or Sherlock had with Minty. Dale admired her unpretentiousness but wondered what really went on inside Michaelson's head because Dale really didn't trust people she couldn't interpret, and Michaelson was about as readable as hieroglyphics.

Dale had barely taken her first bite of lunch when Boehner, just back from her weekend pass, put her tray down next to Dale's and slid in beside her, into the booth. Casually scanning the dining room, Dale saw several near-empty booths and wondered if Boehner was sitting next to her to purposely annoy her. So, Dale thought she would annoy her back.

"Hey, Boner. What's up?"

"The last bunch of females are in," Boehner commented, gritting her teeth.

Damn, she's not taking the bait. "Which brings us to how many?" Dale asked, with the same amount of indifference in her voice. Her mood now bordered on apprehensive because she knew that Boehner had met with Lieutenant Walker already. If

Michaelson hadn't been seated with them, Boehner would have been able to point Walker out so that Dale could have at least had a little time to observe her before the inevitable meeting.

"Forty, I think, but I'm not sure. Have you kept count?" Boehner addressed Michaelson, who shook her head negatively.

"How many came in?" Michaelson inquired. She didn't really sound interested but it didn't seem to be in her nature to be discourteous.

Boehner looked toward the ceiling and counted from memory. "Eleven."

"Then I think that brings us to forty-four," Dale estimated and thought, *Who the hell cares?* "Great. Now that everyone is finally here, maybe we can get down to business and start some actual tr—" She stopped in mid-sentence, shocked by what surely must have been an apparition approaching. Dale's mouth automatically dropped open and a sudden wave of sweat rushed through her entire body. Both Boehner and Michaelson noticed her weird expression and they followed the direction of her gaze.

Michaelson went back to her meal but, in her peripheral vision, Dale saw Boehner react as though she now had a legitimate reason to slap Dale silly. Dale understood that this whole encounter between the two agents was supposed to be nonchalant, two trainees meeting for the first time, no big deal. Dale was sure her giveaway, goofy reaction to someone she was conveniently supposed to never have met was botching it up royally from the beginning. But, like a train wreck, Dale couldn't look away from the woman walking toward her table.

"Oakes, what the hell is wrong with you?" Boehner asked, sounding a little more desperate than she probably should have.

But Dale didn't, *couldn't* answer Boehner right away. She studied the approaching woman in detail. The resemblance was uncanny. She was the spitting image of Shannon Bradshaw, her best buddy from long ago. The closer this person got, the more phenomenal the likeness. Finally, Dale managed to get back to Boehner, but she still kept her eyes on the blonde heading toward the table. "She looks incredibly like someone I used to know."

"Maybe she is," Michaelson spoke up, sipping her tea. "Wouldn't that be a coincidence?"

Dale nodded as Boehner nudged her roughly under the table, finally getting the lieutenant's full attention. "What'd you poke me for?" Dale grabbed her rib and glared at Boehner.

Boehner looked at her, incredulously, then buried her face in her hands, shaking her head, hopelessly.

The woman sat next to Michaelson and smiled a friendly but noncommittal smile at both Boehner and Dale and started to eat.

Dale cocked her head to one side and looked at every inch of this woman's face. She wasn't sure if she could go through four months of hell with someone who looked so much like Shannon. There would be too much of a temptation to make military references this woman wouldn't understand and that could eventually expose Dale as a spy. She just couldn't get past the similarity, it was amazing.

The woman, after four or five minutes of Dale's rude staring, decided to speak to her. Her eyes met Dale's tilted ones. "Did anyone ever tell you that you do a wonderful impersonation of the RCA Victor dog?"

Oh my God, Dale thought and broke out into a huge grin. That voice had given her away. Only one person had a delivery and a compromised New England accent like that. This was too good to be true. "That's cute. I didn't mean to stare like that, it's just...you look like a friend of mine."

"Really? A close friend?" Shannon continued to eat.

"She used to be. I'd like to think she still is," Dale said.

"Oh, I'm sure she still is," Shannon said.

"What's your name, by the way?" Dale asked, still sporting an idiotic smile.

"Shannon. Shannon Walker. What's yours?"

"Dale Oakes." *She must have gotten married.* "And this is Deborah Michaelson and Linda Boner, who is waiting on orders to leave here."

Shannon picked up the cue and directed her conversation to Boehner. "Boner? Man, I'd change that name, especially in the military."

"It's *Bay*ner," Boehner growled, not taking her eyes off Dale.

Dale lifted her tray and Boehner let her out. Boehner sat back down to answer some of Shannon's questions.

"Hey, it was nice meeting you, Oats," Shannon said, dryly.

"It's Oakes. Yeah, maybe I'll see you upstairs, Welker."

"Walker."

"Right, sorry." Dale disposed of her tray and headed up to the picnic table to wait for her long, lost friend. She was definitely going to string up Anne Bishaye when she saw her for her intentional nondisclosure of who Lieutenant Walker really was. And sweet little Karen Henning was going to get blasted, too, because she must have known all along. What a moron she must have looked like, especially in front of Boehner, who was more than likely having a mental field day with what just happened.

"I must say, you carried that off remarkably well," Boehner's sarcastic voice interrupted Dale's train of thought. "If I hadn't known that you two were who you were supposed to be, I never would have suspected a thing."

"I realize what it must have looked like. What you don't understand is that I know her."

"I *know* you know her. That's Walker," Boehner argued, her annoyance unmistakable.

Dale looked around to make sure that no one was within listening distance. "Let me clear something up for you before you talk yourself into a counseling statement, Sergeant," Dale began, quietly, pulling rank on Boehner, a practice she really wasn't fond of doing. "They told me I'd be working on this thing with a Lieutenant Walker. Lieutenant Walker I'd never heard of. That woman downstairs is formerly PFC Bradshaw and we went through basic training and LE School together six years ago. I have neither seen nor heard from her in three years. Now, apparently, she acquired a gold bar and a new name along the way and nobody bothered to tell me. We used to be best friends. She was the last person I expected to see today and until she opened her mouth, I was convinced she was just someone who looked incredibly like her. Now do you understand a little bit of my indiscretion?"

"I'm sorry. I didn't know. I just thought..."

"You just thought that a couple of airheads were assigned to take over for you and it just confirmed your faith in officers...or maybe it's just lieutenants."

Embarrassment burned on Boehner's cheeks as she turned and walked away.

Dale watched her leave. She knew the type. Boehner didn't particularly like her or any officer for that matter and, regardless of how astute Dale or Shannon may have come across to her, her reaction to them was marred by private feelings of failure. Even if Boehner's replacements were smart and professional, Dale knew Boehner would be damned before she'd apologize. Boehner was the sort that was too hardcore to admit her initial impression had been wrong. She had the us against them attitude toward commissioned officers and Boehner was battling a personal conflict that would never be won.

Dale waited for Shannon to come upstairs. There was a lot of inconspicuous catching up to do and urgent questions to be asked that made Dale a little anxious but, overall, she was filled with an overwhelming sense of relief. All her anticipated fears about this Lieutenant Walker stranger were replaced by an almost intimate feeling, as if the next four months were going to be one continuous private joke between Dale and Shannon.

Anne Bishaye had done all this on purpose, Dale knew. More than once in their casual conversations, Dale had mentioned in passing her and Shannon's misadventures and her curiosity as to Shannon's whereabouts. How Bishaye had located her and set this partnership up had to be a case of who knew whom and probably a couple repayments of favors. Regardless, Dale didn't care how she did it or what her motivation was, just that it had been done and was irreversible.

She spotted Shannon ascending the stairs and a smile crossed Dale's face as her mind and body suddenly relaxed at the same time. She knew now that no matter how difficult or complicated this case might become, at least she would have some fun with it.

Holding her two index fingers in the shape of a cross, Dale put them up in front of Shannon.

"What in hell are you doing?" Shannon asked her.

"Isn't that what one is supposed to do when one confronts a ghost?"

"Cut it out, Dale." Shannon lit a cigarette.

"Where have you been, you shithead?"

"I missed you, too."

"If I wasn't so damned happy and relieved to see you, I wouldn't be speaking to you right now," Dale told her, annoyed. "I know America is technologically progressive but the last time I heard, they had invented pens and papers in Korea."

"I left Korea three years ago."

"Oh, really? Well, you never would have known by me."

"What are you so snippy about, anyway? Why should I keep in contact with someone who only writes every other eon?"

"That happened once," Dale protested. "When I was starting classes and my line duty schedule was heavy. Try again."

"Look, it wasn't intentional. I got sidetracked and time just flew. Are you still pissed off?"

"Doesn't it sound like I'm still pissed off?"

"Well, with you, Dale, one never knows. Fine," Shannon shrugged, knowing better. "I'll just go and tell the light bird to take me off the case and we won't have any more problems about it."

"You do and I'll break your arm."

Shannon looked at her and grinned. "So, you did miss me."

"Shan, I was worried about you. Things weren't too friendly in Korea when you were there. And with you up there by the DMZ with Second Division? How the hell did I know what happened? And where did Walker come from?"

"My ex-husband."

"Ex? Divorced?"

"Oh, you're so quick. You should become a detective."

Dale ignored her sarcasm. "So, what happened?"

"He turned out to be a real dog. He was a rock musician, you know, local cover band type, one hell of a guitar player, though. His name is Richard Walker and he was so goddamned gorgeous, I can't even begin to describe him. It was really insane. He relentlessly chased me, which impressed me to death. All the time we were going out, he played it really straight. I had to marry him to find out what a lunatic he was."

"You should have known the minute he asked you to marry him."

Shannon smirked. "You haven't changed one bit."

"I seem to be hearing that a lot lately. Go on."

"He started running from the day we were married."

"Openly?"

"No, not at first."

"Then how did you know?"

"Every time he would play out of town with the band, he would come back and show me a new position."

"Old joke, Shannon," Dale said but snickered anyway.

"Yes, but very close to being true. After a while he wasn't even discreet. He cheated with everybody. Our neighbor, my ex-roommate, my cousin when she came to visit, every groupie who approached him, it was disgusting. His little black book looked like the Manhattan phone directory."

"I can't believe with a fucking firecracker like you at home that he would look elsewhere."

"It was about notches in the belt, Dale. It had nothing to do with me. Or, at least, that's what I keep telling myself. So, I decided I didn't need his companionship, or anyone else's, that badly. Or the humiliation, or the aggravation, so I got rid of him."

"Sounds like a wise move. How long ago was this?"

"Eight months, two weeks, four days and five minutes. Not that I'm counting."

"Are you over him?"

"I don't know...I don't think anyone ever really gets past a bad marriage. You just kind of adjust. I've adjusted. It still hurts, though, that someone could be as insensitive as he was, and I'm angry at myself that I stood back and took it for so long. And I occasionally think about him and I occasionally get lonely for the good times we had, but no matter how lonely I get, I would never take him back."

"You'd never even consider it?"

"No, I think way too much of myself. No man is worth bringing your standards down that much."

"Do you still keep in contact?"

"No. It was a nice, clean break. Can we get off this subject? He really is a waste of my breath."

"No problem. Walker, huh? It's going to be real hard not to call you Bradshaw."

"Just keep reading my name tag." Shannon studied her long, lost friend. "What about you? Did you ever marry that limey you dated on and off for years?"

"My name is still Oakes, isn't it?"

"You could have kept your maiden name."

"True. But no, I didn't marry him. In fact, we broke up last month."

"Oh. Sorry to hear that. How are you doing otherwise?"

Dale made a face that indicated indifference. "I could be better, but I could be worse, too."

"Yeah, a lot worse from what I heard."

"From who?"

"Bishaye."

"She blows things out of proportion," Dale said and waved it off. The last thing she wanted Shannon to think was that she wasn't up to pulling her weight on this assignment.

"Don't tell me that. I was almost getting a respectful opinion of you."

"Yeah, well, don't. You'll be disappointed." Dale smiled at her again. "God, Shan, this is great."

"What's great? I think this whole situation sucks."

"Oh, me, too. I'm talking about you and me working together again in this capacity. Have you stopped and actually thought about the fun we can have?"

Shannon smiled and raised an eyebrow. "Now that you mention it, I guess the only thing to do is sit back and make the most of it."

"When did you get your butterbar?"

"I was commissioned about a month before you were, according to Bishaye. So, tell me, what's going on around here anyway? Brief me on the drill sergeants and the other females."

"Not right now. Someone's coming." Dale nodded her head toward the stairway as a new recruit walked in their direction.

"Oh, that's Wachsman. We met at reception. She's going to be fun to have around here, I can tell."

"Hey, Walker, where'd you go?" Wachsman asked, reaching them. "One minute I'm in line with you and the next, you've completely vanished."

"I wanted to sit with someone who's been here a while to

find out what it's been like so far. I thought you were right behind me."

"I was. Then suddenly you pulled this Casper act and you're gone. Did you find out anything useful? We've been here twenty minutes already and I sure don't feel any different," Wachsman said and shoved her hands into her pockets. "Hi," she said to Dale. "How long have you been here?"

"Too long. Since the twenty-second."

"And...?"

Dale shrugged. "It's not what I expected but I guess I really didn't know what to expect. We haven't really done anything except process in. A lot of paperwork, lectures, rules and regulations. They showed us a few military drills but don't ask me how they're done," Dale said and grinned.

Wachsman and Dale introduced themselves and they remained on the patio and gabbed until their 1300 hours formation. Drill Sergeant Robbins popped in to take attendance and to supervise the afternoon's activities. Dale, Shannon and Wachsman played volleyball with fifteen other women rotating in until it felt as if their feet were going to fall off.

Shannon and Dale didn't get any more time together to talk until right after the evening meal and then it wasn't for very long. Dale knew they would have to be careful, that spending too much time together, especially in the beginning, would put them in a suspicious light among the cadre and most likely spark unwanted rumors among the women in the barracks.

In their second solo conversation, Dale informed her partner of the tension in the bay, but they were joined by several others before she could get down to specifics.

CHAPTER NINE

Women had various reasons for joining the Army and no one's motives were exactly the same. Sometimes, however, even after the reasons were explained, the female in question's psychological whereabouts was still about as clear as tar.

Such was the case with one Emily Zelman, Dale discovered. Zelman had arrived with the second group of women, promptly earned the nickname Dizzy and was continuously living up to it.

If it was just her little quips, such as two's company, three's a menage a trois, her personality might have been bearable. Unfortunately, even her tone of voice was grating. She sounded like a tape player running on low batteries and she looked like a character in a movie that was always a little out of focus.

So far, most of her short time at McCullough was spent shocking the daylights out of youngsters like Creed and Almstead by telling them wonderful little tales of her past, such as being kicked out of college for giving a venereal disease to the dean.

If their eyes got any wider, Dale had thought upon seeing the exchange, *they would not have had any face left.*

Creed and Almstead, however, seemed to have matching overactive gullibility glands as most of the other women passed Dizzy off as a dirty joke.

Dale observed that Dizzy's porno queen act wasn't amusing the hardcore bunch, like Boehner and Lanigan. They seemed deeply disturbed by her explanation of how she became an MP recruit being that it had nothing whatsoever to do with test scores or previous police experience. Unless that experience included bedding down an entire precinct shift in a week to ensure not getting arrested, something else she had freely admitted to doing.

Dizzy quickly became the topic of conversation rather than Kirk, with her loose morals, tight clothes and vacant eyes. Shannon had been up and down stairs on a cigarette break several times and told Dale that each group of women she approached had one comment or another about the bleached blonde upstairs that seemed to descend from the ozone only long enough to get herself back into orbit.

"I'm more worried about Zelman than angry or shocked," Shannon said, as she discreetly spoke with Dale on the south patio, in a corner by the railing. "Not only are they going to assign this bimbo a lethal weapon, they're actually going to place it in her hot little hands and teach her how to use it, too!"

"Clearly, she's not here to succeed," Dale agreed. "And is hardly about to dedicate herself to anything that takes away from her freedom to do sex, drugs and rock and roll whenever and however she pleases."

"No recruiter should have been *that* desperate to make quota." Shannon was disgusted.

"My guess is that Zelman has been sent here to be made an example of. You know that there are recruiters who sometimes alter records and enlist people who are not exactly up to any specific military standard. I think she's here so that the drill sergeants can show the other, more serious trainees what *not* to do their first eight weeks with Uncle Sam."

"Jesus, they're still doing that shit?" Shannon asked.

"They'll *always* do that shit. Mindfuck. It's like a mission. Then they'll use her and any other misfit as examples by freely doling out counseling statements and Article 15s." An Article 15 was a military offense which stayed on a company level, issued by the company commander. The punishment could not exceed fourteen days restriction, seven days loss of pay and demotion to the next lowest rank, which for a trainee, was more than likely civilian status. "Then they'll discharge them for unsuitability."

"It'll be a trainee discharge so at least it won't, or shouldn't, screw up their civilian lives. And, let's be honest, usually people like Zelman earn their counseling statements and Article 15s because they can't grasp the concept of following the rules," Shannon said.

"Still, they waste anywhere from two weeks to two months

in an environment they never should have been exposed to in the first place. And even though they are gone, they certainly won't be forgotten. You know the cadre will use them as training exercises for those who start to turn sour. They will get threatened with punishment equal to whoever had just returned to civilian life before he or she had been bounced out."

"At least we can pretty much eliminate her as a suspect," Shannon said.

They shared a laugh. "Any drill sergeant who would seriously consider a fling with her deserves anything they get."

"Venereal *and* military punishment."

"Most men prefer women from their own planet." Dale said, as she and Shannon exchanged looks and then burst into laughter again.

"We don't serve with most men. We serve with a majority of who would never resist a flirtatious female offering a free evening in the sack."

"If a drill sergeant even remotely entertains the thought of getting any closer to Zelman than necessary, he or she deserves to get caught and punished *and* a trip to the clinic. Zelman is obvious and anyone who pursues her with any unmilitary-like ideas, cadre or trainee, is just begging for trouble."

Shannon put out her cigarette at the sound of footsteps approaching behind her. "Time to go," she whispered, as she turned and nodded a hello to the three females walking toward the picnic table, where Dale was.

"She and Boner almost got into it, too," a short, curly-haired brunette named Charlene Keival was saying. Dale's nickname for Boehner had caught on quickly.

"What happened?" a soft-spoken woman with long, red hair and glasses asked. "I didn't hear anything. I must have been in the bath, pardon me, the latrine."

Her name was Bonnie Kramer, no relation to Brigitte, and she was, thus far, the only married woman in the barracks. Dale and Bonnie immediately hit it off because Bonnie was the only one who admitted to being able to play one of Dale's favorite card games, cribbage. That excluded Shannon, who played cutthroat cribbage, not particularly one of Dale's favorite card games, because when Shannon played, Dale's throat was usually the one bleeding profusely.

"I really don't know. Boner came charging from the other side of the bay, telling Dizzy she wasn't suited for the military, that she shouldn't be here, and it was one or two like her that gave men an excuse to call any woman who enlisted a tramp."

"God...what did Dizzy say?"

"She just kind of smiled at her and said something like, if the shoe fits. Boner just got madder and told her she was a bad influence on all of us."

Who, Dale thought, *Boehner or Dizzy?*

"Yeah," Donna Guierrierre said. She was a pale young woman who sported shoulder-length black hair with a noticeable white streak at her temple. "Then Zelman said that anyone who wanted to die in a blaze of glory by being shot to death in combat couldn't be wrapped too tightly."

Well, Dale mused silently, *that was certainly the pot calling the kettle black.*

"She definitely gives new meaning to the term busybody, doesn't she?" Bonnie Kramer mumbled, more as a statement than a question.

"Have you seen how tight her clothes are? I don't think she has worn one pair of jeans that's even come close to being her size since she's been here," Keival brought up. "How the hell does anyone get into those pants?"

"Probably by buying her a drink first," Dale said, imitating Groucho Marx.

"I'm sure they don't even have to make that gesture," Kramer added. "Foreplay for her is probably, hey, ya wanna?."

Dale would have laughed, except she knew that, most likely, it wasn't a joke.

Upstairs, Shannon settled on the floor by her bunk and began filing her nails. When Wachsman came back from taking a shower. Wachsman sat on her bed, next to Shannon's.

"I couldn't take it anymore," Wachsman admitted. "After listening to that filth, I just had to run in and cleanse myself. I feel much better now. How in hell are we going to live up here with her for three or four months? She's going to drive us all nuts. If she doesn't straighten out and become a nun like the rest of us, I'll never make it. I'll be so hot and bothered from her stories, I'll be attacking the first male who walks through that door...which could be dangerous because what I've seen so far, the pickin's ain't too great."

Shannon, who normally despised gossip, decided to behave like a typical new recruit. "I heard she told Minty that being confined up here with us isn't going to be a problem for her —"

"Meaning what?" Wachsman interrupted. "She doesn't care about gender, just a warm body? We all need to be prepared to sleep with one eye open? What?"

"God, I hope that's not what she meant. Getting adequate sleep is going to be difficult enough without having to be constantly worried about unwanted company. Then I heard that Dizzy told Minty that she'll adjust soon and settle in with the rest of us virgins and we'll never know she's here. She told Minty she's basically a simple girl —"

"And Minty kept a straight face?" Wachsman asked.

"Yes. And Dizzy also said she adapted well to sharing space and all she really needs is enough room to lay her head," Shannon continued.

"And anyone else who stumbles in her path, I'm sure," Wachsman said.

Dale started her regular walk across the patio to the open area before the stairs. Most of the time she had been downstairs, she had been keeping her eye on a young man on the landing outside the second-floor bay door. She knew he was a permanent party soldier assigned to the A company supply room. He was a specialist fourth class, named Ingersol. He had been watching the new Alpha women with what seemed like more than mild interest. He had participated in their volleyball game earlier and had seemingly been studying them come and go out of the barracks, during the last hour, from his perch. Dale guessed he wanted to get at least one or two of them in bed before actual training started and tonight would have been perfect. The patrol of drill sergeants was few and far between and it would have been to his advantage to see which women could be easily swayed. After all, the women were going to be without *it* for quite a while and she was pretty sure it was the least he felt he could do to service as many as possible until their first weekend pass. She was well-aware of the type.

Dale had gotten chilly so she left the patio to go upstairs, alone. Folding her arms across her chest, she spotted the good-looking specialist as he leaned against the steel railing, a choice spot he hadn't left by the women's door since the last volleyball game had ended. She expected him to address her since he was undressing her with his eyes with every step she climbed.

"Hey, beautiful, why don't you come here and lean with me a second?"

Dale stopped and looked at him. "You're kidding, right?"

"Hell, no. Kidding about what?"

Shaking her head, Dale started to walk by him, but he stepped in her way. "Come on, man, I have work to do inside."

"Yeah, but this is important, baby. Come here."

"So is getting my assigned details done before bed check. And I'm not your baby."

"Not yet you ain't," he said and laughed. "Come here." He motioned for her to stand next to him.

Okay, Dale thought, *I'll play your silly little game for a while.* She shrugged and took a step closer to him.

"Look at you. You're way too tense. You need something to help you relax."

"And I'll just bet you think you have that something, don't you?" Dale smiled insincerely at him.

"Yes I do, baby, that's a fact." He leaned in really close, his face possibly a centimeter away from hers, his hand dropping to cup his crotch. "Guess what I'm stroking in my hand for you."

Dale returned his lewd glare and matched his low, throaty, suggestive tone of voice. "Dude...if it takes only one hand to stroke it, why the hell do you think I'd be interested?"

"Oh, a feisty one. I like that," he responded, flicking his tongue almost obscenely across his lips. "So, why don't you come upstairs to my room and let me get in your pants?"

"Gee, that's a real tempting, not to mention suave and debonair offer," Dale told him. "But one asshole in my pants is quite enough, thanks." She left the landing without looking back.

Inside the bay, Dale spotted Shannon, who was chatting

with Wachsman, and decided not to disturb her for a while. Instead, she looked around for Kirk but could not find her, so she ventured into the latrine and then the shower room to see if she was there. She walked back into the bay and was about to ask if anyone had seen Kirk when she noticed that everyone was standing At Ease. Suddenly, alert, she slowed down, making a quick visual search and saw no man on the floor.

From the corner of her eye, she spotted an unfamiliar, scowling face stomping toward her. It belonged to a female dressed in civilian clothes. Dale immediately assumed what was, apparently, the required commanded position but it was too late.

"What's the matter with you, soldier? Can't you follow orders? Didn't you hear the call of at ease?"

"No, I —"

"No *what?*"

"No, Drill Sergeant." It was only a guess that she was a member of the cadre. This abrasive, thick-bodied, masculine-looking woman wasn't wearing any kind of uniform or insignia. Dale's thumb pointed toward the latrine. "I was in —"

"I don't care *where* you were and get your hands back into the proper position!"

Dale returned to At Ease, thinking of a few positions she'd like her hands in, as the two women exchanged glares.

"You're off to a fine start, young lady."

Now, where have I heard that before? Dale thought.

The woman turned to face the other recruits. "Good evening, ladies. My name is Drill Sergeant MacArthur, and I am going to assign you all details for tonight." She pulled out a list of names. "When I call your name, I want you to answer me so that I know I have covered everyone."

Dale listened and watched as MacArthur handed out tasks. MacArthur's appearance and manner of speaking were cold and direct. She sounded unfriendly and unapproachable, the type who took her job a little too seriously as if, regardless of anything else, she was always on duty.

When she placed Dale with a group of females to clean up the orderly room, MacArthur appeared to be memorizing Dale's face with her name. Dale figured that sometime in the near future she would have to purposely mess up to put herself out of the running of any drill sergeant's suspicion as a possible cycle spy, but she hadn't intended to get on anyone's bad side quite this early. Especially anyone with MacArthur's obvious disposition.

The drill sergeant left as quickly and as abruptly as she had entered and other than being a little surprised by her sudden appearance, Dale was not impressed. In fact, her general outlook on MacArthur was that she was going to be more of a pain in the ass than anything else. Even though she tried to be authoritative, she came off sounding more intimidated by the female recruits than they did of her. There was just enough quiver in her voice to knock down any idea that she was any real threat as a disciplinarian.

Wynda Laraway had clearly not been inside the laundry room until now. Shannon had seen many of them before but Laraway, the other female assigned to the detail with her, could not easily disguise her mild shock at the huge concrete cubicle, which stored exactly three washing machines and three dryers.

"Look at this! It's so...so...gray...," Laraway said, stuck in her tracks.

Shannon pretended to be just as surprised. "Well, gee, Wilma, don't you kind of feel like we've been transported back to Bedrock?"

The room was semi-divided in the middle by a long cement table used for various activities such as folding clothes, sorting laundry, ironing fatigues and greens and polishing boots. The walls were concrete, and the floors were cement. There was a small window near the ceiling used for ventilation, a deep double cement sink and a small, open closet area where a broom, mop and a bucket were kept.

"Walker, there are only three washers and dryers in here," Laraway said, astonished.

"I know. I see them."

"That must mean there are only three apiece in the other laundry room."

"That seems like a reasonable guess," Walker said.

"Do you mean to tell me that for a hundred and some-odd people who are going to be in this company, there are only six washers and six dryers?"

The door swung open with a powerful thrust. "What are you waiting for, ladies? This place doesn't clean itself, you know. Now, move it!" It was little Lieutenant Henning with the big Texas voice. As Laraway jumped, startled and nearly tripped on her own feet, trying to get the mop and bucket, a small grin curled the corner of Shannon's mouth. "What are you smiling about, soldier?"

"Nothing, Ma'am."

"Glad to hear it. Have either of you seen Private Kirk?"

"No, Ma'am," they chorused.

"I'm coming back here in fifteen minutes. I want this laundry room spotless. Understood, ladies?"

"Yes, Ma'am," Laraway answered and brought the bucket to the sink.

As she reached around to close the door, Henning's eyes caught Shannon's and Henning winked.

Kotski, Gina Tramonte and Dale had been assigned to sweep, mop and straighten up the Orderly Room and the senior drill sergeant's office. It was then that they met Sergeant First Class Fuscha, who Dale guessed had stayed after hours to help the senior drill sergeant get the paperwork prepared for the beginning of training.

Fuscha was a big man, much taller than six feet and looked like he weighed approximately two hundred fifty pounds. He had thick black hair and a thick black mustache, and he talked with a thick New York accent, using grammar like a person who never got beyond the eighth grade. He spoke gruffly, but Dale guessed that deep down inside, he was as gentle as he was big, that he only had to act tough for the benefit of the newbies.

He showed the trio where the cleaning equipment was and returned behind his desk to continue his paperwork. When Kotski entered Ritchie's office to sweep, Fuscha spoke up again. "Yeah and don't forget to wipe off the glass case in there. And youse girls." He pointed to Tramonte and Dale. "Don't forget the corners here."

"Okay," Tramonte said, not thinking.

"Okay? *Okay?*" He looked up, sharply.

"Yes, Sergeant," she corrected herself.

"Sergeant what? I don't got a name here?" He pointed to the nameplate on his desk. "What's it say?"

Dale looked at Tramonte, who seemed stumped.

"Fuscha. Call me Sergeant Fuscha."

"Is that what it says? Fuscha? Like the color?" Dale asked.

"Yeah, Fuscha like the co–what's your name?" he asked, suddenly seeming to realize that he should be annoyed by all this.

"Oakes."

"You're a smart ass, Oakes."

"Yes, Sergeant Fuscha," Dale said and grinned.

Her expression was infectious and Fuscha returned her smile. "Get to work, smart ass."

The mood was broken all too quickly when Drill Sergeant MacArthur came storming in, ushering Kirk through the first sergeant's office, into the captain's office, slamming the door shut behind her. Henning followed seconds later.

The muted voices rose at a steady pace until they sounded as though they were yelling at Dale instead of Kirk. Even Fuscha looked toward the door when the shouting continued for fifteen minutes. His attention focused on Tramonte and Oakes, who had stopped working and were listening, also. He then looked at Kotski at the senior drill sergeant's office door, leaning on her broom, looking at the closed door on the opposite end of the room.

"All right, youse girls, never mind what's going on in there. Get back to work."

"It's hard to never mind it, Sergeant Fuscha," Kotski spoke up. "They're so loud and distracting."

"It's military business," he snapped. "Now hurry up with your detail or it'll be you in there! Haul it. I don't want to be here all night!"

"Yes, Sergeant Fuscha."

It was difficult to concentrate on anything other than what was being said in the next room but within ten minutes, the women were done. Dale noticed, before she left the Orderly Room to return to the bay, that Kirk was definitely holding up her end of the hollering until the voices hushed in controlled anger to a level where most of the conversation was lost. There was no doubting the gist of the altercation, though, and it was clear that Kirk was in for the fight of her young life.

Less than twenty minutes had passed when Dale emerged from the shower. She was engaged in casual conversation with Pamela Ryan, whose bunk was next to hers, when they heard the sound of the barracks door swinging open and slamming against the wall. A voice called out the command of At Ease and most of the women jumped to it.

A crying Kirk was followed in close pursuit by a snarling MacArthur, who then stood above Kirk as the young woman completely stripped her bunk. When this task was finished, MacArthur told everyone to carry on and then searched the lines of women for one in particular.

"Lanigan!"

"Yes, Drill Sergeant," the holdover responded.

"Get over here," MacArthur barked. When Lanigan reached her, the drill sergeant spoke in a low tone to the new MP, who then returned to her locker and began changing back into her fatigues.

When Kirk passed the undercover lieutenant, she tried to talk to Dale. "Hey, Oakes, don't —"

"Shut up, Kirk, no one told you to speak!" MacArthur shouted at her.

"No one has to, I'm not a goddamned dog!" Kirk screamed back.

"I said shut your mouth! You don't need to tell her anything!"

Now the attention became focused on Dale, who smiled sheepishly at some of her cellmates. When her gaze went back to Kirk's, it was intercepted by MacArthur, who looked anything but pleased. Dale's smile disappeared quickly.

"Well, you certainly are keeping yourself inconspicuous to a tee," Shannon mumbled, after the drill sergeant and her ward had left the floor. Shannon was on her way outside for her last cigarette break of the evening.

Dale shrugged. "She trusts me."

"Lucky you. Do you know what's going on?" Shannon

glanced around them to make sure they were alone.

"I'm not sure. There was a lot of hollering downstairs when I was there."

"Is there a reason you find that unusual in a basic training environment?"

"They're making her sleep in the senior drill sergeant's office tonight and Lanigan has to guard her...or that's what I heard Lanigan just tell Boner. I don't know. I don't like this shit."

Shannon waved it off and made a face. "I wouldn't worry too much about it. They're just trying to show her she's not in the neighborhood anymore. She's playing with the big boys here."

"Oh, come on, Shan, there are other ways."

"Of course, there are other ways but we're not talking about trained psychologists here, either."

"Well...exactly my point."

Deborah Michaelson exited the latrine, nearly bumping into both Dale and Shannon. "You two still trying to figure out which life you met in?"

It was the first time either lieutenant had seen Michaelson really smile and it was disgustingly dazzling. It also had some mischief behind it. However, both Dale and Shannon took it as a cue to move apart.

"I'm going to bed," Dale yawned.

"I need a smoke," Shannon mumbled and laughed.

Shannon moved outside where she was immediately confronted by the same GI Dale had encountered earlier. Shannon lit a cigarette and tried to ignore him. She looked out into the clear, starry, Alabama sky but his incessant staring caused her to break out into a grin.

"Hey, baby, where'd you get that smile?"

"From my orthodontist. And I paid plenty for it."

"Yeah? Well, smiley, how 'bout if you and I head up to my room and have some fun?"

"No, thanks. If nothing better comes along though, maybe I'll look you up." She still had yet to look at him.

His lascivious little snicker told Shannon that her attitude had not put him off. Oh, if he only knew that she was not a naive, confused little trainee. The specialist 4th class, leaning on the railing, inched closer. "Now, that was just cold."

"I never was known for my tact." Shannon was halfway through her cigarette and if he made it impossible for her to finish it, she was going to kick his huge ego all the way down to the Orderly Room and report him.

"Come on, baby, let's go up to my room right now. No one will ever know."

"I'll know," Shannon told him.

"Darlin', if you just got to know me..."

Shannon looked at him for the first time and her expression was not one of interest. "Oh, please. I know all about you. You were a nookie bookie before you got caught and the cops said Army or prison."

He stared at her, speechless. "Now, how could you have known that?"

Stabbing her cigarette out on the railing, Shannon brushed off the black mark it left on the metal. "It's written all over you."

Seconds later, she returned inside the bay, and it was more than obvious that smiley was not very happy.

"What happened to you?" Wachsman asked.

"A moron named Ingersol."

"Oh, Christ, he tried you, too?"

Nodding, Shannon opened her locker and removed her nightshirt. "Has he been there all night?"

"Yeah. He stopped me on my way up from mopping the patio. Told me something like he'd be my last chance for a while and don't worry he'd use protection."

"How considerate. What did you say?"

"I told him if he didn't leave me alone he'd need all the protection he could get. He's got to be about the dumbest son-of-a-bitch I've ever run across. Forty women have turned him down flat and he's still out there. Wouldn't you have gotten the idea after about oh, say, the tenth or so rejection?"

"I'm surprised Dizzy hasn't hopped his bones."

"She's the only one he's ignored."

"Men. I'll never understand them."

"Yeah. I think we're going to find that military men are in a league all by themselves," Wachsman added, rather prophetically.

CHAPTER TEN

Shannon hated shots. It didn't matter how big, how brave or how strong she pretended to be, or how well she tried to disguise her cringing, she hated the thought of that little hypodermic pinch. Especially after she had just had her arm punctured several times five days earlier and had to have three more holes put into her body so soon.

She and the other women lined up at the reception center as one medical specialist stood to their left and started to shoot away at bare upper arms, quickly and impersonally, with no regard for soreness or people's feelings.

Tracy Travis strolled toward the plastic chairs in the back section by the soda and candy machines. She read her shot record with interest. "What does A Vic/B-HK-Flu mean?" she asked Mroz.

"I'm not sure. Sounds like some type of flu shot."

"That part I figured out. What do all these letters before it mean?"

"God knows, Travis, it's probably something for VD. I hear it runs rampant in the military."

Travis gave Mroz a sobering look, then scanned the room. "Where's Dizzy? She'd know."

Mroz looked around the waiting area. "I don't see her. She's probably still being inoculated."

"Inoculated? Is that a new code word for having her brains fucked out?" Travis asked.

"If it is, then she should be back out here in no time," Shannon, who sat behind them, quipped.

"God. She's probably off somewhere with an overanxious medic. And if that's the case, I hope he stands in line for his shot afterward. It'll be just my luck, if he doesn't, that he'll be the first man I'll meet on my first weekend pass," Travis mumbled.

She went back to studying her shot record. "Hmmm...what does meningococcal 0.5cc mean?"

"I don't know that, either," Mroz responded. "Why don't you ask one of the medics?"

Travis laughed. "Have you seen them? They don't know. They look stoned. Probably from snorting too much meningococcal fluid. Whatever the hell that is."

"I'm not used to being up and about at this time of the morning. I'm usually just going to bed." Dee Tierni asked and yawned. "What's happening next, does anybody know?"

Kay Verno, who was stretched out in a chair, resting, spoke up. "Someone said they've got to set up again and give us a polio shot and then we're going to be vaccinated in the corridor."

"I will be vaccinated in the arm or not at all!" Travis joked, with mock indignation.

"Well, somebody informed you wrong," Deirdre Snow announced, not looking up from her shot record. "The polio serum is taken orally."

Shannon watched with interest as Travis looked like she was about to respond with something crude but then seemed to think better of it and snapped her mouth shut. Snow was not Travis' favorite person at this point as they had already clashed a few times in the barracks.

Travis wasn't the only one Snow's attitude had rubbed the wrong way. Shannon hated being patronized by anybody and twice already, Snow had incorrectly contradicted Shannon in front of others. She was quickly earning the nickname of Prof which came close to what she was as a civilian, a teacher. But her tone of voice was unreasonably condescending, and she had an arrogance she wore on her sleeve toward what she clearly felt about being surrounded by people she considered inferior.

Snow had spoken to Shannon as though Shannon had just fallen off the turnip truck. Shannon could have corrected her both times but decided against it. Sounding too knowledgeable at this stage of training would have consequently come back and slapped her in the face, especially with someone as sharp as Prof. Shannon's time would come with Snow, there was no doubt in the lieutenant's mind, she just hoped the mounting aggravation could be held off until the end of the cycle.

The day seemed to last forever. As if shots first thing in the morning weren't bad enough, immediately following that, the new recruits were confronted with their first military dentist. Shannon decided that the fact that he had food particles stuck between his crooked, yellow teeth and breath that smelled remotely like a Georgia pig farm did not frighten her as much as the new-fangled dental x-ray machine.

One after the other, the recruits were fastened into a chair, their faces pressed into an unyielding chin strap, forcing them into a paralyzing position that appeared to defy human design and then they were told not to move. The scanner started at their left ear and moved excruciatingly slow around their jaw to their right ear, the imaging device coming so close to their skin it was obviously chasing an amoeba.

"We should have enlisted back in the days of George Washington," Shannon mumbled to Travis, as they waited in line to see the eye doctor. "We wouldn't have had to worry about dentists."

"Right. Just termites and knotholes," Travis commented.

Their conversation was interrupted by a deep, male voice in the eye examination room. "You're not supposed to pronounce the words, young lady! It's an eye chart!"

The women were allowed a break in the rear of the building while they awaited the bus that was to take them back to Alpha Company for noon chow. Shannon had just finished her session with the optician and bought a Dr. Pepper for Dale. As she passed the room she just left, from within, a voice boomed out, "Read the chart, please!"

Then she heard Dizzy's midwestern drawl reply. "What chart?"

Shannon handed the soda can to Dale as Dale spoke with Kirk. It was hard not to hang around her close friend and renewed colleague, especially when her curiosity was heightened. Shannon felt it would do no harm to unobtrusively listen in, as she stretched out in a chair, directly behind Kirk, and pretended to be bored.

"Wait a minute, wait a minute," Dale stopped Kirk who, the more she got into the story, the faster she talked. "They wouldn't let you speak to your sister?"

"They handed me the phone," Kirk said, slowing down. "And I had it to my ear just long enough to hear hysterical crying and they told me to terminate my conversation and then they hung up for me. I had the phone in my hand exactly five seconds. I don't even know who the hell I was terminating my conversation with. Shit, man, I wasn't even conversing."

"Why did they do that? Did they say?" Dale asked.

"They told me that I had pre-arranged the call. Remember that day we all went to the PX?" Kirk continued. "They thought I had called home and told someone to call me here and act like there was some sort of emergency in the family. But, Oakes, you were with me! I didn't even get near the damn phones!"

"Did you tell them that? That someone was with you who could verify that?" Dale asked.

"Yes, but they didn't believe me. They're not even being sensible, Oakes. My family forced me to enlist in the first place, why would they do me any favors? Especially try to get me home. I don't even know which one of my sisters tried to call me. I didn't get a chance to recognize her voice. In fact, I had to take their word for it that it even *was* one of my sisters."

"How many sisters do you have?" Dale asked.

"Nine."

"Do you have one named Marva?"

"How do you know that?" Kirk tilted her head, questioningly.

"I don't. Do you?"

"Yes."

"It was her."

Kirk immediately got visibly upset. "How did you find that out?"

Shannon opened one eye and looked at Dale inquisitively. How *did* she find that out?

"I was cleaning the office last night...well, me and Tramonte and Kotski. I saw it written down on a phone message pad. It said Kirk and that was circled. Then underneath that was written Marva called, will call back and that's it."

I haven't seen Marva since I was eleven years old." Tears streamed down Kirk's face.

"Calm down," Dale soothed. "Maybe Marva heard about what your folks did to you to and tried to call you to see if you were okay."

"Crying like that?"

Dale shrugged. "Maybe she was upset that your own parents could do that to you."

Kirk appeared to consider this possibility as she took a sip from her can of soda.

"How's MacArthur about this whole thing?"

"A fucking bitch."

"What about Henning?"

"She's trying, I guess, but she keeps telling me her hands are tied. I think she's just too fucking scared to buck the system. But she did finally get me in to see the chaplain. He wasn't any help, though. He seemed to close his ears to anything I had to tell him and he kept telling me God would help me through this. I told him the only God I know is freedom. Then they took me to mental health, but those freaks are worse than MacArthur."

Dale glanced at Shannon quickly, then returned her attention to Kirk. "You know what I would do if I were you, Kirk?"

"What?"

"Request. No, demand to see the senior drill sergeant or the company commander. Tell them you're gay," Dale said.

"I'm not gay!" Kirk stated, defensively.

"I didn't say you were. Just tell someone in charge that you are."

"I'm not going to do that." Kirk was shaking her head, stubbornly.

Shannon mulled Dale's recommendation to Kirk. The controversy of armed forces females being lesbians was still a universal rumor, supported by the fact that an ample fraction of service women were. The military seemed to be a common meeting ground, especially for women who believed in career first, family second. But the service didn't make it easy. Women and men got discharged for the indiscretion of not concealing their homosexuality and though the military had relaxed some of the pressure, they still would not tolerate blatant physical or verbal displays. Caution was an excellent exercise to practice unless, of course, someone wanted it known for the reason of getting out and then even dropping a subtle hint spread like wildfire. Dale was right. This *would* most likely be the easiest way for Kirk to get her separation papers.

"You want out, don't you? It won't go on your record. If you're released in the first levels of basic training, unless you've murdered someone, you're going to get a simple trainee discharge. You're a thousand miles from home. No one will ever know."

"How do you know all that?" Kirk asked, her expression revealing that she was already considering Dale's suggestion.

"I read up on that kind of stuff before I came in. Just in case."

"What about my recruiter? Wouldn't he get a reason for my discharge? He'd say something to everybody, I know him."

"I believe the only reason he'll get is that you were unsuitable. That you couldn't adjust to military life. So, tell whoever it may concern that you're gay and you don't trust yourself upstairs with all us beauties. I know there would be a couple of our barracks-mates willing to come forward and make statements if it will help you get out. I know I will."

Kirk skeptically contemplated the thought. "You really think it'll work?"

Dale shrugged. "Well...I can't guarantee it but really, Kirk, what have you go to lose?"

After chow, the women filed back on the bus, all except Kirk, who stayed behind to pitch her coached confession to the phantom Captain Colton.

The bus wheezed its way into the parking lot of the central issue facility, or CIF. CIF was an old, large warehouse-type building where all facets of the Army uniform were issued, received and stored.

The recruits stepped off the vehicle and moved to the entrance, waiting for Specialist Harriman to lead them in, as if they were fifth graders being taken to the museum by their teacher. Dale took one final look at everyone in their civilian clothes. It would be the last time they would all be dressed as individuals for a long time and she wished there was a way she could have hugged her jeans goodbye without appearing a tad insane.

After finally being escorted inside, Dale sat in the waiting area with the other females. They listened to a middle-aged woman's instructions as she handed out nametags that were to be sewn on the right breast pocket of their fatigue shirts, plastic name plates that were to be worn with the Class-A uniforms, and two round pieces of brass about the size of a quarter. One with the letters US on it and the other with the head of Pallas Athena on it. Athena was the Greek goddess of wisdom, skills and warfare and she was the official symbol of the Women's Army Corps.

The female military police trainees would wear the Athena insignia on their dress uniform until they graduated from Law Enforcement School, whereupon they would receive an identical circle of brass with crossed pistols on it, which was the emblem of an MP.

They were also handed two dog tags and two chains. One tag went on the long chain and the other went on the short one. The information on the small, metal plate included last name on the first line, first name and middle initial on the second line, social security number below that, blood type on the next line and religious affiliation on the last line. They were told that these dog tags were to be worn at all times.

The women were then directed to another room where they were issued four sets of female fatigues, one ball cap, three white undershirts, two sets of winter underwear, five pairs of socks, one female fatigue jacket, two leather glove shells, four wool glove inserts and one wool knit scarf. No one laughed when Travis asked if chastity belts, one each, olive drab green in color, were to be handed out also. The issuers did not find it amusing and the new Alpha trainees did not want her to give them any ideas.

Dale mechanically accepted it all, recording what she did or did not get on her personal clothing request sheet, form 3078 which she was handed in the first room. Instead of paying attention, her mind was on Kirk's plight. If the company commander and/or the senior drill sergeant went for Kirk's lie, it wouldn't be long before the young woman would be on her way back to Detroit.

In the next section of CIF, the women were measured for their dress green Class-A uniform and their summer dress uniform, referred to as cords. Both garments had to be altered before being issued and given out at a later date. The trainees also received dress gloves, dress hats, one raincoat, one dress winter coat, one black leather handbag and one white scarf.

The women shoved their new wardrobe into their duffel bags, except for one set of fatigues, which they now wore. They dragged the heavy, cylindrical canvas case into the next and final room, where they were to be fitted for combat boots and one pair of dress oxfords or low quarters, usually referred to as granny shoes.

On the bus back to A-10, Dale silently lamented about not being given a second pair of combat boots, as CIF advised her they only had one pair of her size in stock. Knowing how that worked, Dale figured she most likely would not see her other pair of boots until it was time for her to leave McCullough for good. It was imperative that a trainee have two pairs of boots to switch off and on every other day. The average foot, new to such footgear, lasted much longer by alternating boots and Dale's tender left ankle would suffer more stress by wearing the same ones.

Dale did not see Kirk anywhere, either upstairs when she and the other trainees secured their gear in their lockers, or downstairs on the north patio, where all the women were told to wait for further instruction.

Dale looked around at the other women in their brand-new uniforms and boots. She remembered how she felt wearing fatigues for the first time, a complete loss of identity, like a clone. Dale found Shannon and wandered close to her, while they waited for a drill sergeant.

"I can't believe how comfortable these fatigues are!" someone in the sea of green exclaimed.

Shannon eased herself away from the group and strolled out toward the sidewalk. She took a cigarette out of a pack in one of her utility pockets and lit it. She looked at Dale and cracked a half-smile. "Wait until they wash a set," Shannon said, quietly. "Comfort will go right out the window,"

The female uniform, which differed from the permanent-press male uniform in every way except color, was not a very welcome sight just out of the dryer. It usually looked as if it had been balled up in a corner of someone's closet with an anvil resting on it for about three years. And even with spray starch, it never ironed out the way a drill sergeant liked to see it at morning formation.

"Yep. They'll learn to hate them as much as we did," Dale agreed. "Especially after a long training day and not being able to get a washing machine until close to bed check and then having to cautiously stay up long past lights out to iron them."

"And it doesn't matter how good they try to make these things look, they won't return the compliment," Shannon said and chuckled.

The pants buttoned on the left side and also sported two droopy utility pockets with flaps that fastened on each side of the leg that made the women look like they were wearing saddlebags. The shirts had two buttoned breast pockets and one small pocket on the left sleeve, just below the shoulder. Also, the tops of the female uniform were required to be worn outside the trousers, as opposed to the male fatigue shirt, which was worn tucked in.

Shannon finished her cigarette, pushed the remaining tobacco out with her thumb and forefinger and placed the butt in her pocket.

Wachsman approached the two women. "Okay. Well, I finally feel like I'm officially in the Army now."

The Alpha females were distracted by the sound of two military buses that entered the company area. Shannon, Dale and Wachsman returned to the patio, rejoining the others. Silently, they all watched as ninety-eight men raced off the vehicles as if they were on fire. The males stood at Parade Rest and looked terrified, as the buses then pulled away from 10th battalion. Some of the men were almost recognizable from that first day at the reception center, even with their lack of hair and wearing identical clothes.

It seemed like the men stood there for hours, not moving,

not flinching, but it was only, in fact, ten minutes or so before someone came out to help them. In the meantime, Drill Sergeant Robbins continued to go in and out of the Orderly Room and stroll by the females, ignoring the new male trainees. Every time he walked within ten feet of the women, the group of fifteen who stood or sat by the picnic table jumped to Parade Rest after they all took turns yelling, *At Ease!*

Also, every time Boehner walked by them, they were at Parade Rest, waiting for a drill sergeant to tell them to Carry On. As a smirk crossed her face, Dale knew Boehner could have told the women they did not need to remain in that rigid position when no drill sergeant was around. Dale guessed Boehner decided they needed to learn on their own just like she did.

Finally, Robbins came back out to help the men, but he was not alone. With him was another drill instructor, a sergeant first class most of the women, including Dale and Shannon, had not seen before. Word spread quickly that this was the company Senior Drill Sergeant, James Ritchie, and he was a force to be reckoned with, even though he was a physical contradiction of his reputation. Ritchie was a man who just barely qualified for the military height regulation at five feet, six inches tall, stocky in build but clearly in shape. He reminded Dale of a bulldog, not just in appearance but in temperament, with his beady little eyes that snapped when he growled instead of speaking.

He had dark brown hair cut in a strict, short, regulation style and a square face that had an almost cherubic quality to it. He wore thick, black-framed military-issued glasses and when he smiled, he bared tiny little teeth. Though they were straight, Dale guessed, they had more than likely been worn down through plenty of gritting and grinding in aggravation about the new soldiers.

Dale disliked him immediately and it looked as though Shannon felt similarly. Dale hoped her impression of him as a little dictator was wrong. Yet, from the moment he opened his mouth, he proved to her that her instincts were sharper than ever.

"Okay, you mealy-mouthed, sorry bunch of dirtbag fuckups, listen up! You sissies have been pajama partying for a few days now, you should know one another intimately. Everybody A through D line up here." He pointed to an area in the parking lot.

"E through Mc here," he said, indicating another place. "And ME through Z here. *What are you standing around for? Fall out! Double-time into those lines, you scumbrain asswipes!*"

The men nearly trampled one another as they got into alphabetical ranks. Ritchie made them stand at attention until he got their names in order and then he proceeded to go from man to man, making sure that everyone was present and accounted for and that all his information was correct. The whole process took at least forty-five minutes and by the time he was through, the males appeared to be ready to drop. It was obvious that Ritchie deliberately took his time and even Robbins looked a bit annoyed with him because he finally left the senior drill sergeant alone with the group and walked back toward the CQ office.

"*At Ease!*" Mroz shouted, so loudly even Ritchie jumped, as Robbins reached them.

"Carry on," the drill sergeant told them, preoccupied, and disappeared inside the Orderly Room. Seconds later, as Boehner walked by the group again, Robbins stepped back outside the office and somebody else yelled, *At Ease* with almost as much vigor as Mroz.

"Look, ladies, if a drill sergeant is going to be constantly in your immediate area, like Sergeant Ritchie or myself, you don't have to keep jumping up like this, okay?" Robbins said.

"Yes, Drill Sergeant," the group of women answered.

"Carry On," Robbins told them, and they all returned to their former positions of either standing or sitting.

"Drill Sergeant Robbins," Ritchie yelled. "Take these lowly recruits upstairs and get them squared away."

"Yes, Sergeant Ritchie." Robbins said and smiled. He ordered the men to follow him, which they did, promptly and gratefully.

Ritchie walked back by the females who, honoring Robbins' instruction, did not respond to his presence. Infuriated, Ritchie slammed his fist down on the picnic table and spoke in a tightly controlled voice. "You'd better get up off your lazy, fat asses when a drill sergeant walks by you! Am I understood, trainees?"

Everyone jumped to Parade Rest. "*Yes, Drill Sergeant,*" they chorused.

"I can't hear you!" he told them, his pointy little nostrils flaring.

"*Yes, Drill Sergeant!*"

"Drill Sergeant?" Mroz spoke up.

"Identify yourself, young lady," he snapped. He stepped closer to her and stood less than an inch away from her, apparently fulfilling his wish to intimidate her.

"P-Private M-Mroz, D-Drill S-Sergeant."

"What is it, P-Private M-Mroz?" His expression showed that he actually thought he was being comical. Unfortunately for the women, none of them found him humorous, only obnoxious. This made him angry that nobody even cracked a smile because now he had no one else to yell at.

"Drill Sergeant Robbins said —"

"I don't give a good goddamn what Drill Sergeant Robbins said. *I* said you'd better stand, or your ass is mine, Mroz. You got that?"

"Yes, Drill Sergeant."

"You all got that, you sorry excuses for women?"

"Yes, Drill Sergeant."

"I can't —"

"*Yes, Drill Sergeant!*" They cut him off.

A smile curled the right side of his mouth and he walked away from them, never telling them to Carry On. They were too inexperienced to know they did not have to stay at Parade Rest, something Boehner could have told them a half hour earlier, which would have saved them a lot of aching and anxiety. Thirty minutes later, with some of the women near tears from holding the inflexible position, a scoffing Ritchie walked back by them with MacArthur in tow.

"They're almost as stupid as that weepy, wimpy Kirk, don't you think, Sergeant MacArthur? They're the worst looking bunch of candy-assed females I have ever seen. They'll never make MPs. I think they should give it up and go home right now." By that time, most of the women would have gladly obliged. "Carry on, men. That's obviously what you want to be," Ritchie said, and went inside the orderly room. MacArthur followed, like an attention-starved puppy, shaking her head.

It was at that moment, Dale surmised, that Ritchie earned the deserved nickname of senior drill prick.

CHAPTER ELEVEN

The women spent the rest of the afternoon and early evening learning how to arrange their lockers under the supervision of the holdovers with an occasional visit or two from the individual platoon sergeants. The female trainees had already met MacArthur, McCoy, Kathan, Robbins and Ritchie and had come to recognize certain voices on the bitch box.

"Another one of those drill instructor creatures is running around downstairs," Wachsman said to Shannon while they were rolling their clothes to fit in the drawers in their lockers. "Keeps his hair sheared closer than Kojak's. I think his nametag said Putnam. He's probably a jerk just like the rest of them."

"Robbins isn't that bad, though," Shannon said.

"Don't say that to the girls who were sitting at the picnic table earlier. They feel completely betrayed by him. He's on thin ice," Wachsman commented. "Oh, and have you noticed how many women seem to be taking numerous cigarette breaks since the guys have got here?"

"You mean including the ones who don't smoke?" Shannon said with a grin.

"Right. Just as an excuse to go to the downstairs patio to become better acquainted with the males."

Wachsman stopped and both she and Shannon listened to the excited buzz about prospective romances. "Maybe this co-ed thing wasn't such a good idea," Wachsman said.

"I don't know. I think the drill sergeants will put a kibosh to anything that will divert our attention from training but, since this is a first, who knows?" Shannon could only hope that the cadre could keep everyone focused. That would make things a lot easier for her and Dale.

Following a quick, quiet chow, the women filed back

upstairs to finish their lockers and hang around to await further instruction. Some used the latrine, some were in the shower, some sat on their bunks – something both lieutenants knew was a future no-no, but no one had informed them yet so neither could say anything. A few women wrote letters or played cards while others got involved in conversations about their past or the new male recruits.

Regardless of what they were occupied with, when that bitch box light came on and somebody saw it and shouted *At Ease*, those women, all except the holdovers, immediately jumped to the position of Parade Rest. Dale and Shannon eyed one another, and Dale wondered if she should speak up and say something but, again, decided to play ignorant. She was relieved when, after the announcement was made on the intercom and the red light went dark, Boehner walked through the relaxing ranks. "It's a goddamned light, what are you doing?" she asked, incredulously.

"Huh?" Lesley Minkler said in all her natural eloquence. She was a dark-haired woman who seemed in a constant state of bafflement.

"The bitch box light, every time it comes on, everybody snaps to a position. It's only a light, it's not a camera. All somebody has to say is at ease and listen. Nobody has to move."

"Usually when somebody says, At Ease, it means there is a drill sergeant in the bay," Claire Steele, a slender woman from Michigan, brought up. "I think I'd rather play it safe and look stupid instead of taking it for granted and end up humiliated and deaf from being screamed at."

There was a chorus of agreement with Steele and Boehner shook her head as she sauntered back to her bunk. "I guess you'll just get to a point where you'll be able to tell the difference in the tone of voice of whoever calls it out."

"You could have been as forthcoming this morning when our backs were breaking down on the patio!" Travis called after Boehner, who just laughed and turned to look at her. "As much as I hate to admit it, I have to, at least, partially agree with Boner. I can see what Steele is saying but that doesn't mean we have to stay in that position once we find out it's only the bitch box."

"Exactly," Boehner said. "And its *BAY*ner," she corrected, searching out a smugly smiling Dale to give her a nasty look.

Travis stuck her tongue out and made a rude noise at the former cycle spy when the door opened with a bang and the words *man on the floor* were called out by a deep, male voice. The command At Ease was hollered by Boehner, which had to mean business.

This drill sergeant was a big man, overweight but not the least bit flabby. There was something gentle about this man who wore crisply starched fatigues, military-issued glasses and a thick mustache. He possessed a speaking voice somewhat more delicate than the others, Dale thought. *It only would have registered a six on the Richter Scale.*

"Good evening, ladies, my name is Drill Sergeant Audi. I would like you all to come to the left side of the bay. I have a few announcements to make. You holdovers can go on with whatever you're doing." He waited until everyone had gathered around him. "I'm going to divide you into platoons. After I do that, you will change to the bunk and locker that is in accordance with your platoon. You will be placed alphabetically."

"But, Drill Sergeant, we just barely got our lockers straightened out," Minty drawled.

"Are you whining at me, Private..." he leaned in to read her nametag. "Minty?"

"No, Drill Sergeant," she said, defeated. "I was just —"

"Don't *just* anything, Private Minty, because whatever it was you were just going to say or do, bear in mind, you do not have that privilege yet."

"Yes, Drill Sergeant."

"All right. I would prefer not to be interrupted again unless it's an emergency...and that better be an act of God. First platoon is mine and Sergeant MacArthur's. Sound off and go stand by your new bunk when I call your name. Almstead."

He pointed to the first bunk on the left side of the bay entrance. "Beltran." He pointed to the second bunk. "Brewer," he continued, checking their names off as they answered with, *here Drill Sergeant*, and went to stand beside the following bed in the row. "Caffrey. Creed. DeAmelia. Ferrence. Guerrierre. Hewett. Jaffe. Keival. Kirk..." He looked up when no one answered. "*Kirk!*"

"She's still downstairs, Drill Sergeant," Quinn Brewer volunteered.

He made a notation on his sheet and moved on to the next bunk. "Kramer, Bonnie. Kramer, Brigitte. And Kotski, you're all in my platoon." He walked back down between the beds until he reached the entrance again. "Second platoon is Sergeant McCoy and Sergeant Kathan's." He pointed to the first bunk opposite Alexis Almstead's newly assigned one. "Laraway. Lehr. Mackey. McKnight. McTague. Michaelson. Minkler. Minty. Mroz. Newcomb. Oakes. Ryan. Ryder. Sager. Saunders."

Audi walked around the lockers and moved to the right side of the bay. He led a dwindling party back down to the entrance and started with the left group of beds. "Third platoon will be Sergeant Robbins and Sergeant Putnam's. Segore. Sherlock. Snow. Steele. Swinegar. Tierni. Tramonte. Travis. Troice. Verno. Wachsman," he pronounced it Washman. "Walker. York and Zelman. Did I miss anyone?" No one spoke. "I want you to change bedding and lockers now and have it done and in order before lights out at 2130 hours. You holdovers will be placed in the remaining bunks in the fourth row. Any other questions, I'll be downstairs."

Drill Sergeant Audi left the women standing there a little dumbstruck by the amount of work facing them. Dale was confident there would be a fair amount of bitching taking place after the barracks door shut. There always was. That's why drill sergeants always waited until the end of the evening when they knew the new soldiers had spent all day neatly squaring away their lockers according to regulation to come upstairs and rearrange bunk assignments. It was all a part of the game.

Prepared for it though they were, neither Dale nor Shannon welcomed another hour's work. They were tired and, along with everyone else, they had hoped for an early night.

Shannon, eager for a last cigarette before lights out, headed downstairs while Dale was putting the finishing touches on her locker. When someone tapped Dale on the shoulder, she turned around to see Kirk, who had clearly been crying.

"They didn't believe me," Kirk began, her eyes welling up again. "I sat there all afternoon and when I got the chance to see the senior drill sergeant, all I got to say was yes, Drill Sergeant, and no, Drill Sergeant. What a fucking son-of-a-bitch that Ritchie is!"

"Yeah, we kind of got that impression this afternoon. So, did you at least get to tell them?"

"Yes. For what it was worth. But I told you they didn't believe me."

"How do you know?"

"Because I'm still here."

Dale smiled, patiently. "Did you actually expect to walk in and say, hey, guys, I'm a lesbian and have them back up in horror and hand you your discharge?" Kirk's expression told Dale that was exactly what she had assumed would happen. "I'm telling you, this is a macho shithead organization. It's still a man's Army. They don't even like women here, much less women who openly desire other women."

"How can you say that? There are more dykes in this room than in the Netherlands."

"You don't know that for a fact. And neither does Uncle Sam. But a female who is admittedly gay is usually booted out, one way or another."

"I don't know," Kirk said, skeptically.

"Don't give up. Stick to that story, it will get you out of here in the end, I'm sure of it. It'll take a while, but it will work. Cheer up. At least they didn't make you sleep in the office again."

"Hey, I'm not knocking that. At least I had some privacy there."

"Kirk! Get over here and clean out your locker, would ya?" The voice shouting across the bay belonged to Toni Sherlock. "It's my locker now and I want to get my shit in it and get to bed."

Dale responded to Kirk's look of confusion. "Oh, yeah...a new drill sergeant, Audi, was up here earlier and he switched everyone around alphabetically, according to platoons. You're in first platoon, by the way."

"Oh joy, oh bliss," Kirk mumbled, mirthlessly.

"Come on," Dale grinned, "I'll help you transfer everything and get it in order."

"Nah, it's okay, I need something to concentrate on. But I appreciate it."

The undercover agent nodded and decided to head downstairs for some fresh air before lights out. Reaching the north patio, which was directly under the women's bay, Dale ambled up to a small group of women that included Verno, Laraway, Boehner and Shannon just in time to hear the young Hawaiian ask Boehner a question. "Is it Attention we stand at when a drill sergeant walks by?"

Verno heard three adamant negations.

"Good God, we have to stand at Attention for enough as it is," the ex-cycle spy told her. "You'll find out that your back can handle more when it is in a little more relaxed position, so don't go giving the Department of the Army any ideas that we might possibly like standing at Attention any more than we already do," Boehner laughed. She really did laugh. Shannon and Dale were so shocked that they laughed, too, and they knew for a fact standing at Attention unnecessarily for long periods of time was no joke.

Laraway brought up that afternoon. "I'll tell you, for the length of time Ritchie had us standing at Parade Rest, I'm not so sure that's any better."

"It is, though, you'll see."

Dale yawned intentionally and stretched. "Man, it's almost time for lights out."

Verno glanced at her watch. "I'm glad. I'm bushed." As if on cue, Laraway and Verno said their goodnights and headed for the stairway.

Shannon pretended to start another conversation with Boehner while the two were still in hearing range. "What did you mean when you said the drills try to be hardcore in the beginning?" When the area was clear of traffic, the party of three moved into the laundry room where their presence together and conversation could not easily be detected or eavesdropped on. "Anything?" Shannon asked Boehner, getting right to business.

"No, not really. The big topic of conversation is this Kirk thing. Did you know she's a dyke?"

"You say that like none exist in the military," Shannon said.

"I told her to say that." Dale spoke up.

"Why?" Boehner looked surprised, if not somewhat annoyed.

"Because she wants out. She was trapped into enlisting, and she shouldn't be here. As an ex-training officer, I know the last thing other troops need for morale is someone who's an instigator. If she's forced to stay here, there's going to be problems."

Right," Shannon agreed. "If she pushes the gay thing and some of the girls back her, she should be out in no time."

"You hope," Boehner intoned.

"Well, let's just say in any normal training company that's how it would be, but it seems to me like they're putting her through a lot of unnecessary bullshit here," Dale stated.

"Oh, come on," Boehner said, a little provoked. "You mean to tell me in all the training companies you put through, you never had the cadre mess with a trainee who decided they didn't like it and wanted to go home to Mommy? I can't believe that."

"We had plenty of trainees who never dealt with any form of discipline and changed their minds about wanting to be a soldier *after* a taste of Army life. Yes, the cadre played games before the trainee was sent home. This is not what I find questionable here. This girl said, before she even got off the bus, that she shouldn't be here. It has nothing to do with being a little baby who can't take it after she tastes it. She hasn't even tasted it and she hardly wants to go home to mommy when mommy is responsible for her predicament. Her being here is a direct result of fraudulent enlistment." Dale was worked up. "And then you have morons like Ritchie..." She let her sentence trail off.

"Yeah, what's Ritchie really like anyway? He seems like the type who eats trainees for breakfast," Shannon commented, picking a stray string off her fatigue shirt.

"He's —"

The laundry room door opened and slammed against the wall. The term, speak of the devil took on a figurative meaning as they really weren't sure if Ritchie was possessed or not. The senior drill sergeant and MacArthur stood in the doorway and wore identical smirks.

"At Ease!" Boehner called out, unnecessarily, as the three had already snapped to Parade Rest. It was a good thing that military movement was instinct because an authentic first week trainee would have frozen from unmitigated fear at being caught with an off-limits person while one of them was committing an infraction. At that point, any improper response to any command would have iced the cake. The two cycle spies could have played dumb but neither felt like taking on the extra misery that would have caused them.

They waited for either drill sergeant to tell them to carry on. Fortunately, the women did not hold their breath. Ritchie approached them, as if in slow motion, followed closely by his smug little shadow. He circled them and a long two minutes passed before he spoke.

"Boehner, what did I tell you about openly talking to the new trainees?"

"You told me not to, Drill Sergeant."

"Get out of here, Boehner. I'll deal with you later."

"Yes, Drill Sergeant," she hastily answered and wasted no time leaving.

Ritchie made another circle around Shannon and Dale and watched to make sure their eyes stared straight ahead. He spoke, finally, when he stepped in front of them. "You both reacted very fast. That would tell me that you both pick up quick. And talking to Boehner out here in the open when you were repeatedly told not to and when you could easily converse with her up in the barracks unobserved would show me that you are incredibly pig-headed and blatant in your disregard of obeying orders or very eager to learn all you can about the military. And learn it before we can teach it to you." He stepped closer to Shannon than Dale, reading her nametag. "Are you that interested in the Army, Private Walker?"

"Yes, Drill Sergeant."

"So interested that you also ignored the rule about smoking in the laundry room?" His voice was dripping with sarcasm.

"Yes, Drill Sergeant."

"You are? Well, good. Now I'll let you learn discipline. Field strip that butt and get down and knock me out fifteen, Walker."

"Fifteen what, Drill Sergeant?" Shannon knew very well what his statement meant and she faced it with dread. Dale knew

Shannon had always equated doing push-ups with being vaporized. She extinguished the cigarette and forced the tobacco out of it with her thumb and forefinger.

"Fifteen push-ups, you idiot!" MacArthur yelled, in a burst of confidence. "Hit it, Walker."

"Yes, Drill Sergeant," Shannon responded, with not much enthusiasm. She assumed the position and started pushing up.

Ritchie waited until she got at least five done. "I don't hear you counting them off, Walker."

Having to play stupid was going to get a little hard on her muscles. "One...two..."

"One, Drill Sergeant. Two, Drill Sergeant," Ritchie corrected her.

Shannon held a front leaning rest position to get her bearings and started again. "One, Drill Sergeant, two, Drill Sergeant..."

MacArthur monitored Shannon's slow progress while Ritchie's attention moved to Dale. "You find this funny, Oakes?"

"I wasn't laughing, Drill Sergeant."

"You shouldn't be, Oakes. I understand that you and Kirk are real close. Is that true, Private?"

"Yes, Drill Sergeant."

"Are you queer, too, Oakes?"

"Well, I've been told that I can be a bit odd at times but —"

Ritchie stepped up to Dale, his face barely an inch away from hers. "Don't you ever fucking get smart with me again, Private! Do you understand me?"

"I apologize, Drill Sergeant. I thought it was a legitimate question."

"That's what you get for thinking. Now get down and knock me out twenty-five for having a smart mouth and another ten for being late for lights out, which was two minutes ago. Walker, give me five more and I want to hear you both count off."

While Shannon struggled to do five more push-ups, Dale knocked out thirty.

"Now do you know what to say?" Ritchie asked them both.

"No, Drill Sergeant," they both lied.

"Isn't this exciting? Next time you'll know so you won't

have to corner Boehner about it. You say, Drill Sergeant, thank you for conditioning my mind and body. Private Walker and Private Oakes request permission to recover." He waited while the two women repeated it, about to lose the use of their arms.

Ritchie and MacArthur exchanged the type of smiles that made Dale feel like Hansel and Gretel being prepared for the oven.

"I think we should leave them like that, Sergeant Ritchie."

"It's a thought, isn't it, Sergeant MacArthur? Nah, let them up."

She hesitated, then shrugged. "*Recover!*"

Shannon and Dale jumped to their feet and stood at Attention.

"Dismissed," Ritchie commanded. "Get your worthless asses upstairs and don't let me catch either one of you walking."

"Yes, Drill Sergeant," they said together and ran out the door. They double-timed across the patio and neither said a word until they reached the stairway.

"Boy, that was close," Dale said, taking two steps at a time.

"A little too close for me, thank you," Shannon let her know, keeping up with her. "We're going to have to be more careful, Dale."

"I do not like him one bit," Dale stated, reaching the landing.

"She doesn't thrill me, either," Shannon said, referring to MacArthur. "I never met anyone suffering from such a terminal case of irregularity before. She better get herself to a doctor. Nobody is that openly miserable all the time without a reason." She opened the barracks door. "Christ, I cannot do push-ups! I hate them! Son-of-a-bitch..."

Behind the closing barracks door, the voice of Drill Sergeant Kathan could be heard. "Late for lights out, ladies? Get down and knock me out twenty-five."

CHAPTER TWELVE

The next morning at 0520 hours, Alpha company, 10th battalion, learned how to get into morning formation.

The company's six drill sergeants stood before their respective platoons, alphabetizing the men and women as a group and setting them up in ranks. The NCOs then showed the trainees what the drill Dress Right, Dress at close interval meant. The new GIs were then put through the paces of learning those instructions several times. In addition, the trainees were taught how to count off and how to align the squad. After repeated practices, trial runs were made with the individual platoons.

"Fall in!" each platoon sergeant yelled. "At close interval, Dress Right Dress!" They waited five seconds. "Ready, Front!" It wasn't too horrible for amateurs.

The drill sergeants went on to show them Dress Right, Dress at normal interval, and then at double interval. When that was completed, the trainees tried to master Open Ranks.

On an individual basis and at a closer range, the drill sergeants looked friendlier but not entirely harmless. One still approached with extreme caution, especially those with a fear of the unknown. Some, after a trainee got used to their presence, became less overwhelming than when they had made their first brutish entrance.

Nevertheless, it would be a while, if at all, before the drill instructors dropped their defenses completely and there was a good reason for that. With all those fraternization charges so freely flying around, anyone who valued his or her career could not be too careful about the company he or she chose to keep or even converse with.

Then there was the problem with the discussion remaining in neutral territory. Most of the recruits had not been in long enough to decide whether or not they liked the Army so their dialogue was flecked mainly with civilian chit chat, whereas the drill sergeants felt uncomfortable when not talking about anything military to a trainee, unless they were talking about sex which, under any circumstances, was very, very unwise.

At 0600 hours, the senior drill sergeant stepped up to the platform in front of the company. The platoon sergeants put their troops to work, and it wasn't half-bad the first eighteen times they did it for Ritchie, however, with his growling insults at them, it wasn't easy to find the inspiration to continue.

Ritchie made an intimidating little speech to the new Alpha trainees about them being the most laughable bunch he'd seen to date and more than half of them, females mostly, wouldn't even make it to AIT, then he giggled himself off the podium and returned to the orderly room. Dale knew those uninspiring words weren't abnormal for the first few weeks of basic training. Ritchie didn't need to be quite so gleeful about the sentiment, though.

The drill sergeants then took charge of their platoons and marched them to different locations for physical training, or PT. One platoon stayed on the patio, another platoon moved to the open area between the patios and the third platoon, nicknamed third herd by Robbins, assembled in the parking lot. Once in formation and the Open Ranks command was given, the front rank took two steps forward, the second rank took one step forward, the third rank stood fast, and the fourth rank took one step backward. The next command was at double interval, Dress Right, Dress. The soldiers extended both arms until their fingertips touched those of the person standing next to him or her. The following action was to remove their fatigue shirts and ball caps.

The first exercise was the side-straddle hop. Dale always wondered why they just didn't call the exercise by its slang name or at least tell the trainee that it was a simple jumping jack.

During the next twenty minutes, the new trainees tackled bent leg sit ups, squat thrusts, the bend and reach, the knee bender, the high jumper and the four-count push-up. There was a reason why they saved the push-up for last. After that one was correctly done, most of the women couldn't move their arms, with the possible exception of Creed, who knocked them out as if she had springs for forearms and hinges for elbows.

Each platoon went on to do a half-mile run, singing cadence with the drill sergeant for the first time, feeling an odd sense of belonging. Everyone kept up, mainly because the first day wasn't that strenuous and nobody would have had the nerve to fall out, regardless of cramps from improper breathing habits or turned ankles from setting a foot down unexpectedly on a stone or uneven ground.

The company marched to morning chow directly after PT. Following that, the trainees did their assigned details then returned to the bay to make sure their lockers and the barracks were flawless for an inspection that never took place.

It was not uncommon for a rumor of an inspection to be planted by a drill sergeant to keep everyone on their toes, just in case the company commander decided to conduct one on his own without forewarning.

After Dale looked at her clothing issue form again, she decided to tell Kathan what she had *not* been provided at CIF. Long john tops were a necessity and Alabama winters were so unpredictable, she didn't know how soon she'd need them. Kathan instructed her to wait downstairs on the patio and they'd get the company driver to take her to WacVille's CIF. Dale wanted to protest that she did not have the proper paperwork to request what they had neglected to give her but because she wasn't supposed to be familiar with any of that, she had to keep her mouth shut.

Incompetence annoyed her and if Kathan had checked her clothing document, he would have realized that CIF wasn't going to give her the time of day. Kathan, unfortunately, never seemed to swim into consciousness until at least noon and Dale was perturbed that she would have to unnecessarily miss whatever was going to happen or be taught not only during her absence that morning but also whatever day she'd have to return to CIF when Kathan, or whoever, removed his head from his rectum and saw to it that her paperwork was properly taken care of.

Also waiting downstairs for the jeep was Kotski, who had

not been issued any long johns at all nor was she heading off to WacVille with appropriate documentation, either. Disgusted, Dale had to shrug it off. If nothing else, at least she'd get a chance to visit with Kotski, something she hadn't had an opportunity to do since the first two processing days.

"Nobody told you yet?" Bonnie Kramer said, as she pegged two points on the cribbage board,

"Told me what?" Dale moved her peg up one notch for the last card.

"Oh, that's right. You and Kotski were gone all morning. Did you get your stuff?"

"No. Neither did Kotski. We didn't have the right paperwork. Told me what?"

"It's your crib," she pointed to the four cards left face down on the floor. "That Kirk tried to run."

"*What?* When?"

Uh oh…this can't be good…

"This morning. But they caught her and brought her back."

"Well, where is she?"

"I don't know. I haven't seen her. Maybe they locked her up for attempting to go AWOL."

"Who caught her?"

"Some MPs. The captain sent them after her. Boy, he's really nice looking."

"Fifteen two, fifteen four and a three-card run for seven," Dale said and pegged out. "Who's nice looking?"

"Colton." Kramer put away the cribbage board and cards. "The company commander. We caught a glimpse of him today standing outside the orderly room."

"What does he look like?" Dale asked, with mild interest.

"Oh, he's tall, nice body, black hair, mustache, beautiful blue eyes, perfect white smile, gorgeous long eyelashes, straight nose, hair nicely styled for its length..."

"Boy, that was some glimpse."

"I was the closest."

"The only married lady in the bunch and you get to stand the closest to someone who looks like that? That figures," Dale said and laughed. "What was he doing outside the orderly room?

Catching a glimpse of you?"

"No. Talking with Ritchie about Christmas company. Then Ritchie had Audi explain it to us."

"What is...Christmas company?"

"It's for those of us who don't want to or can't, for some reason, go home for Christmas."

"They're sending us home for Christmas?" Dale asked, incredulously.

"Oh. Yeah. Gee, you missed that, too. They're sending us home on the fourteenth of December and we have to be back on the second of January. Practically the whole post is leaving except for a certain permanent party. They're calling it Christmas exodus. Those who aren't going will be moved to the Delta-12 barracks for the duration of the leave and then be brought back on January second with everybody else."

"God, I can't imagine anyone not wanting to leave if they have the chance," Dale made a face. "What will they do here?"

"I don't know. Probably work. You know, details and stuff."

"Well, that certainly sounds exciting." Christmas company also more than likely meant that one of the two cycle spies would have to remain behind with the few who decided to stay. *Oh, goody.* Dale looked at her watch. Snoopy indicated it was almost time for noon chow. "Did we have that inspection today?" She stood up and stretched.

"No. Not yet, anyway."

"Good. Probably none of us would have passed anyway."

"Why do you say that? We all worked very hard on our lockers and I think they looked great."

"Somehow on a first inspection, I don't think that would have made a difference. We couldn't be good the first time no matter how perfect our lockers were."

"I don't believe that," Kramer said.

"No, think about it, Bonnie, really. What would they have to yell at us about if we did things right the first time?"

Kramer contemplated Dale's words. She shrugged. "I suppose you've got a point, although I would hope they would give credit where credit is due."

"I have a sneaking suspicion that things don't work that

way. I don't think they're paid to give us credit. I think they're paid to be a pain in the ass."

"Well, they certainly do that well," Kramer agreed, smiling.

"At Ease!" someone in the bay shouted.

"Fall in downstairs for PT," stated a disembodied voice from the ceiling.

After lunch, off-key and off-time, Kathan marched those who wanted to go, to the main post exchange, for a forty-five-minute visit. The PX was approximately ten minutes walking distance from the company area. Unless a female had incredibly long legs, she usually had to put it into fifth gear just to keep up with the normal thirty-inch steps the males took. The women ordinarily marched in fifteen-inch steps but since they were training with the men, they marched to the standard set for the males. It was not easy at first but, eventually, it would become second nature.

Dale lined up, waiting to use one of the pay phones. When it was her turn, she made sure that the people around her were engaged in conversation, inserted her dime and then dialed the number to LTC Anne Bishaye's office. Her back to her peers, she asked for Bishaye in a barely audible, yet clearly understandable voice.

After being informed that the battalion commander was in, Dale turned around to keep an eye on things. Her stomach fluttered a little knowing that Bishaye was literally right around the corner from her at A-10 but it really started to tremble when she heard the honey-rich voice of the colonel on the other end of the line.

"Colonel Bishaye," Anne announced, sounding very relaxed.

Dale imagined Bishaye sitting at her desk, that sexy grin splitting her face, those hot blue eyes burning up whatever they fell upon in that room. The undercover lieutenant then remembered the colonel's last words to her before she left Vermont. Again, dare she hope...? She had to stop this! She took a deep breath and in the sweetest tone of voice she could muster, she said, "Hello, Mother."

"Hello, dear." Anne always seemed to enjoy their banter.

"If I ever get through this, I will never speak to you again."

"Promises, promises."

There was a definite smile in Bishaye's voice and it thrilled Dale to know that it was just for her. "Anything up?"

"Nothing yet. Nothing on Carolyn Stuart's murder, anyway."

"Hmmm. Too bad. Hey, I've made some friends here."

"Congratulations! I'm impressed. I guess there's a first time for everything."

This made Dale snicker. "You're not funny."

"Yes, I am. You're laughing."

"Yeah? Not for long. One of my friends is from Detroit. Her name is Jascelle Kirk."

"Yes, I already know this. Why are *you* telling me?"

"Because she is having real problems here."

"Is she connected with this case?"

"No. I mean, I highly doubt it."

There was a distinct change in Bishaye's tone. She was suddenly very professional. "Then what is your interest in her, Dale?"

Instantly defensive, Dale responded. "Nothing other than I believe she's unnecessarily being given a hard time."

"I met with her today. I have my eye on it."

"Thank you."

"Look, Dale, she's not your concern. Back off from her. She's our worry and we'll deal with it. I feel she is distracting you from your mission."

"Wait a minute —"

"I mean it, Dale. I said I've got my eye on it," Bishaye sternly cut her off. "We are looking into Articles 83 and 84, Fraudulent and Unlawful Enlistment. She should be out by the middle of the month. However, for you," she emphasized. "The Kirk case is closed. I do not want you involved any more than you already are. Are we clear on this?"

Dale hesitated, then sighed. "No, I think you need to bludgeon me with it a little more." She bit the inside of her cheek in thought and wondered why the shift in attitude. Regardless, she felt it was wise to change the subject. Bishaye being annoyed with her was never pleasant. "Give me some news about the old man."

"You mean Colton? Hasn't he talked to you yet?"

"Nope."

"Well, be patient. He will."

"I can't wait," Dale said, unenthusiastically. "So far I don't have such a hot opinion of him. Neither does my friend."

"Kirk?"

"No. My other friend."

"You have *two* friends?" Bishaye gasped, feigning shock.

The smirk returned to Dale's face. "Heh…you're cute for a bit—"

"Ah! Don't say it," Bishaye warned, playfully. "I gather you mean Lieutenant Walker."

That's the one. Why didn't you tell me?"

"I did. I distinctly remember telling you Lieutenant Walker would be on this case with you," Anne told her.

Yeah, you told me a lot of things that day, little of which I understood. Focus, Dale, focus.

"I looked like an idiot when I spotted her."

"I told you to act normally…you were just following orders."

"Why do I feel like a straight man to you today? Look, seriously, it wasn't funny, this whole thing was almost history right then and there."

"I thought she would be a pleasant surprise."

"After the initial shock, it was."

"Good. You two still get along, yes?"

"Better than ever, it seems. That reminds me, we learned about this thing called Christmas company today."

"One of you will have to stay," Bishaye said.

"I figured that. I'll talk about it with my friend."

"I hope you're referring to Lieutenant Walker. I couldn't handle it if you told me you'd made three friends," Bishaye joked.

"Fuck you and the horse you rode in on," Dale said with a smile on her face.

"Hey…what'd that horse ever do to you? And watch your language. You're supposed to be talking to your mother. You're beginning to sound like a grunt."

"I can't help it. It's all that low crawling in the mud I have to look forward to. I'm going to feel like one. I'm just getting into practice. Also, you need to know that the natives are getting restless."

"Does that surprise you?" Bishaye asked.

"This soon? A little." Dale glanced up at the people who were in line to use the phone. A few pointed to their watches. "Okay, well, my time's up. I'll check in again soon. Any words of wisdom before I go?"

"Yes. Remember you only get out of it what you put into it."

"Bullshit."

"What was that?"

"I said, is that it?"

"Yes. Call me again soon."

"Sure. Whenever the master gets out our leash and takes us for another walk."

"Be a good little puppy and he will."

Dale growled and then said, "Take care, *Mother*. Anything comes up, let me know."

"Yes, dear. Goodbye, Dale."

"Bye," the CID agent said, quietly, as she hung up the phone. Anne Bishaye. So close, so far.

Dale and Kirk sat on the floor, leaning up against Kirk's bunk. "So, what happened after they got you to mental health?"

"Some major interviewed me and then told me to behave myself and go back to the barracks," Kirk said.

"Well, that certainly was therapeutic. So where did you go when you left there?"

"They brought me back here and sent me to see Colton and he told me he was sick of my crybaby routine and if I wanted to run to go ahead."

"He actually told you to run?" Dale was a little surprised by that.

"He sure did. Then the son-of-a-bitch came after me with the MPs, o'er the hills we go, laughing all the way. He acts like this is all a big joke."

"Yeah, I heard he was pretty smug. So, then what did you

do?"

"They took me to see the battalion commander," Kirk said.

"No kidding? Did he rake you over the coals, too?" Dale asked. She pretended she had no knowledge of who Anne Bishaye was.

"He is a she, and no, she didn't. In fact, I liked her a lot. She made everyone get out of her office and we just sat and talked. She made me feel like she was really interested in what I had to say, not like these clowns around here."

"How long were you in there?"

"I'm not sure. It seemed like a long time because we talked about everything from my life in Detroit to my hobbies to my problems here. She didn't make me call her ma'am and she never even brought up the gay thing," Kirk said, She sounded really hopeful since the first time she arrived at McCullough. "She asked me if I thought I'd gotten anything out of being here so far."

Did she laugh when you told her a terminal headache?" Dale threw out.

"I thought of saying something like that, but she seemed so on the level with me that I didn't want to be a smart ass with her. She also wanted to know if I felt like I'd made any friends here and, I hope you don't mind, but I told her about you and that you seemed to be the only one here who even tried to understand my side."

"No, I don't mind, unless she asked for my description and social security number, too," Dale smiled. She was pleased that the colonel had impressed Kirk but she was also confused as to why Anne had been so adamant about Dale keeping her distance from the situation. "So, is she going to resolve this?"

Kirk nodded. "When she was called to another meeting, she got up off the edge of her desk and said, and I quote — she modified her voice to try and imitate Bishaye — 'young lady, if you don't want to be here, I don't want you here, either. These things take time and until I can cut through the red tape, I would appreciate it if you would try to act like a soldier, if nothing else, for the benefit of the others who want to be here. Your situation isn't any easier on them, you know, but I am going to do my best to have you out of here by Christmas exodus on 15 December.'

She walked to me, shook my hand and that was it."

Dale's smile broadened. "Seriously? She said you were going home?"

"Within the next three weeks!" Kirk said, excitedly. "Why didn't they send me to her the first day I got here?"

"I don't know. It probably has something to do with going up the chain of command."

"I'd like to shove it up the chain of command."

"Wait until December fifteenth. By then I'm sure you can find a lot of people to help you."

Karen Henning may have looked fragile because of her size but, as Dale and Shannon both could attest to, her presence was strong, and her confidence seemed unshakable at times.

Her demeanor was personable yet firm when she called the Alpha women to the far picnic table that afternoon. She introduced herself as the company training officer and then proceeded to establish her merit with them. She continued with her woman-to-woman pep talk.

"You're going to hear this a lot while you're here so let's tackle this before we move on to anything else. I cannot stress enough the importance of you being the first females to take basic training with the males and the first company to make it continuous training through the end of LE school. You are making history and the success or failure of this experiment rides solely on you. Because of that, I really must emphasize the purpose of you women pushing for one another and the need to stick together and work as a team. You will get my total support as long as you earn it, as long as you make a pact to work with one another and not against one another.

"That being said, let me warn you against playing delicate during that time of month, or any time, for that matter, because it would result in a contradiction of your bid for equal training." Henning then grinned. "If I have to play soldier during my period, you have to play soldier during yours."

It broke the ice and some of the women relaxed a little. "Also, you need to make sure from this point on that you refer to yourselves as MPs and not WACs because there is indeed a difference."

Dale raised her hand. "Ma'am?"

Henning kept her expression neutral. "Yes, Private?"

"Oakes, Ma'am. Could you explain to us what the difference is?" Dale believed the women should know that it wasn't an elitist issue.

"Of course," Henning answered, easily. "No basic training course is easy, especially if you are not used to strict discipline and teamwork. In WAC or Women's Army Corps basic training, the females learn pretty much everything the males do so with the exception of combat training and instruction. Since women are not allowed in combat, the military sees no sense in wasting time and money on training they will never need or use."

"Ma'am, I thought women went through bivouac," Minty said.

"They do. But they don't learn combat, they learn how to survive in field conditions. You will essentially be going through male basic combat training right alongside the men. The fact is that, hopefully, by the end of this course, you will be qualified as military police officers and that is an importance not everyone will be able to say they've earned. And, even though, women are not officially allowed in combat, as MPs, you just might end up in a conflict or situation where you have to be just as instinctive about the use of your weapons as any infantryman.

"To identify yourselves as LE trainees, or MPs, when you graduate, will distinguish you from others. And if you make it, it is, indeed, an honorable distinction." She looked around as the crowd soaked in her words. "We will not take it easy on you. You must understand what will be expected of you and you must make the effort to get it done. What happens here, during the next four or so months, will let the Army know just exactly what women are made of and what they are capable of."

When Henning was finished, she individually shook hands with everyone in the group and welcomed them to the Army.

Later, when it was discussed in the bay, Dale was pleased to know that the meeting itself had come off favorably, to the extent of several positive attitude changes. The Alpha women seemed to feel that they had an ally in Henning now and because of her professed allegiance to them, they must reciprocate loyalty. The women had found a mentor and several of them openly vowed to drive themselves to their maximum potential so as not to disappoint their lieutenant. At least she gave them a goal, which was their first positive motivation since they had arrived there.

CHAPTER THIRTEEN

Dale picked a vacant spot on the hard, damp laundry room floor and plunked herself down next to a rather pale, irritatingly thin, young man named Gilbert Hibbon. She took out her can of Kiwi shoe polish, cotton balls and a diaper, then unlaced her combat boots and removed them. She stood back up, walked to the sink, filled the top section of the can with water and returned to her place on the cement.

Gil Hibbon had been selected as their squad leader by McCoy because of his ROTC training in high school so Dale felt she should get to know him. Otherwise, she never would have chosen to spit shine her boots next to him. At least not yet.

The new Alpha members would discover during basic training, that when there were any free moments, trainees could usually be found reading their smart books, polishing insignias, doing laundry, ironing fatigues or polishing boots. *Especially* polishing boots, which was what five other people were doing in the laundry room, not including Dale and Hibbon. There were two men doing laundry, two people waiting to do laundry and one-man ironing fatigues.

One of the men who waited for an available washer was a rather immense person named Robert Gauthier, who was also in Dale's platoon. He didn't have to worry about polishing his footgear because there was really no way to spit shine canvas. Gauthier, or Bigfoot as he was not-so-affectionately nicknamed at CIF, couldn't be fitted to a pair of combat boots because there wasn't one his size in existence. His boots had to be specially ordered and until they came in, Bigfoot opted to wear an unusually large pair of sneakers. Dale almost envied him. She knew how those boots were going to affect her feet.

"What'd you think of MacArthur this afternoon?" Hibbon asked Bigfoot.

"Jesus, that bitch." The tall man shook his head. "How much longer you gonna be?" He asked Thomas Lark, who had hogged one washer for what seemed like way too long.

"I'm on my last load."

"Good. I'm sick of waiting. You never should have sneaked that second load in there."

"What'd MacArthur do now?" Dale asked Hibbon.

"She walked into the barracks, unannounced, caught Bigfoot coming out of the shower and dropped him for ten."

"No shit." Dale looked up at Bigfoot. If she bent her neck any farther, it might have snapped backward. "Were you naked?"

"No, I had a towel around me when I came out into the open, but it fell off when I was on my fifth or sixth push-up."

"That must have thrilled her," Dale commented, and shamelessly wondered if everything on his body was in proportion to his size and build. Her second thought to that was *Ouch*.

Hibbon couldn't contain himself and was almost to the point of giggling. "MacArthur told him he shouldn't be a GI, he's too feminine."

"Compared to her, I am," Bigfoot groused.

"I thought that, as a female, she couldn't go into the male bay without announcing female on the floor," Dale said.

"She's not supposed to, but she can get away with it on a technicality, which is we're not sure whether she's a female and neither is she."

Dale shook her head. "Aw, guys, that's not nice."

"As if she's any better to you women," Hibbon said.

"She probably is," Bigfoot commented. "She likes women."

"Is that so? She doesn't sneak into our barracks and try to catch us naked," Dale told them and looked directly at Bigfoot. "I mean, you don't know that for sure. Maybe she's just got an attitude because she's been fucked by too many men." She felt defensive and then wondered why. She couldn't stand MacArthur, either.

"That's strong talk. You sound like one of those women's libbers or something," Bigfoot said.

"Yeah, I have feminist tendencies," Dale admitted. She

hated labels, however, she had to admit, the older she got, the more feminist she became. "But I'm not radical about it, okay? I know my limitations. On the other hand, I'm not here so that I can learn to cook and clean while I stand barefoot and pregnant in some man's kitchen."

"That sounds radical to me," Hibbon told her.

"Come on, guys, don't pick on me. You're just upset because you have to train with women," Dale said.

"Sure, that's upsetting. You women being with us means they're gonna bring our standard of training down so that you can keep up," Bigfoot told her.

"That's an unfair assumption," Dale said. "Hopefully, most of us were accepted into this program because they felt we *could* keep up. Anyway, what makes you so sure they're not going to be twice as hard on us?"

"Please. Don't make me laugh," Bigfoot said.

"Look, there's no point in arguing about this until training actually starts," Lark stated. "So far, I don't see them taking it any easier on the women or bringing down the standard for the men. So far, I don't see them doing much at all. I've never wasted so much time in my life."

Dale stopped polishing and stared at Lark. His words echoed hers a week ago and she felt that was an odd statement for someone to make who was supposed to have no previous knowledge of what training should be like at this point. "Why do you say that?"

"He was in the Navy for four years," Hibbon explained. "He and a few other prior service guys try to fill our heads with horror stories."

"How many prior service are in our company?" Dale inquired. The thought of prior service members gave her new ideas. She'd have to kick her thoughts around with Shannon to see if there might be some basis for suspicion in that area.

Shannon had scarcely picked up bits and pieces of conversation, nothing that meant much, except for Minty who took a poll on whether or not any of the other women had experienced a bowel movement since their arrival.

"I don't know about the rest of y'all," she drawled to

anyone who would listen to her. "But I haven't taken a shit since I've been here. How's everybody else doin'?"

"Oh, come on, Katherine, we have very little privacy and dignity left, must you insist on our making public how often we move our bowels?" Belinda Ryder asked.

"Don't call me Katherine. I've always hated that name. My nickname is Koko."

"Koko?" Travis asked. "Koko Minty?"

"Yeah, what's wrong with that?" Minty challenged.

"You sound like a peppermint patty or a new flavor of ice cream," Travis told her.

"I'm a new flavor, all right," she said, slyly and then glared at Caffrey. "But none of y'all will ever get to taste me." When there was dead silence in response to her statement, she continued with her other subject. "No, I'm wonderin' if it's the food they're feeding us. All that starch and all? I'm bound up or somethin' but I'm not in pain. Is it just me? Come on, girls, am I the only one?"

After several exchanged glances and unintelligible mumbling, almost every female spoke up and agreed with the tall Oklahoman.

"You happy now, Minty? You've finally exposed us all," Quinn Brewer said. "Do me a favor, ladies? When you do take that first shit of the cycle, I, for one, do not want to know about it."

"Does that mean you don't give a shit, Brewer?" Travis hollered to her. The response was many groans and a few pillows tossed at her head.

Shannon's bunk was opposite Lanigan's, so she used that placement to strike up a conversation. "Hey, Lanigan, is that normal? Did your cycle have a problem with that?"

"I really don't remember," Lanigan answered and yawned. She prepared for bed. The holdover had just emerged from the shower and wore a yellow terrycloth bathrobe with a hood and a white bath towel wrapped around her head.

"Was your cycle this slow in getting started?" Shannon asked, off-handedly. She removed her jacket and put it into her locker.

"You writing a book?" Lanigan snapped, as she curiously

regarded Shannon.

"You taking courtesy lessons from Boner? Lighten up, I just want to know how come we seem to be wasting so much time or if this is normal."

Lanigan smiled apologetically and unwrapped the towel from around her head. "I didn't mean to jump at you. I'm just nervous about talking to any of you. Boehner's been caught twice and yelled at. I've got a good record and I'd rather not have Ritchie on me when I have so little time left here."

Shannon nodded, looked around the bay and back at Lanigan. "I think you're clear to talk."

"I'll give you two weeks. You won't be kidding around like that. These drill sergeants find out everything that goes on in these bays."

"How do they do that? Eavesdropping through the bitch box?"

Lanigan shook her head negatively before Shannon had even stopped speaking. She leaned in closely and whispered. "Spies."

Shannon took a sharp breath, covered her mouth and looked utterly horrified. "Up here?" Shannon whispered back. "With us?"

Lanigan nodded, seriously. "You'd be amazed. It's the only way the cadre can find out all that they do."

"Who does the spying?" Shannon kept her voice low and hoped desperately that Lanigan hadn't planted this notion in anyone else's head and wondered who the hell had planted it in hers.

"It could be anyone. All these girls are supposed to be new recruits, right? Well, one of them is assigned. It's the same with the guys. One of them was assigned, too."

"Assigned from where?"

"Okay, it might be different now because you're in this Oshit training program and you're all starting out together, that's why your spies are probably here already. With us, a lot of us came to LE School from different basic training companies, so our spies didn't come in until AIT began, and they were people who were already in the Army, prior service or inserts from another field."

"This all sounds a little James Bond-ish to me. How do you know all this?"

"Boehner found out. She overheard it one night when she was on CQ and came up and told me. It made sense. There's no other way that the drills can find out all they know."

Shannon was stuck between being bewildered and furious. What the hell was Boehner up to? She tried to cover her confusion by shrugging off Lanigan's suggestion. "We're talking about Boner here. This sounds like something she'd think up."

"I've spent two and a half months with Boner...I mean, Boehner. I think I know her a little better than you."

"Do you? You spent two and a half months with all the other women, too. Do you know which one was the spy?"

"No." Lanigan hesitated, thoughtfully. "But why would Boehner make up something like that?"

"Why does Boner do anything?" Shannon moved to Lanigan's area and leaned against her locker. She made sure her back was to a majority of the bay. She wanted no one to overhear this conversation. "We're talking about a girl with a real sense of the dramatic here. I'm fully convinced that she does not have both oars in the water, anyway. I mean, did she or did she not stand right about here the second night we were all here and announce that she wanted to die young by being shot to death in combat? Let's face it, Boner is a little extreme. Nobody should be that dedicated. She should be a *marine*, for Christ's sake. Think about it...what do we need to be spied on for? We're basic trainees, not prisoners of war. We don't have any secrets that could be of any possible value to the U.S. Army. I can't believe Boner would even suggest something about spies and I can't believe that you would actually believe her."

"It just sounded logical," Lanigan mumbled, as she apparently felt the sting of Shannon's reasoning. "So how come the cadre knows so much? How do they find out what they do if there are no spies?"

"It's just a guess," Shannon started and tried not to sound too sarcastic. "But I would assume that they don't enlist and automatically become drill sergeants. They had to go through basic training and AIT and when they graduated and met other people who'd been through AIT, I'm sure they compared and collected stories. They've probably heard it all and done most of it. I'm sure the charming ones can always get the naïve recruits to inform for them, especially with a promise that such a service can further one's military career. To me that makes much more sense than some wild story about...spies." She practically spit that last word out as if it left a bitter taste in her mouth. She then shook her head. "I don't know about that Boner."

The holdover looked as though she felt she'd been made a fool of, and Shannon couldn't tell if she were angrier at herself or Boehner.

"Listen, Lanigan, you didn't say anything about this spy business to any of the other girls, did you?"

Lanigan told her that she had not. "You're the first one I've felt like I'm not going to get caught talking to."

"Good, because we have enough problems getting along as it is. Being cooped up for as long as we're going to be with each other, the last thing we need is some crazy rumor floating around that one of us can't be trusted. You know what I'm saying?"

Nodding, Lanigan put her robe away. "I won't say anything. But I think I'm going to kill Boehner."

Stand in line, Shannon thought. "She probably can't help saying things like that and, what's worse, she probably believes what she says. Let's not talk about Boner anymore."

"Gladly."

The blonde agent relaxed and turned so that her back rested against the locker. "I know you can't tell me what basic training with the guys is like, so what's LE School like?"

"It's interesting if you're into learning a lot in a short period of time. You've really got to stay alert because some of the classes are very complicated and they expect you to learn that material in a one- or two-day period."

"Like what?"

"Like triangulation." Lanigan made a face.

"What's that?" Shannon asked, mildly surprised that Lanigan had found that particular course difficult. She'd never had a problem with it.

"It has something to do with measuring traffic accidents, you know, distance of vehicles from fixed objects. I thought I'd never get through that class. But on the whole, I felt I had to cram a year's worth of learning into seven and a half weeks and I'm not sure I'll remember everything when I get to my permanent duty station."

You won't, Shannon wanted to assure her, *and it wouldn't matter even if you did because your first month on the road will be like school again.*

"As long as you don't believe everything you hear, you should do okay." She winked and decided to change the subject again. "There are a lot of male drill sergeants here. I'm surprised it's not more balanced out. MacArthur must be in her glory."

"Would we recognize it if she was?" Lanigan said and laughed.

"Probably not," Shannon agreed, thinking a full-blown smile from the dour female sergeant might actually burn her retinas. "What about male drills? Do we have some harassment or being hit on to look forward to with them?"

"I seriously doubt it. Two girls in my cycle pressed charges against two drill sergeants for that kind of stuff. I guess it's not the first time it's happened either, so these drill sergeants are real careful about what they do, what they say and how they behave."

"Which drills were they?" Shannon inquired. She leaned in close, as if she were about to receive a juicy tidbit of forbidden gossip.

"Oh, they're not here anymore. They left or were sent away. I'm not sure. All I know is that it was really hushed up."

"God, they must have really pushed those girls. I certainly wouldn't bring a drill sergeant up on any kind of charges my first few months with the Army and then expect to have any kind of successful military future...especially if it wasn't true."

"Me, either. One of the girls graduated immediately afterward and was stationed in Maryland and the other was discharged and sent home. If it hadn't been for Stuart, I would have thought they might've exaggerated but Stuart was gay, so it didn't make sense."

"Sounds like your cycle had some excitement."

Lanigan sighed. "Yeah, we sure did. If you're lucky, though, you won't have to go through anything like that. The drill sergeants were unbearable afterward. Forget possible fraternization, they weren't even civil. I don't wish those last couple of weeks on anyone."

"They seem okay now."

"That's because most of the drills are new. Ritchie, McCoy and MacArthur are the only ones left from last cycle. The rest are either brand new or transferred in from other companies."

"Robbins is new?"

"Well, he came in after Drill Sergeant Halpin left. There were only about three weeks left in the cycle, so he's relatively new."

"He seems like a flirt."

"I don't know about that. They all acted pretty much the same, especially after the second incident with Halpin. If he's a flirt, he just started, and if he just started, either he'll be sorry, or your cycle will."

Shannon was seated at the far picnic table, which faced the laundry room and looked out across the parking lot. The nights had gotten damper and cooler and though the temperature hovered around the lower fifties, there was no doubt winter was on its way. That was a thought neither Shannon nor Dale welcomed with much enthusiasm.

She had just lit her second cigarette when Dale exited the laundry room in her stocking feet, holding her freshly spit shined boots in her hand. Dale spotted Shannon and padded to her. She scanned the area to make sure they were alone.

"What a coincidence you're out here. I've got to talk to you," Dale told her.

"I've got to talk to you, too." Shannon exhaled smoke. She stood up and they both walked to the railing that was the farthest from hearing range. As Dale glanced at her watch, Shannon spoke quietly. "We've got ten minutes before lights out. Keep your eyes open for Ritchie."

"Boy, winter's really beginning to make its presence known, isn't it? I can feel it in my foot."

Shannon blew out another long stream of smoke. "What's your news?"

"What do you think about prior service?"

"They give me indigestion if they're not cooked long

enough," Shannon cracked. "Prior service pertaining to what?"

"This case."

Shannon pondered this. "Hmmm, well, prior service would know their way around and that would make it easier to set things up but we don't have any prior service females."

"I thought of that. What about prior service males using the females to set up the drill sergeants? They could be more mobile within the company and a lot of obvious things they could do would go unnoticed because they are not as closely watched."

"The same could be said for inserts."

"Right. If this is a grudge thing, whoever has the grudge could be finding prior service guys and getting them to work with him on this. Why don't we see if we can get a list of prior service members from Henning and watch them. Which prior service male takes up with which female could prove to be very interesting. What do you say we start monitoring their behavior?"

"Can't hurt. I mean, we're sort of grasping at straws right now, anyway. It's a start," Shannon agreed, with renewed interest.

Dale, who had her back to the railing, saw the laundry room door open. Bigfoot and Hibbon emerged and nodded in the direction of the two CID agents. They both said goodnight and headed upstairs. "So, what did you want to tell me?" Dale checked her watch again.

"Don't trust Boner."

Dale blinked at her, seemingly momentarily at a loss for words. "Hopefully, you have something better than *that*, because not trusting Boner is already a given in my book." Dale folded her arms and waited for what else Shannon had to offer.

"I don't know what she's up to and I haven't had a chance to ask her yet, but something prompted her to fill Lanigan's head with tales of cycle spies and barracks espionage."

"What?" Dale whispered harshly.

"Yeah. My reaction exactly. I strike up a casual conversation with Lanigan and she gets all secretive and tells me we have spies in the bay. She said Boner told her."

"Oh, my God. Has she told this to anyone else?"

"I have no idea if Boner has or not, but Lanigan told me that she had not and I finally talked her into thinking it was a ridiculous idea. I don't know how you want to handle it but as far as I'm concerned, Boner is the enemy."

"You won't be twisting my arm to see things your way. Why the fuck would she do something like that? We need to go find her and ask her."

As Dale started to push herself away from the railing, Shannon stubbed her cigarette out and stuck the butt in her pocket. "Wait. Let's think about this. Confronting her might not be a good idea. I know you'd like to do it to watch her squirm but that might be counterproductive."

"You better have a damned good reason for spoiling all my fun."

"Boner was really pissed off when we were brought in. I don't think it really has anything to do with us, personally, I think Boner is angry she didn't get her job done and found out she was being replaced in the same company for, well, practically the same assignment. And look at some of the stuff she's done since we've been here, for example, that insane statement about dying in combat."

"That was said for shock value."

"Was it? She's been acting like GI Jody ever since we've been here. You know, the ideal soldier, everybody look up to me, and obviously she must have been acting that way long before we got here because Lanigan hasn't mentioned anything about a drastic personality change."

"So, what are you saying?"

Shannon arched an eyebrow and grinned. "Too many Kiwi fumes or what? You're usually right on top of things. A normal cycle spy is supposed to blend in, not draw unnecessary attention. Who has consistently been the center of attention since we have been here?"

"Kirk and Zelman."

"Well, yes, them, too, but then who?"

"Boner. Naturally she's going to draw a little more attention. She's a holdover and the women are curious."

"Agreed. But think, Dale. It's just Boner. Lanigan, St. John, Rossi and Cornish are holdovers, too and hardly anybody knows their names. *Everybody*...even a lot of the guys...know Boner.

She just doesn't draw attention, she demands it."

Dale cocked her head. "My goodness, Miss Marple, you're right."

"My point is, if we hit her with this incriminating little piece of information, will she crack a little more than she already has and blow our cover out of resentment? That would set this investigation back at least six months."

Dale took a deep breath. "Good point. I'd like to tell Anne, I mean, Bishaye, and have her get it out of Boner the day she leaves. And I'd like to be a fly on the wall when she does. Let's just hope Boner hasn't said anything to anyone else."

"I don't think she has. I'm sure we would have heard a rumor like that flying around. These women are not the height of discretion when it comes to gossip."

"Ah...true soldiers already," Dale smiled. "Let's drop a dime to Henning and let her make the decision as to whether to tell Bishaye or not."

"That sounds like a plan to me. Come on, let's get upstairs. I'm not taking any chances tonight."

CHAPTER FOURTEEN

Fate intervened for Dee Tierni when Lieutenant Henning decided to make the morning run. The women as a group had vowed to make a special effort to excel and a majority had succeeded but Private Tierni had to fall out with not even an eighth of the run left because she was too queasy. Dale was a few yards behind Tierni when Henning had reached the trainee and attempted to get Tierni to breathe properly after Tierni had just deposited last night's meal at the base of the parking lot.

Dale knew Tierni was probably upset enough about dropping out of the run, so she was sure the last thing Tierni needed was the senior drill prick running up one side of her and down the other. "You get that fat ass moving, Private, or you'll really know what it's like to be tired, you candy-assed —"

Henning snapped upright, turned around and glared at him. "Lay off her, Sergeant Ritchie. I mean it!"

Ritchie's face registered more than mild shock. Dale presumed that, at first, all he saw in his approach was another GI's body bent next to Tierni's, so he undoubtedly assumed it belonged to another trainee. Second, he was noticeably embarrassed because Henning not only stuck up for a lowly trainee, but she also reprimanded him in front of a few of them on top of it. Ritchie sputtered and ran off to most likely tattle to Captain Colton, Dale guessed. Ritchie definitely seemed the type.

When Henning and Tierni returned to the south patio, everyone was already in formation, sweating and panting and standing at rigid Attention. The company's concentration was devoted to Ritchie, who stood at the podium before them. Henning wisely decided to keep Tierni next to her until Ritchie's speech was over. Henning's fear was probably the same as Dale's, that if he saw the trainee walk alone into the already formed ranks, he would unleash a more focused humiliating personal attack against the already mortified private.

"What I saw this morning was absolutely pathetic! I cannot think of enough words to describe how disgraced I was to have passing troops know that you belonged to my company," Ritchie yelled. "You're nauseating! Those other companies we passed were laughing at us and they will probably use us as a bad training example. You men run like a bunch of sissies! And you women —" He paused and shook his head, dramatically. "At this pace, you'll never be good enough. Now, I would hate to have it on *my* conscience that *I* held these men back. Do you understand what I'm saying, ladies?"

Everybody understood. It was very clear to everyone from that point on that Ritchie did not approve of female MPs and resented being a part of a unit that had to train them. In fact, from what Dale had already heard, Ritchie rebelled against females being in the Army, period. His mantra seemed to be a woman's place is in the home, not the barracks. Dale knew and had worked with too many men like Ritchie. Their goals seemed to be the more females they could personally eliminate, the better they would feel.

"If you ladies cannot keep up, you will be placed in a special class," Ritchie continued. "If you cannot keep up with that, you will be recycled to another MP training unit. If you fail there, you will be put into a slot where you can handle the job and I guarantee it won't be your choice this time. Your performance today was way too far below standard. It was unacceptable for this unit, and I will tolerate it no further. I will not have you dragging down A-10! Maybe you ladies better think seriously about being MPs because you'll never make it at the rate you're going. We're supposed to be the cream of the crop not the bottom of the barrel. I'm very disappointed, ladies, *very* disappointed."

He left the podium abruptly and stomped into the orderly room.

Each sergeant turned to face their troops, a little dazed. It was customary for someone in a position of authority to deliver a severe speech to let the trainees know that they needed a lot more work. Even so, such an agitated assault that mostly singled out the women clearly took the cadre by surprise, especially considering how obvious it was that most of the women had put out one hundred percent on the run.

The command of At Ease was called for all three platoons. The somewhat provoked trainees were ordered to attend to their assigned details until they were called down for chow. After Attention and Fall Out were commanded, the soldiers returned to their respective bays.

When noon chow was over, the women were given free time, but they were restricted to the barracks because one by one they were scheduled to meet with their platoon sergeants. They had to be available to go to the drill sergeant's office when the female before them, alphabetically, had completed her personal interview so it was better to keep them all in one place beforehand.

The women were still ruffled and grumbling about being degraded that morning and quite a few stated they were determined to bring that matter up with their drill instructors, one way or another. They had discussed it as a group and, even with Tierni falling out, they didn't feel they deserved that kind of verbal abuse.

Everyone was also understandably nervous about how to act, including Shannon and Dale, who hoped they could play up their alleged military naïveté to the hilt and not say anything that might give them away.

"What do you think they'll ask us?" Tramonte inquired.

"Probably stuff like why we enlisted, why the Army instead of the other services or why did we choose military police. I mean, those would seem like logical questions," Minty said, as she assisted Sherlock in aligning third platoon's bunks.

"Hey, I've got an idea," Travis said, suddenly. "Let's turn the tables. Let's go in there and say, why did you become a drill sergeant and why did you enlist, or how about, what concentration camp did you find Ritchie running and why didn't you leave him there?"

Tierni spoke up, calmly, breaking her silence from that morning. "Ritchie's the type of man who has to be handled with extreme care. Clearly we're dealing with an insensitive, high-strung, fucked up little maniac."

The women were still too wounded themselves not to nod in agreement with Tierni.

If Ritchie was listening to them downstairs, he would have to have been a bigger fool than Dale had given him credit for to come upstairs and confront them on any of their remarks. He would have to have been a complete idiot not to realize that if he set one foot in the barracks that afternoon, a justifiable lynching may have taken place in the shower.

At the beginning of her meeting that afternoon with her drill instructor, Shannon smiled inwardly as Robbins tried to be his most charming self, but her amusement soon turned to frustration as their interview continued. He couldn't have cared less how she answered his questions, he never even heard her answers because he was trying so hard to impress and flirt with her. On anyone else, it might have worked but Shannon was all too prepared for it and was very disappointed that Robbins had fulfilled her expectations. It was times like this she wanted to expose her true rank and reason for being there.

Robbins would definitely have to be watched more closely.

Dale, on the other hand, enjoyed her meeting with McCoy. She was careful in how she handled herself with him, not intimidated but careful enough not to blatantly let him know it. Unlike Minty, who had come back to the barracks after her interview, bragging about how much she thought she could wrap McCoy around her little finger. Dale remained respectful and inquisitive and let him lead the meeting.

The subjects, not surprisingly, remained military and Dale clearly hit a nerve when she asked McCoy to compare the differences of the Army now, when she had, supposedly enlisted, to then, when he had.

"Honestly? The military is my life. I love the Army but ever since they got rid of the draft, I'm dismayed by the decline in discipline and morale."

"But, Drill Sergeant, wouldn't it be better because the people enlisting want to serve? I would think that would make a difference in morale, as opposed to the people forced to be here because of a draft." Dale really was interested in his answer.

"I see what you're saying, Oakes, but to make the Army inviting to today's able young person, Uncle Sam has had to lower his measure of eligibility, therefore accepting individuals way below standard. The freedoms of the New Army make them virtually untrainable. It's not now, nor ever will be again, like it used to be. If we go to war with the caliber of people presently being inducted we'll probably never get to the front line."

"If you really believe that, Drill Sergeant, why are you here?" Dale made sure her tone stayed more on the curious side than the judgmental side.

"I've got eighteen years in. It would be foolish for me to get out now," McCoy said and shrugged. "Besides, I do what I can to ensure the recruits under my control are turned into the best trained soldiers possible."

"What do you think of this program, Drill Sergeant? Do you think it's a waste of time for us to be trained alongside the men?"

"Not at all. As far as I'm concerned, you're a soldier and my expectations of you are the same as anyone else. Just listen to us, to your field and class instructors, keep your head up and your nose clean and there is no reason why you can't be successful."

"Some of the other drill sergeants seem to have made it clear that they don't feel the same way, Drill Sergeant," Dale commented.

"But I'm your senior platoon sergeant, Oakes, and my attitude will shape your attitude. If I have confidence in the program you're in and that everyone will be given a fair chance to make it through, what the others do should not be your concern."

The soldier part of her liked and respected that answer, even though her own experiences contradicted McCoy's apparent nobility. The cop part of her wondered if he was so discouraged about what he thought was a serious downfall in standards that he might try to send a misguided message to battalion. Even though she sincerely doubted it, she was smart enough to never say never.

"I'm just wondering if he acted the same way with the other

women as he did with me," Shannon said, exasperated, her voice as quiet as possible, but not without its conviction. They were sitting on the picnic table on the north patio. "I could have told him that I was a Russian spy and that I enlisted to learn any and every American secret I could get my hands on, and he would have said, that's wonderful, young lady, you have such initiative. I like him but really, Dale, we're going to have to watch him. He's inviting trouble with a capital T."

"Right here in River City," Dale finished, smirking.

"It's not funny. I wanted to smack him."

Dale swallowed her smile as her partner was obviously not in the mood to play. "Sorry. I didn't have that problem with McCoy. He's here to put a cycle through and he's strictly business, there is no mistaking that." She decided not to openly speculate yet on her fleeting thoughts of his disappointment in the Army prompting conspiratorial behavior. Although it was in her nature to think that way, she really did not get that feeling from him. "He's going to be strict but he's going to be good. Who knows? Maybe Robbins is just looking at the crops."

Shannon threw a glare to Dale. "Crops? You're calling me a crop?"

"You know what I mean. And just don't take him up on any invitation to do squats in his cucumber patch."

Shannon rolled her eyes and lit a cigarette. "What do you think we should do about him?"

Dale shrugged and said, "I don't know. Maybe he's all insinuation. I think we should just monitor him for the time being. Maybe tell Henning about him, too."

"Oh, speaking of Stubby, I ran into her before I went in to see Robbins and I had a chance to tell her about Boner. She said she would tell Bishaye and not to worry about it, she'd take care of it."

"Stubby?"

"Like I'm sure she hasn't heard that before. Anyway, she was a tad pissed off and I somehow imagine that sooner than later, the strange Sergeant Boehner will have deep teeth marks on her butt."

Instead of taking that in the context in which it was intended, Dale had to fight to keep a lewd smile off her face as the visual of Anne Bishaye playfully biting her on the ass danced in her head. Then she pictured the colonel chewing on Boehner and it shrouded her with an entirely different mental picture. She shook that imagery from her memory banks. "Regardless, it won't be a pleasant experience. I've been there. Boner is in for a real treat."

They sat in silence for a while and enjoyed the noticeably mild November night. "Were you in the bay when MacArthur was telling everyone about fireguard, KP and CQ?" Shannon asked.

"No but I heard about it. Everyone's confused, probably because they've never done any of it before. When does it start?"

"Fireguard starts tonight. There's a list of names posted for the next two weeks. I believe I saw your name up there somewhere within the next couple of days."

"If MacArthur made up that list, I'm surprised it's not Kirk and me every other night. What about KP?"

"That starts tomorrow. There's a list up for that, too, and I believe your name is on that for the very near future. And CQ starts tomorrow night at five...excuse me, I mean 1700 hours."

CQ stood for charge of quarters, and it was pretty much self-explanatory. The soldier designated to work that time slot was in charge of the company and all four bays. In reality, he or she was the representative for the commanding officer after he signed out for the day and left the company area. The CQ and CQ runner's job was primarily to insure that the unit stayed safe and sound and that any problems be taken care of or reported immediately to the staff duty NCO who was the CQ on the battalion level. The soldiers worked out of the orderly room, or CQ office, where their immediate priority was to answer the telephone, keep a log, perform several checks of the company area and conduct reveille. For a basic trainee, CQ was a royal pain. It took away four hours of precious sleep time, eight hours on weekends, that could not be compensated for the next day.

KP, or kitchen police, had changed during the past years. Privates didn't spend hours peeling potatoes anymore. In fact, the last Dale had heard, the kitchens had been invaded and captured by civilian cooks and trainees were no longer allowed to touch the food that was being cooked or prepped. All they were expected and assigned to do now was to be a cleanup crew. They washed dishes, pots and pans, swept and mopped floors, cleaned tables, made sure there were enough trays and silverware and they helped to take in food deliveries, if there were any, while they were on duty. It made for a very long day, but it usually wasn't a hard day anymore.

Fireguard was exactly what the name implied. A trainee guarded the barracks against fire and kept his or her fellow trainees safe from any such impending disaster. It lasted from nine o'clock at night until five o'clock in the morning. It was an eight-hour shift divided into two-hour time slots with four trainees working as fireguard at night. Again, it took away valuable sleep time, but a lot of people made good use of that two hours they were supposed to stay awake. They polished boots, ironed fatigues, shined brass, studied or practiced what they had learned during the day, together with making sure that no threatening flames popped up anywhere.

The one thing that wasn't allowed was sleeping. If a fireguard was caught sleeping on her shift by a fellow trainee, she was at the mercy of her buddy who had the choice of keeping it to herself or reporting the violation to her superior. If the fireguard was caught taking a snooze by a roving drill sergeant, being sent to the gallows may have been quicker and less painful.

An Article 15 would definitely ensue, meaning a temporary absence of one type of green that a trainee would sorely miss, not to mention the ridicule that would accompany the punishment which would, no doubt, include extra duty.

To discipline a fireguard who had fallen asleep while on duty with extra detail was redundant. If she wasn't so exhausted from everything else she had to do during the day, she wouldn't have been so apt to nod off. To assign a worn-out fireguard more to do wouldn't solve the problem, it would just compound it. However, fining her a couple weeks' pay might better underline the importance of the mistake.

A fireguard had to constantly be on her toes anyway because a drill sergeant always appeared unannounced sometime during the night. It was inevitable. It wasn't bad enough that the trainee had to see him or her the first thing in the morning, the last thing at night and every waking hour of the day but even in a few of the unwaking ones, too.

The drill sergeants got all bent out of shape, or so they claimed, when the females were caught in various stages of undress yet if nature had felt it urgent enough to wake one up in the middle of the night, that was hardly the time to stop and fiddle with opening one's locker to get a bathrobe. It never seemed to fail, though, if a female trainee got up to use the latrine improperly attired, she was guaranteed to run into a drill sergeant who was pulling a surprise bed check and she was dropped for disciplinary action before she could relieve her bladder. If she wanted to be prepared and draped a robe or something proper to slip into across the bottom of her bunk, she was written up for having a security violation.

Most of the time, being in basic training was a no-win situation.

CHAPTER FIFTEEN

The weekly weigh-in was initiated that evening and was, at the very least, a humiliating experience, especially for most of the women. The drill instructors didn't let a single female pass without emitting some devastatingly sarcastic remark about her weight and that ranged from the slightest of build, such as Debbie Ferrence, to the heaviest, like Dizzy Zelman. Or, from the most physically fit, Deborah Michaelson and Shelley Creed, to the most out of shape, Dizzy Zelman.

Stepping on the scale and the wit and wisdom of the platoon sergeants didn't put the women in the best of moods. They moped around their last hour before lights out and some took care of last-minute details like locker neatness while others spit-shined boots, ironed fatigues and shined brass.

Some of the females had already forgotten about the remarks and had gone to bed. Some felt they didn't need this kind of harassment and still secretly wished they could go home.

"Ritchie told me that if I didn't go on a diet they'd have to let out the mess hall," Dizzy drawled to Verno and Linda York, who regarded her blankly. "You know, like you let out a dress?"

"Oh. Funny," York responded, unsmiling, and yawned. "Well, maybe they have Army diet doctors they can send you to."

"Well...I really don't see myself as being that fat. Although a diet doctor could maybe give me some of those nice little black capsules that keep your eyes open...for days. My metabolism really seems to like those. You know what I mean?" She shrugged. "Oh, well, if they can't, I know where I can get some, but I'd rather not have to pay for them." It didn't seem to surprise anyone that she had already located a company or battalion pusher.

"Aw, Dizzy, don't feel so bad," Bonnie Kramer said, as she

strolled from the other side of the bay. She had also been chastised for being eleven pounds overweight. "We've only been here a week and just barely started any physical training. They've got to give us a chance to get in shape. Just think of it this way," Kramer soothed. "When there's a universal famine, you and I will be the only two prepared for it." She patted her tummy. "We could live off the land for a while."

"Personally," York said "I don't think any of us should be criticized for our appearance when we've got MacArthur as a role model." There was an overwhelming chorus of agreement.

"Drill Sergeant MacArthur isn't that big," Snow spoke up, sternly. She never bothered to look up from studying her ranks and insignias.

"She's not?" Travis challenged. "I swear Captain Ahab came through here last night after bed check, looking for her."

Snow shook her head. "You're insensitive, Travis."

"And you're an open nerve ending, aren't you, Snow?" Travis shot back.

The former teacher looked up from her reading material with a very shocked look on her face. She recovered quickly from this attack of insolence. "You may take this whole thing as a joke but some of us feel that this phase of our life is extremely important and I, for one, don't feel I have to be barracks clown at someone else's expense." Her voice was coolly even. "I believe Drill Sergeant MacArthur is doing her best. It can't be easy for her. I caution you not to ridicule her in front of me again."

The silence in the bay was deafening.

Travis approached Snow's bunk. By Travis' own admission, she was not a physical person, rarely getting into fights even in her worst tomboy stage, but neither was she one to back down from being bullied, whether it be verbal or physical. "You *caution* me? What's going to happen if I do?" Her fists rested on her hips as she leaned in close. "Are you going to make me write, *I will not insult MacArthur in front of Snow* five hundred times?"

"Very funny, Travis." Snow did not act threatened by Travis' looming.

"Sounds to me like you've got a crush on her, Prof," Travis

stated, amused.

Snow bounded off her bunk and right into Travis' face. "Shut up, Travis!"

"Ooooh, I think I just might have hit one of them thar open nerves. Well, for what it's worth, if you do have the hots for her, I think your taste is in your mouth." This comment drew several obscene sounding snickers from a few of the women around them.

Snow then pushed Travis who pushed back. Tramonte and Tierni, who were the closest, grabbed Travis and held her back as Sherlock got a firm grip on the former teacher.

"Knock it off, both of you!" Belinda Ryder, the fireguard, yelled. "If a drill sergeant comes up here because of a fight, I'm going to kick the living shit out of both of you and I am not, by nature, a violent person!" She stomped to Snow. "Tone down that attitude of yours. We all had to be smart to get here and just because you have a degree doesn't make you any better than any of us here, so stop acting as if this is all so beneath you."

She then turned to Travis. "And you stop your instigating. This infighting has got to stop or we are going to be proving that son-of-a-bitch Ritchie right when he says we shouldn't be here. Let's stop acting like a bunch of high school girls and start acting like soldiers."

The embattled women let Ryder's words sink in and when they were let go, they quietly returned to their areas. Soon, the atmosphere in the bay had gone back to normal.

Downstairs, as new trainee John Urso entered the Orderly Room to ask questions about the responsibilities he was to undertake as CQ the next night, Private Mroz couldn't have left the Orderly Room any faster if she'd been shot out of a cannon. She had overheard Colton and Henning tell Ritchie that they were going to have a surprise fire drill later on in the evening and she had to get upstairs and warn the females. She hit a dead run two steps beyond the CQ office but was collared by Lieutenant Henning, who was waiting for her at the door of the first sergeant's office, which was ten steps beyond the orderly room.

"Private Mroz, I believe?"

"Yes, Ma'am," Mroz stood at Attention and stared straight ahead.

"Private Mroz, you were in the Orderly Room when Captain Colton and I were discussing something with Sergeant Ritchie?"

"Yes, Ma'am."

"And you heard something about a fire drill, is that right?"

"Uh..." Her hesitation confirmed that she had.

"Look at me, Private Mroz."

Mroz's eyes met Henning's. "Ma'am?"

"You heard nothing, isn't that right?"

"About what, Ma'am?"

"Exactly. You learn fast." She stepped back from Mroz. "Carry On, Private Mroz."

"Yes, Ma'am. Thank you, Ma'am."

Mroz threw the barracks door open five seconds later. The sound of it slamming against the wall startled everyone but it wasn't quite as alarming as Mroz' behavior. She silently ran up the aisle of one side of the bay and down the other. She looked in every nook and cranny, even eyeballed the underside of a few beds. She then stopped abruptly, craned her neck and stared at every inch of the bitch box. Seemingly satisfied with its darkened stillness, she then took off for the latrine and checked each cubicle and shower stall individually. By this time, she had drawn a curious crowd of twelve. Chris Steele was the first to speak.

"Mroz...what are you doing?"

"Checking for drill sergeants."

"I'm sure we don't have any. We sprayed for them an hour ago," Travis said. "Pesky little buggers."

"Why would you be checking for drill sergeants as if they were hiding?" Boehner asked. "They're really very good about letting you know that they are around."

"I'm not supposed to say anything, but I think you should know," she addressed them all. "I overheard the captain and the lieutenant tell Ritchie that we're going to have a fire drill later tonight."

"Oh, no. When?" Tramonte asked.

"I don't know. Sometime after lights out and bed check,"

Mroz said.

"Then you'd better be prepared," Boehner warned. "Wear something warm because you don't know how long they'll keep you out once you get out there. I suggest sleeping in your long johns. Don't sleep with your fatigues on, that will look too suspicious. When the fire bell rings, wrap your dust cover around you and the women on the right side of the bay should run out that fire exit door," she pointed. "The one located by Zelman and Walker's bunks. The women on the left side of the bay should go out the regular door and save time and avoid a major clusterfuck. Try to remain calm and as orderly as possible because they'll be looking for that."

"Do we have a lot of these fire drills?" Tierni asked.

Boehner shook her head. "My cycle only had one."

The word spread as quickly as if the barracks really had been in flames, and everyone suddenly became preoccupied with trying to remember Boehner's instructions.

At almost exactly midnight, a horrible noise that could have only been described as a fire alarm went off and, even though they were prepared for it, the silence broken suddenly by the screaming alarm, caused palpitations in just about everyone.

Downstairs in the parking lot, every platoon sergeant, senior and junior, was there to monitor the behavior and quickness of their troops. In the open area between patios, Ritchie stood, barking orders and standing next to the north and south fire exits were Henning and Fuscha, respectively, in civies, directing traffic. All of the cadre exchanged looks with one another, clearly suspicious that the females seemed to be wise enough to be warmly attired and wrapped in one of their wool Army blankets. Henning gave Mroz an intimidating, knowing glare when the private passed her.

While they ran into line behind their platoon sergeants, several men, who figured the women must be scantily dressed if they were wrapped in their dust covers, tried to step on the blankets the females wore around their upper bodies to see what the women were, and hopefully, were not, wearing underneath their green wool capes.

Once everyone ended up where they were supposed to be, Ritchie lumbered around, while Henning paced behind him.

"What I just saw was the most disorderly exercise I have ever witnessed!" he screamed at them. "In a real situation, a lot of lives would have been in grave danger or lost!" Ritchie, again, found a way to blame the non-success of the drill on the women and proceeded to make the entire company, some in bare feet, stand in the parking lot in thirty-degree weather for close to twenty minutes. After he made sure they were wide awake, chilled to the bone and wired from anger, he allowed them all to return to their bunks for, hopefully, the rest of the night.

After PT, morning chow and details, the trainees were issued steel pots, helmet liners, male fatigue jackets, for the women, because they were lined and heavier than the female fatigue jacket, web gear, or LBJs as they were more commonly referred to, a canteen and holder, a first aid pouch, an ammo pouch and a pistol holder.

The trainees were instructed to wear their LBJs over their fatigue jacket and to place all of the equipment on the designated places of the heavily constructed suspenders connected to a belt woven with nylon and canvas. The web gear was restricting and heavy, but it had to be adjusted to quickly.

The drill sergeants, from that point on, continually reminded their troops that they were not celluloid soldiers. When Caffrey was caught wearing her helmet improperly, she was immediately reprimanded in front of the entire company. "Don't wear that steel pot on the back of your head like that! Who do you think you are? Jane Wayne?"

The trainees were then marched to an empty classroom in the LE School. The company stood at attention next to what resembled school desks until Lieutenant Henning, dressed in her Class-A uniform and looking exceptionally pretty, entered and walked up to a dais in the front of the room.

"Take your seats!" Drill Sergeant Kathan yelled.

Everybody simultaneously sat in the desk next to them. It wasn't good enough, so they did it again. And again. It wasn't fast enough so they did it faster. Then faster yet.

The drill sergeants didn't seem to care that every time the trainees sat down, this new, bulky, awkward equipment they were wearing would get caught on the desk or the attached chair and prevent them from a smooth and quick descent. When they finally did it to someone's satisfaction, probably Henning, who was getting impatient, standing up front in uncomfortable pumps, they were able to stay seated and listen attentively as the lieutenant passed around leave request forms and explained again about Christmas exodus.

It was boring and hot with all that gear on, but Henning was amusing. She put the trainees at ease, which, clearly, put a majority of the drill sergeants on edge. Dale could feel the resentment they had toward Henning coming off them in waves. Dale was sure they took umbrage at the rapport that seemed to have formed between the company training officer and the new soldiers. This was basic training, not summer camp, the trainees weren't supposed to feel like anyone was on their side.

It didn't matter that the new members of Alpha company really tried their best for her sake. It didn't matter that she motivated them, gave them support and consciously improved their morale. It looked as though as far as a majority of the drill sergeants were concerned, Henning was just another uppity little female who had better watch her step because she was interfering with their carefully laid plans of molding new GIs by degradation and terrorization.

They had been dismissed for noon chow and some green clad figures were now haphazardly on the north patio about fifteen feet from the Orderly Room door, nervously awaiting one o'clock formation. The trainees had been forewarned that the commanding officer would address them that afternoon and everyone was a little anxious. The men wanted him to be their answer to Lieutenant Henning. The women were at the point where they just wanted him to be fair.

At 1300 hours, everyone was called into formation on the south patio where each platoon faced the podium, with second platoon directly in front of it. First platoon stood perpendicular to second platoon's right, facing the podium's left side and third platoon stood opposite first platoon.

There were a little less than fifty soldiers in each platoon but together it added up to almost one hundred fifty wide-eyed cherries, a nickname experienced soldiers called virgin trainees.

This was the kind of crowd most company commanders loved to make speeches to.

"*Company!*" Ritchie called out.

"*Platoon!*" Each drill sergeant screamed in unison, over their shoulders as they stood in front of their respective squads.

"*Ah-ten-hon!*" Ritchie commanded.

Everybody, including all of the drill sergeants, snapped to the position. An extremely handsome man in his late twenties appeared from nowhere, followed closely by Henning and a tall, distinguished-looking, silver-haired gentleman, who wore a set of crisply starched fatigues with a lot of stripes on his upper arms. As the captain stepped onto the platform, the trainees studied him as much as looking ahead would allow.

He had jet black hair, militarily cut but fashionably styled, a full black mustache that just barely passed regulation, a bronzed complexion that made his straight white teeth seem even brighter and his aquamarine eyes a deeper color. His muscular build, teamed with his obvious good looks, apparently captured every female's attention, including the ones who normally would not look at men that way. His youthful quality and the way he appeared to command respect from the older snakes, like Ritchie, seemed to have caught the awareness and envy of most of the males.

The senior drill sergeant saluted Colton, then dropped his hand when the captain returned the gesture. Ritchie did an About Face so that he stared straight ahead between McCoy, Kathan and second platoon.

"*Company!*" the senior drill sergeant yelled.

"*Platoon!*" The three senior platoon sergeants responded.

"*Stand at —!*"

"*Stand at...!*"

"*Hease!*"

"*Alpha-10, first and best of the LE School, sir!*" Voices thundered and echoed around the south patio as the company snapped quickly to the commanded position.

"Thank you, Sergeant Ritchie," Captain Colton said, in a well-modulated voice. He turned to look at the entire company. "Good afternoon."

"*Good afternoon, sir!*" The floor almost vibrated.

PERMISSION TO RECOVER

"Did you hear something, Sergeant Ritchie? I didn't." Colton addressed the crowd.

"The captain can't hear you!"

"Good afternoon, sir!" It really wasn't any louder, but the cadre could never let the first time be acceptable.

"Better. Not great but it will have to do for today. My name is Captain Rory Colton. Welcome to Alpha company, 10th battalion. I am your company commander. I know you've already met Lieutenant Henning, your training officer," he nodded toward her to his left. "And this is your top sergeant, First Sergeant Fleece," he indicated the silver-haired gentleman who stood behind Henning. "I know you've met Sergeant First Class Ritchie, who is your senior drill sergeant, and your individual platoon sergeants.

"First of all, I'd like to welcome you to Fort McCullough, Alabama, home of the United States Army Law Enforcement School and Training Center. For most of you, this will also be your home for approximately the next four months, where you will take your basic and MP training.

"I'd to explain to you what the Army says basic training is. Army regulation – AR-350-1 states the purpose of basic training is to convert non-prior service enlisted personnel into well disciplined, highly motivated and physically conditioned soldiers who are qualified in their basic weaponry and drilled in the fundamentals of soldiery.

"What I say about basic training is this. We intend to work you physically and mentally to the point where you know right from wrong. Right is anything we tell you. Wrong is anything you think outside of that. You cannot become disciplined soldiers if you think like a civilian." He paused to let what was just said sink in and watched the crowd absorb his words.

"Let me tell you a little bit about OSUT. It stands for one station unit training, and it means instead of taking your basic training at one installation or in another company at McCullough and then moving here for your AIT, you'll do it all in one shot right here in A-10. So, get used to those handsome, smiling faces and sunny dispositions standing in front of your platoon. You're going to be with them for a while.

"I'm sure you've already been informed that you are

participating in a first. This is the first time we have attempted training men and women together in a basic training environment and then continued it into AIT. It's only experimental and we're not sure how good a plan it is yet. This company and the following 10th battalion cycles of Bravo, Charlie and Delta will give us a pretty good idea if it'll work out or not. We hope it will. We hope you women will prove to the outside world that you can be as good as the men because we will not take it easy on you. This is what the women's movement is all about and equality is what you are going to get. If you buckle, you are out. It's as simple as that. Same with you men. Don't think you can kick back just because there are women in this unit.

"This is the first and we are going to make it the best. We are going to work you and run you so that, by the time you leave basic training, you are a soldier and by the time you leave AIT, you are an MP. Not everybody here is going to make it and you won't make it because you won't try hard enough and then you'll give up, which is unfortunate. Your failure is our failure, and we don't like to fail but we also don't like people who give up when the going gets a little tough. We don't like losers. That being said, we are going to work our hardest and do everything we can to not only make you succeed but make you the best soldiers possible.

"You have seven weeks of basic combat training and seven weeks of advanced individual training. Now, fourteen weeks may not seem like a long time, but after your first full week of actual training, it's going to feel like an eternity. Yet, when you get to the last week, it'll have felt like it took no time at all. You won't think so now but you'll look back on this moment and realize that time in basic training really flies." His smile was almost convincing.

Of the average four million, two hundred thirty-three thousand, six hundred seconds one spends in basic training, time is the last thing that flies, Dale thought.

"Think we're going to be rough on you? Damned right we are. I don't want any babies in my company. You get out of it what you put into it. You put in one hundred percent, that's what you'll get out of it and that's what we'll expect from you. Yes, it will be a challenge. Your mind and your body will experience things you've never experienced before but you're not alone. The person next to you is going through it with you. And these drill sergeants have all been through it, too. So, let's work together and make this the best cycle Alpha-10 ever put through. Okay, troops?"

"Yes, Sir!" Everyone answered in unison.

"I can't hear them, Sergeant Ritchie, are they talking to me?"

"The captain can't hear you!"

"Yes, Sir!"

"What was that?" Colton asked, again.

"Yes, Sir!"

"Outstanding. Okay, Sergeant Ritchie, you can have them back now."

"Company!" Ritchie bellowed.

"Platoon!" the drill sergeants yelled.

"Ah-ten-hon!"

Colton turned the command to the senior drill sergeant. Henning, Bobby Fleece and Colton left the south patio and headed to the orderly room.

Following Colton's speech, which Dale found capriciously disappointing, the trainees marched to Raburn Hall, a building on the WAC side of McCullough where a lot of the non-physical basic training courses would be taught. It was an old, three-story brick building which housed several classrooms and one small lecture hall. Ritchie delivered the trainees' very first class, which was a spiel on the dos and don'ts of basic training.

The session more or less reiterated what the company commander had just told them. The senior drill sergeant tried extremely hard to be comical and he did make an occasional witty remark but if he hadn't already been so widely despised, his jokes might have been more appreciated, even laughed at. Saying James Ritchie was not a popular man was an understatement and his attempt at humor and congeniality failed miserably. It clearly left him in a very bad mood.

That evening, the women's floor was all excited about the dashing captain. A majority of the females openly romanticized about him, and it made Dale want to stick her finger down her throat.

"Don't you think you're overreacting a little?" Shannon said and chuckled. She shook her head at Dale's grousing. "Just because he looks like a Playgirl centerfold doesn't mean he's a bad CO."

They were sitting on the picnic table while Shannon finished her cigarette.

"Until he proves otherwise, I have no reason to believe he is a good one, either. That speech was an unimaginative, standard crock of shit. The recruiters told everybody the same thing he said before the trainees even got here. Then Henning reiterated it. How many times have you heard that exact same prattle?"

"That's you and me. How many times have we been down this road previously? I am sure that speech was inspiring to those who have never done this before. I mean, did you see them up there? Pretty soon, these women are going to be fighting over him like Cinderella's ugly stepsisters."

"That's not inspiration, that's horniness. Where the hell has he been, Shan? That fire drill last night? He should have been there. If this new, experimental program is supposedly so important to him, why hasn't he been around?"

"Maybe he's had things to do."

"Yeah...ten to one it was a blonde named Comet or something equally as...cosmic."

"You're not picking on us blondes, are you?" Shannon teased. "Look, let's reserve judgment on him for now, okay? We've got enough facing us. I don't want to have to think about dealing with an irresponsible CO on top of everything else. Let's give him a chance first."

"You give him a chance. I've got a bad feeling."

"It's probably just gas. From all the starch they feed us."

Dale sighed and decided that Shannon's deflection of the subject was probably wise. They did have enough to deal with without the added complication of an indifferent commanding officer. Dale followed her partner upstairs and knew she'd have to turn in right at bed check as she had KP first thing in the morning. She decided that maybe that was what her nasty mood was all about.

CHAPTER SIXTEEN

Vanessa McKnight, who was fireguard, woke Dale up, along with Tramonte, Troice, Pam Ryan and Kirk at 0415 hours so that they could be down at the mess hall by five o'clock.

When they got there, they met three sleepy males, Matt McKeighan, Scott McNulty and Joey Overton, who were assigned KP duty with them. They all stood outside the kitchen, yawning and stretching, too tired to really converse, until the NCO in charge unlocked the door and let them inside. He was a hefty, African American man, average looking, in his early thirties, named Melvin Crosby. He was an E-6, staff sergeant, and he made it clear that he was their boss for the day, but he also let them know that he was a lot more relaxed than everyone else had been lately.

Crosby showed them around the mess hall and explained to them that they had to wear their ballcaps any time they were in the kitchen area because heads were to be covered at all times. "Also, you're not allowed to touch the food with the exception of emptying the trays into the trash cans, and dividing the contents into edible and non-edible garbage."

The cooks were all busily at work as they made breakfast and the smell of it woke everybody up.

"One good thing about working KP, you get to eat a hot meal before everyone else," Crosby told them and smiled. That perked everyone up.

"So, you two," he pointed to Tramonte and Ryan. "Get two brooms and start in the kitchen and work your way out here to the dining area. "You two," he pointed to Troice and Overton. "Get a bucket and two mops and follow them." He nodded toward Tramonte and Ryan.

"You two," he indicated Dale and Kirk. "Wipe off all the tables and make sure they're set up with napkins and salt and pepper shakers and you two," he said to McKeighan and McNulty. "Come with me. There's last night's trash to haul."

Dale and Kirk got sponges from the supply closet and started cleaning the tables and booths.

"How did it feel spending another night upstairs, in your bunk, with the rest of us? You're on a streak. What does this make, the second or third night in a row?" Dale asked.

"I'm not knocking it. Ever since I talked to that lieutenant-colonel, things have gotten better. These tables aren't even dirty."

"We'd better keep wiping. He'll just find something else for us to do." Dale studied Kirk. "Hey, did you ever think that things have gotten better because your attitude has changed?"

Kirk gave a non-committal shrug.

"Maybe now that things are going more your way, you've relaxed and it's really not so bad, huh?"

"Hey, whose side are you on? It's still bad. It's just that I know I'm leaving and that's what I was fighting for. I don't need to fight anymore."

Dale nodded and continued wiping.

At 0530, all eight trainees sat down with a tray of hot food and ate. They had to be finished and cleaned up by ten minutes before six, where the guys would then wait in the kitchen to empty trays and the women would wait in the room with the sinks for the dirty trays, pots and pans and utensils. That took them from 0615 until 0800 hours. Then they had to redo everything they had done before chow had been served, only switching details so that everybody got a chance to do everything. After those chores were completed, Crosby brought them through the kitchen to an outside delivery door where they stood on a dock, awaiting a bread truck.

The van rolled in moments later and backed up as close as possible to the platform. Overton and McKeighan jumped off the dock to unload the racks, but Overton's foot turned on landing with a resounding crack and he crumpled to the ground in pain. Crosby jumped down to the injured GI and ordered Tramonte to get help.

Gina Tramonte returned moments later followed in hot

pursuit by MacArthur, who amazed everyone when she hoisted Overton up into her arms and carried him upstairs so that the duty driver could transport him to the hospital.

After the bread was taken care of, the remaining trainees followed Crosby into the kitchen, where they each grabbed a cup of coffee, and proceeded into the dining area. They congregated, then sat at two adjoining tables in the middle of the room.

"What do you think will happen to Overton?" McNulty asked. "Do you think they'll discharge him?"

Crosby shrugged and took a sip of his coffee. "I don't know. Maybe. Depends on the extent of his injury, I guess."

"Maybe I should have thought of jumping off the dock," Kirk mumbled.

"I can't get over MacArthur just lifting Overton up like he was a loaf of bread himself. I didn't know she was that strong," Troice stated. "I'm going to have to be more careful around her."

Crosby snickered. "Honey, you'd better be careful around her anyway. She likes your type."

Troice looked at him, cautiously. "What's my type?"

"Female."

McKeighan turned to McNulty with a triumphant laugh. "Pay up."

McNulty groaned and shook his head. He looked at the NCOIC, pleadingly. "Is that true? I mean, some of the guys —" He looked pointedly at McKeighan. "They thought that she was but..."

"But now you know," Crosby clarified.

"Do you know that for sure?" Dale asked.

"Do you mean has she ever done anything in front of me? No. I just know, that's all." The staff sergeant then looked at the undercover lieutenant intently. "Do I look like the type of man who goes around spreading rumors, Private?"

Dale returned his probing stare. "I don't know, Sergeant. I don't particularly care for Drill Sergeant MacArthur. Personally, I keep looking for her flying monkeys to show up but in all fairness to her, something like that shouldn't be spread around about her if it isn't true. Rumors like that can ruin careers."

While the others held their breath, clearly surprised by Dale's boldness, expecting a severe reprimand because of it,

Crosby said nothing. It was almost as if Dale's candidness impressed him. He sat back and smiled. "It's true. I've met some of her *friends*."

Everyone exhaled and then, seemingly relieved, nodded in comprehension.

"What about Bradbury?" Tramonte inquired, referring to Bravo company's female drill sergeant, the one who nailed them for talking their first day in the mess hall.

"I thought she was obvious," Crosby said. He laughed and shook his head. "She walks a fine line, that one. At least MacArthur messes with consenting adults. Bradbury goes after trainees. She used to be a permanent party in WacVille. Then she got caught last year during bivouac with about four female trainees and it wasn't the first time she'd done that. It's just the first time she got caught. She has a habit of sending females to the arms room and meeting them there, whether they want to be met or not...if you get my drift."

"You mean she forces herself on them?" Troice asked, obviously horrified by the thought.

"Not really," Crosby admitted. "In the end, the females she targets really don't seem to mind. She kind of has this uncanny radar about them."

"How does she get away with it?" Ryan asked. "How come she's still a drill sergeant?"

"Because she never gets turned in. Just turned on," Crosby told them.

"Is it ever a case where the trainee comes on to her?" Dale wondered.

The staff sergeant focused on Dale again. "Looks like you've definitely chosen the right MOS. It's possible. But it's mostly her."

"What about our company? Does she ever stray into our territory?" Troice still looked a little spooked about the thought of possibly being attacked by Drill Sergeant Bradbury in the arms room.

"No. Well, she hasn't yet, anyway."

"What about the other drill sergeants? From our company?" Ryan wondered. "Do they ever do anything like that? You know, go after trainees?"

Dale leaned in closer, curious to hear his answer to this.

"I don't think you're going to have that problem this cycle. Your drill sergeants will be too scared."

"Why is that?" Dale asked, innocently.

"Because it seems your company has a jinx on it. A couple of drill sergeants got caught with their hands in the cookie jar and the cookies came forward and pressed charges."

"Did they really put the make on them or was it just a case of a few women with overactive imaginations?" Dale asked, hoping she didn't sound too much like an investigator. Obviously she didn't, as Crosby didn't seem at all surprised by the question.

"Well, it's like this, since you seem to be so stuck on fairness, Private Oakes. A lot of the stuff is hearsay and, by the time word got around, everybody involved had been cleared out. But I knew a couple of the NCOs who were accused and, as far as I'm concerned, it was cut and dried. The drill sergeants were careless. No question about it."

Dale liked Crosby's openness but for the life of her, she could not understand why he not only told these stories to *trainees* but also talked to them as if they were all drinking buddies, sitting around the NCO club, having a beer. She also found it interesting that he was the only one, so far, who did not think it was all a set up.

Shannon had been assigned Orderly Room detail that afternoon to cover for Tramonte and Dale. She had finished sweeping the deserted first sergeant's office and knocked on the door to the CO's office, ready to clean in there. It was Saturday but she waited for Colton's voice to tell her to come in anyway. When she heard nothing, she figured he was probably out somewhere, possibly trying to charm leaves off trees.

She had been quick to jump on Dale about not giving Colton a chance, which wasn't quite fair, due to the fact that his first impression annoyed her, also. That was an unusual reaction for Shannon who normally liked to wait and see before she formed an opinion whereas Dale often spoke out impulsively before she thought.

The undercover agent quietly found everyone threatening in

one way or another. Experience had made her that way. She, like her partner, had a knowledge of caution that surpassed her chronological age. It was in her nature, like Dale's, to expect the worst and she recognized this fault and tried very hard to control it. It wasn't always easy.

She entered the hallway, swept the doorway and then closed the door behind her. She kept her head down and moved the broom to the far corner.

"Please, come in," a smooth, baritone voice broke the silence.

Shannon nearly jumped out of her skin and whirled to see Colton seated behind his desk, smiling at her.

"Don't do that!" she admonished him, her hand up near her throat. If her heart didn't fall back to its normal resting place within the next few seconds, she was going to have to push it down manually.

His tone of voice immediately became indignant. "I beg your pardon, Private..." he squinted to read her name tag, "...Walker. Oh. Walker. I've been wanting to see you."

"How come I didn't see *you* when I came in?"

"I was bent down, looking in my bottom drawer for a file."

"Why? Are you going to try and escape?" she laughed, amused by her own quirky sense of humor.

Colton looked puzzled. "Huh?"

Even with that moronic expression, he was incredibly handsome. At closer inspection, Shannon could feel the magnetic pull of his charisma and alerted to that, wondering just exactly how he used that gift.

"Forget it," she said and cleared her throat. She looked away, then glanced back up at him to catch him openly scanning various parts of her anatomy.

"So...where's the other one?" he asked her, as his eyes finally met hers.

Shannon immediately dropped her head and focused briefly on her chest. "They're both here. I know they're small but, Jesus..."

He shook his head. "No, no, no, I meant Oakes."

"Oh," she nodded in comprehension. "KP."

"So, what have you and Oakes come up with?" he asked.

There was a smugness to his tone that set Shannon's teeth on edge.

"Well, Sir, I think —"

"Don't think," he snapped. "Answer my question."

Now it was Shannon's turn to get indignant. "Hey, I'm sorry. I'm supposed to be enlisted, not an officer, remember? I just can't talk without thinking anymore," she told him.

"Well, I'll tell you what I think," he began, apparently not even realizing she had insulted him. "I think your being here is a waste of time and a waste of money and a waste of, if you'll pardon the expression, manpower. This is not a conspiracy, Walker. Frat charges come up every cycle in every company. They're just not reported. Lonely women without *it* for six weeks or so...they get hungry. And when the drills don't respond or respond a little *too* enthusiastically, the trainees get embarrassed or scared or whatever it is that propels them and they report the incident. It's called looking out for number one in my book. It's human nature. Lieutenant Henning jumped the gun."

She could barely hold her temper. His arrogance astounded her. "Colonel Bishaye agrees with Lieutenant Henning," she said, through gritted teeth.

Colton smirked, patronizingly, and shrugged it off. "That's the way you women think. Life is like a Harlequin romance mystery. You have this wild imagination and thing for adventure. If you didn't, you wouldn't have joined the Army."

Shannon bristled even more and approached his desk, almost menacingly. So much so that Colton slid his chair back against the wall in anticipation of her crawling across the desk to strangle him. "Is that so? Well, Captain," she spit out each word. "My wild imagination and thing for adventure would like to inform you that Carolyn Stuart was murdered a couple of weeks ago. I bet she just had a ball getting her head shot off for having had such a wild, adventurous time here in your company." Shannon walked to the door and placed her hand on the knob. Before exiting, she paused to look back at his startled expression. "Put that in your Harlequin romance mystery and smoke it!"

The thing that disturbed Rory Colton the most was not that his theory had just been trounced on by a brutal truth, but that Shannon Walker ignored military courtesy and did not address him as sir and did not request permission to be dismissed.

He really didn't like females in the military unless they were in what he felt were subservient positions. In his opinion, females should only be accepted in fields of clerical, medical, food service and supply. All these other opportunities seemed to be leading women to forget their place.

He wrote himself a note to speak to the battalion commander about reining in her agents. He hadn't met Oakes yet, but he automatically assumed she would be as insubordinate and disrespectful as Walker.

He knew he probably wouldn't have to wait long to find out.

Later on, still incensed, Shannon pulled Dale out to the landing and relayed the incident to her in an exasperated whisper. Dale was too exhausted from KP to look up at her partner and say, *I told you so.*

"I couldn't believe him! What a fucking asshole!" Shannon continued, irate. "How does someone like that get to run a company and someone like you or me doesn't even get considered?"

"Because that thing between our ears will never compensate for that thing missing between our legs."

"You need to talk to Bishaye about him."

"I will. You should have heard Crosby today. He was saying shit that never should have been said in front of trainees."

"You said he didn't think it was a set up?"

"Yes but his opinion doesn't hold any water for me. He sounds like a man who just likes to hear himself talk."

"What do you make of what he said about Bradbury?"

"I think he's jealous of her sexual prowess and the fact that she probably gets more women than he does. And I think Bradbury should be watched on a peripheral level but only if she decides to expand her stable into our area."

"And MacArthur?"

"MacArthur's odd but I really don't get the feeling that she's involved in any of this. She's too easily intimidated. Hey,

PERMISSION TO RECOVER

I'm ready to go inside, how about you?"

Shannon stabbed the butt of her cigarette out on the railing and wiped away the black residue with her thumb. "Me, too. Dale...thanks for not saying I told you so about Colton."

Dale sighed. "I was kind of hoping I was wrong about him." She waved it off and dragged herself back into the bay where she fell into bed, almost unconscious at impact.

CHAPTER SEVENTEEN

Sunday was another unproductive day and, more or less, a repeat of Saturday. Heavy PT was conducted in the morning and even though it took place an hour later than usual, it still didn't make it any easier. After the morning exercise and chow, the trainees milled around the barracks and voluntarily tested one another on enlisted grades, commissioned officer grades, military time and the phonetic alphabet. Only a rare few had already memorized them as the rest held tight to their little laminated booklets marked *GTA 21-2-26, April 1973*, which also explained saluting, general orders and special orders.

Later in the afternoon, any female who wanted to go was marched to the PX by Audi. They were given one hour to do whatever they had to do as long as they steered clear of the snack bar. Some women got their hair cut and others replenished their personal items and cigarette supply. Dale tried to phone Anne Bishaye at home and was disappointed to find no answer.

That evening, an informal inspection was performed by the individual senior platoon sergeants. It was quite relaxed, and mistakes were pointed out then constructively corrected. The drill sergeants explained to their troops what the inspecting officer or NCO would look for and what the trainees should expect if they didn't comply.

Following the inspection, there was plenty of free time before lights out and bed check but, with being restricted to the barracks, the north patio or the laundry room, most of the women decided to do their boots, laundry or brass or just hit the showers and go to bed.

Monday morning officially became known as sheets day. The trainees stripped their bunks and stood in formation on the south patio. Single file, by platoon, they carried their dirty linen to Sergeant Carey to exchange for clean sheets.

Down on Delta company's north patio, Dale threw her dirty linen into a pile and moved slowly in line toward two young privates who handed out freshly laundered, folded sheets. She heard her name being called and searched for the source of the voice.

"Oakes? Private Oakes?"

Dale agent saw MacArthur stroll up the ranks of trainees. The sergeant dangled on her fingers the metal tag that the Alpha women had hung on the foot of their bunks so that their personal areas could be identified when the trainee was not present for one reason or another.

"Private Oakes?"

"Here, Drill Sergeant." Dale reluctantly stepped out of line and faced MacArthur. She assumed the position of Parade Rest.

A tiny smirk curled the corner of the drill instructor's mouth when her eyes met Dale's. "Is this your name tag, Private Oakes?"

"Yes, Drill Sergeant." Dale observed the plate carefully. It was definitely her handwriting on the cardboard square in the middle of the tag.

"You left your locker open, Private Oakes," MacArthur told her, almost too pleasantly.

No way, Dale thought. She was too conditioned to securing her personal items. "Are you sure it was my locker, Drill Sergeant?"

"Your name is Oakes, isn't it?" MacArthur yelled, all geniality now gone, furious at being questioned.

"Yes, Drill Sergeant," Dale nodded and looked straight ahead. *The mindfuck continues*. She sighed inwardly. *Neither rain nor snow nor dark of night...*

"Do you realize leaving your locker unlocked is a security violation?" Everyone had stopped what they were doing, and all focus was on the drill sergeant and the trainee. MacArthur so enjoyed being the center of attention as long as she was in control.

"Yes, Drill Sergeant," Dale answered. It was difficult enough being made an example of in front of one's own company, but Bravo-10 had gotten in line also which made it twice as humiliating. The frustrating part was that Dale knew she had not forgotten to secure her locker, and this was obviously MacArthur's way of singling her out for her association with Kirk. It was nice to see MacArthur finally acting like a drill sergeant but not at Dale's expense.

"You realize, of course, that I could give you a counseling statement...or...I could do any number of things to make you understand the significance of not letting this happen again."

"Yes, Drill Sergeant."

"What do you think I should do, Private Oakes?"

"I don't know, Drill Sergeant."

"I'll let you off easy, Oakes. Hopefully, this will be the only time we'll have to go through this. Get down and knock me out twenty-five."

"Yes, Drill Sergeant." Dale proceeded to assume the front leaning rest position and began pushing up, counting off as she completed each repetition. "Drill Sergeant, thank you for conditioning my mind and body. Private Oakes requests permission to recover," she said, when she was done.

"Recover, Oakes," MacArthur allowed. "Get back in line. I just want you to know I locked your locker back up."

Dale stood up and stepped back in line. "Yes, Drill Sergeant. Thank you, Drill Sergeant." She knew everyone was looking at her, but no one spoke. Dale didn't know whether to be proud of MacArthur or hire a hit man. If she had actually done something wrong, she could have accepted being disciplined but she knew beyond any doubt that she had shut and locked her locker before she even stripped her bunk. She tried to forget the incident. She may have been the first one to be put down in front of the entire company but she knew for a fact that she would not be the last.

They returned to the barracks, remade their bunks and then they lined up in formation, ready for PT. The calisthenics were unusually heavy again and, as basic training had officially started at 0520 hours that morning, a majority of the soldiers had made up their minds that from then on, the exercises were going to be more strenuous.

It had been a cold morning, so far the coldest they had spent

PERMISSION TO RECOVER

there and, after PT and chow and before 0700 formation, most of the trainees made sure they dressed warmly. They wore long johns, fatigues, a male utility jacket, wool glove liners, leather glove shells, a steel pot and liner and their LBJs and though it caused limited mobility, it did help ward off some of the chill.

The trainees were marched to Raburn Hall where they sat and listened to Audi tell them about the history of the U.S. Army. They were given a ten-minute cigarette break which they took outside then they reassembled in the auditorium where they watched MacArthur explain and give a demonstration of military uniforms.

They returned to 10th battalion for noon chow and then they marched back to Raburn Hall where Ritchie treated them to a filmstrip and lecture on MP duties. Following that ninety-minute class, the trainees were marched to the PT field where they removed everything but their long johns, fatigue pants and boots, which made it uncomfortably chilly for Dale and Kotski, who still had no long underwear tops. All clothing was folded and placed neatly on the ground in front of them.

Led by the senior drill sergeant and supervised by the company commander, the trainees were put through the most vigorous exercises they had experienced to date. Rolling around in the cold, moist, damp Alabama clay for two hours did nothing for morale and Ritchie appeared to be playing a game to see how long it would take everyone to get soaked and frozen. He also had to let them know how badly everyone was doing and how they were all shaming him.

Ritchie's brand of motivation was, at the very least, provoking. It motivated the trainees to the point of wanting to poke his eyes out. One could not be inspired to drive himself to do better by being called a filthy maggot whose mother was a slut and whose father was a faggot and who would never get anywhere in life because he or she looked like such a fucking scumbag. Ritchie liked to brag that he got to be such a good drill sergeant by being sarcastic, contemptible and insensitive. He did for the motivation of troops what Attila the Hun did for universal charm.

"My God," Wachsman exclaimed as she stood by her bunk

and peeled her fatigue pants away from her long john bottoms. The red clay had soaked through both materials straight to the skin and some of it had semi-dried and was caked on. "We'll be lucky if we don't all catch pneumonia. That man should sue his brains for non-support."

"How come they're called grass drills when there's no grass?" Esperanza Beltran wondered out loud.

That evening *everyone* stood in line for a shower. It wasn't so much that they wanted to clean the dirt off them, they had been chilled to the bone and the hot water felt great. Some of the women even skipped chow so that they could get into the showers first and then go to bed. The PT had completely worn them out.

Later on, the women who didn't pass out, relaxed and discussed the day's events. The biggest topic of conversation, other than MacArthur's singling out Dale that morning, was the female drill sergeant's fashion show.

"Coco Chanel would be so jealous!" Wachsman stated in a mock serious tone, to which Travis grabbed the sheet off her bunk and wrapped it around herself. She strutted down the aisle like a runway model. "And here comes Drill Sergeant Virginia MacArthur now, looking quite sensible in percale..."

Wachsman then began hopping around Travis, pretending to be a photographer snapping pictures. "Ginger! Oh, Ginger, look here! Look this way, Ginger!"

"Very funny, ladies. Very mature." The icy voice belonged to none other than the sensible one herself, who had sneaked in, unannounced. "I will be sure to remember this in tomorrow's details."

"At Ease!" Snow yelled, a little belatedly. Some had wondered if she had held back on purpose, since she was fireguard and had obviously seen MacArthur enter the barracks. Everyone who was conscious, responded. Most of the women stared at the floor, embarrassed, except for Travis and Wachsman, who glared daggers at Snow. Prof could not hide her smirk.

A good drill sergeant would have had everyone out of bed and into the front leaning rest position or something equally as punishable. Dale wondered if MacArthur would act as bravely as she had that morning. One didn't make fun of a drill sergeant when they were anywhere in the immediate area. However, MacArthur appeared to be living in a state of continual frustration and confusion and never seemed quite sure as to how to deal with the trainees on a consistent level.

"Oh, lay off them, for Christ's sake, they were only having a little fun," Kirk spoke up. She had been seated on her bunk, blowing her nose. She was obviously coming down with a bad cold and felt miserable. "Maybe if you went out and got yourself laid, you could relax and loosen up."

"You shut your filthy mouth, Kirk!" was all a stunned MacArthur could manage to say.

Kirk stood up and walked toward her in a casual fashion. "Make me, Bitch."

"At Ease, Private Kirk!" MacArthur commanded.

The women stopped looking at the floor and total focus was on Kirk and MacArthur, until Dale jumped in Kirk's way, facing her.

Drop it," Dale urged. "Don't even think about taking a swing at her."

"Get back to the position, Oakes," MacArthur ordered. "Let her spend the next two weeks in the stockade before she leaves."

Kirk started to walk around Dale. "You know, punching you out just might be worth two weeks in the stockade. At least I wouldn't have to look at your ugly puss every day."

"Your ass is mine, Kirk!" the female drill sergeant spit out.

"In your wildest dreams."

"Kirk!" Dale said, sharply, to get her young friend's attention.

MacArthur walked around Kirk. "Oakes, I suggest you take your buddy into the latrine and calm her down before the whole bay gets disciplined."

As the drill sergeant exited the barracks, Kirk reluctantly allowed Dale to push her into the stall area. "Oooooh, I really want a piece of her!"

"What the hell is the matter with you? Do not fuck with MacArthur, Kirk, I'm telling you. Not only can she postpone you getting out, but she can make your life even more of a living hell than it has already been for the rest of the time you are here."

I can take anything that bitch can dish out."

"You think so? You think she's an amateur at this and you're the first trainee she's ever messed with?" *Even though she might act like it,* Dale thought. "I wouldn't want to test her. Now cool off, okay? Because if you hit her, you'd better have your walking papers in your hand and your running shoes on your feet. Battery is still battery and it is still a punishable crime and you can kiss your discharge and your freedom goodbye."

"Is that true?"

"Yes. And as pissed as MacArthur is right now, she'd have you strung up above an open fire...if she could get somebody else to do it. I understand she gets under your skin, but you should have never confronted her like that with an audience. She's probably on her way downstairs right now to cry to Ritchie."

"Oh, shit..."

"Yeah, oh, shit is right. Look, you're getting out. That's what you want. Don't make it any harder on the rest of us. Just play the game and keep counting the days until the fifteenth."

Kirk exhaled and shook her head then looked back up at the CID agent, almost desperately. "Oakes, what are you doing here? You're too smart for this."

Just then Minty stuck her head around the corner. "She's gone. Thanks a lot, Kirk. We'll be lucky if she doesn't get Ritchie up here and we're up working all night."

"Hey, I wasn't the one making fun of her," Kirk countered.

"No, but you did threaten her." Minty then stepped fully around the corner. "You've got what you want now, okay? Do us a favor and keep your mouth shut so that the rest of us can get what we want. If you don't, you're headed for a blanket party."

"Oh, now who's threatening?"

"I'm not threatening, I'm promising. And I don't think I'll have any problem finding a party crew. So, keep that in mind, the next time you think about opening that big mouth of yours."

Dale sat on the floor in the hallway between the bay and the latrine. She massaged her foot and ankle and rested her head on her knee. She was extremely tired and pulling fireguard from

0100 to 0300 hours did nothing to improve her state of exhaustion. She thought fireguard duty was an important job until it came to her turn and then she thought it was bullshit. *Get with it, Army,* she thought, *install smoke detectors. They do the job just as well and will allow me at least twenty more hours of sleep in basic training.*

Shannon appeared around the corner and watched Dale rub her foot. "The dampness in the weather beginning to bother it?" Shannon's voice, though barely a whisper, shattered the stillness and startled Dale.

"Hey. What are you doing up?" She raised her head to look up at her partner.

Shannon shrugged and smiled. "Nature called. I don't care how unconscious one gets, one cannot sleep on one's stomach with a full bladder. You weren't nodding off, were you? I'd have to report you."

Dale laughed and stood up slowly. "And you would, too." She accompanied Shannon into the latrine. "Was anyone else up?"

Shannon checked all the stalls to make sure they were empty then entered one and locked the door. "Are you crazy? Everyone is comatose after Ritchie's workout."

"I know. My body's running on remote control. Not only that, it's just barely thawing out."

"Jesus, you and Kotski must have been like popsicles without your long john tops. Speaking of that, did you see those Neanderthals in our company during PT yesterday when Stubby walked out in her T-shirt?"

Dale rolled her eyes. "I know it. I haven't seen behavior like that since the eighth grade." *Although she did look quite...nice in that tight little T-shirt.*

There was the sound of a toilet flushing and a door unlocking. Shannon exited the stall and leaned against the sink with Dale. "I mean, what are we? Dog food? Look at some of the women in our company and the boys get orgasmic about her?" Shannon had brought a cigarette with her. She stuck it in her mouth and lit it. Smoking was prohibited anywhere in the barracks, but trainees often took chances and smoked in the latrine anyway. The lieutenant doubted any drill sergeant would be in at that time, so she felt it was safe to fire one up.

"You say that as if you don't like her."

Shannon waved her hand as she inhaled. "Stubby's a good shit. I like her and all, it's just why do men always desire things that are inaccessible?"

Boy, isn't that an understatement? Dale thought. *Especially lately.* "Don't we all, though?"

"Yeah, but we women have a way of getting what we want." She exhaled a stream of smoke. "Maybe our goals are more realistic so that our dreams aren't so unattainable."

Dale smiled. "How profound. And what do you do with all this midmorning logic?"

"I sell it to fortune cookie factories."

"Why, you must be worth a mint. How is that you come to grace us lowlifes with your wealthy and worthy presence?"

"I felt like slumming it for a while."

"Christ, now you sound like Ritchie."

"Hey, don't ever use that sniveling little snotrag's name and mine in the same paragraph, even jokingly. At least I have a conscience. He had his removed...minor surgery and all that."

Dale started to laugh as she remembered a quote from the first time she and Shannon had traveled this military path together. "Whereupon a butterfly kicked him in the head..."

"And he completely lost his mind," they finished together. "Did you talk any sense into Kirk tonight?" Shannon asked. "We're very lucky Ritchie wasn't up here with his whip and chair."

"She's a kid. She's not used to being treated like she's been treated here."

"Doesn't sound like she had it too good before here, either. She's got to remember she's not in the neighborhood anymore. Let's just hope she keeps her mouth shut for the rest of the time she's here. It's not that I don't sympathize with her but enough already."

"As long as she doesn't get provoked again, she should be fine."

"What Army do you belong to? You honestly don't think she'll be left alone, do you? Besides, this time she did the provoking."

"I know. I spoke to her about that. Oh God, I wish I weren't here," Dale sighed.

"And you think I do?" Shannon took a long drag off her cigarette.

Dale looked at her partner seriously. "Yes. I thought you did."

"They said they ran over your foot, not your head. What person in their right mind would enjoy going through this shit more than once?"

"Boner would."

"Operative phrase was right mind. Listen, I don't know if it means anything or not, but I don't think it should be ignored, either. Standing in formation today, I overheard Perry Sargent say that while he was on CQ last night, Beltran, who was his CQ runner, used the autovon line to call Fort Devens." Shannon stated.

"Fort Devens? That's interesting. Isn't Beltran the self-proclaimed *chola* from L.A.?"

"Yeah, that's the one. Walks around here like she's stumbling in her own personal smog bank."

"Why would she be calling Massachusetts, a post three thousand miles from her hometown and the post closest to where Carolyn Stuart was murdered? Why is a trainee using something that she is not supposed to know anything about yet, number one, and number two, she's awfully bold to do something like that so openly when we've been told again and again we haven't earned any privileges yet."

"Using the autovon line is a privilege we won't earn anyway until we get to our first permanent duty station, remember? There could be a very reasonable explanation for it all. I just think it should be questioned."

"If you get the chance, tell Henning and if I get the chance, I will."

"I'd rather not have Private Sargent—" Shannon stopped and thought about that name. "If he makes it and gets promoted to E-5, his troops are going to have a ball with calling him Sergeant Sargent. Anyway, I'd rather not have him made out to be a snitch and the only way Henning would be able to find out is through word of mouth through Sargent."

"But Beltran doesn't know that."

"We hope."

"Well, how Beltran learned to use the autovon line is her secret for now. I'm sure it is in the CQ instruction book somewhere so let's say, for argument's sake, she just got real studious last night. Let's count on her being naïve. Henning could tell her that all basic training unit phone lines have to be cleared by a code and the ones that aren't are automatically logged through battalion as a violation."

"And Stubby could let her off with a warning if Beltran tells her who she made the call to and why and how she learned to use the autovon line."

"It would be too good to be true to crack this wide open this soon." Dale had a hint of anticipation in her voice.

"Yeah, well, I've dealt with the women who have been tied up in this thing. I've talked to them at great length and although Beltran is sneaky, she doesn't even come close to being in their caliber of deception so don't even get your hopes up."

Dale yawned. "That's a shame." She checked her watch. "I have to go wake up Segore." She started to leave and stopped, looking back at Shannon. "Put that cigarette out or I'll have to report you."

The blonde agent took her last puff and flushed the butt down the toilet. "No problem. I was finished, anyway."

CHAPTER EIGHTEEN

Tuesday was to be the longest day of the cycle and, unfortunately, it had nothing to do with the training schedule. Nobody knew it when the wake-up call came so, naturally, no one was prepared. Especially not Dale.

It started out as treacherous as a Monday with heavy PT, morning chow and a cold march to Raburn Hall. Once inside, the trainees listened to a lecture on bugle calls, flags, ranks and the responsibility of a soldier by Robbins. Following that, there was a drug and alcohol speech from Putnam while Henning paced in the back of the auditorium. She and Putnam were the drug and alcohol coordinators of Alpha-10, which was why she was present during that class.

Outside for a ten-minute break, the topic of conversation was the weather. At twenty-three degrees, it was starting to be much colder than anyone had anticipated, and the big complaint was how the damp wind blew right through the uniforms to the skin. The boots were becoming annoying, also. For some, they were either starting to rub against the tendon or bothering the arch, which made it aggravating while marching. Things didn't appear to be going much like anyone thought they might.

Back inside the auditorium, everyone was introduced to Captain Harrison, the battalion chaplain, a smooth talking yet comical fellow who gave an interesting speech on faith. It began with a long story involving the names Abigail, George, Sinbad, Ivan and Slug which made absolutely no sense to anybody but him. Upon completion of his tale, he looked out at a sea of blank and confused expressions and decided to continue on a more traditional note.

The final class before noon chow was a film and lecture on the Uniform Code of Military Justice, or UCMJ. The UCMJ was the military's bible, a constitution of military rules that were not to be broken. If any of the articles were violated, the punishment fell under the jurisdiction of the UCMJ, such as Article 85 for desertion, Article 86 for AWOL, Article 91 for insubordination, Article 92 for failing to obey an order or regulation, etc. The movie and speech only covered minor ground, an introduction to mainly open new eyes, not so much as to why one should follow regulations and be a good little soldier but what happens to one if one doesn't.

Jascelle Kirk had remained in the company area that morning. She had not felt well and had reported to sick call. Troop medical clinic number three was the facility that handled all of 10th battalion's soldiers. It was located on the left side of the gymnasium which was approximately four minutes walking distance directly opposite the Alpha-10 parking lot.

Kirk, who had started battling a deep cough and felt the distant thundering of bronchitis in her chest, and Michaelson, who had an upset stomach which prevented her from eating breakfast and kept her throwing up in the latrine while everyone else was begrudgingly participating in PT, were the only two females to go on sick call that morning.

Some of the other women would have gladly traded places with the beautiful blonde trainee but Michaelson really enjoyed exercising three times a day and having to miss any sessions greatly discouraged her. Anyone who caught a passing glimpse of Michaelson just out of the shower could clearly see that she took very good care of herself. Her muscles were femininely defined but hard as a rock.

Up until PT became a tri-daily ritual, one could always find the subtly captivating blonde working out on her own in the morning and before her shower at night. Had it been anyone else, the other women might have made jokes or poked fun but the fact that Michaelson's face and body just seemed so perfect and she kept so much to herself, hardly any comments were made...at least not of the negative kind. There weren't too many women with Michaelson's discipline and determination and even if there were, there were few who ended up with the same overall results.

To top it all off, she had scored one hundred sixty points on her military entrance examination so there were some definite brains behind that beauty.

At approximately 1030 hours, both Kirk and Michaelson returned to the barracks with their respective signed sick slips. Michaelson's was marked quarters which meant she was restricted to the bay for the day. They had given her something to settle her stomach which tasted like and had the consistency of blackboard chalk, but it seemed to be working. Kirk had been given a cold pack, which was a little brown paper bag that contained a bottle of cherry-flavored cough syrup that only seemed to make her hack more every time she took a swig and cold pills, which did absolutely nothing for her because an hour after taking the two tablets, she felt no difference and was still incredibly congested. Her slip was marked back to duty but when she reported to Sergeant Fuscha, he told her to wait upstairs because the captain was going to call for her momentarily.

Momentarily turned into two hours later, just after the troops marched in from Raburn Hall for noon chow and back out again. She could only imagine what Colton wanted now and impractically hoped that it was news of an earlier release. She figured what he really wanted to speak to her about was that incident with MacArthur, which she realized now had been a mistake. Oakes was right, Kirk was going to have to keep her lips zipped from now on. The battalion commander had told her to behave and, if nothing else, she should do it for Bishaye simply because the lieutenant-colonel was giving her what she wanted.

At twelve-thirty, she entered Colton's office. Once inside, she was given the military command to Report.

"What for?" She wasn't trying to be defiant, she just didn't understand why they kept trying to turn her into a soldier when she was leaving.

Colton slammed his open hand down on his desk and stood up. "Listen to me you ignorant, ungrateful little nobody! You're in deep enough trouble as it is, I'd advise you to do as you are told!"

A little stunned by his vicious verbal attack on her, she swallowed her natural instinct to strike back. "Look, *Sir*, I understand what I did last night to MacArthur was wrong. I understand that when I'm with others I should act like them but I'm getting out real soon, why do I have to practice this stuff when I'm by myself? I'm not going to use it back in Detroit."

"After your behavior last night, you may not be going back to Detroit."

"What? Wait a minute —"

"No, *you* wait a minute, Private Kirk. We're doing you the courtesy of getting you out of here just like you want, and you return the favor by abusing one of my sergeants? And disrespecting her in front of other troops? How are my drill sergeants supposed to get any respect or cooperation from those trainees now? Why should you be allowed to get away with such conduct without being disciplined just because you think you're due to be discharged?"

"I'm sorry. I told you that I know now I was wrong. It won't happen again."

"You're goddamned right it won't happen again! The only way I can punish you that will have any impact on you at all is to postpone your release date. So, as of now, you will be spending another thirty days here. Maybe longer, depending on your attitude."

"On whose authority?" Kirk raised her voice, throwing any previous semblance of respect for him away.

"On *my* authority! I do still command this unit."

She unraveled before his eyes and instead of him clearly recognizing it for what it was and being concerned, his unmistakable arrogance pushed him blindly forward and plainly energized him with power Kirk was positive he didn't have. "You can't do that, you can't take my freedom away from me again," she pleaded.

"Don't you dare tell me what I can and cannot do!" he shouted at her. "Your total disregard for this system is what got you into this mess. You are still in the Army and until you are officially discharged, you will obey the oath you took at AFEES in Detroit!"

Kirk had tears streaming down her face as she angrily confronted him. "Listen to me, you fucking son-of-a-bitch, I will not play games with you! I am going to leave her on the fifteenth of this month like I was promised! You can't go against the battalion commander!"

"The colonel lied to you, all right?" His voice rose above hers to get her attention. "She said it just to pacify you. She doesn't have the authority to release you unless it's okayed by a psychiatrist, which it was not," Colton told her. "The psychiatrist at mental health said there was nothing wrong with you, that you were just passing off responsibility, that you were obviously used to being indulged and that the worst thing we could do is indulge you further by letting you out. He said that if we kept you here long enough that you'd eventually give in and fall in with the other trainees. He said you have the potential of becoming a dedicated, responsible soldier." He smiled at her in pompous triumph, as though his posturing should have automatically changed her mind.

"You're lying! You're fucking lying, man! Did he also tell you that if you motherfuckers didn't let me out of here that I would kill myself? Did he also tell you that?" By this time, she was yelling her intense desperation at him which was, evidently, a total waste of energy.

"And you are bluffing," Colton scoffed. "You are trying to manipulate me, Private Kirk, and nobody does that."

"You're trying to manipulate *me*, Colton! I want to see the battalion commander right now!"

"Permission denied, Private Kirk. You will go back to duty now. That's an order."

"Fuck you, Colton," she hissed. "Fuck you, fuck the battalion commander, fuck the Army, fuck everything!" Suddenly her voice was very calm. "You think I'm bluffing? I'll show you a bluff."

She opened the door and ran out of his office, through the first sergeant's door to the patio and headed upstairs to the bay. She heard him yelling after her.

"Kirk! Get back here! Sergeant Ritchie!"

Michaelson watched Kirk with growing apprehension as the young African American woman threw open her locker door and removed a small, plastic container and ingested the entire contents. Kirk then pulled out a plastic baggie from her tampon box containing approximately twenty-five dark orange and peach colored capsules and swallowed all of them. Michaelson knew that the pills had been purchased from Ingersol, the second-floor landing lizard who had more success pushing drugs than himself on the ladies. God only knew what kind of narcotics they were or what kind of effect they were going to have.

"Hey, Kirk, come on, what are you doing?" Michaelson approached her, fully willing to force her finger down the younger trainee's throat if she had to. After the throwing up Michaelson had been doing all morning, seeing another pile of vomit wouldn't have bothered her all too much, especially if it resulted in saving Kirk's life.

"It doesn't matter anymore, Michaelson. They aren't going to let me go. Colton told me so. He said it all was a lie, that Bishaye lied to me about everything. I can't stay here. I can't do this Army thing...this...it was a mistake. It was all a mistake."

"Kirk, it'll be okay. You have to go to a hospital, though. This isn't going to solve anything."

"It will for me."

"Kirk, it's not worth it, you —"

"I don't care anymore! You understand? I'm fucked either way." She started to cry again.

Ritchie burst through the door and ordered Michaelson to go back to her locker.

"But, Drill Sergeant, she —"

"I said return to your area, Private Michaelson, this does not concern you."

"She took some pills," Michaelson told him, quickly.

"Is that true?" Ritchie asked, turning to Kirk.

The young woman sat on her bunk with her arms folded across her chest, obstinately ignoring him.

"How many pills, Michaelson?" Ritchie inquired.

"A lot, Drill Sergeant, I can't say for sure."

"Snitch," Kirk spat out at Michaelson.

The word stung but Michaelson remained by Ritchie's side, unyielding, as she and Kirk locked stares.

"I've had it with your bullshit, Kirk. Let's go. We've got to get you to a hospital, although I can't, for the life of me, think why anyone would want to save your worthless soul."

The distraught young woman stood up and thrust her middle finger practically up his left nostril. He grabbed her hand with such force, he almost broke it. With lightning speed, Kirk reached around with her other hand and punched Ritchie directly in the groin. Ritchie released Kirk instantly as he fell to his knees in what looked to be agony. Kirk then spit on him as she ran out of the barracks.

"Michaelson!" Ritchie barely got out in between grimaces and groans. "Go downstairs and tell Sergeant Fuscha to call the MPs."

"Yes, Drill Sergeant." Michaelson said.

At five o'clock, while the Alpha-10 trainees were standing in formation at Parade Rest on a patio they had just marched onto, Dale noticed an overabundance of official looking strangers and uniformed MPs running around. That was curious enough but then the senior drill sergeant ascended the podium and made a statement. "Please cooperate with MPI. They will be asking questions about Private Kirk. Company..."

"Platoon!"

"Ah-ten-shun!" The trainees snapped to the commanded position.

"Fall Out!"

For the first time since she had seen him, Ritchie looked rattled, if not downright terrified. When he stepped down from the platform, he hurried back to the orderly room, accompanied by two plain clothed investigators.

Dale raced upstairs, trying to quell an unpleasant gut feeling that was seizing her brain. Her heart was pounding so rigorously, she wouldn't have been surprised if the motion could be seen outside her uniform. She threw open the barracks door and immediately found Michaelson, who she knew had been in the bay all day.

"Michaelson..." Dale called, out of breath. "What the hell is going on around here."

"It's Kirk. She's dead," Michaelson threw at her, without any build-up or warning.

Dale stepped back, her hand automatically going to her

chest. "Wh—? Dead? What are you talking about...she can't be dead."

"She's dead, Oakes." Michaelson faced her, squarely, obviously shaken. "I was told not to talk to anybody about this, but I know you really tried to help her and cared about her, so I'll try to get this out before the others come in. She overdosed. They found her in the weeds across from 12th battalion."

"No, no, no...what happened?" Some of the women were starting to enter the bay, so Dale and Michaelson walked up the aisle toward the windows at the far end of the bay, as Michaelson lowered her voice.

"All I know is that when we got back from the TMC, she was sent for by the captain. She came back upstairs, out of control. She started swallowing all this stuff she had in her locker, saying Colton had told her that the battalion commander lied and they weren't going to let her out. She said nothing mattered anymore. Then Ritchie came to get her. She got away from him by punching him where he thinks and, let me tell you, it couldn't have happened to a nicer guy," she told Dale, sourly. "He definitely escalated the situation. I ran downstairs and told Fuscha and he called the MPs but it was too late. By the time they found her, about thirty minutes ago, she was gone."

"Son-of-a-bitch! Son-of-a-motherfucking-bitch!" Dale shook her head, her eyes flashing with anger. She reeled from Michaelson and headed outside, running directly into her partner who was entering the barracks.

Shannon held onto Dale, pulling her to the side of the landing. She could feel Dale's rage and waited until the area was clear.

"Calm down," was all Shannon could get out.

"Calm down? She's *dead*, Shan!"

"I know. I just heard. I'm sorry."

"Goddamn that fucking Colton," Dale's voice was coolly controlled but there was no mistaking her fury. "He told her they weren't going to let her out. That was the final straw."

"That's hearsay, Dale."

Dale looked directly at her partner for the first time. "It came from Michaelson who was witness to Kirk's last words."

"He was playing commander, which is what he's paid to do.

He must not have felt the situation was that desperate. He tried something, it didn't work—"

"*Didn't work?* She is *dead*, Shannon. I think that constitutes a little deeper issue than just not working. He can't get away with this. I'm going to talk to him right now."

"Dale, stop it!" Shannon looked around to make sure they were still free to talk. Her voice was quiet but urgent. "You cannot say anything without letting on to the cadre or MPI who you are. You *will* blow it. Kirk's death is a terrible tragedy, and it definitely should be investigated but not by us. It has nothing to do with why we are here. I'm not trying to be cold, but I think we should use this situation to our advantage. The women are going to be scared now and so will the cadre for a while. For people like MacArthur, who seem to be on the verge of a nervous breakdown anyway, the pressure might be too much, and she might talk...if she has anything to talk about. And those set up girls, if they are here in this cycle, aren't going to be so willing to play along if these people in authority here can drive somebody to the point of taking his or her own life. We may be able to clear this thing up really fast and get the hell out of here."

Dale windmilled her arm and broke Shannon's grip. She glared at her friend and partner, incredulously. "I don't believe you! Do you realize what has happened here? A human being is dead because of our company commander. A seventeen-year-old girl who had her whole life ahead of her."

"She committed suicide."

"She was driven to it," Dale countered, not at all pleased by Shannon's almost indifferent attitude. "What has happened to you? You can't tell me that none of this bothers you, that because this isn't connected to our case that we should just ignore it and the implications that go along with it...I can't believe you are that insensitive."

"I am not insensitive!" Shannon said, defensively. "I have as much compassion as you do. I keep forgetting that you don't have as much to lose as I do. This would suit you fine, wouldn't it?"

"What does that mean?"

"Blowing our cover wide open. That would be perfect, wouldn't it? Then you could go back to Vermont, do whatever it was you were doing and never have to deal with Uncle Sam again. Unless, of course, they made a mistake on your disability check."

"That's unfair, Shannon, and it isn't true," Dale argued, hurt by Shannon's personal attack.

"The hell it isn't! Well, go ahead, hot shot, go down to Colton's office and explode! Bishaye's down there, too, make sure she gets an earful!" Before Dale could respond, Shannon stormed inside the barracks.

Completely enraged now, Dale ran downstairs and pounded on the Orderly Room door. MacArthur stepped outside, looking dazed. "Go back upstairs, Oakes. Kirk's death does not concern you. If MPI needs to talk to you, we will notify you."

"Her death doesn't concern me? She happened to be my friend and, for your information, Drill Sergeant," Dale was biting off every word. "Her death concerns everybody."

"Go upstairs, Oakes! Now!"

"Drill Sergeant," Dale said, evenly, trying to control herself, "I understand the battalion commander is here. I would like permission to speak with her."

MacArthur looked at her as if she had lost her mind. "Permission denied. The colonel is much too busy a woman to comfort you because you lost your friend."

"Then I want to see Captain Colton."

"Permission denied. Go back upstairs, now, Oakes, or you're in big trouble!"

"*Knock it the fuck right off, MacArthur, I want to see somebody now!*"

"How dare you speak to me like that! Who the fuck do you think you are?"

Before Dale crossed that line and told MacArthur exactly who she was, they were interrupted by Anne Bishaye and Rory Colton stepping outside the CQ Office to see what the disturbance was.

"I want to talk to you!" Dale pointed directly at Bishaye while MacArthur called them both to Attention.

When Dale did not obey, MacArthur panicked at what would appear to the battalion commander as a lack of control of a trainee.. "*I told you Attention, Oakes!*" MacArthur yelled at her.

"At ease, carry on," Bishaye calmly told the female drill sergeant, overriding her. Bishaye then looked at Dale, inquisitively, making a deliberate point to look at her nametag. "Private Oakes? Are you the young lady Private Kirk told me about who befriended her?" she asked, clearly for MacArthur's benefit.

"Yes, Ma'am," Dale responded, fixing her gaze on the usually warm, blue pools of the lieutenant-colonel's eyes, which now looked like ice chips. Bishaye was obviously disturbed but still composed and in control.

"Captain Colton, would you please escort Private Oakes to my office? I will join you in a moment."

"Certainly, Ma'am." Colton put his hand on her arm to usher her away and Dale shook him off, still scowling at Bishaye.

"Private Oakes, please," Colton urged.

She finally went with him, ignoring any attempt he made to speak to her on the way to battalion headquarters. Dale didn't know what Bishaye would say to MacArthur to cover Dale's behavior but Dale knew if anyone could smooth someone's actions, it was Anne. Of course, at the current moment, Dale didn't care if her cover was blown or not.

CHAPTER NINETEEN

Colton and Dale were waiting in Anne's outer office when she got there. Unlocking the door, Bishaye let them inside, shutting the door behind them.

"You look like you could chew nails and spit rust," Anne observed, facing Dale. "This better be good for you to risk blowing your cover."

Dale stared at Anne, agape. Who was this cold woman standing before her and what had they done with her beloved Anne? Had she become so much of a fantasy that Dale had completely deluded herself to the reality of this woman? Suddenly, she was not so attracted to her anymore. "What the fuck is wrong with you? A girl *died* today, Anne!"

Unaware of the relationship between the lieutenant-colonel and the CID agent, Colton flared up. "You address her as Ma'am or—"

"Or what?" Dale whirled toward him, menacingly, with enough attitude to make him take a step back. "Or what? Tell me exactly what you'll do, you lying sack of shit!"

"Dale..." Anne warned.

Ignoring her, Dale continued with Colton. "I've known this lady forever, I will call her whatever I goddamn well please!"

"Dale!" Anne hissed at her to get her attention. "Keep your voice down! I don't need the entire battalion hearing everything that is said in this office. Now I ordered you to keep your distance from this young woman. Why did you stay connected? Was she a lead?"

Looking back at Anne, Dale shook her head, almost laughing out of frustration. "You are unbelievable. No, she was not a lead, she was exactly what she said she was. I'm furious because her death was senseless. She was a nice kid. I liked her. I considered her my friend." She looked at Bishaye, accusingly.

"Were you going to let her out or weren't you?"

"Yes. I told her that. She didn't belong here."

Narrowing her eyes, Dale cut Colton a look that should have made him wet himself. "Then why did you tell her she wasn't going to get out?"

Obviously this was news to Anne. "What?" She took her focus off Dale and put it onto Colton.

"Where did you hear that?" Colton asked, visibly shocked by information that had reached Dale so quickly. Dale figured with all the confusion and chaos, Colton had not had time to conjure up a cover story and the MPI reports were not available yet so Dale's knowledge of this was most likely unexpected. The hesitancy in his voice hinted at his guilt.

"Deborah Michaelson gave a sworn statement saying that Kirk returned to the barracks provoked and unhinged, suicidal, because you told her she wasn't getting out, that Anne, here, had lied to her."

"You told her *what?*" The frosty tone in Anne's voice was positively glacial. "Is that true?" At that point, Colton may have actually wet himself. Dale didn't blame him. She'd been on the receiving end of Bishaye's wrath, and it was not an experience she wanted to relive.

"She's lying! I never said that!"

"You ball-less, gutless coward! Own up to it!" Dale yelled at him. "What reason would Michaelson have to get you into trouble? That doesn't make any sense!"

"No, I never said that to Kirk!" He said, defensively. "And besides, Michaelson isn't supposed to be telling you or anybody else anything until the investigation is over," Colton said, uneasily.

"She didn't," Dale told him. She did not want to get the trainee in trouble for talking when she was told not to. The poor woman was dealing with enough, as it was. "I overheard two MPIs discussing it. Michaelson wasn't speaking to anybody about it, as is procedure. Why did you tell Kirk that, Colton?"

"I didn't!"

"Then what *did* you say to her to incite that kind of reaction?" Anne pressed, clearly knowing, as Dale did, that Colton was the one doing the lying.

"Well...uh...it seemed...uh...I was, you know, trying to use psychology on her...to get her to stay in..." He looked up at Bishaye, knowing he didn't answer her question, almost pleadingly. "She wasn't supposed to react that way!"

Anne bowed her head in disgust. She then refocused on Colton. "Well, that is just great."

"Why, you sorry son-of-a-bitch..." Dale began.

"Dale —" Anne grabbed her arm before she could charge him.

Dale did not shake the battalion commander off. Under different circumstances, the feeling of Anne's hand on her bicep would have been pleasantly welcome. However, now Dale knew that if Anne didn't have a good grip on her, anchoring her in her place, her rage might push her into doing great bodily harm to Colton. Not that he certainly didn't deserve it, but Dale did not want to go to jail for what she considered a justifiable beating.

"Where's your license to practice? Huh? It's not there because you're not trained, you asshole! You didn't earn one! So why the hell are you trying to use psychology on anyone? It's not about that. You just get a kick out of exercising your power by telling anyone who darkens your door, who might not happen to agree with you or want to do things your way, anyone who's troubled to knock it off and double-time it back to duty. And why? Not because you believe it, because you *can*. The power of manipulation is an intoxicating thing, isn't it, Colton? Well, bully for you. I hope you're happy and real proud of yourself, you just manipulated a seventeen-year-old child right into an unnecessary grave!" Dale was seething.

He was unable to hide his outrage. Dale guessed it was because a female was speaking to him so disrespectfully, especially since she was a subordinate. He turned to Bishaye. "Are you going to allow her to talk to me like that?"

"Yes, I am. How dare you cross me like that? How dare you tell a trainee, or anyone, for that matter, that I lied to them? You and I will discuss possible disciplinary action for that later. But, for now...do you realize what you've done? We not only have the suicide of a seventeen-year-old girl on our consciences, we're going to be damned lucky if the whole battalion isn't brought up on charges for gross negligence and full responsibility for her death. Alpha-10 has a swell black eye already because of the fraternization charges, we didn't need this on top of it."

Dale removed Anne's hand and took a step away from her, staring at her, contentious and puzzled. "I cannot believe that you are talking about this so goddamned clinically! She was a person, a living, breathing human being as real as any one of us standing here! She was not a troublemaker or a bad person, she was a *misfit!* Why is it that the Army trains its authoritarians to have such a killer instinct?" She glared at Colton. "Why can't people like you let go without stripping somebody completely of any dignity, self-confidence and self-respect? Why must you carry on a tradition of punishing people for not fitting in? You don't solve the problem, you intensify it. Jesus Christ, send them home! Save us all a lot of money, headaches and time. Beating them until you have bled them dry may be great fun for you, but it is total hell on the morale of the other trainees. Why do you think the UCMJ has AR 635-212? It's for unsuitability. It's there for a reason."

"Look, if we let them off that easy, we'd have everybody wanting to go home after the first week," Colton argued.

"Bullshit. You can tell the ones who are really out of place from the ones who are just a little shell shocked. I can. Kirk needed to be left alone until she was let go, you should have recognized that, then none of this would have happened." Dale said.

"Kirk needed to be disciplined!" Colton protested. "I couldn't let her get away with her antics with MacArthur. What kind of message would that have sent, not only to the other trainees but to the drill sergeants, as well?"

"First, if MacArthur was any kind of a drill sergeant, the *very least* she should have done is had every single one of us down in the front leaning rest position and she should have taken control of the situation right then and there. She should have gotten everyone out of bed, made us do push-ups until our arms shook and then kept us up half the night with an impromptu GI party. *That* would have made an impact. *That* would have sent a message. She would not have had to single out Kirk to discipline her, the other women would have taken care of that. So, when you start talking messages, maybe you need to look in your own back yard."

Shrugging, Colton cleared his throat. "Well, you do have a point there."

"I'm not done," Dale spit out. "Second, yes, I agree that Kirk's defiance and insolence with MacArthur should have been addressed but you could have picked a punishment more suited to the crime. You knew she was acting that way because she wanted to get out of here as soon as possible, did you honestly believe you would cure her hostility by making her stay longer? By constantly making an example out of her? You drive people to commit a crime, bait them into it, like AWOL, assault, murder at times...suicide, in Kirk's case...and then you punish them for it. It's like a fucking game to you, man, and it's obviously gotten so that it doesn't matter how high the stakes are just as long as you win."

"Okay, Dale," Anne put her hand up to halt the conversation. "That's enough. It's done."

"It's not done —" Dale stated, ready to continue to her verbal onslaught of Colton.

"It's done." Anne's elevated voice silenced Dale and then she softened her tone. "There is nothing anyone can do to bring Kirk back. We obviously made a monumental mistake. I am sorry. I will do my best to make sure this never happens again in my battalion as long as I'm here. But I don't think taking this personally and avenging the death of someone you barely knew is wise, Dale. I believe you are too upset to think clearly. Just bear in mind that if you blow your cover, it doesn't just involve you and you could make it miserable for a lot of people, including me. I suggest you go back to the bay, think it through and get a good night's sleep. Tomorrow, if you decide that you have really had enough, come to me. Don't just announce it to whoever will listen. I expect you to do me that courtesy. I know for a fact that you are much more professional than you are acting right now."

Still fuming, Dale sighed. "I don't have to think about it. I'll stay. I owe you that. But I want something done about him," she nodded to Colton. "Can't you bring him up on individual charges or something?"

"Charges?" The captain looked startled.

"I'm going to hold off on doing anything for the moment. At least until the investigation is finished and I have more information."

Dale and Colton exchanged looks of pure venom. "You're the boss. But if he gets away with this—" Dale began.

"Dale, Jesus Christ!" Anne ostensibly reached her breaking point. "The fault is not all his and I am not going to make him the scapegoat. Yes, he was wrong in what he did, but we all have a hand in this from her parents to me. Now you settle down and think about this less irrationally than you have been doing. You have points well taken about the system, but you are not going to change it. I do realize just because things are the way they are doesn't make them right, but now is not the time for a crusade." She took a breath as Dale glanced downward. "You have exactly eleven days until Christmas exodus and then you can have some time to yourself again. Just hang in there with me until then."

Looking back up at Bishaye and then at Colton, she bit her lip trying to keep herself in check. Returning her attention to the striking battalion commander, knowing her feelings for the older officer had changed, although she wasn't quite sure how yet, Dale replied. "Like I said, you're the boss." Turning to leave, she stepped around Colton and opened the door. "Just keep him the fuck out of my way."

The door slammed and the room went silent for a moment. Colton finally looked at Anne. "She and her partner aren't real big on waiting to be dismissed, are they?" The expression on his superior officer's face told him that Dale's insubordination was the least of his worries. At any other time, he enjoyed being in her presence because she was gorgeous and what red-blooded male did not like to be in the company of a beautiful and, yes, sexy female? And, despite the fact that she was a woman, she was an admirable and inspiring officer, one that he had often employed the fantasy of being intimate with but right now he would have settled for being in her good graces, which he, obviously. was not. "Is, uh, she always like that?" he asked, cautiously.

"Like what?" The battalion commander crossed her arms.

"Angry."

"I think she has a right."

"Is she a dyke?"

Anne's eyes flew open at the brazen question. "Excuse me?"

Oblivious to her indignant astonishment, he repeated it. "Oakes. Is she a dyke?"

"What does that have to do with anything?"

"Well, she comes on really strong and tough and Kirk admitted that she was a dyke and she and Oakes were obviously close and —"

"And that automatically means they were the apple of each other's eye? You amaze me. I think the last thing you should be thinking about here is Dale Oakes' sexual orientation. Dale cares about people. She may not always express her compassion well, but she can be very emotional. This girl's death was very avoidable, and it hit way too close to home. When you look death in the eye, like Dale has, and just escape it by a hair, I think you might be a little overboard in this situation, too. She's acting so tough and strong because she's scared and that's how she filters her fear."

"She sure doesn't act scared to me."

"You don't know her."

"Not sure I want to either."

"Now...whether she is or isn't a lesbian is none of your business."

"But it *is* the Army's."

"Where are you going with this, Rory?" she inquired. "Did nothing she say to you in here sink in? You don't like her because she stands up to you, challenges you. She isn't impressed by your good looks so that obviously makes her gay? So, if that was the truth, instead of accepting her and leaving her alone to do what she needs to do to finish out her career, you want to make as much trouble for her as possible, maybe even be responsible for getting her discharged just because she goes against your grain. You should be ashamed."

But he wasn't. "She wants to have me brought up on charges!"

"Because you actually did something wrong!" she told him,

exasperated. "Listen to me...Dale Oakes is a good officer and a better cop and if you want this case solved then you should be damned grateful that she is here. She is very good at what she does. You, on the other hand, are skating on very thin ice. Clean up your act, Rory, it may be your last chance. Now, have a seat and make yourself comfortable. We're going to discuss you calling me a liar."

All activities were cancelled the next day and anyone who wanted to visit the chapel was allowed to at ten. Kirk had not been very popular during her short stay, but no one had wished her the unfortunate and untimely ending she came to. And, for some, it was proper alleviation of their own guilt to pray for Kirk's worthless soul, as Ritchie had called it, to reach the arms of God and finally find the freedom she had been so desperately seeking.

No one in the barracks, with the exception of Michaelson and Dale, knew any of the details surrounding Kirk's death and the cadre reacted as though struck mute when Kirk's name was mentioned.

Dale had decided not to attend church. She felt it was just ironically sanctimonious that the Army had driven Kirk to her death and then graciously held a memorial service for her. She wanted no part of it.

Yesterday's meeting with Anne and Colton lingered in Dale's head. What the hell had happened to Anne Bishaye? Of course she held a higher rank and a lot more responsibility now but...was the pedestal Dale had placed her on now crumbling? She used to be relaxed and amusing and enjoyably unpredictable and, although her presence had always been commanding, she never let her power go to her head. Now she seemed strained, which was only natural under the circumstances, but she had appeared uneasy and anxious even in Vermont. She had arrived as Anne and left as Colonel Bishaye, apparently unbothered by the fact that she had used her position and influence to get what she wanted. And now an innocent girl had paid for the mistake of being imperfect with her own life and Anne seemed only concerned with the political repercussions not about Kirk and what she must have gone through and that, to Dale, were the actions of a stranger.

Dale stood on the north patio, leaning on the railing, looking at Bravo-10's barracks when Shannon leaned in next to her.

"Hi."

"Hi." Dale answered, without looking at her.

"Heard some big brass was in Bishaye's office this morning."

"I'm not surprised. The post commander and all his mafia have to figure out a way to cover all this up. Nobody needs bad press. Especially not Uncle Sam. He got enough of that with Viet Nam." She looked at her partner, who lit a cigarette. "I see you're still talking to me."

"I see we're still on the case."

"Shan...I'm sorry about yesterday. I just...I just don't understand this..." she let her voice trail off as she pounded the railing with her fist.

Leaning in closer, Shannon touched her shoulder to Dale's. "Yeah and I said some things I shouldn't have, too. You're a lot more emotional than you used to be."

"No. I'm as hard as a rock when I have to be. When it's necessary."

"I really am sorry about Kirk. I never meant to sound as if I wasn't. How did your talk with Bishaye go?"

Dale shrugged. "Let's just say it went."

"In other words, she was about as sympathetic as I was."

"In other words, she was very military."

"I was in the CQ office this morning, sweeping up, and I heard that Bishaye met with a representative from the Army legislation liaison. They're going to issue a statement saying proper care and caution were exercised in regard to Private Kirk's treatment."

"Oh, gee, how noble of them and I am sure if further statements need to be issued, they'll parade out the long-standing personality problem bullshit. I just can't swallow the injustice of this."

"I guess, so far, either the press has been kept at bay or they really don't care. So...what really happened, Dale?"

"Colton played a vicious head game with Kirk and it backfired. Ritchie got into the act but by then she had swallowed a slew of pills which Michaelson thinks she bought off Ingersol. Anyway, she got away from Ritchie by punching him in the balls. By the time he got up off his knees, she had booked."

Shannon shook her head. "God. So, Michaelson thinks she bought them off Ingersol. Why is that?"

"She told me this morning that Ingersol tried to entice her up to his room by offering her a baggie of pills just like the ones Kirk took."

"Interesting. Do you know if she told anyone else about that?"

"I don't think so." They exchanged glances.

"Don't you find that odd that she wouldn't tell anyone except you?"

"Not really. I was the only one she told what really happened, so no, I don't find it unusual."

"But why not MPI?"

"As we have already observed, Michaelson keeps very much to herself. And I think being thrust into the spotlight like this made her really uncomfortable. She seems the type that wouldn't want to point a finger at Ingersol if she wasn't positive. Besides, who wants the tag of snitch their first few weeks of basic training? She may be doing us a favor by not mentioning it. That way, we can keep an eye on him, see who he hangs out with, who buys from him...if nothing else, maybe we can weed out the possible druggies in the company."

"What good will that do?"

"Well, you know, it's just my opinion, Shan, but I believe that if one is serious about law enforcement, one does not break the law. I find it rather hypocritical that someone would get stoned or purchase illegal substances either before or after a shift and then go out and bust somebody for doing exactly what they do. And, to me, anyone who has a drug habit is not serious about enforcing the law. So, anyone who buys one of Ingersol's little baggies should be watched closely because I, personally, wouldn't want somebody like that as my partner and I don't feel like standing around, watching that person being trained and turned out and permanently assigned and end up being partnered with someone like me."

"You want to hop down off that soap box for a second? I

agree with you to a point but let's face it, Dale, the pressures of police work are unlike that of just about any other job. Just because I may go home after a hard shift's work and occasionally fire up a joint or two doesn't make me any less of a cop that you are."

"The point is, it's still breaking the law."

"So is drinking and driving. Tell me you're not guilty of that."

"All right," Dale conceded, holding up her hand. "So we all have our guilty little vices. You know what I'm saying, and I wasn't leveling any accusations at you. Maybe we have fallen into some bad habits, but we never did it in training."

"And never on or before duty and to set the record straight, I have never busted anyone for pot except one sleazeball on a grade school playground on post. Now, let's get off this discussion. One fight a week with you is about all I can handle. One a day with you would be too much for even Henry Kissinger."

Dale smiled. "Okay. Back to Michaelson."

"Do you think she's protecting Ingersol?"

"If she was, she wouldn't have mentioned him to me at all," Dale said, shaking her head. "I think she wanted at least one other person to know what she knew in case it does turn into something. And I think she's scared."

"Of what?" Shannon asked.

"You join the Army to train for law enforcement, you're not even in basic training one month and someone practically kills herself right in front of you? Wouldn't that spook you a little? This kind of stuff doesn't ordinarily happen."

That's true but what's going to happen when the real investigation starts? About how Kirk got the downers to kill herself?"

"For all they know, she could have come in with them. Nobody checks that closely. She could have been hiding them in a tampon box since she got here, you know how the male drill sergeants avoid those as if they were electrified. She could have packed them, brought them in on her, who knows? Unless they already suspect Ingersol or someone else of pushing pills around here, I imagine they are going to assume she brought them from home. Regardless, you can bet your ass we're going to have a massive shakedown within the next twenty-four hours and they're going to check for anyone else's drug supply," Dale said.

"I bet Ingersol is sweating bullets as visions of criminally negligent homicide charges dance in his head." Shannon took the last puff on her cigarette, field stripped it and stuck it in her pocket.

"Well, maybe not that particular charge, but..."

"Probably no charges at all knowing this place, but it gives me great pleasure to think the son-of-a-bitch is shaking in his boots. I'm going back upstairs. You coming?"

"Naw. Not yet. Being outside feels good." Dale took a deep breath of the crisp late autumn air and heard the bay door close as Shannon walked to the stairs.

Dale sensed someone walking up behind her, but she thought it might have been Shannon coming back to tell her something she might have forgotten.

"Oakes," the voice said. Dale recognized it as belonging to MacArthur.

"Yes, Drill Sergeant," Dale responded, unenthusiastically moving to Parade Rest.

"What's your second special order?"

"I will obey my special orders and perform all my duties in a military manner."

"What's your fourth special order?"

"There is no fourth special order, Drill Sergeant."

"Very good. You didn't even have to stop and think about it." MacArthur was obviously impressed. "At Ease, Oakes, Carry On."

Dale returned to leaning on the railing and MacArthur leaned with her, studying her. *Great*, Dale thought, *a tête à tête with the third of her four least favorite people in the world right now.* What could MacArthur possibly want to talk to her about?

"Why aren't you at the service? I would think you, of all people, would be there."

"I'm not big on church, Drill Sergeant." Dale said, half smiling.

"That's too bad, Oakes, you'd be surprised what a good

religious reprimand could do for you."

Shaking her head, Dale thought, *this is not an area you want to go with me.* "I'm beyond God's help, Drill Sergeant. That's why I joined the Army."

They locked stares and MacArthur broke out into a grin, which made Dale blink in astonishment. A smile changed this woman's entire appearance.

"I'm sorry about yesterday in front of the battalion commander, Drill Sergeant," Dale told her reluctantly. She hated apologizing to a person who should have been openly sharing the responsibility of the tragic circumstances that brought them together at that very moment but if she was going to continue this charade, she knew she was going to have to make nice with MacArthur. She exhaled a deep breath, trying to sound her most confused and puzzled for the drill sergeant's benefit. "I thought everything was starting to go so smoothly for Kirk and then bang, it's over. Like she never existed. I, uh, know things were wild for you yesterday, too, but I needed answers...maybe not really answers, we'll probably never get any real answers. I guess I needed some direction. And I didn't think you could give that to me, and I thought the battalion commander could. So...I just wanted to say I am sorry if I made you look bad, Drill Sergeant."

I'm sorry you made yourself look bad, Drill Sergeant.

"I'm sorry too, Oakes," MacArthur admitted, sincerely, her voice and attitude softening. "I'm not used to this kind of stuff myself."

"Has something like this ever happened before, Drill Sergeant?" Dale's curiosity sounded genuine.

"Suicide? Not since I've been around here but there's been a lot going on in this company that doesn't add up." She seemed to be talking to herself more than Dale.

"Like what, Drill Sergeant?"

"Nothing," MacArthur said. "Nothing that I have a right to discuss with you, anyway."

"We, upstairs, have heard stories...rumors..." Dale prodded.

"That's all they are, Oakes. Every cycle starts on things they heard about the cycle before them. It's all garbage."

"Well, whatever is going on around here, it looks like it's

beginning to take its toll on you, Drill Sergeant. Maybe you should take a vacation." *Please oh please oh please...*

"I plan to. I have sixty days coming to me and I really need this leave. Especially after this."

Blinking, Dale took a moment to vocalize. Wishes like this weren't usually granted, especially this quickly. "Are you going to be going before our cycle ends?"

"I'm leaving in January."

"Are you coming back at all after Christmas exodus?"

MacArthur looked at her, smiling. "Why? Don't tell me you're going to miss me, Oakes."

Returning her grin, Dale said, "Drill Sergeant, honestly, if I ever saw any of your faces again, it would be too soon for me. But I am starting to get used to you."

This made MacArthur laugh. "Your honesty kills me, Private." She cleared her throat and straightened up, becoming official again. "Well, I suppose I have said enough. Again, I am sorry about Kirk. I honestly didn't think she'd end up this way." MacArthur sounded as though she truly meant it. "I hoped the colonel helped clear things up for you."

"She was really nice about it, Drill Sergeant." Dale's senses were suddenly overcome by the memory of Anne's intensity, and it sent a shiver of pleasure down her spine. Well, so much for her being pissed.

"She's a terrific leader. We're lucky to have her here," MacArthur stated, her voice thick with hero worship. "Well, you'd better get back up to the bay and study your three special orders."

"You sure you don't want me to study the fourth one, Drill Sergeant?" Dale asked, teasingly. "Are we having a test on them?"

MacArthur shrugged and donned a guilty expression. "You never know. I wouldn't take any chances if I were you."

Dale didn't really want to leave the patio yet, but MacArthur's request was much better than an order. Double-timing it across the concrete floor, Dale took two steps at a time to get upstairs to the bay.

CHAPTER TWENTY

"MacArthur's leaving," Dale told Shannon, who was polishing her brass. Dale sat on the floor, leaning her back against Wachsman's bunk. Wachsman was on the other side of the bay, talking about Kirk's death with some of the other women, which left no one around the two lieutenants. Most of the others were doing their best to avoid Dale. They had heard about her scene with the battalion commander, and they no doubt felt awkward about approaching her until they felt she'd had time to cool down.

"Why? Did this Kirk thing get to her?"

"I think it contributed. But she has leave time coming in January. She's taking two months."

"Not soon enough or long enough for me," Shannon commented, tipping the bottle of Brasso onto a cloth diaper and polishing her brass insignia with it. "How'd you find out?"

"She told me. She came out and talked to me after you left."

"Lucky you. How did she treat you?"

"Fine. You know, I think she'd be okay if she just got that wild hair out of her ass. You were right about her being wound up. I tried to get her to talk about it, but she said it was nothing she could discuss with me."

"Well, we know what that means. I wonder who will replace her."

"With our luck, it will be someone who knows us both."

"Jesus, that's all Bishaye needs." Shannon inspected her nice bright, shiny brass and then set it down. "What do you think is really going to happen with this Kirk case?"

Pinching the bridge of her nose to relieve the tension built up in her head, Dale replied. "Probably nothing. She signed all her insurance papers but because of the suicide clause, somebody is out twenty thousand bucks...hopefully her folks.

Anyway, I doubt her parents will press charges. They didn't want her around in the first place so, in a roundabout way, they got their wish."

"Yeah, but that's just the type who would make a big deal out of it."

"Doesn't matter. Uncle Sam will cover it up. If they can cover up the bullshit that still goes on up around the DMZ in Korea, as you well know, they can handle a simple suicide. They'll foot the bill for the funeral expense and burial and it'll be over. No repercussions. Nice and clean. And by the time the cycle is over, Kirk will have become a faded, misty, bad memory. If anyone remembers her at all."

Shannon nodded at Dale's changed attitude of dubious acceptance. "Life goes on," she concluded for her.

"And on and on, regardless. Jesus, I'd love a drink right now," Dale said.

"Mmm. Hey, I heard Boner is leaving tomorrow."

"Now, there's some more good news. For a while I thought she might be going through it all again because she doesn't trust us."

Shannon picked up her Athena insignia with the cloth so as not to get fingerprints on it. She recapped her bottle of brass cleaner and stood. "Wouldn't surprise me. However, I think things are getting just a little too hot for her to handle. Has she given you anything, by the way?"

"Other than a perpetual migraine? Nothing. You?"

"Hell, no. Maybe she's talked to Stubby."

"No, I think she's avoiding Stub...Henning, too. Henning spoke to her about the cycle spy rumor she tried to start. Boner said she told Lanigan that shit to throw any attention off herself."

"Bullshit. It was to draw more attention to her. I'm glad she's going. Let's just hope she doesn't make one grand gesture before her dramatic exit."

Dale stood up and yawned. "I'm going to bed early tonight. I didn't sleep too well last night."

"I'm sorry this Kirk thing has hit you so hard."

"It's not just that...when I did finally get to sleep last night at God knows what hour, I dreamed I was marching all night long. I'm exhausted."

"And so it begins," Shannon thought out loud, grinning. "She starts dreaming like a full-fledged trainee."

0200 hours. The barracks door flew open, slamming up against the wall. The lights went on and there were various shouts of *man on the floor!* and *At Ease!*

Everyone was rousted out of bed one way or another and the women stood at Attention by their bunks until their platoon sergeants got to them. At that time they were commanded to open their lockers and watch while all of their belongings were meticulously searched and then thrown on the floor. Nothing in the locker was sacred or left untouched and some of the women felt naked, standing there, in their sleeping attire which, for a few, consisted of nothing more than a T-shirt and panties.

Captain Colton oversaw the whole project and shadowed the drill sergeants as they conducted their shakedown. When McCoy reached Dale's locker, Colton stood directly in front of her, glaring into her eyes. She tried to look straight ahead, as was required for the position of Attention but her eyes eventually strayed to meet his.

"Don't look at me, Private," he snarled. "You're breaking position. If you're staring at anything other than my upper lip, you're wrong."

Dale said nothing but made sure he knew before she resumed looking straight ahead that the feeling of contempt that passed between them was undeniable.

The shakedown easily took an hour and putting their belongings back into their lockers neatly and according to regulation took almost as along as the drill sergeants' surprise visit. Most of the women, jolted awake, forgot about going back to bed and just got themselves ready for morning formation which was to take place in forty minutes.

The outcome of the impromptu inspection resulted in three of the trainees getting counseling statements for not wearing their dog tags and also in the first Article 15 of the cycle. It was awarded to one Emily Zelman, who had minor quantities of drugs shoved in every nook and cranny of her locker. By the time 0520 formation came around, Dizzy had not only lost her pills, she'd lost two weeks' pay and gained two weeks of extra duty. As with everything, Dizzy took the news with a shrug. She clearly didn't care. Dale knew Dizzy could get more whenever she wanted. Dizzy had stated many times that as long as she had a roof over her head, three meals a day, a warm bed to sleep in, clothes to wear and all the sex she wanted, she was happy.

Dizzy had told Wachsman, who told Shannon, who told Dale that at night, after bed check, when everyone else was unconscious, Dizzy would arrange her duffel bag in her bunk to look like a sleeping body. Then she would sneak out the rear fire exit door, stick a penny down by the base to hold it open just enough not to lock so that she could get back in and then run one flight up where the men in first platoon slept. They kept a penny in their fire exit door, too.

Dale thought they probably should have reported it, but they didn't want to be labeled snitches, either. At present, her actions didn't appear to be interfering with their mission, so they agreed to leave it up to the drill sergeants to discover and deal with.

Boehner left at 0600 hours, bag and baggage, with an uncharacteristic whimper instead of a bang. She told everyone she was being stationed in Germany, that her orders came in yesterday and Dale secretly wished her tower rat duty but shook her hand along with the rest of the women, thanking her for her help and advice. Even when she told Dale good luck, Dale refused to feel guilty for hoping Boehner might eventually draw a tour of seclusion on a missile base in Iceland and she waved with the rest as the bus Boehner was on drove out of their field of vision. Both undercover lieutenants breathed a collective sigh of relief and returned upstairs to attend to their details.

Kerrie Hewett, an innocent young Mormon from Utah, used a buffer for the very first time in her life after all the other details were done and pissed everyone off by not being able to control the contraption. She switched it on, and it careened away from her like a horse stung by a bee, knocking all of the beds out of alignment on the right side of the bay. Several women spent an extra fifteen minutes showing the mortified woman how to handle that nasty old machine with a mind of its own and then putting their bunks back in line. If Hewett hadn't heard obscenities before that point in her life, she certainly was familiar with them before 0700 formation.

The temperature hit zero degrees that morning with a wind chill factor of eight below, so the trainees were transported to their classroom. They stood down by the bus stop at the end of North building in formation for at least thirty minutes waiting for the repainted school bus to arrive. In that allotted time, they could have marched or double-timed to Raburn Hall and stayed warm through the exercise of it.

The trainees were scheduled for two tests that morning on the UCMJ and on court martials. Before anyone even took their seats, Henning singled out Shannon and spoke to her so everyone could hear.

"Private Walker, I would like to see you outside to discuss your barrette. It's not very military," the Alpha-10 training officer said, crisply, stepping off the low platform in front of the classroom and walking toward the door.

"Yes, Ma'am," Shannon responded, following her outside.

After making sure the coast was clear, Henning spoke. "Where did you find that thing you are wearing on your head? From Peter Max? That's disgustingly civilian."

Shannon grinned. "It got your attention, didn't it?"

"It got everyone's attention. That thing could land aircraft. What's up?"

"Well, first, on behalf of all us lowly trainees, I'd like to thank you for that shakedown last night. We didn't need sleep anyway. Dale and I expected it but that doesn't mean it was appreciated. I mean, you could have pulled it earlier so that we might have had the option to go back to sleep. Now anything anyone tries to learn today will be a waste."

"Welcome to the Army, Private Walker." Henning smirked with an almost coltish tone to her voice then she got serious. "We just wanted to make sure no one else had any drugs like Kirk did. One overdose death is enough."

"I bet Zelman's supply surprised you...although I don't think she has suicide on her mind. In fact, we're not even sure she has a mind yet. I'll be curious to see if she behaves any differently after her Article 15."

PERMISSION TO RECOVER

"How's Dale?"

"Better. Still pissed off. How close did she come to blowing it?"

"Not at all. General consensus was that she reacted in a normal, angry, confused manner. No one would have thought a cycle spy would have put herself out in the open like that. And, she did *not* make a friend out of Captain Colton."

"Well, *that* should cheer her up. She doesn't hold much regard for our illustrious CO. Can't say he's exactly my cup of tea, either. Did you find anything out about Beltran?"

"Yes. She's a sneaky one. At first she denied the whole thing, told me that Sargent doesn't like her and is trying to get her into trouble but then I used that call logging crap Dale came up with and Beltran freaked out. I guess she's not used to getting caught. Anyway, she called her boyfriend, who is stationed at Devens. He taught her about using the autovon line. She gave me the number and I confirmed it. It's legit. She's on warning and knows if she tries it again, she will be severely disciplined."

"I hope she believes you."

"If she doesn't, she's going to exhaust herself by polishing one of those three-foot tall silver bullets in the stones near the stairwell between the patios."

"And Boner and the cycle spy rumor?"

"Colonel Bishaye took care of that."

"Oh...Dale told me that you had."

"I spoke to her about it, but Colonel Bishaye took care of it...if you get my drift."

"Ouch."

"My sentiments exactly. Boner – I mean, Boehner did tell me to keep an eye on Minty, Sherlock and McTague, though. She said they seemed more interested in becoming wives than soldiers."

"Sherlock and McTague, maybe but Minty knows where to draw the line. I think because she's an Army brat, she's testing how far she can step on that line before she crosses it and gets reprimanded. And I've watched them myself. I don't think their behavior warrants any special concern. What's going to happen with the Kirk thing?"

"Basically nothing."

"Yeah. That's what we thought. I hope this incident shook enough people up so that it doesn't have to happen again."

"Hard to say. I mean, how much more has to happen to this company, period? How's Michaelson doing? She refused counseling."

"Michaelson seems to be very strong, emotionally. I think she's fine, under the circumstances. She keeps very much to herself, anyway. Dale and I will keep an eye on her, though."

"Good. Anything else?"

"Not at this time."

"All right. You should probably get back inside. And don't wear that barrette again or I'll have you polishing that other silver bullet with Beltran."

"No task too tough, Ma'am," Shannon said, flashing Lieutenant Henning a grin.

"Oh, God I *hate* that Ritchie," Travis stated, walking in the bay door and leaning against her locker. "If I hear one more time, *Far be it from me to start any shit...* , I'm going to jam my disgustingly dirty combat boot right up his nasty, funky ass." She had been singled out by the senior drill sergeant for having soiled footgear after her platoon had just finished doing PT in the dusty field. Everyone else had blemished combat boots, too, but for some reason, known only to Ritchie, he picked on Travis.

"I know it. He's just so arrogant and obnoxious," Tierni grumbled, removing her LBJs and field jacket.

"I think he's just the opposite," Travis remarked, slamming open her locker door. Tierni looked at her. "I think he's obnoxious and arrogant," Travis clarified.

"The fucker provokes me just by saying my name," McTague said, still stung by receiving a counseling statement earlier that morning.

"I can't believe they actually gave us tests today. I had to concentrate just to keep my eyes open. Who could think about answering questions? I'm surprised any of us passed," Melanie Mackey yawned as she put her fatigue jacket away.

"I think I got the highest mark..." Tramonte spoke up. "Of anyone who failed."

PERMISSION TO RECOVER

On the other side of the bay, the conversation was dominated by the subject of Drill and Ceremony. That afternoon, the trainees had been taught and practiced Left Flank, Right Flank and Rear March.

When the trainees first started learning Drill and Ceremony, it was much like being children on a playground. Incorrectly anticipating commands and not being able to read a drill sergeant's mind, resulted in several wrong turns, bumping into one another and what the drill sergeant's described as clusterfucks and open confusion. Dale and Shannon purposely followed directions so badly that it was almost as if they had always done it wrong.

The drill sergeants also increased the trainees' education by adding two predominantly military acronyms SNAFU, situation normal, all fucked up, and FUBAR, fucked up beyond all recognition, to the trainees' vocabulary. Both terms were used repeatedly that afternoon by the drill sergeants in reference to their troops.

But flanking movements were difficult for anyone, even some experienced soldiers, so the drill instructors didn't ride their platoons too heavily about it. Most of them remembered all too well their own failure to immediately pick up the drills.

Michaelson ran into Dale coming out of the shower and they stopped in the hallway. "How are you doing, Oakes?"

"Oh, I'm okay, thanks. How about you?"

"Still a little shaken, I guess, but I'm getting past it. I didn't know Kirk that well."

Have you had to make any more statements?"

Michaelson shook her head. "No. They were satisfied with what I had to say and I'm glad because I would have hated to have implicated Ingersol in anything and then be wrong. The last thing I need is a jerk like him trying to get revenge on me."

"I can understand that. I don't think I would have mentioned his name, either."

She nodded. "Well, I just thought I'd ask how you were. They've been keeping us so busy I haven't had the chance."

"Yeah. I know. Thanks."

"You're welcome."

Dale watched Michaelson walk away from her before Dale

headed to her bunk. She missed Kirk, truly surprised at the physical loss that she felt, especially since she really didn't even know the young woman. She often looked at Kirk's stripped bunk and pictured her on the rare occasions when the seventeen-year-old genuinely smiled. Small pangs of guilt tugged at the undercover lieutenant, wondering if there might have been anything she personally could have done to help Kirk. But as time progressed, Dale would also come to think of Jascelle Chandani Kirk as nothing more than a faded, misty, bad memory.

Friday morning the trainees marched to one of the many rifle ranges to see a demonstration of the M16A1 rifle. They were joined by troops from WacVille, who were also there to observe.

While standing in formation, waiting to be seated on the bleachers, Dale saw Helen Zerby, the young woman she had flown down to McCullough with, standing with Delta-2. But Private Zerby failed to notice the undercover lieutenant. The Delta drill sergeants made their troops sound off and do a few cadences to impress the MP trainees but as soon as the females' shrill voices pierced the air, the Delta women appeared more embarrassed than their cadre.

Dale looked around to see Lieutenant Henning trying to hide behind third platoon, facetiously pretending to be a cheerleader as the Delta women chanted. She was comical, regardless of how unprofessional and, when she was caught by McCoy, she stopped immediately and smiled meekly, her eyes avoiding his amused stare and she said, sheepishly. "I guess I'd better stop before I become Private Henning."

During the ceremony, which demonstrated the power and range of the M16, one of the shots ignited a small spark in the area where a truck had been camouflaged and that section of the range suddenly erupted in flames. The trainees sat, mesmerized, while they waited for a fire truck to arrive and extinguish the burning area.

The flustered range commander cleared his throat. "You won't see that happen again in a hundred years."

Ten minutes later, after the small blaze had been doused and the demonstration resumed, the cluster of camouflage caught fire again.

"A hundred years, huh? Boy, time sure flies around here," Dale cracked to Gil Hibbon, who was seated next to her.

At that point, the presentation was cancelled, and the Alpha drill sergeants rounded up their troops, marching them five miles, in a roundabout route, to the company area, just to kill time. Once back at 10th battalion, they practiced Drill and Ceremony.

Staff Sergeant Walter Van Es conducted the class on the M16A1 rifle that afternoon. He seemed to be a no-nonsense man who tolerated no violations of his classroom code which included no talking, no disrespectful behavior and definitely no sleeping.

He strolled around the room, lecturing the trainees on engaging targets during daylight, loading and unloading the rifle, prevention and correction of malfunctions, how much time was allotted during qualification, in what positions they would be learning to fire and weapons safety.

He touched on the subject of night fire and how the recruits would be practicing primarily the same fundamentals of day fire, keeping in mind the problems of limited visibility. This meant, he informed them, that they would have to be so familiar with their weapon that they could disassemble and reassemble their M16s in the dark if they had to.

In the middle of a sentence, Van Es stopped abruptly, reached into his pocket and pulled out a marble which he proceeded to throw with great skill and velocity at Private Freddie Swan, who had fallen into a deep sleep. The instructor's aim was impeccable, and the marble ricocheted off Swan's head, into a wall and out the door, waking the private up with such a start, he wondered if he had been shot.

The behavior of Van Es had alarmed the other trainees to the point of forced alertness. He must have had a radar directed at sleeping trainees. He hadn't even so much as glanced Swan's way, yet he instinctively knew Swan had been asleep. Not wanting to suffer the same demeaning, if not somewhat painful, fate, everyone made a mental note to keep an eye on Van Es' whereabouts at all times. To Dale and Shannon, the action brought back a flood of almost forgotten memories. It seemed some rituals never changed.

Both Shannon and Dale felt the temperatures in the classrooms were purposely high enough to rival saunas so that the trainees would immediately nod off during the sessions which justified the NCOIC to wake the poor, unsuspecting souls up by throwing at them anything the instructor could get his hands on. Sometimes the instructors liked to try the skipping stones on the lake pitch and hit as many dozees as possible in one throw. It must have been interesting going home on leave and boasting about a battle scar that was made, in all actuality, by a blackboard eraser.

Dale knew that after a while, hearing someone's head slam against a desk because he or she had fallen asleep or hearing a solid object whizzing by nearly fracturing a skull or two became familiar sounds.

Van Es, satisfied with the overall reaction, continued the class by explaining how to maintain the M16. Displayed on his desk was a general-purpose cleaning brush, a bore brush, a chamber brush, a pipe cleaner, a cleaning rod, bore cleaner, rifle patches, clean dry rags and LSA lubricating oil.

"You will constantly be cleaning your weapon free of water, carbon, grit, rust, dirt and gummed oil," he told them. "You will disassemble your weapon and clean it by parts— the magazine, bolt carrier group, upper and lower receivers and the barrel." As he spoke, he took apart the rifle and demonstrated, naming each section as he laid it on the desk. When he was through, he reassembled the M16 and the class with Van Es, for that day anyway, was over. Then it was back to the company area for one more hour of PT.

CHAPTER TWENTY-ONE

Dale knew when she woke up that morning that it was cold outside. Her foot told her. It took her at least five minutes before she could get out of bed and stand on it but, once up, the pain became tolerable and, as the day went on, almost non-existent.

The temperature hit twenty-four degrees Fahrenheit, with a wind chill factor of four and was becoming more wintery as the morning progressed. The company was issued weapons cards and they lined up in front of the arms room to be assigned their M16s. Each weapon had an identification number on it and each trainee was ordered to memorize it.

The company marched to the range where they became reacquainted with Sergeant Van Es. He briefly reviewed what they had learned yesterday and then turned the class over to Sergeant Glass who stood in front of them, cocky in his red and white range cap, demonstrating for what seemed like the hundredth time how to hold and fire an M16A1 rifle. He pointed to a sign that had the initials B-R-A-S in bold letters. Next to the letter B was the word breathe followed by R for relax, A for aim and S for squeeze. Glass pointed to the B.

"Breathe in a regular manner, then inhale deeper than normal, exhale partially, complete your aim, then fire."

He moved his pointing stick down to the R. "Relax. Try to avoid tensing your muscles. You'll have more of a chance to hit your target."

He then pointed to the A. "Aim. Aiming means aligning the target placement with the sights. The front and rear sight posts are aligned. The front post should line up in between the middle of the back post, evenly across. The best possible insurance of knocking that target down is to aim your front and rear sight posts just below the middle of the white square on those pop-up targets and fire your weapon."

His stick touched the S. "Squeeze. Squeeze the trigger, don't jerk it. It doesn't take much pressure to fire this weapon. If you remember BRAS - these four steps - you should have no problem. Breathe, relax, aim, squeeze."

The trainees were then marched to a range where they were to zero their weapons and shoot four rounds of live ammunition hopefully into graphed targets designated for their individual lanes. All the drill sergeants were interested in at that time was getting the trainees used to firing the M16, not in any particular score.

Given their firing orders, thirty trainees lowered themselves into cylindrical-shaped foxholes that covered their bodies up to their armpits. They were the first firing order. The second firing order stood fifteen feet behind each foxhole to observe how it was done as the rest of the company waited in a huge warm up tent off to the side of the range, getting used to the smell of gunpowder.

Shannon was in the first firing order and after she had put in her issued earplugs, she jumped down into her foxhole, removed the wool glove and leather shell on her right hand and placed the M16 in its firing position. The temperature was so low that she felt as if her finger was going to weld itself to the metal of the trigger.

On amplified instruction, Shannon flipped the selector level to safe after checking to make sure the hammer was locked. She then looked to see if her bolt was open, keeping the weapon pointed down range. She inserted a loaded magazine, slapping the bottom of the metal bullet container to insure that it was positively engaged and then she pulled the charging handle back as far as it would go, releasing it so it slammed back into position. She struck the forward assist and took aim at the target some fifty meters in front of her. She placed the selector level on semi.

"Now...watch...your...lanes...," the voice boomed from the loudspeaker in the watchtower.

When the firing orders and ammunition had been exhausted, the trainees remained on the range to police up spent cartridges, which seemed to take forever to do. It was, perhaps, worse for the first firing order who had to wait around until the fifth firing order was completed.

To keep those bored trainees occupied while everyone else blew away targets and assorted vegetation, the range crew had them pick up rocks from one pile and make another pile with them. When that was done, the drill sergeants made them put the rocks back where the first pile had been.

Members of the consolidated mess hall and a two-and-a-half-ton truck, affectionately referred to as a deuce and a half, brought chow out to the trainees for their lunch. The meal consisted of goulash, mashed potatoes and mixed carrots and peas. For beverages, coffee, tea, fruit punch and milk were available and chocolate cake was offered for dessert. It all sort of tasted the same and though everybody complained, they ate ravenously.

Following chow, the platoons were put into formation where their drill sergeants strongly advised them to keep their weapons with them at all times. Loss or misplacement of an M16 would be dealt with severely. They were told to never leave their weapons unattended because if a drill sergeant found a rifle without an owner, the weapon would be gone. They were also informed that from that moment on, the M16 was to be referred to as a trainee's weapon, not his gun.

According to a drill sergeant, you fight with your weapon, you sleep with your gun. Anyone thereafter who called it a gun would be making an intimate acquaintance with Mother Earth.

Alpha-10 marched back to the company area where they cleaned their M16s for the first time. It was not as easy as Van Es made it sound.

After disassembling the rifle and cleaning it, each platoon's drill sergeant had to inspect the weapon to see if it met his or her standard of cleanliness. A majority did not. Most M16s had to be cleaned at least three times before they met with a drill sergeant's approval and could be turned in to the arms room.

The rifle cleaning supplies were put away and the trainees were marched to the PT field, given extensive physical training and then they were lined up to do wind sprints.

Supervising the exercises, Lieutenant Henning noticed that

Dale appeared to be having difficulty running. Strolling to Dale when she was at the back of the line, Henning gently grabbed the CID agent's arm and pulled her out of hearing range of anyone else.

"What's wrong?" Henning asked, pointing to Dale's foot.

"Nothing. Really. It's the weather turning so cold all of a sudden."

"Are you telling me the truth?" Henning looked concerned.

Dale always thought that was a silly question. Did anyone ever answer No? "Yes. It's the dampness. It's not just me, everybody's complaining. It probably has a lot to do with getting used to the boots again. Don't worry about me. If it does get bad, I'll see somebody."

"But you're limping now."

"But I won't be in a few days," Dale argued.

They were approached by McCoy. "Is there a problem, Ma'am?"

Henning glared at Dale and then looked up at the drill instructor. "No problem, Sergeant. I just noticed that Private Oakes seemed to have trouble running, that's all."

McCoy turned to Dale. "Something wrong, Oakes?"

"No, Drill Sergeant."

"Are you calling the lieutenant a liar, Oakes?" he asked, playfully.

"No, Drill Sergeant."

"You're fine, then?"

"Yes, Drill Sergeant."

"Well, then, if the lieutenant is through with you, get back in line, Oakes."

"Yes, Drill Sergeant." Dale looked at Henning, who nodded. She double-timed it back into line, being careful not to let her aching foot show.

The Army barracks were obviously decorated by the same people who designed the terminal ward of a hospital and about as exciting to live in as sick call. In fact, by the second full week, the way most of the women coughed and wheezed, it could have easily been the clinic just cleverly camouflaged as troop housing.

PERMISSION TO RECOVER

The trainees were not quite used to running themselves that ragged yet and the hours they were keeping were beginning to take their toll, so much so that Travis had nicknamed them the Green Bay Hackers. Only the most physically fit seemed to be barely escaping the viruses and run down feelings that were rampantly traveling through the barracks.

Shannon awoke Sunday morning after what seemed like an hour of sleep. She and half the female population of Alpha-10 spent most of the night coughing and continuously adjusting their body positions so that they could breathe properly without having excess fluid from their noses dripping on their pillows or down the back of their throats. Unfortunately, they still had to get up for formation and they still had to participate in physical training.

Between Fall In and PT, the trainees caught up on their laundry, writing letters, their boots, their brass, their rules, regulations and drills and being marched to the PX to replenish their personal supplies. Since Shannon and Dale took turns doing one another's laundry, Dale played cribbage with Bonnie Kramer again while Shannon was flirted with madly in the laundry room by Private Jason Zachary.

Sunday night, after the evening meal, the trainees were thrown their first GI party. It was not a festive experience. They moved their bunks around, stripped all the old wax off each tile, then mopped, re-waxed and buffed the floor. They completely scoured the bathroom and showers with toothbrushes so that they not only shined, they glistened. They finally made it to bed a little after midnight and the fireguards enforced that only one toilet and one shower was used until the morning inspection.

Monday morning, after the ritual of changing the sheets and an inspection where the barracks and the trainees' mental stability was torn apart, the company drew their M16s from the arms room and marched back out to the range. The temperature was lower than it had been on Saturday and Shannon made the comment that it was so cold, one could probably get ice cream straight from the cow. Several times while waiting to zero their weapons, there were rumblings from soldiers fearing the loss of feeling in their fingers and toes. The wind chill factor was ten degrees below zero.

Everyone who wasn't in the present firing order stayed close to the warmup tent. It was still cold in there but at least the canvas warded off the wind.

It would not have surprised anyone to discover that there were days when Ritchie was possessed by demons. His behavior was so erratic and immature at times, no one could really believe he held such an authoritative position. There was never any flattering conversation in the barracks about him, which Dale and Shannon knew was dangerous because of the bitch box's traitorous power to unsuspectingly transmit. However, even that knowledge did not prevent them from occasionally joining in the denouncement of his credibility or questioning his sanity. Planning fictitious ways for his untimely demise was done as often and with as much vigor as polishing boots and scrubbing toilets.

The senior drill sergeant circled the exterior of the tent as a majority of the three remaining firing orders crammed themselves inside to keep relatively warm. He was scanning the shape of the canvas for small bulges that may have been caused by unattended M16s. After making a mental note of what could possibly have been four rifles leaning against the inside of the tent, he began his mission to snatch and degrade.

Dale was the first one to spot the ungloved hand reaching under the tent from the outside, feeling around for the butt of a rifle. If she hadn't known what the humiliating outcome would be if the hand took possession of an M16, it would have been an entertaining sight. Fortunately, before the sneaky hand could connect with Private Eddie Belden's weapon, Belden picked it up to take it with him outside so that he could smoke a cigarette. He was totally unaware he had just escaped certain mental agony.

Shannon passed Belden as she entered the tent and nodded to Dale, who was seated on the ground near the wood stove that had been providing heat.

"Walker, come here," Dale called to her, standing up. As Shannon approached her, sniffing back her cold, Dale saw the hand come sliding underneath the tent again, reaching for another M16. "Which drill sergeant is outside the tent?"

"The only one I saw was Ritchie."

A smile that could only have been described as sadistic crossed Dale's face. She crooked her index finger at her partner. "Come with me."

By this time, Dale was not the only one who had noticed the mysterious hand. Private Kerwin Cross, who had his rifle leaning up against the tent, almost lost the weapon. Just when Ritchie's hand was about to triumphantly grab the unattended M16, Dale picked it up and handed it to Cross.

Shannon and Dale followed in the direction they knew Ritchie was going, toward the next M16 that was leaning up against the tent. The two women were also beginning to take on a quiet but appreciative audience. They silently alerted Private Caffrey that she was about to have her weapon stolen and she picked it up and laid it across her lap seconds before Ritchie would have had it.

Dale motioned to her partner to play along as they got their telepathy going. The dark-haired lieutenant quickly walked ahead and leaned her M16 up against the tent so that the outline of it could be seen from the outside and Shannon stood poised, with her weapon ready. Just as they had anticipated, Ritchie took the bait and his hand sailed underneath the tent for a quick steal but before he could get a grip on it, Dale pulled it away and Shannon slammed the butt of her rifle onto the back of Ritchie's hand with such force, it would not have surprised or upset anyone if she'd crushed every bone halfway up to his elbow.

At the sound of Ritchie's eight-octave scream, all of the trainees occupying the warm-up tent scrambled back to inconspicuous locations, just in time before the furious senior drill sergeant burst through the entrance displaying a throbbing, swollen hand. He was followed closely by Audi and McCoy. The tent occupants jumped up to the position of Parade Rest.

"Who did this?" Ritchie screamed, his face beet red and his pointy little nostrils flaring. He definitely had everyone's attention.

"Did what, Drill Sergeant?" Jimmy Judd, a six-foot, five-inch bean pole from Tennessee, drawled. It was an act of extreme control for the occupants of that tent not to even crack a hint of a smile. McCoy, who was standing behind Ritchie was grinning, but he was allowed.

"*This*, you hillbilly!" He held up his pain-riddled hand. "I want to know who battered me with their weapon and I want to know *now!* Battery is a serious offense and I want the cocksucker who is responsible for this!" Each word was being spit out because of the excruciating agony Ritchie was experiencing.

A specialist fourth class named Raymond Haviland stepped forward. Specialist Haviland was prior service, having spent four years in the Navy, and wasn't quite so intimidated by Ritchie's rampage. "Excuse me, Drill Sergeant..."

"What is it, Haviland?" Ritchie snapped.

"Somebody tried to steal my weapon from the outside of the tent about five minutes ago," Haviland lied. "I don't know who that somebody was, but it is my understanding that if somebody tries to take my weapon and bring undue hardship to me that I can do what I have to keep my weapon secure. None of us knew that was your hand, Drill Sergeant, it could have been anybody's. Whoever did that to you was protecting his weapon from a thief."

"Jim, I think you should get to the hospital right away and have that looked at," Audi suggested to Ritchie, soothingly. The two of them exited without another word.

McCoy turned to Haviland. "Nice job, Swabbie."

"Thank you, Drill Sergeant, but it wasn't me who did it."

"It doesn't matter. Whoever did it was in the right, and you backed them up." He looked at everyone. "Carry On."

As McCoy left the tent, there was a lot of easier breathing, laughter and backslapping.

On the warpath when he returned an hour later, Ritchie, whose hand had miraculously not been broken but was severely bruised and was now wrapped, was becoming unbearable even for the other NCOs to deal with. At one point when he had been out of view of the range commander or Alpha company cadre, he kicked Private Scott McNulty in the helmet while McNulty was in a foxhole. Ritchie had asked the private a question and McNulty, not used to wearing earplugs, did not acknowledge the senior drill sergeant. Ritchie let go with a sharp kick that connected with the private's steel pot, which knocked McNulty violently against the side of the circular trench.

Immediately looking around to see if anyone of importance saw it, Ritchie was relieved to find all the NCOs preoccupied. However, both Dale and Shannon had witnessed the action and, unbeknownst to him, that may have been Ritchie's biggest mistake.

At the end of the day while Henning was rodding weapons, Shannon moved toward her and announced loudly, for the benefit of everyone else. "No brass, no ammo, Ma'am!"

"Fantastic," Henning responded, bored with shoving the metal stick down the barrel of the M16s. "How'd you do?"

"Could have done better."

"Then stop fooling around and get to it, Private." Henning winked.

"Right." Shannon looked behind her, seeing no one close by. "If you see any knuckle skin on the butt of my weapon, take my word for it, it was deserved."

"I figured it was either you or your partner who did that. No one else would have had the guts."

"That man is going to get himself into some deep shit if he doesn't straighten his act out real soon."

"What else did he do?" Henning sighed, but before Shannon could answer her, another trainee came up behind them both, holding his weapon at such an angle that, if it went off, it would have removed the left cheek of Henning's rear end. *"Point that weapon down range!"* the training officer yelled at him. "Moron," Henning sputtered, as Shannon grinned and moved on.

CHAPTER TWENTY-TWO

That night, directly after evening chow, the trainees were called into formation and given a be careful while on leave speech. Henning informed them that this was the first time they would be in public as members of the U.S. Army, and they were to represent the military with respect and pride. Her lecture lasted approximately fifteen minutes and then the trainees were ordered to line up by platoons on the north patio by the CQ office where they were to be individually paid before leaving for Christmas exodus. Henning was in charge of that, too.

Dale could tell when she went into the Orderly Room to get her leave money that the training officer was provoked about something. Two armed MPs stood by the desk, Fuscha ran around like a chicken with its head cut off, an assorted number of NCOs came in and out of the first sergeant's office and barracks gossip Charlene Keival was on wide alert at CQ duty. It wasn't a good time to ask why.

Dale attempted to break the mood while Henning searched for Dale's name on the list and pointed to a small, sparsely decorated spruce twig that sat on the corner of the desk. "Excuse me, Ma'am, is that a Christmas tree or did somebody bring you a holiday corsage?"

Henning looked up in the direction of the six-inch tall tree. She looked back at Dale, her expression unchanged. She made no comment, then she returned to scanning the list.

"Don't bother the lieutenant," snapped a particularly burly MP, who wore an E-4 rank on the upper sleeves of his dress uniform. Dale looked at him, surprised.

Henning kept her eyes on the computer readout sheet. "You're out of line, Specialist," she admonished, not even glancing at him. "Here you are. Oakes." Henning handed Dale a slip of paper. "Make sure all the information is correct."

Dale checked the sheet of paper that informed the Army where Dale was going to be from December fifteenth until January second and how they could get in touch with her if they needed to.

"It's correct, Ma'am," Dale told her and handed the paper back. Henning counted out the cash and held it out to Dale.

"Count it again, Private Oakes and make sure it is the correct amount."

Dale recounted the money. "It's all here, Ma'am."

"Outstanding. Sign right here.," She pointed to an area on the receipt and Dale scribbled her signature. "Thank you, Private Oakes. Have a nice holiday," Henning said, blandly, disinterested.

"Thank you, Ma'am. Same to you."

"I'm beginning to really hate you," Shannon said to her partner. "You get to get out of here for two weeks and I have to stay."

"Come on, we decided reasonably and intelligently. It's not my fault you called tails," Dale reminded her. "Besides, you said you'd be better off not going back home for a while."

"Look, just because my landlord and I are at odds right now doesn't mean I don't have other places I could go," Shannon protested, blowing her nose.

"You have a cold. You would be miserable wherever you went, anyway. I'm not giving in. I've got work to do, too, you know. My time isn't going to be spent entirely in Vermont, partying."

"You don't understand. You get to go out and drink and dance and have wild, uninhibited sex," Shannon emphasized, almost whining.

"That's right," Dale confirmed. "And I'll probably do it all with the exception of the wild, uninhibited sex. I told you I broke up with Keith, remember?"

"Big deal. It never stopped you when you two were together, why should it stop you now?"

Dale laughed and shrugged, "Let's just say I've changed." And nothing seemed to have brought that home more than the past month. Before her partner could question that statement too deeply, Dale gave Shannon a gentle shove. "Just think how much more aware you'll be of the other trainees who are also staying."

"Don't try to pacify me with a sense of duty."

"Come on, it's only two weeks."

"*Only* two weeks? We've *only* been here a little more than two weeks. Can you honestly say it doesn't feel like two years?"

"It's been three weeks and don't be a poor sport. Think of all the money you'll save by being stuck here. Let's get off this subject because you are not going to make me feel any guilt what-so-ever."

"Selfish bitch."

"Thank you. Feel better? Now that you've aired that, do you happen to know why our industrious training officer was in such a wonderful mood when she was giving us our pay tonight?"

"Yes, I do. Town crier Keival said Stubby had to ride all the way back to the company area with a ten-thousand-dollar payroll and absolutely no armed or unarmed escorts. Just her and Silva, the duty driver."

"How the hell did that happen?"

"Ritchie never arranged for an escort and Henning waited as long as she could. When she and Silva got back to the company area, the MPs were here instead of where they were supposed to be. I guess she couldn't contain herself and blew a fuse at Ritchie. They took it into Ritchie's office and closed the door, but Keival said she did overhear Henning ask him if he had any knowledge of any inbreeding in his family, which led to one hell of a shouting match."

"What is the matter with that man? I don't believe him. I especially don't believe that incident this afternoon. Thankfully, McNulty was just dazed. Ritchie could have really caused him a serious injury. Somebody better put a bug in McNulty's ear to go to the IG. He's supposed to be protected against crap like that."

"It makes me even happier about what we did to him in the warm-up tent this afternoon. Does Bishaye know what kind of lunatic she's got as a senior drill sergeant here?"

"Well, you would think, wouldn't you? But I doubt it because she didn't give us a head's up. I sincerely hope I have opened her eyes about Colton, if I start in on her about Ritchie, she's going to think it's me."

"Thank God both Colton and Ritchie leave in two days for two weeks."

"Where are you lucky dogs moving to until we get back?"

"Delta-12"

"Who else is staying?"

"Caffrey..."

"See?" Dale winked at her. "You don't have to be lonely if you don't want to be."

Shannon ignored her and continued. "Michaelson..."

"Michaelson is staying? I can't believe that such a perfect specimen of womanhood has no one to go home to or anywhere else to go."

"Maybe she flipped a coin, too," Shannon said, sourly. "Quinn Brewer is staying. Swinegar is staying..."

"She is? How come? She was so enthusiastic about leaving."

"She says she's fallen in love. She and that holdover, Ribak, have apparently struck up a romance. He's staying because he is waiting on orders so now she's staying. I think she really wants to stay because Robbins' going to be NCOIC and we all know, including Robbins, how she feels about him."

"And who knows? Maybe Robbins volunteered to be in charge because Michaelson is staying. We all know how every normal, red-blooded male feels about her. Crushes are a terrible thing, aren't they?"

"As long as it stays a crush."

"Well, there you go. Now you have a reason to stay."

"Yeah. Right," Shannon sniffed. "People fall in love so goddamned easily around here and I don't understand why. This is not my idea of a romantic environment. Let's face it, between you and me, falling in love in basic training is just silly."

"Silly? This is Alabama, I think love between you and me is illegal."

Shannon glared at her. "Thank you, Henny Youngman."

Dale chuckled. "So Swinegar is staying because Ribak is staying or so she wants everyone to believe."

"Right. And Tierni isn't going home for exodus. She's

going to spend two weeks in Atlanta with Silva."

"Oh, you know, I thought something might be developing between those two. I'll make it a point to get to Atlanta sometime before January first."

"How will you find them?"

"I'll just ask Tierni where they plan on going and take it from there. Or I could always just hit the Atlanta Underground. Everybody eventually ends up there."

"Maybe you should just leave them alone. I would be very surprised if either one of them knows anything about why we're here."

"You just don't want me to go to the Atlanta Underground without you. But you do have a point so I will leave them alone. Good for Tierni. That Silva is a cutie." And though she meant it, it no longer implied she found him personally attractive or wanted him for herself.

"Yeah, he is. He's a good company driver, too. I think they're going to miss him when he comes down on orders. What do you say we go upstairs? We've got another long one tomorrow."

"And another cold one, I heard."

"Peachy," Shannon commented, blowing her nose.

It was now Tuesday, and Christmas exodus was only two days away. The company was starting to get anxious about returning temporarily to civilian life and most of them acted as though they were convicts about to be paroled.

Imminent release was approximately forty-eight hours in the future and there was still serious military business to attend to. The new soldiers were warned not to be caught daydreaming and, in all honesty, if a trainee's thoughts drifted to anything other than the matters at hand, they drifted to how it was going to feel to be in a nice, warm, comfortable bed, regaining that lost art called a decent night's sleep.

The temperature was also higher than it had been the past week and the extra clothing the members of the company had worn to fight the cold weather now made them sweat. This was a distraction they did not need.

Once they marched to a different range, the trainees were

again placed into firing orders. They now dealt with shooting at pop-up targets, at intervals of fifty, one hundred, one hundred fifty, two hundred, two hundred fifty and three hundred meters for trial classification. They had to fire four times, ten rounds each, in one order in two positions, prone and standing foxhole, and each position allowed sixty-five seconds to engage and hit ten targets.

Everybody was understandably nervous about the next two days. Their performance on the rifle range, especially tomorrow, determined whether they would continue training with Alpha-10 or be recycled to another company that was a week behind them in schedule. It was the first phase of the process of elimination. If a trainee didn't hit at least seventeen out of forty targets that would qualify them as the military classification of *marksman* by the end of training day tomorrow, he or she would be moved out of Alpha-10 and reassigned to Bravo-10, where he or she would have another chance to qualify after Christmas exodus.

Dale hadn't done all that well on the practices because she knew, when it came time to qualify, that she would be able to pull off a decent score. She also knew she could knock down all forty targets, with the exception of the lockups, if she fully concentrated. She found the M16 a weapon that had been easy to master the first time she picked one up. It was just something that had come very naturally to her, which surprised her because before she entered the Army, she had never handled a rifle or associated with anyone who was familiar with one, despite the fact that her hometown was known for being deer hunting country. Dale's uncanny ability for accuracy astounded her. She did almost as well with a pistol.

Not as well as Shannon, though, whose father, uncles and brothers were cops so she grew up around handguns. When her brothers would shoot at homemade targets, Shannon would be right there with them, unloading one of their .22s, .38s, .357s or .45 caliber pistols into plastic, water-filled, gallon jugs or tin coffee cans. Rifles, on the other hand, got the best of her and when it came down to actual qualification, she rarely hit higher than a *sharpshooter* designation which was twenty-four to twenty-seven targets out of forty. Dale never scored lower than an *expert*, which was twenty-eight up to forty targets.

Until it counted, Dale was intelligent enough to look like a scared amateur. In fact, she did so poorly that morning that Kathan made it a point to make an example of her in front of her firing order in the warm-up tent. He also told her that he doubted she would be continuing her training with Alpha-10 if she kept shooting in a similar fashion.

It was not a pleasant moment, but it was one that Dale could tolerate because she knew she still held high esteem with her peers for what she and Shannon had done to Ritchie the day before. Most importantly, she also knew there was no chance of her being recycled even if she couldn't improve her score.

Henning, who had just arrived at the range, patrolled behind some of the foxholes, supervising, when she finally decided to sit down and observe. She was surprised to see the name Walker printed in laundry marker ink on masking tape on the back of the camouflaged steel pot of the scorer and even more surprised to see the name Oakes on the rear of the helmet of the soldier in the foxhole.

"How did you two manage this?" Henning shouted, to make herself heard through Shannon's earplugs.

"We didn't. It just worked out this way. She makes me sick on the rifle range," Shannon told her. "I'll get even on the pistol range though."

"Have you fired yet?" Henning asked.

"Last order. I got fourteen."

"Fourteen! How did that happen?"

"There were snakes in my foxhole. I had to use twenty-six cartridges to kill them. The rest went into the targets."

Henning stared at her, incredulously. "Are you serious?"

"The weather warming up must have brought them out. I hate snakes. It was them or me."

"How many were there?"

Shannon held up two fingers.

The thought of two snakes plugged with thirteen bullet holes in each made Henning laugh out loud. "What kidders, you two are. You're joking, right?"

Suddenly, a blood-curdling scream pierced the air from two foxholes up.

PERMISSION TO RECOVER

Henning returned to Shannon's scoring table five minutes later, after she investigated the ruckus. Henning made no comment to her.

"What's all the commotion?" Dale yelled to her. She turned around and removed her earplugs.

"Sherlock had to be given another firing position," Henning answered, not looking at Shannon. "She refused to stay in a foxhole with what was left of two snakes."

Dale shook her head, turned around and faced the targets again. As she put her earplugs back in, she commented. "Shannon, the reptile annihilator. She doesn't just kill them, she obliterates them. Nothing left but a bloody shadow. Coming this Christmas to a theater near you."

"I can't help it. I hate snakes," Shannon said to no one in particular.

After checking the flags to make sure everyone was clear down range, the order was given from the tower to fire. On the second shot, Dale's rifle jammed.

Henning was about to wave her flag which indicated an improperly functioning weapon when Shannon stopped her. "She knows what to do."

Immediately, Dale placed the selector level on *safe* and released the magazine. She pulled the charging handle to the rear and the jammed live round popped out. She then placed the bullet back into the magazine and slapped it back up into the rifle and moved the selector level back to *semi*. Even though the entire process took only eleven seconds, she only got to shoot at two more targets but was too distracted to regain her concentration.

"Cease fire...cease fire..."

As Dale pushed the selector level back to safe, she pounded her fist into the ground in frustration. On command from the tower, she lifted herself from the foxhole and into an unsupported prone position.

"That's too bad," Henning said to Shannon. "She missed those points."

"Not really. She'll do extremely well when we qualify tomorrow, and it actually looks better if she keeps her scores lower today."

"I suppose that's true but what happens if her weapon jams during qualification?"

"Then she'll call a safety NCO over. It would have been a waste of time today."

Dale was heard muttering, "Shit, shit, shit," to herself as she lined up her sights.

That afternoon, the drill instructors marched the trainees back to the company area to learn more Drill and Ceremony, mainly Inspection Arms, Order Arms and Port Arms. After the trainees' own arms felt like they were useless and devoid of any strength from holding the M16s in an outstretched position for most of the drills, the troops were assembled on the south patio to listen to a speech from Colton.

The company commander actually told the women how proud he was of them and was surprised at the way they were holding up. As if that wasn't enough of a shock, Colton then admitted that so far, he felt that they had outshined the males.

Dale had to bite the inside of her cheek to keep from grinning at the clearly painful admission from the company commander. No matter how agonizing it was for Colton to have to say, it seemed even more excruciating for Ritchie to hear. The expression on Ritchie's face was one of ultimate betrayal and his stare at the handsome captain indicated that now he had gone too far. But knowing Colton as Dale thought she did, she realized how difficult this speech must have been for him and she accepted it, as did everyone else, in good faith.

Some of the male trainees were plainly disappointed but a majority of them had to agree that the females were really putting forth a noticeable effort. Some suggested later that the CO's speech was said to boost morale for qualification on the range in the morning. Everyone openly anticipated the next day, some eagerly, some nervously, some with dread and they all hoped, if nothing else, the weather would be on their side.

It had started to get chillier again toward the afternoon. By evening chow, the temperature had hit below the freezing mark again. Being that it was one of the coldest winters Alabama had ever experienced so far, the weather did nothing to encourage even the most inspired troops and one of the main reasons for that was the combat boots.

As with any new footwear made from leather, it took time for the material to soften and mold comfortably to the foot. Since the military budget had been cut back, the labor and materials used to manufacture the boots had been cut back also and the quality of leather and the construction of the boot itself weren't what they used to be. Therefore, this brand-new leather combat boot was fitted to the foot, accompanied that foot outside and because of the colder climate, immediately froze, becoming brittle and shaping itself incorrectly and harmfully to that foot.

It was not uncommon for some troops to have problems with improperly fitting footgear in any type of weather, particularly females with arches too high or arches that had fallen but this cycle, especially, ran into problems. Uncle Sam's Christmas gift that year was contusions, abrasions, swollenness, blisters, callouses, shin splints, tendonitis, stress fractures, bone spurs and bursitis. Podiatry problems that would stay with some of these trainees the rest of their lives.

Everybody took extra-long showers that night and let the hot water beat on their already abused feet. A few of the women even bribed some of their barracks mates for foot rubs. Ordinarily, a patrolling drill sergeant observing such activity between the females would have made mental notes on who was doing whose feet but under the circumstances, not only did the drill sergeant who performed the bed check fully understand, he acted like he wanted to get in line.

Shannon had lucked out. Possibly it was that she was used to combat boots but so far her feet still felt pretty good, considering. Although the onset of what resembled shin splints from marching on hard surfaces, such as concrete, paved roads and sidewalks, crept up from just above her ankle bones and she was not pleased about that, but she knew how to work with her feet to lessen the pain.

Dale, on the other hand, wasn't faring as well. Not only did the boots wreak havoc on her left foot but the dampness mixed with the cold made the arthritis that had obviously settled in nearly unbearable. Four times in the past week, Dale had awakened in the middle of the night with the kind of paralyzing pain that was reserved for charley-horses. She had to bite her tongue not to make any noise and rouse anyone else out of their well-needed slumber but, within minutes, after being able to completely straighten out her foot, the throbbing tapered off to a dull ache.

Every time this happened, Dale promised herself she would go to the TMC. By the time 0730 rolled around and sick call time came, Dale had decided she could live with it, at least until Christmas exodus, and forged ahead with her day's training. By 1500 hours, she usually regretted the decision. And by 1700 hours, as Henning had noticed, Dale had a slight limp. If any of the drill sergeants had seen it, they had neglected to mention it.

Pamela Ryan, who was in the bunk next to Dale's, had noticed and offered to massage Dale's feet as she exited the shower.

"No, but thanks anyway, Ryan," Dale acknowledged her, appreciatively. "The heat from the water seemed to do the trick," she lied. It was a sweet gesture but at that point, any inexperienced hands pushing and pulling on Dale's already aggravated injury may not have been the wisest idea. "How about you? How are your feet?"

"Great. York did mine about ten minutes ago. She should go professional. My feet feel like they just had sex."

Dale perked up and looked in York's direction. She saw two people sitting on the floor, waiting for Linda York's magic fingers. Dale climbed into bed. "Thanks anyway, Pam."

"Sure. Good luck on the range tomorrow."

"Thanks. You too," Dale said to Ryan.

"Thanks. I'll need it."

Ryan really didn't need luck. She had paid attention to her instructors and had shot well all along. She told Dale she had this underlying fear that her qualifying score was going to be analogous to high school. She had always done fine in class but when it came time to be tested, it was as if the questions were in a foreign language, and she always drew a blank. This time, however, her fear was unfounded. She had been picked for the first firing order and had scored twenty-five out of forty, which qualified her in the second category of *sharpshooter*. She would have preferred *expert* and if she had knocked down three more targets, she would have made it, although she wasn't disappointed. Of her firing order, she had tied with Perry Sargent for the highest score, and, at the end of the day, her score was average with most of the males but higher than most of the females who barely made *marksman*.

Both Dale and Shannon had been selected for the third firing order, Dale fit snugly in foxhole number twelve and Shannon three away from her in foxhole number nine. For the most part, it went well, except when Shannon's first ten round magazine jammed and forced her to shoot an alibi round. That broke her concentration and hindered the outcome of her score, but even with the alibi round, she had hit enough targets to qualify as *expert*. Everyone else acted impressed but as familiar with the weapon as Shannon was and as many times as she had qualified, she felt she should have done better than knocking down thirty targets.

Henning, who rodded, saw Shannon's discouraged expression as she approached. "That's not an encouraging look."

"No brass, no ammo, Ma'am," Shannon announced.

"Great," Henning responded, as she shoved the metal rod down the barrel of the M16 and pulled it back out again. "How'd you do?"

"Thirty, Ma'am."

"Don't sound so depressed. That's better than I ever did. From what I hear, you and Annie Oakley shot higher than a majority of the guys."

When all the results were in, at thirty-eight out of forty, Dale Oakes had scored the highest on the rifle range. Drill Sergeant Kathan was noticeably offended. For two and one-half weeks, he had insulted her, berated her and tried to pound into her head what a no good, lazy shitbird she was so how dare she have the audacity to not only challenge his insight but downright defy him by being the best...in anything. At first, he accused her scorer of cheating but when he checked who had signed her fire performance test sheet, he saw it was Haviland. Haviland would have had no reason to alter her score. As prior service, he knew better, he knew what the consequences of those actions would have been if he'd been caught. The lockups were correct. Kathan had to face facts...it must have been a fluke.

News traveled fast. While she cleaned her weapon on the patio with her platoon, every drill sergeant with the exception of Ritchie congratulated Dale. Second platoon males were happy because she put them in first place in the unofficial and playful competition between the three platoons. The women were proud of her because she was a female and had scored the highest. They didn't care what platoon she was in.

"Show-off," Shannon said, later, as they stood outside by a picnic table, overlooking the parking lot.

"Damn right," Dale answered, proudly. "My grandfather always said, what comes easy, let go free."

"My grandfather used to say that, too. Except he was referring to farting."

"Come on, Shan, if your M16 hadn't malfunctioned, you would've been real close to my score. You've gotten as good with that rifle as I have gotten with a pistol. In six weeks, I'll be the one pissed off because you scored higher. How's your cold?"

"It's better. Last night Stubby sneaked me a cap full of some cold medication that is supposed to not only help you sleep but it works on your cold while you sleep. It's potent stuff. She gave me another hit of it tonight when I cleaned her office. I'm getting a little drowsy."

"Maybe you should get to bed then. Take advantage of its effect." Dale yawned, also feeling tired. "Twelve people didn't qualify. They're going to be taken back to the range tomorrow. I heard they are going to be kept there until they score a seventeen or over."

"So much for the idle threats of recycling. Who are they, do you know?"

"Well, there are three women, Hewett, McTague and Newcomb. I'm not sure of all the guys but the ones I do know are Muscatello, Stavis, Swan, Jasinski and Drago."

"I wonder what happened to McTague? She did okay in the practice qualifying. She must've choked."

They had reached the landing to the female bay. "I can't believe they're giving them a second chance, though. Things

PERMISSION TO RECOVER

sure are being done differently here."

"I know. I wish they'd make up their minds whether they were going to be hard asses or not. It's very confusing."

CHAPTER TWENTY-THREE

The next day after details were done and everything was in order, the soldiers received their Class-A uniforms and their individual scoring badges. The people who had gone back to the range had returned by noon chow, all with a qualifying score. A group of fifty trainees left later that afternoon and night for the beginning of exodus. The rest who were going would leave the next day.

That evening, for the remaining trainees, the mess hall was set up with snacks and refreshments until twenty-one hundred hours and the new GIs were allowed to dress in civilian clothes and mingle freely with the soldiers and cadre from the other three tenth battalion companies. The only Alpha cadre members in attendance were Putman, Audi and Henning.

It was a relaxed but well-controlled little affair with the drill sergeants keeping a close eye on everyone, ensuring if anyone disappeared, it was to return to the bays. They didn't want any of their trainees, especially the ones remaining for Christmas company, to make any impulsive trips to the green grass motel. All cadre members were wise to neglected raging libidos and were used to all kinds of excuses from trainees to sneak off for some emotionally detached sex at the first available opportunity. They certainly couldn't control what the trainees did while away from the post but, under supervision of the company's authority figures, no one would be getting pregnant on their watch.

Upstairs in the female bay, Dale packed as Shannon sat on her partner's bunk and watched her. "So, are you going to make up with your boyfriend?"

Dale shrugged, noncommittally. "I doubt it."

"Don't you miss him?"

"Yeah, to a point. I got used to having him around. Or him being around me."

"Wow. That sounds like true love at its best."

Dale smiled at her. "True love it wasn't. I never had any illusions about that." She closed her small suitcase. "You'll never guess who came up to me about fifteen minutes ago and congratulated me on my score."

"Colton?"

"No. Bradbury," Dale said and laughed.

"Why, you dog! Where's your first stop going to be on your leave? The arms room? No, seriously, maybe she's trying to improve a score of her own."

"That occurred to me, but she seemed very sincere. And, incidentally, fuck you." Even though Dale was losing the battle in fighting an attraction toward women that was slowly beginning to overwhelm her, Staff Sergeant Jane Bradbury was not someone who entered her realm of interest. The Bravo-10 drill instructor was too obvious, too stereotypical, too on the make. Pretty sure now that these urges and feelings Dale was having were not just a phase, either way she knew that the overly horny Bradbury was not someone for whom she was willing to risk what was left of her Army career.

Thankfully, their conversation was interrupted by the odd sound of change jingling. Lots of it. Clearly oblivious to the noise she was making, Travis passed Dale's bunk, munching on something chocolate, her pockets literally loaded with coins. Less than a minute later, Tramonte entered the bay, strolling by Dale and Shannon, also snacking and jingling. The two CID agents exchanged looks as curiosity got the better of them.

"Hey, Travis, when's the sleigh ride?" Shannon inquired.

"What do you mean?" Travis asked, her mouth full of her last bite of candy bar. Instead of looking genuinely puzzled by the question, she and Tramonte, whose bunks were next to one another, looked caught.

"You two come up here sounding like jolly old St. Nick, we were just wondering if maybe Dale and I missed a floating poker game somewhere."

Travis and Tramonte glanced at one another again and shook their heads negatively. Since no one else seemed to have noticed or, if they did, obviously didn't care, the two trainees decided that there was no harm in confiding in the two lieutenants. Especially after some of the stunts *they* had pulled. Travis made a gesture with her head that indicated she wanted them to step a little closer. Dale and Shannon complied.

"You know the candy machine downstairs?" Travis began.

"The one that's off-limits?" Dale asked.

"That's the one. Well, I wanted some gum, so I asked Putnam if he minded if I got a pack since everything else seemed okay tonight. He said sure, go ahead, and he left to go downstairs. So, Tramonte and I put the money in and pushed the button for the gum to come out and with the gum came the money we put in plus more change. So, on a whim, we tried it again and doesn't our change come back in triple. So we kept on putting money in and getting more money back."

"Didn't you have to buy a candy bar to make it work?" Dale wondered.

"Sure, at first, but then we ran out of candy. So, we tried putting a quarter in the slot, pulling on the change return handle and we'd get two quarters and a dime or two back."

"It felt like we were in Atlantic City for a while," Tramonte added.

"How much money did you end up with?" Shannon asked.

"About seventeen dollars out of the machine."

"What did you do with all that candy?" Dale wanted to know.

"We took it up to the male bay and sold it to the guys for twenty cents apiece." Tramonte said and looked at Travis.

"We made about five dollars off that, which we split," Travis admitted.

"What would you have done if one of the drill sergeants had answered your knock on the bay doors?" Dale inquired.

"We made sure Putnam, Audi and Henning were downstairs before we went up," Travis said, confidently.

"What about McCoy?"

"What about him?" Tramonte said and shrugged.

"He's the staff duty NCO. He's floating around, making sure no one's taking advantage of the Army's hospitable behavior tonight."

"He is?" Tramonte asked, a little weakly. Both she and Travis were terrified of McCoy.

Shannon nodded. "You must have just missed him 'cause he was in here about ten minutes ago."

"Oh, shit," Travis sighed, relieved, and they all started to laugh. "He's probably wondering why all those candy wrappers are in the trash."

"He's probably not wondering at all. He's just a nicer guy than we all think," Dale commented.

"Yeah. And my future husband is Robert Redford," Travis said and rolled her eyes.

Shannon looked at Dale. "You keep saying things like that and we're going to change your last name to Snow."

Tramonte sat down and unloaded all the change from her pockets onto her bunk. "Where is Snow, anyway? Did she leave already?"

"Who cares?" Travis said, opening her locker.

"Well, I just hope she didn't sneak up on this conversation without us knowing and go running out to find the first Smokey the Bear hat she sees."

"Wasn't she downstairs in the mess hall when we looked?" Travis paused, trying to recall.

"Maybe what she was really looking for is an Australian bush hat," Shannon stated, a smile curling the corners of her mouth.

"MacArthur's not here," Tramonte said.

"But...Bradbury is," Dale volunteered.

"Bradbury, who's that?" Travis asked.

Dale and Shannon remembered that not everyone else was as aware of the surrounding personnel as they were. They didn't have to be. "She's that wench from Bravo," Shannon explained.

"Oh..." the two trainees chorused, immediately knowing who she was.

Good pair," Travis nodded. "I hope they live happily ever after."

"Bradbury's probably not interested anyway," Dale spoke up. "I don't think Snow presents much of a challenge."

"But we know who does, don't we?" Shannon winked at Dale.

"So, Walker, why aren't you leaving for exodus?" Dale said, glaring at her.

"Yeah, Walker, are you insane?" Travis inquired, taking quarters out of her pocket by the handfuls and putting them into stacks of four.

Shannon shot Dale an *okay, we're even* look. "I just figured if I had a chance to leave here, I might not come back. Tell me all of you haven't thought that at least once."

"I haven't," Dale lied. "If I'd wanted to stay in Vermont so badly, I wouldn't have enlisted."

"Then why are you going back for a visit?" Shannon challenged for the benefit of Travis and Tramonte.

"Because I have family and friends and a boyfriend or two there."

"If they mean so much to you, why did you leave?" Shannon prodded.

"They weren't enough. The jobs I had were too boring. Spending every night out drinking in bars gets a little old after a while. I couldn't care less if I ever live there again but that doesn't mean I don't ever want to see the people again."

"Well, I made the break and I'm sticking to it," Shannon said.

"Then why don't you leave here and spend Christmas somewhere else, somewhere new?" Tramonte wondered.

"I can't afford it. Besides, I don't have anywhere else to go. Just think how much more conditioned I'll be when all of you get back."

"Brainwashed, you mean, don't you?" Travis clarified, counting her profits.

"Obnoxious," Dale countered, smiling.

"You don't really think I'm obnoxious, do you?" Shannon asked Dale, later, as they sat on the picnic table closest to the north laundry room.

"About as obnoxious as you think I am," Dale replied.

"Wow. That bad, huh?" Shannon exhaled cigarette smoke.

Dale pushed her a little too playfully, causing her to fall off the table. Regaining her balance, Shannon sat back down.

"I don't think you're obnoxious. It's just that sometimes you sound a little too gung-ho about the Army."

"Sometimes I feel very strongly about it."

"Think you're going to be a lifer?"

"You mean a career-oriented soldier?" Shannon corrected, then cocked her head, thoughtfully. "I don't know. I've been in six years now. I see no reason to get out yet. When it stops being a job and starts consuming my whole life then I'll look into something else. I think if what happened to you hadn't happened, you wouldn't be getting out."

Dale paused, pensively. "Maybe."

"So, really, what are you going to do with your two weeks?"

"Hmm...let's see...I will give myself a day or two to catch up on my rest and then I'm going back to Manchester, New Hampshire and spend a couple of days with a friend of mine, Tess Burke. She's a career counselor at the AFEES there and someone I was stationed with at Fort Lost-In-The-Fog. She also knew Carolyn Stuart on a social level so I am going to nose around and see what I can find out. If that turns into a dead end, I will come back here a few days ahead of schedule, stay with Anne Bishaye and her husband and try to get what I can on our current drill sergeants. Then Anne will drive me to the airport early on January second and I will hang around the USO room and come back here with a bus load of trainees."

She had not really thought about spending time with Anne Bishaye until that very moment when she vocalized it. She was still ambivalent about her feelings toward the woman, angry at her for her handling of the Kirk situation and acting so damned clinical about it and still powerfully attracted to the older officer's knockout presence and sizzling magnetism. Part of her hoped they would get some time alone and part of her prayed that Jack would not leave her side during her visit. She was torn between wanting and not wanting to be tempted.

"Are you sure you're not going to see your British bartender while you're home?" Shannon's voice interrupted her internal musing.

"Keith? No. I told you, it's over." It made her sad and relieved to admit that. Keith had become such a part of her routine, had become such a safety net for her that suddenly realizing he was no longer there for her to fall back on, to...well, *use*...was a revelation she wasn't sure she was ready for. Yet she knew this was how it had to be. She had to start being true to herself, whatever that meant at this point in her life, and that truth did not involve her ex-boyfriend. In fact, it may have no longer involved any man and although the sense of freedom in that thought was refreshing, it was also frustrating.

For years she had fought against the rumor that just because she was a female in the military, that she was gay, that just because she was a cop who also happened to be a woman, that she was gay. It wasn't that she was offended at being thought of as a lesbian, it was the *assumption* that irked her, her being considered a stereotype just because she had entered into an environment and a profession that was predominantly male-oriented and dominated. Because she desired to do and was good at an occupation considered manly, she was automatically thought of as too tough to tame and not feminine. She could have had sex with the most masculine male in the town square and had it broadcast on the evening news and people would have still suspected her of being a dyke. She couldn't escape it.

Dale had lost her virginity at seventeen to a boy she didn't even know. Her carnal history for the next seven years was strictly heterosexual, and she indifferently satisfied her needs when the urge struck her with whoever was available, and reasonably attractive, at the time. Sex didn't really mean much to her and it became more of a bodily function than a pleasurable reward. It was only now, since she had begun questioning her orientation, that she understood why she may have been so determined to sleep with a lot of men and that was not so much to prove to everyone else that she was straight but to convince herself, as well. The funny thing was it hadn't worked. She had really convinced no one, least of all herself.

But this realization that was now washing across her, that she might, indeed, be a lesbian and not just curious, this epiphany was a double-edged sword. The acknowledgement of her desire for the fairer sex, which might now finally set her free, would also now keep her imprisoned as Uncle Sam would not be quite so enthusiastic about her acting upon who her heart and loins told her to gravitate toward. If she decided to make a career out of serving in the military, she would have to hide any personal happiness she might find along the way. Although she really loved what she did for the Army, she hated the catch-22 of not being able to openly enjoy a whole life like her straight colleagues did. While she agonized about coming to terms with this realization, she wondered how anyone could think homosexuality could actually be a choice. Sure, one could choose to closet themselves and hide their true inclinations, but that was almost impossibly distressing, and one could also opt for coming out and face ridicule, ostracization, family banishment, religious exile and violence. Who, in their right mind, would openly choose that?

"So, it really is over, huh?" Shannon again broke Dale's introspection.

"We want different things." Boy, *that* was an understatement. "We have for a while. Better to make a clean break than to let it linger, you know?"

"Well, thinking about my own failed marriage, I agree."

"What do you say we head upstairs and hit the rack? We both have a long day tomorrow. Well," Dale grinned, impishly. "I do, anyway."

"Yeah. My heart bleeds for you." She took the last drag on her cigarette, extinguished it and flicked it out into the parking lot.

"You should have field stripped that butt and put it in your pocket. You'll just be picking it up tomorrow."

"My hope is they'll make you do police call before you go."

"I wouldn't make any bets on that." They were just about to leave the patio to return to the bay when they heard the sound of a car driving up. Caution being second nature to both of them, they waited, hidden by the east wall of the laundry room, holding their breath, hoping to catch a piece of conversation. All they heard was the muffled sound of two female voices.

After the car door shut and they heard the sound of the vehicle driving away, Dale and Shannon scrambled around the side to the north wall. Shannon peeked around the corner to see Snow head upstairs. She turned to Dale who had quietly moved to a better position to see who was in the car. When they heard the door of the female bay open and close, Dale joined Shannon.

"Who was it?" Dale asked.

"Snow. Who was in the car?"

"Bradbury."

"Why am I not surprised? I'm beginning to agree with you, except I think Bradbury gives all women a bad name, not just gay women." Shannon leaned back against the wall, staring at the ceiling. "Well...what do you think about this?"

"I think it was just sex, Shan. I think they both got the itch, found an opportunity to scratch and took it. I think if it had anything to do with why we are here, it wouldn't have happened so soon, and it would not have been so stupidly blatant. *Anyone* could have seen them, from here, from the second and third story bay windows, anyone."

"Anyone could have...but you and I *did*. I'd like to not say anything to anyone about this. Just keep it between the two of us and eventually, the four of us. When the time becomes right."

Dale eyed Shannon suspiciously. "What have you got up your sleeve? Blackmail?"

"Blackmail is such an ugly and illegal word, Dale. I mean for incentive. And convenience. Bradbury obviously couldn't care less but this is just an ace up my sleeve for Snow."

"I understand your motivation but what if it turns into something bigger."

"Do you honestly think it will?"

Dale thought about it, seriously. "No."

"Me, either," Shannon smiled, rubbing her hands together in glee. "I can't wait for the opportunity to wipe that smugness right off her face." She glanced at her watch. "Oops...light's out was five minutes ago."

They jogged to the stairway and up the steps to the second-floor landing. "Let's go in quietly so that she's not sure who came in after her."

Shannon nodded. "God, how depressing. Even *she's* getting some."

"Oh, don't sound so deprived. Ingersol is more than willing and I'm positive that Ms. Bravo company nineteen seventy-seven would be more than happy to accommodate you."

"Please. If I was a knothole, they'd both accommodate me, too."

"And you'd probably be cruel enough to leave those slivers in Bradbury's tongue and Ingersol's —"

"You're very ill, Dale," Shannon said, laughing.

PERMISSION TO RECOVER

"You wouldn't want me any other way," Dale teased as she opened the door. "See ya in two weeks," she whispered as they slipped inside.

# PART TWO

PERMISSION TO RECOVER

CHAPTER TWENTY-FOUR

January 1978 Georgia

Two weeks felt more like two days and before Dale knew it, she was flying into Marietta, Georgia. It was a neutral airfield where Anne Bishaye told Dale she could pick her up and not run the risk of being seen by anyone from Tenth Battalion. The original plan was for Dale to come back to Averill a day or two early, brief her boss on what she had accomplished while away, and then relax before she had to return to her mission at McCullough. Now the strategy was that Anne would drive Dale to Atlanta where she would drop the lieutenant off on the north side of town and Dale would take a cab to the Hartsfield-Jackson Atlanta International Airport, meeting up with a few of her fellow trainees for the bus ride back to post.

Dale conveniently came up with an excuse to get out of having to spend the night at the Bishayes' home. She had no idea what could or could not have transpired between her and Anne, especially with Jack not in the house. Maybe nothing at all would have happened but she was not willing to run that risk.

The thought of actually being confronted, face to face, with her feelings for the enchanting colonel in an environment that could have very well included a bedroom and an absent husband, was too much for Dale at this point of her self-discovery to deal with. Dale, knowing she wanted Anne Bishaye sexually, and knowing she couldn't, shouldn't have her, was enough for her not to tempt herself unnecessarily. She had not forgotten what Bishaye had said to her just before leaving Vermont but since the colonel made no move to act upon the intent from those words, or clarify them, and probably wouldn't, especially after the Kirk incident, Dale was left in carnal limbo.

The notion of being with Anne alone thrilled her while the

idea of not being with her alone relieved her. If Dale could have gotten out of meeting Anne at the airport that day, she would have. But she knew she had to get together with her boss, her friend, to not only brief her but to receive a report on what had been going on, if anything, during her absence.

Still, when Dale got into Bishaye's car and studied the beautiful woman behind the wheel as unobtrusively as possible, Dale suddenly wished she had returned the day before. And that the sensual colonel had seduced her, as she had so many times in her daydreams.

"You look exhausted," Anne said and smiled easily, as they drove out of Marietta toward Atlanta. "Some vacation. Walker looks more rested than you do."

"She probably is," Dale commented and stretched out as much as the front seat would allow. "So, while we're on the subject, what is going on with Walker?"

"Absolutely nothing. She said you would understand this. Swinegar has turned into a lovesick fool, which, fortunately, has gotten her nowhere. Michaelson plays a mean card game and Caffrey is a sore loser. Other than that everyone has been well behaved, minded their own business, done details and hung around. I understand the part about Michaelson and Caffrey...what about Swinegar?"

"Probably nothing. She's got a wild crush on Drill Sergeant Robbins and we thought since she stayed for Christmas exodus, we'd keep an eye on her behavior with him, being that he was one of the NCOs in charge of Christmas company. We figured if a move was to be made on him by her, or vice versa, this would have been the ideal time to do it. Obviously, nothing happened but I have a feeling that wasn't from lack of trying on Swinegar's part. But let me say again that both Shannon and I feel Swinegar's overzealous crush on Robbins is nothing more than just that."

"So, what about you? I can only assume you came up empty, too."

"Nothing else has happened since I called you last week."

"Was that you that called me last week?" Anne kidded. "I thought I was speaking to someone at a lumberjack convention during dueling buzzsaws."

"I had a problem with my phone," Dale told her.

"Really? I thought it might be that you didn't want to talk to me."

Dale looked at her, feeling caught. "Why wouldn't I want to talk to you?"

"Dale...I never know with you. So, refresh me since I could barely understand you."

"I talked to Stuart's girlfriend, who is helium from the shoulders up, did I mention that before? It was like conversing with a vacuum. She gave me absolutely no information. Then I spoke to everyone in Stuart's art class and that came up a dead end, too. She had only been in class for two weeks so nobody really got to know her. And nobody admitted to giving her a ride home after class at any time. Everybody had alibis for the night of the murder. I spoke with Stuart's mother, who let me look at all of her daughter's personal effects and that was a complete No Go, too."

"Was her mother able to help you with anything?"

"Nada. Zip. I guess because of Carolyn's *lifestyle*, she and her mother were barely speaking. They footed the bill for Carolyn's funeral but neither she nor her husband attended. In fact, they had already tossed a lot of her stuff. I was lucky to be able to get my hands on what I did. Not that it helped me."

"Were you able to track down the people who did show up at the funeral?"

"I was able to talk with a few but not all. Theresa Burke is looking into the rest of the crowd and said she would let me know immediately if she comes across anything odder than what we already have."

"Burke is a sharp woman. If something looks out of place, I am sure she will spot it."

"Which reminds me, could you possibly talk her into staying in? What a waste if she gets out."

"I agree. We discussed her situation when she called me about Stuart's murder. It was very unfair, how all that transpired with her, and I told her I would make a few calls. I've done that and, hopefully, it will make a difference."

"Well, I just saw her and she hasn't changed her mind yet."

"These things take time, Dale. You are too impatient."

"She doesn't have a lot of time left."

Anne sighed. "All right, I'll call her again and give her one of my patented Fort Ord MP pep talks."

"Jeez, don't bore her to death. I want you to give her incentive to stay, not remind her of why she wants to leave," Dale said with a grin.

The colonel arched an eyebrow and glanced across Dale. "Remind me again why I like you?"

"I can't, for the life of me, think of one reason." This is what Dale most enjoyed and now missed about her friendship with Bishaye, the playful banter that dangled between insult and flirting. Now, every word that left either of their lips seemed suspect for double meanings and the ease with which they traded barbs before had now become strained. With a confused Dale realizing the exact intent of her true feelings for the colonel, she felt she needed to be cautious with what she said and just how she said it. She did not want to make a fool of herself with this woman who caused her insides to shudder with pleasurable and unrequited anticipation at just the mere thought of her. She was startled out of her contemplation by a slight slap on her arm.

"Where'd you go?" Anne questioned.

"Nowhere. Just thinking."

"About?"

"The case," Dale lied. "So...I then contacted the same people Shannon saw before training started and nobody gave me anything new. Everybody's story is still the same, so that was a big waste of time." She yawned and stretched out again. "What's going on with the cadre? I really need a heads up here because they certainly are a mixed bunch."

"MacArthur and Kathan have been replaced. So I have no idea what to tell you to expect there."

"What about the others?"

"What do you want me to say, Dale? Yes, they are all a little crazy but they have a right to be with everything going on right now."

"Some may be a little crazier than others and I think they were that way long before this case developed into what it is." Greeted with dead silence, Dale decided to confront the one work issue she knew was throwing a wedge between them. They had danced around it long enough and she wanted it out in the open. "You know, you really pissed me off about the Kirk thing."

Anne immediately bristled and fired back. "Yeah? Well, you pissed me off, too. When you saw she was going to be trouble, you should have stayed away from her. But no, not you. Trouble is like a magnet to your steel. If you are anywhere near it, you draw it right to you."

Dale was a little taken aback by Anne's sharp tone and how her temper had flared so quickly on this subject, but backing down now was not an option. "I don't often tell you that you're wrong, Anne, you know that, because you rarely are and I respect your judgment. But you and your tight little cadre miscalled that one. I'm not sorry I got involved with her situation and I would do it all again except if I could have foreseen the outcome, I would have walked her out Main Gate myself in the middle of the night," Dale argued, her voice elevated to match Bishaye's.

"You disobeyed a direct order from me, Dale!" the colonel yelled at her, gripping the steering wheel so tightly that the color drained from her hands and fingers.

"What the fuck is with you and this order bullshit? You got past ordering me to do things a long time ago."

"Really? When was that?"

"When you started asking instead...when I thought we had become friends." Dale was sounding resentful now. "You know, *Colonel*, it's not just the Kirk thing, it's your whole attitude lately that pisses me off! And don't tell me that it's your job. Sell that crap to a farmer for fertilizer."

Bishaye suddenly pulled to the side of the road, far enough away from the flow of traffic where they would not be a hazard to any passing vehicles. She threw the car into park.

When she turned to face Dale, Bishaye placed her hand on the back of Dale's seat.

"You have no idea what goes on at my rank, what is expected of me in the position I have!" Anne spit out. "I'm already tired of your innuendo that I've somehow become as cold and as calculating as some of our superiors who are only trying to make rank so that they can conquer the world. I have an impossible job and you know it doesn't matter how many connections I have, I will forever be under a microscope doing what I do because I am a female in an experimental position of power! If I fail, you know it doesn't just look bad for me, Dale, it looks bad for all women in the military! You *know* that. You know that I am expected to walk on water and not even get the bottoms of my feet wet because if I do — despite the fact that my competitor, Colonel Joe Shit The Ragman, has made repeated obvious mistakes — if I make one, one mistake, he will be promoted and I will be removed and disgraced. Because in the DA's eyes, I couldn't cut it! And all because I'm a female, no other reason. So, don't you dare sit there and give me shit about things about which you can only scratch the surface!" Her eyes flashed angrily at every syllable.

"Fine! That part I can understand but what about us? You used to be able to talk to me about this stuff, you used to share with me the frustrations of your job, you used to tell me to watch what you were doing because some day I would learn from your mistakes and be able to do it better! You used to call me every once in a while to show me that you cared. You created me, Anne. You built me from scratch to be your perfect little soldier, molding me into exactly what the Department of the Army wanted, with you singularly being in charge of me. It was great when I was making you proud and you were getting all the attention when my apprehensions were valid and lauded. Then I get hurt and I'm no longer a priority. I know you were there in the hospital when I woke up but after that, you moved on. And I barely saw you or heard from you, like I was no longer useful so I was no longer important to you. And that hurt. A lot."

There. It was out. And the truth in Dale's words and the pain in her voice silenced both of them. When Dale looked into Anne's eyes, the depth of emotion was intangible. Before either woman could reply to what Dale had presented, Anne grabbed a fistful of the younger officer's jacket and yanked her closer, threading her other hand through Dale's hair and pulling their faces together. Before Dale could even contemplate pulling away, Anne seized Dale's lips, pressing against them with a passion that matched her previous anger.

The action surprised and stunned Dale. At first, she was too shocked to react any other way than rigidly, trying to comprehend the less-than-diplomatic moment and then it hit her. Anne Bishaye was kissing her! *Anne Bishaye!* The woman who occupied all of her lesbian fantasies. While the soft lips bore down on hers, it only took another second for Dale to snap out of her awkwardly confused state and begin responding to this dream come true.

Sliding one hand behind the colonel's head and the other cupping her face, Dale tried to adjust her breathing as she let instinct take over. Tentatively, she moved her lips against Bishaye's, afraid Anne would come to her senses and break contact and come up with a logical reason why what was happening, had happened. But then, since Dale never did anything tentatively and couldn't have cared less about logic at that point, she advanced more aggressively, daringly outlining the colonel's lips with her tongue, causing a low moan to emanate from Anne's throat.

Anne took that invitation and lightly sucked Dale's tongue into her mouth, dancing with it once it was there. Seconds, minutes passed as the kiss intensified and the women continued to explore a new side of their relationship, a different kind of need Dale never thought they would acknowledge, much less satisfy with one another.

Dale reveled in the softness of Anne's lips and, at the same time, was captured by the sheer virility behind the kiss, the power alone in the gesture. But she expected nothing less from Anne Bishaye as the colonel never did anything timidly. With every sexual nerve ending tingling, Dale knew they would either have to stop now or they never would. Even though they were somewhat isolated from the main road, they still could get caught and, as difficult as it was to put the brakes on, she really didn't want to run the risk of being discovered doing what they were doing.

Reluctantly breaking contact, Dale put pressure on Anne's shoulders to gently push herself away. Both women's eyes were closed, the only sound in the car being heavy breathing.

"I shouldn't have done that," Anne gasped and put her hand over her eyes as though she were shielding her exposed emotions from Dale.

"I shouldn't have let you," Dale responded, feeling just as vulnerable as Anne sounded.

"You couldn't have stopped it." The beautiful colonel moved her hand and looked at Dale.

"I know." Dale had still not opened her eyes yet.

"Dale...I —"

"Don't, okay?" Brown eyes met concerned yet still desire-filled blue ones. "We'll just sit here and analyze everything and it will ruin the moment."

"Because it can never happen again..."

"I know," Dale admitted and nodded. She finally got her breathing somewhat regulated. She knew it could never happen again but she still needed to know why not. "Why?"

"Why? I think you know why. I'm married. I'm your boss. The military would destroy us. I'm not gay. You're not gay. How many reasons do you want?"

"How can two people who aren't gay kiss like we just did?"

Anne looked down and conveyed embarrassment. "I don't know. I can't and won't try to explain it. You and I have something...primal...going on between us, we always have. It defies reasonable explanation. I know you usually take your cue from me and I'm sorry. I should have had better command of my actions."

Dale had to rein in the urge to scream, *I wanted you to kiss me!* Maybe she should have thrown herself at the woman who had just ignited her blood as Anne seemed to be desperately wrestling with her own self-control. Or maybe she would just let the older officer believe that Dale was strictly a lesbi-*Anne* and drop it.

Anne obviously still reeled from what had just taken place between them and took a deep breath, finally releasing her grip on Dale's jacket. This prompted the lieutenant to slide back toward the passenger side of the bench-type seat, away from the object of her affection.

"All this and you still talk like a CO." Dale smiled. She looked back up at the colonel and saw an expression that now seemed distant and sad. Dale interpreted it also as remorse and regret. "Hey, don't sweat it. We don't ever have to talk about it again."

Anne ran her hand through her auburn hair and focused her attention out the driver's side window. "I'm sorry."

"Yeah. You said that. So, why, uh, don't we get back on the road?" She watched as Anne put the car into drive and merged the vehicle back into traffic. "I guess this is it for private meetings for a while."

"Unless it's absolutely necessary, yes." She glanced at the younger woman next to her. "Especially in lieu of today—"

"Look, either we're going to talk about it or not. If we're not, stop referring to it. If we are, let's talk," Dale told her, annoyed.

"I just don't think talking will accomplish anything."

"Fine. Then how 'bout them Yankees, huh?" Dale was a little disturbed that she had started to choke up. She swallowed the lump in her throat and focused her attention out the passenger side window.

Anne couldn't help but laugh. "It's football season and you're living in Alabama, you'd fare better asking how them Crimson Tide are doing."

"Okay, I just want to say one thing and then I'll drop it. I know I get on your nerves. I know I can be unorthodox and hard to handle, but you could and you can always deal with me. Regardless of the obstacles put before us, we always worked through it. Whatever this is, whatever happened today, whatever has been happening to cause this friction, the good and the bad, which I'm not too sure aren't related, we will work through this, too. Because, despite my bitching, I believe in you, Anne. Always have and always will and I never want to do anything to let you down. And...um...if my letting you kiss me and then kissing you back in any way let you down, I apologize. I will readily admit that I suffer from a little bit of hero worship."

The colonel did not seem prepared for this confession from her young friend. After a minute of silence, Anne responded. "Well, you shouldn't."

"I know. I know better," Dale said and smirked. "We've changed, haven't we?" She received a nod from Anne and continued. "I thought it was just you at first but it's both of us, isn't it?"

"Unfortunately. You've changed, too, Dale. When you first

came to my company at Ord, you were wide-eyed and innocent, very respectful of rank and position. You were your average, every day, small town girl coming to the big city and very eager to please."

"I'm still eager to please. Except now it's me I'm eager to please and not the Army. You can't fault me for wising up." She knew it was more than evident that she was also eager to please Anne.

"You're cynical. You have every right to be, of course, but you're only twenty-four years old. You're too young to be so disillusioned. I just wish you hadn't been so Goddamned good at this."

Curiously, Dale studied her. "Why is that?"

"Because then you wouldn't have gotten hurt. And maybe if you hadn't been so good at it, you wouldn't have been MP material and you would have been recycled. Or you just would have been bounced to another MOS. I mean, who knows? You'd probably be out by now, doing something unrelated and productive and searching for your happily ever after."

"I don't believe in happily ever after." *Especially now*, she thought, *with you so close yet so very far*.

"No, you don't *now* because you have jaundiced ideals."

"If you had left me alone in Vermont, maybe I could have gotten my life on track but you forced me back into service and not even for something good. I mean, if I had to come back in, why am I being wasted on this chickenshit assignment?"

Anne paused momentarily. "I wouldn't call a what's happening in Tenth Battalion and a murder exactly chickenshit."

"Anne, we still don't know if Stuart's death is even related and we have no clue whether whoever is behind all this is even going to strike again. I can't speak for Shannon and I really don't know what her CID experience is, other than being on loan to them for a few months when she was an XO at Bliss, but my experience and skills are way beyond this. This isn't a sure thing and Shannon and I could spend another three months here and have no more than we have right now. In the meantime, I'm not sure my foot is going to make it. I'm not pleased with the idea that I could end up in a wheelchair for something that isn't even going to produce results. If you had intel that told me, for sure, that this cycle was going to be affected then I would have no problem doing my job and waiting for it to happen and then do what I had to do to get the case resolved. I should have been assigned to Stuart's murder before I was put here."

"A lot of things should have been done differently but they weren't. And..." Anne sounded almost wistful. "I thought it would be nice that you and I could be together again."

"Then why can't we be?" Dale asked bravely, squeezing her eyes shut.

"Please, Dale," Anne said, quietly. "You know I didn't mean it like that and you know why."

No, she didn't. Not really. Dale knew that Anne would reiterate about marriage, career, orientation but, as for why, when they both obviously felt the way they did, they could not discreetly meet every once in a while to get whatever this was out of their systems. "So, we're just supposed to ignore our feelings for each other? Forget that kiss?"

Silently, Bishaye drove and just stared straight ahead. Then, finally, she spoke. "I won't ever forget that kiss, Dale, and you have no idea how much I would love to do it again. But we can't. I love my husband and it could only end badly for you and me."

"Why? I know you love me, too." She had wondered when she had gone from sounding strong to sounding pathetic.

"Yes, Dale, I do. But not enough to give up my career for," she admitted, bluntly.

As difficult as it was to hear Anne's words, as much as it hurt, Dale knew it was harder for Anne to say them. Dale, knowing the colonel as well she did, would have been shocked if Anne had been anything less than direct. In a way, it was better that Anne was candid now than to let Dale think there was even a chance for them to be together. She wanted to ask Anne about her parting statement at the Rutland airport but realized that her boss would probably tell her that, like today, she had a lapse in judgment and leave it at that.

I've really missed you. More than you'll ever know or I want to admit. I need you near me, Dale, I can't lose you again, she'd said.

Even at face value, that was an odd declaration for someone

who didn't want to be more than friends.

The rest of the ride to Atlanta was uncomfortably quiet as both women became lost in their own thoughts, their own private regrets. Finally, the colonel brought Dale out of the hypnotizing hum of the car and the rhythm of the road seams being crossed by the tires.

"Are you okay?" Her tone was one of concern.

"I will be," Dale assured. "I'm a big girl."

"I know you are."

"I'm just confused about us, that's all."

"So am I," Anne admitted, truthfully. "But I have to keep things in perspective and the bottom line is an affair would not be productive for either one of us. It may be momentarily gratifying but the end result would be disastrous for us both. As tempting as the thought is..."

"So you have thought about it?"

Anne glanced at Dale, who was staring at her, then returned her attention to the road ahead of her. "It's crossed my mind. You have an energy, an intensity, that is very hard to resist and you and I connected from the moment we met. We have, basically, fulfilled all of the other's needs except sexually. Before you, the idea of being with a woman was very, very foreign to me. Not that I think it's wrong, it's just something that never interested me. I love men, I love Jack, but you drew something out in me that was different, unexpected. I have wanted to kiss you for a very long time, to know how your lips felt on mine, to feel that link which could not be accomplished any other way."

"Then why did you stop us?" Dale asked, as the memory of that kiss flooded every part of her body.

"Because I could see it spiraling out of control very quickly and I don't think either one of us would have been ready for that."

"And you have to be in control. Always. Don't you?" There was a hint of bitterness in Dale's tone.

"Yes," Anne responded. "I do."

Dale wanted to scream out of sheer frustration. If Anne only knew. She decided to change the subject. "When are you going to give your battalion speech? You're overdue."

"Tomorrow," Anne replied, apparently surprised by the abrupt change in conversation.

"Just make it more compelling than Colton's. What a piece of shit CO he is."

Anne sighed. "He isn't that bad. You two just got off on the wrong foot."

"Wrong foot? I could be a centipede and I still wouldn't have enough feet for us to get off on the right one."

Pulling up to the curb, around the corner from the taxi station, Anne spoke again. "Gee, Dale, I would love to rehash this with you but you need to get to the airport so that you can ride back to McCullough with your fellow trainees. Who knows what you could learn on the trip back?"

Dale laughed. "Oh, you're a crafty one, Mrs. Bishaye." They exchanged a tentative look, wanting to hug goodbye but neither daring to touch the other. "Well...I'd better get going then."

"Yes. I, uh, will see you tomorrow in the vast sea of trainees."

Getting out of the car, Dale removed her suitcase from the back seat. She walked around to the driver's side. "Thanks for the ride, Colonel. It was very...educational."

Anne nodded. "We're okay, right?"

"As okay as we can be under the circumstances, I guess." Dale put her hand up and made a small waving motion. "See you tomorrow."

"Bye," Anne said, and Dale knew the colonel was watching her. She could feel Anne's eyes on her.

CHAPTER TWENTY-FIVE

The ride to the airport by bus was short but incredibly lonely. Dale was numb and could not help but feel a loss at what had happened between her and the woman of her dreams. But every now and then, out of nowhere, a smile would split her face as the full recollection of the kiss would attack her brain and surge unabashedly to her groin, reminding her that she had kissed Anne Bishaye. It didn't matter that she could not tell anyone or that it would go no further between them. The memory would always be hers.

She had waited an hour in the USO room with other McCullough soldiers until the announcement was made that the coach had arrived to take the military personnel back to McCullough. Dale immensely enjoyed the bus ride back. In the two-and-a-half-hour trip, she, Kotski, Tramonte and Tierni shared a lot of tales from their exodus from post and two forbidden bottles of Mad Dog 20-20. Alcohol was not allowed on the vehicle but that didn't stop the only four women on the bus from partaking, when Tierni had produced the contraband from her large purse. If ever Dale needed a drink, it was now. Before they reached the main gates of McCullough, Tierni placed the two empty bottles back in their brown paper sack and stuck them in the overhead luggage rack. She ignored Dale's suggestion of dropping them down the commode, where even the most persistent drill sergeant would not have stuck his or her hands.

When the commercially chartered bus pulled up to the on-post depot, the riders were divided up and put onto two military coaches. The vehicles remained stationery, idling, while the civilian bus drove away. Within moments, an angry-looking male drill sergeant stepped purposefully onto the bus in which Dale and her pals were now seated. All conversation ceased as the staff sergeant established his presence by just standing there. He then raised his left arm high into the air and in his grasp was a familiar looking paper bag.

"Oh, shit," Tramonte whispered, closing her eyes.

"Whose whiskey bottles are these?" he roared, and slapped the bag with his right hand, causing the empty bottles inside to clink together. He hesitated a reasonable amount of time. *"I asked you a question! Whose whiskey bottles are these?"* Again, his query was greeted with silence. The drill sergeant scanned the blank faces. "Well, I'll tell you what...until somebody confesses, these buses ain't goin' nowhere. I don't care if your sorry asses sit here for three days and you all get nailed for AWOL."

The occupants of the bus returned his stare, looking genuinely innocent, with the exception of the only four females in the back,.

"Well, we weren't drinking whiskey, we were drinking wine, so maybe those really aren't ours," Tramonte whispered to Tierni, trying not to move her lips.

"They're ours," Kotski whispered back.

They all watched the staff sergeant exit their bus and climb aboard the next bus. Reading his obviously shouting lips, the Alpha women saw that he was repeating the same message to the other group.

"What are we going to do?" Tierni asked, in a desperate whisper.

"Nothing," Dale advised, under her breath.

"You should have shoved them down the toilet like Oakes said," Kotski intoned, softly, through clenched teeth.

"Shh, he's coming back," Tramonte warned.

The drill sergeant boarded the bus again and began walking down the aisle, stopping at each seat, bending down so that he could smell everyone's breath as they answered. "Are these your whiskey bottles?"

The response of *No, Drill Sergeant* was repeated until he reached Tramonte and Tierni. He studied them intently and then looked behind them at Dale and Kotski. Passing the four of them, he proceeded to ask the remaining males on the bus. When he got back up to the front, he instructed the driver to carry on.

PERMISSION TO RECOVER

As the driver pulled away, everyone breathed a sigh of relief.

"Why do you think he skipped us?" Tramonte asked.

"He probably didn't think we had the guts to do it, being female and all," Kotski commented.

"I never thought I would say thank God for discrimination, but I am so glad we got out of it," Tierni said.

So am I, Dale thought, not wanting to imagine having to be hauled before the battalion commander *now*.

January 1978, Alabama

The Alpha trainees were dropped off at Tenth Battalion at twelve-forty in the morning and signed in at 0045 hours. Dale was the last to file into the Orderly Room, surprised to find Shannon on CQ duty. As Dale signed in, Shannon had a knowing smile on her face.

"What?" Dale asked, wondering about her partner's smirk.

"You four smell like a still."

"Seriously?"

"Yeah," Shannon laughed. "You stink. What was your pleasure?"

"MD 20-20."

"Whoa! Good luck getting up tomorrow."

Dale smiled. "No task too tough. How are things here?"

"Real quiet."

"What time are you off CQ?"

"One."

"See ya upstairs?"

"No way. I'm going to bed. We've got a big day tomorrow, including a scheduled speech by the battalion commander."

"Oh, great," Dale said, sounding annoyed for Shannon's CQ runner's benefit. "That will probably be like a sedative, as if we need that."

"That's why I want to get my rest tonight, so that I am not tempted."

Dale grinned at her. "I wish I could be as dedicated as you," she stated sarcastically.

"Fuck you very much."

"You're welcome."

***

Morning came much too early, especially for those with a fermented grape hangover. Dale was not one of them, however. She fell asleep instantly and woke up clear-headed, her dreams overrun by having sex with Anne Bishaye. That was not going to make her seeing the delectable colonel that day any easier.

Surprisingly, no one missed formation, where the buzz was that one female did not return. Bonnie Kramer was officially AWOL at 0520 hours. Kramer was the last person Dale thought would not come back. She seemed pretty settled with the military lifestyle and the plans she had for herself and her husband when she got permanently assigned. Three females disobeyed orders and got married during exodus without permission from Battalion or prior counseling from the chaplain, which was required now that they were in the Army. Lesley Jaffe was now Lesley Flack, Lesley Minkler was now Lesley Horan and Tracy Travis was now Tracy Novak. The drill sergeants, however, refused to recognize their marital status and would still refer to them by their maiden names.

It was obvious that no one, with the exception of Christmas Company, had continued to do PT during the fifteen-day break. The platoon sergeants were clearly conscious of this and drilled their troops extra hard that morning, some to the point where eating chow afterward would have sent them reeling immediately into the latrine.

At 0800, all four companies that made up Tenth Battalion marched to Quigly Auditorium, directly across from the Alpha Company area, where they were to be addressed by the Battalion Commander. Command Sergeant Major Hernan Soledad was the first to appear on stage. He was a short but hefty man with a salt and pepper buzzcut. He stood stage right and spoke to the soldiers without the unnecessary aid of a microphone.

"*You will stand when Colonel Bishaye comes onto this stage and when she leaves! You will yell a good morning, ma'am that will be heard around this post! Am I understood?*"

"Yes, Sergeant Major!" the mostly male audience responded.

PERMISSION TO RECOVER

"What did you say, ladies?"

"Yes, Sergeant Major!"

"That's better." He looked across the crowd of maybe five hundred GIs, give or take a few drill sergeants. *"Battalion! Attention!"*

The entire auditorium was on its feet and at rigid attention when Anne Bishaye walked on stage and stood behind the podium. Dale couldn't see her with everyone standing but her heart traitorously fluttered when she heard the colonel's voice when she spoke into the microphone. "Good morning, Tenth Battalion,"

"Good morning, Ma'am!" the chorus of voices yelled, literally shaking the auditorium.

"Nice," she nodded and smiled. "At ease, take your seats," she commanded. In unison, her audience sat.

Dale observed Bishaye scanning the crowd and felt a mixture of sadness and pride. She turned to look at the reaction of her peers and she noticed a majority of the men were staring at Bishaye with startled admiration, obviously shocked at the colonel's movie star appearance. Dale suppressed a smile, knowing that they would soon find out that their battalion commander was as strict as she was gorgeous.

"I'm sure each one of you is finding your stay at sunny Fort McCullough a real pleasure," Anne began amid a lot of snickers and throat clearing. "I hope each of you had an enjoyable holiday. I know I did. Now, down to business. First, let me apologize if I repeat some of what you have already heard from your company commanders. However, some things are important enough to say again.

"I'd like to begin with the importance of OSUT. One Station Unit Training means you are going through Basic Combat Training and Advanced Individual Training — LE School, in your case — in the same company, with the same drill instructors, with the same trainees. This is also the first time that men and women will go through basic combat training together. It's all experimental and it's going to depend on all of you to prove to the Department of the Army whether or not it is possible to continue this kind of co-educational training, altering the entire program so that everyone, not just MP trainees, can go through BCT together with no separation for male and female except for billeting. That would mean the Women's Army Corps would be phased out, altogether, and we would just have one Army, which I personally think is long overdue."

Her smile was so contagious that most of the trainees found themselves smiling, too.

"Let me point out that Tenth Battalion's cadre is the finest. Believe me, for everything new that we're all going to be experiencing, we need them. Converting you from civilians to disciplined, proud soldiers is not as easy as it sounds. Your drill sergeants are not tough just for the hell of it. They don't get on your case just because they feel like picking on you. They are tough to make you tough, to make you the best and most highly motivated soldiers you can be. I have no doubt you won't remember your drill sergeants with a great deal of affection but I guarantee you *will* remember them and you will respect them. If you went through basic training as many times as they have, maybe you'd even understand them, but it isn't their job to make you like them.

"We're aware of how you are going to be feeling during the next couple of months. You're going to be more than tired. We realize that for some of you, your schedules have been completely reversed. You are getting up at 0445 instead of just getting in at that time." It wasn't a coincidence that she was looking directly at Dale when she said that and Dale wondered how she could always pick her out in a crowd so fast. "That's a considerable adjustment for some of you. Exhaustion is inevitable. But I would strongly advise you not to fall asleep during your training day.

"The PT — Physical Training — as you already know, is not like a high school gym class. It's hard, it's damned hard but you need it and don't expect it to ease off as training advances. If nothing else, it will only get harder. Also, you are going to march until your feet are ready to fall off. But keep marching. And you are going to run until you think you have no breath left. But keep running. When the PT tests come around, you'd better push yourselves to the max. It's all there inside of you, it is just a matter of attitude.

"I will also strongly suggest that nobody gets into a fight

while they are here. That might cost you time you don't have and some money and you are not making that much to lose. Besides, if you have the extra energy to fight, it's saying to your drill sergeants that you are not getting enough PT. Knowledge that your drill sergeants will then have to take out on your entire platoons and that's not going to make you very popular in the barracks.

"The chow will not conjure up memories of a home-cooked meal but the food won't kill you, either. The menus have been much improved during the past few years and, personally, I feel this battalion has one of the best mess halls on post. I eat here many times and I'm not dead yet, or at least that's what my husband tells me," Bishaye smiled, winking at the front row, causing a few sets of hands to be folded over crotches.

"I am sure you realize by now that you have selected one of the hardest times of the year to go through training. The dead of winter. Summer has its disadvantages, too, with snakes and poison ivy, heat prostration and sunstroke. But you are faced with pneumonia and frostbite. You will survive. You're not the first to deal with this kind of weather and you will not be the last.

"Most of you are here by choice. And this particular Military Occupational Specialty, or MOS, is your choice, also. However, just because you've opted to serve in the Military Police Corps doesn't mean that's what you'll be spending your entire enlistment time doing. You first have to get through basic training and then LE School and continue to maintain a certain standard to remain an MP.

"Being a military police officer is a privilege. It is one of the Army's elite corps and you must constantly earn the right to stay there. You will continually be on display. People will be listening to everything you say and watching everything you do. You cannot break the law, military or civilian, and expect to be forgiven because of what you do. You cannot break the law and bust somebody for the same things you do when you are not on duty. And off-duty is just a figure of speech to a member of the Military Police Corps, as you may not always be on the schedule but as an MP, you are always on duty. You will discover, especially after you have reached your first permanent duty station, that you must maintain the highest level of military standards at all times. You cannot be a hypocrite. You cannot preach what you do not practice. Am I getting my point across? You cannot be perfect but you'd better damned well try." Her tone was much more emphatic now.

"If you are prejudiced, you will have to get over it. It's hard to hold a prejudice against someone you are forced to eat, drink, sleep, and train with. You are literally spending twenty-four hours a day with each other. You will either learn to love them a lot or hate them intensely. Just remember, the person you may be making racial slurs against is getting the same training you are and may have to save your life one day. It's always good to have a clear conscience if that day ever comes. Not that you won't make enemies, you will, but the enemies you make in the military will fade away when they are out of your sight. But the friends you make are forever." Once more, she searched out Dale's face, zeroing in on it. "I can personally attest to that statement."

Dale's heart leapt in her chest again and she wondered why Anne was doing this to her. Was she purposely trying to torture her? Dale looked around, uncomfortably, hoping the people directly in front of her, beside her and behind her were wishing that affectionate expression on the colonel's face was for each one of them. She knew, unless anyone was privy to their situation, no one would have a clue that Anne was singling out Dale and, squirming in her chair, Dale wished the colonel would quit it.

Bishaye looked at the crowd and continued. "In closing, no applause, please, I need to mention AWOL. It means Absent Without Official Leave. It's not kid stuff, it's a federal offense. If you have a serious problem, talk it out with your drill sergeant, your senior drill sergeant, your first sergeant, your training officer, your company commander, your chaplain...even come to me, if no one else will listen. Allow us the chance to help you resolve your issues before you reach a point where you act irrationally. An AWOL offense is something that will follow you around the rest of your life. It will never go away and you *will* go to jail. So, I am asking you to think about your options very carefully before you make the choice to desert.

PERMISSION TO RECOVER

Unfortunately, as most of you already know, there was an incident in Alpha Company before Christmas exodus which made us all realize that sometimes your problems are much deeper than we think they are. Although these were unusual circumstances, we are not blind to the fact that it could happen again.

"Most of you, especially in this group, are here because you want to be, you freely enlisted, no one forced you to raise your right hand and you need to take responsibility for that choice. We understand that nothing will feel like it is going right at first but it will all fall into place, believe me. The Army is an entirely different environment than civilian life. It takes some getting used to, some major adjustments for those of you who have never been around the military. You are not in prison or a concentration camp. There is a method to our madness and it is all for your personal improvement, so suck it up. We want you to get through these cycles as much as you want to get through them. The end result is just as important to us. With that said, good luck during the next three months. You're going to need it. I hope to see you all at your graduations."

Everyone seemed to be so mesmerized by Bishaye, they didn't see Command Sergeant Major Soledad enter the stage and stand off to the colonel's right. When his harsh voice boomed out, it visibly startled the first six rows.

"Battalion! Attention!"

Recovering quickly, the members of Tenth Battalion were on their feet.

"Good morning, troops," Bishaye smiled.

"Good morning, Ma'am!" Tenth Battalion responded.

CHAPTER TWENTY-SIX

As third platoon stood in a haphazard formation on Range 24, they observed members of first platoon given instructions on how to operate a grenade launcher. The range commander joked that if anyone could hit the deuce-and-a-half truck they had camouflaged in the woods at the edge of the range, approximately five hundred meters away, he or she would be rewarded with a case of beer. With that incentive, everyone was aiming for that particular clump of trees with the exception of one of the male trainees and Private Caffrey, who were evidently trying to shoot down aircraft.

The temperatures had reached zero and clearly nobody wanted to be outside, including the instructors. Shannon, who was standing next to Tierni and behind Snow, folded her arms tightly across her chest and moved back and forth from one foot to the other. "Jay-zus, it's colder than Drill Sergeant Bradbury out here," Shannon commented loudly to Tierni.

Snow turned around quickly and glared at Shannon, who confidently competed with Snow's domineering gaze. She seemed puzzled by Walker's *I've got a secret* smugness that appeared to result in a rare feeling of intimidation. Snow turned back around and faced the range again, her cheeks burning a deep red. Her expression said it all. Was Walker guessing or did she somehow know something incriminating? Just for a moment, her self-satisfied air disappeared. But only for a moment.

After noon chow, no one being a case of beer richer nor an expert at launching grenades, the trainees were marched to a different range. This particular area was extremely close to Tenth Battalion. In fact, in the evenings, if any of the Alpha or Bravo trainees looked out their bay windows, the effects of night training were obvious because of the tracer rounds being shot.

PERMISSION TO RECOVER

Range 4 was the area used to teach trainees the fundamentals of Night Fire. As one of the courses that had to be passed in order to move on in the cycle, the trainees were going to familiarize themselves with how to engage and fire at targets in limited visibility. As with qualification on the M16, this event was also considered a Go or No Go situation. If a trainee achieved a particular level of requirement for whatever the training subject was, he or she received a Go. If the trainee failed to attain the minimum goal set for that training subject, he or she received a No Go. In most cases, depending on which exercise, a trainee was allowed three No Go's before being recycled, except the Alpha trainees were not informed of that at first. They were under the impression that if they No Go'd once, they were out of Alpha-10. Dale and Shannon knew differently but said nothing on this one, as they felt maybe it would be an added bonus if the trainee felt he or she was being given a second chance. On the other hand, maybe it produced more motivation if they felt they absolutely had to do it right the first time.

The range instructors talked about Night Fire to the trainees as the three platoons sat and shivered on a set of bleachers. They were advised about the three principles of this course, which were one, allow one's eyes to adapt to the dark or low levels of illumination. Two, be able to scan the area around the target every four to ten seconds and still be able to engage the target. Three, have the ability to look at a target at a six-to-ten-degree angle and still see it, called off-center vision. While the range NCOs discussed the rest of the day's and evening's activities, Dale looked around at the cadre. McCoy had shaved off his mustache, which gave him a much softer appearance and Ritchie was trying to grow one, at which he was failing miserably.

Second platoon had been introduced to Kathan's replacement that morning, a staff sergeant named Jay Holmquist. It was hard to tell what kind of drill instructor he was or would be because he had mainly observed, not yet opening his mouth. Regardless, it could not be denied that this new member of the Alpha-10 cadre was a handsome man. He looked to be about thirty years old, at the most. He was approximately five foot ten and weighed about one hundred sixty-five. He not only appeared as though he was in good physical condition, he also looked well fitted to his crisply starched uniform. His fair complexion was complimented by light brown hair, a mustache to match and piercing green eyes. He was quiet and it seemed as though he preferred to watch and take everything in. The more Dale studied him, the more she thought her concern needed to move from Robbins to Holmquist. If there were to be a setup, this would be the man who would be the target...especially if he was a flirt.

That night, at dusk, the Alpha company trainees were back out on the range for the Night Fire course. The weather had become increasingly worse, the ground frozen solid for at least two hours and the long johns, fatigues and regular winter issue did nothing to protect the trainees' unprepared bodies against the stinging wind and the biting cold. They had been on the range all day and, by their collective behavior, still had not acclimated to the bitter temperatures. Shannon smiled to herself, now glad she had stayed during Christmas exodus because it didn't help the others that had to get up at such an ungodly hour following two weeks of civilian merrymaking, or that they were run into the ground by unmerciful drill sergeants at morning PT.

Then there was that long speech in an auditorium that was warm enough to put them to sleep but given by a woman they visibly fought to stay awake for. After that, they spent the afternoon zeroing in, practicing, shooting at the targets, only to be advised that firing at night would be entirely different. That all the rehearsal in the world couldn't prepare them. Shannon heard a majority of trainees all grumbled the same thing *then what was the point?* No one of importance heard them.

Receiving their firing orders at 2100 hours, they consecutively took their positions as the range commander started broadcasting the rules on the public address system attached to the watchtower.

"The sun has been completely down for an hour and a half so your eyes should have gotten used to the dark by now," the disembodied voice told them.

Shannon listened closely, just wanting it to be over with so that she could get back inside the barracks where it was warm.

The voice continued. "...Rounds of 5.56 ball ammunition

and two rounds of caliber 5.56 tracer ammunition. You must get twenty hits out of eighty in order to obtain a Go. Your respective drill sergeants will keep score."

Bullshit, Shannon thought. This was just a formality. The drill sergeants could see as little, if not less than, the person who fired at the targets.

"Keep your weapons pointed down range. Ready on the right?" He received an okay signal with a flashlight. "Ready on the left?" Another okay flash. "Ready in the middle? Lock and load your first ten round magazine."

Shannon was in the second firing order and sat on the cold, hard ground behind Robert Snow, no relation to the Prof. He was in a prone position and watched his lane for pop-up targets. It was kind of like watching for shadows in a dark room. Shannon liked to see the tracer rounds as it reminded her of the laser fire from *Star Wars,* however, she would have gladly foregone the thrill to be in a nice, hot shower at that moment.

Wachsman, who was in the third firing order, sat five feet behind Shannon. It was the interval every firing order was placed in but Wachsman didn't stay there very long. "Oh, my God, my ass has turned into ice," Wachsman wailed and crawled to Shannon as the firing began.

"Mine, too, but it beats standing. We've been standing most of the day," Shannon reminded, her voice shaking.

"You don't seem to understand. I'm going to be shitting ice cubes and peeing icicles." Wachsman blew on her gloved hands. "Did your recruiter tell you it was going to be this cold here?"

"No, he said have a nice winter with a smile like a good bookie. God, I hate this. We have seven firing orders, it's nine-thirty already, we've been up since four and probably won't get to bed until two or three and they'll expect us to be up again by four. With no complaints. And we'll have to be wide awake," Shannon shouted so she could be heard above the gunfire and through earplugs.

"Tell me about it. I'll probably fall asleep while I'm firing. As if it will make any difference in my score."

"Wachsman, why aren't you five feet behind Walker?" It was Drill Sergeant Robbins, who had moved up on them, unnoticed.

"I'm sorry, Drill Sergeant but if I talk to Walker, it takes my mind off freezing to death. And we were discussing military subjects," she threw in for good measure.

"Is that so? Like what?"

"Like why is this necessary, Drill Sergeant?" Shannon interrupted and tried to sound annoyed. "No one can even see those targets, much less engage them. Isn't it a waste of time and ammunition?"

"Well, Private," he paused to sip steaming hot coffee out of a Styrofoam cup. "Your concern for the Army's resources touches me deeply." His tone was pleasant, not cutting, which let Shannon know he understood her real reason for griping. Exhaustion and possible frostbite. "But this exercise is necessary in your training. We need to see how you function during adverse and limited conditions. Wars don't end just because the sun has gone down or it gets cold and this *is* combat training."

"But, Drill Sergeant, we're training to become MPs, not front-line infantrymen," Wachsman protested. "I don't see where this is essential to our law enforcement careers."

"Private Wachsman," Robbins explained, almost gently. "Your actual police training does not start until you have completed basic combat training and basic combat training prepares you for the essentials of combat warfare. That's why this is necessary. Besides, it never hurts for any soldier to be so familiar with his or her weapon that he or she could operate it blind. No matter what MOS. Night fire helps you to do that."

Both Wachsman and Walker reluctantly nodded because he made sense. Robbins smiled, triumphantly, and he handed Wachsman his coffee cup, still three-quarters full.

"You two share this and don't tell anyone I gave it to you." He turned and walked away. Both women looked at one another, stunned.

"I think I'm in love," Wachsman stated as she watched him leave. She took a good swallow of the warm but quickly cooling off beverage and handed the cup to Walker, who looked at her for a few minutes. "What?" Wachsman asked in response to Shannon's odd stare.

"Nothing. Not really. I guess the wind chill factor is beginning to freeze my brain. I was about to say something enormously profound and witty like, one cup of coffee doesn't a good relationship make."

"Well, thank God you didn't."

Shannon nodded and handed the cup back to Wachsman.

Later, at eleven-forty-five, when the company was back in the barracks, Dale approached Shannon's bunk and stopped. "How do you think you did on our all-important Night Fire?"

"I'm not sure but I think I almost shot Ritchie."

"What makes you say that?"

"Because I aimed."

"Ah. Well, better luck next time."

CHAPTER TWENTY-SEVEN

The next morning, after extensive PT, Alpha Company marched to the troop medical clinic for another series of shots before they double-timed to Raburn Hall to attend one more Human Relations class. Dale paid little attention during the lecture as her thoughts seemed to be wrapped up with a certain colonel and the recollection of a very hot kiss. *Damn that woman for screwing me up even more with this issue*, Dale thought, yet still not able to keep the smile from her face.

The soldiers occupied their afternoon by cleaning their individual M16s on the two patios in their company area. During the couple of breaks the trainees took, two incidents occurred and both involved Private Vanessa McKnight. It confirmed suspicions Dale and Shannon had about her before Christmas exodus, that she was a conniving little snitch. First, she let it *slip* to Putnam that she heard a couple of the guys in Christmas Company lost money on bets they had made on the Rose Bowl and *excuse me, Drill Sergeant, but isn't gambling illegal*, she inquired, big Bambi eyes batting away at him. This conversation was overheard by Bonnie Saunders, as she emerged from the bay after using the latrine.

Second, McKnight asked Silva, the company driver, in front of Audi and half of First Platoon, how he enjoyed his leave in Atlanta with Tierni. Silva, embarrassed and a little shocked by her lack of honor among trainees, told her she must have mistaken him for someone else since he had spent Christmas in South Carolina with his family. McKnight apologized for her *faux pas*, but by then the damage was done.

Later that evening, after most details were completed and McKnight was taking her shower, Minty and Saunders took four cans of shaving cream that were willingly donated by the guys and filled McKnight's bunk with it. Then they remade the bed very carefully so McKnight wouldn't notice until she got in it. After discovering the deed fifteen minutes later, McKnight ran down to the Orderly Room, only to find two other trainees, Dave McElroy and John Pickett in charge. She ignored them and knocked on the First Sergeant's office door.

Karen Henning listened patiently as McKnight wailed and whined about how mistreated she was by her barracks mates and how she was trying really hard to get along with everyone. A few tears later, for effect and permission to be escorted down to the linen supply room by Pickett for clean sheets, McKnight left, feeling satisfied that justice would prevail.

Dale opened the door between Colton and Henning's offices, where she had finished her sweeping detail, the second Vanessa McKnight was safely away.

"Did you hear any of that?" Henning asked, her voice at a hush so McElroy wouldn't hear her.

"All of it."

"What about it?"

"She's got diarrhea of the mouth. She deserves anything she gets."

"I thought so." Henning rose, straightened her paperwork, and stood very close to Dale. "Do you think she might be in the running?"

"It's possible. But I am leaning more toward thinking she's just your average snitch you get with every cycle. The type who thinks it makes her look good by trying to make everyone else look bad. She's already got too many enemies and that includes some drill sergeants, I'm sure. They don't like her particular species any more than her fellow trainees do."

By the time Dale got back upstairs, McKnight had changed her linen but was still pouting. Koko approached Dale when she got to her bunk. "McKnight said Henning is reporting us all for agitating and provoking. You were down there. Is that true?"

Dale sighed and shook her head negatively and pulled her towel out of her locker. "Can't speak for the XO but it looked to me like she thought McKnight was a big baby."

"I knew it. She's too smart not to see through that bitch."

The temperatures were again below the freezing mark in the

morning. The dampness in the air, especially in the wind, cut right through to the bone and made it difficult to concentrate on any given command. Even exercise did nothing to warm the trainees up, in fact, sweating only seemed to intensify the bitter cold.

Upon his arrival to the company area at 0515, Colton heard about the McKnight situation. Henning innocently related the incident to him and included the retaliation by two second platoon females, thinking he would get as much of a kick out of it as she did.

He didn't.

When the platoons were divided for PT, Colton and Ritchie approached McCoy and Holmquist and spoke with them, off to the side. After a few minutes, McCoy returned to his platoon and ordered the males to Fall In to a different formation. When that task was completed, he made the females close ranks and intervals and he and Holmquist marched the men to another location.

Ritchie took charge of the second platoon females. He put them in the front leaning rest position while he castigated them *again* about their not getting along and lack of respect for one another. He further cautioned them against taking disciplinary action into their own hands. It wasn't their job. Ritchie kept them on the cold patio floor in the first count of the push-up exercise until even Michaelson's arms shook. For the next forty-five minutes, he drilled the women mercilessly, while Colton silently supervised. If anyone could not continue the particular exercise they were doing, Ritchie put her back into the front leaning rest position until everyone else completed the last repetition. The females who couldn't keep up were ordered to attend a special remedial PT class that evening after the training day was over.

Finally, he made them do a two-mile run. By that time, there were only two females, Michaelson and Ryder, who made it all the way. The rest fell out due to either injuries or not being able to breathe. Even Michaelson hacked and wheezed and was bent at the waist when the run ended.

Dale hadn't dropped out. She had fallen behind but she was driving herself to make it, favoring her bad foot. Every step she took brought her great agony but she refused to give in to Ritchie or Colton.

"You look hurt, Private, are you okay?" Holmquist jogged beside her.

She nodded, not having the breath to answer him, wondering where he came from.

They both glanced up to see Colton run back toward them. "Just don't push yourself too hard out of pride. It's not worth any permanent damage," Holmquist said.

"I'll handle this, Sergeant," Colton told him, crisply, after he had reached them.

"Yes, Sir," Holmquist said. He left them and caught up with the group Ritchie led.

"These women will soon realize that we will not tolerate this kind of behavior. Maybe Sergeant Ritchie can break a few of them in the process. Why couldn't you or Walker say something to calm them down? Or don't you have enough confidence to do your job?"

Dale stopped running and gawked at him as she limped around in a circle, walking off the run. She tried to catch her breath. "Listen to me, shit-for-brains," she panted. "This punishment is not going to stop McKnight from being how she is and, if anything else, this is going to bring more hostility toward her. And rightly so. She's a crybaby and a troublemaker. Just what is it that you expect my partner or me to do, anyway? We're trainees, you hemorrhoid, we have no power in the barracks! We're acting like we normally would if we were actually doing this for the first time. So get off my back."

She started to walk away from him with the intent to catch up with her platoon when Colton grabbed her upper arm. She easily got out of his grasp, surprising him with the fluidity of her movement. "Next time you lay a hand on me, I'll knock you on your fucking ass," she warned him.

"You watch your mouth, Oakes." He thrust his index finger at her. "Remember something, young lady, I am still your superior officer. You will address me accordingly." It was all he could manage to say as his machismo was backed down by her glare.

"Are you deaf as well as fucking stupid? I have been

addressing you accordingly this entire conversation. Don't try to play head games with me, Colton, you are way out of your league. Besides, you can't touch me, get it? I answer to Anne Bishaye and only to Anne Bishaye. Your threats are a waste of breath. I can walk out of this assignment any time I want to and if you're the cause of me leaving this case then *you* will answer to Anne Bishaye and I really don't think you want to do that." Dale could feel his hatred like an icy heat on her back as she walked away.

The rest of that morning was taken up with Drill and Ceremony, which the women almost mastered. They made sure they weren't going to be put down for any more push-ups that day. The men, however, weren't that fortunate. They were dropped every time one of them was caught messing up. By noon chow, the male trainees were almost as tired and sore as the females.

First Aid classes at Raburn Hall took up the afternoon. The trainees learned how to perform artificial respiration, which most of the second platoon women no doubt thought was about four hours too late, plus how to splint fractures and dress wounds. Holmquist was quietly monitoring the class from the back of the room. When everyone partnered off, he joked with Henning, and asked her if he could practice his heart massage technique on her. He was so disarming about it that it seemed Henning clearly felt more amused than disrespected.

The dreaded remedial PT class was cancelled after evening chow for unknown reasons. Even while Dale nosed around when she swept the Orderly Room, she picked up no hints as to why. She then assumed that the drill sergeants, who had exhausted themselves running their platoons ragged, were too tired to supervise another hour of physical training. On her way upstairs after she finished her detail, Dale ran directly into Henning, who had just left the First Sergeant's office and was on her way to her car.

Dale rendered the proper greeting and after Henning returned the salute, she pointed to Dale's combat boots. "I want you at Sick Call tomorrow and have that foot checked out. No more fooling around, Private Oakes."

PERMISSION TO RECOVER

Dale knew Henning had witnessed her hobbling back to the company area that morning. Dale nodded and played the part of the dutiful trainee for the benefit of others who still milled around. "Yes, Ma'am."

Damn it, she thought, *I don't need this right now.* She knew Henning wasn't aware of the can of worms she could open by insisting Dale report to the troop medical clinic. If they had someone competent x-ray her foot, Dale would be placed on profile, which was a military medical term principally based upon a soldier's physical condition and how that directly related to him or her being medically qualified and able to adequately perform his or her military duties.

If that happened, she would be taken out of training every morning for at least two weeks to be put through physical therapy but, more than that, if the podiatrist really did his or her job, an investigation would probably be launched into why anyone with an obvious foot injury like hers had ever been allowed into basic training in the first place. Bishaye would have to step in and then at least one more person would have to be let into this little private circle of knowledge. It wasn't worth it. Dale would have to find an excuse not to go and then call Anne to tell Henning to back off on the foot issue.

***

The women around Minty's bunk planned an unhealthy immediate future for McKnight. When Dale walked by the group, she was stopped.

"Hey, Oakes, you want in on this?" McTague asked.

"On what?"

"A major blanket party for McKnight," Minty said.

"No, thanks," Dale shook her head.

"Why not? It was her fault none of us could breathe half the day," Saunders said.

"Look," Dale said, not unpleasantly but in no mood to be congenial. "The last time I got involved with somebody's problems, she ended up dead. I sympathize with you but no thanks."

The sober reminder of Kirk's death silenced the group of

five. At least until Dale left their area. On her way to Shannon's bunk, Dale heard a small group discuss the First Aid class.

"I'm sorry," Travis stated. "But a sucking chest wound sounds like something out of an X-rated horror movie."

The remark made Dale smile and she relaxed for the first time all day. Once she got to Shannon's bed, she saw that her partner was sound asleep. Dale sighed, stretched and headed for the shower. As the hot water streamed over her body, she couldn't get her mind off Anne Bishaye. She closed her eyes and unintentionally moaned at the recollection of Bishaye's soft lips on hers.

"Hey...I don't know who's in the last stall but cut it out!" Travis shouted from the first stall.

"Oh, behave. It's just my reaction to the temperature of the water on my aching muscles," Dale shouted back and mentally slapped herself for not having better control.

"Either you're lying or I'm taking a shower all wrong," Travis responded.

CHAPTER TWENTY-EIGHT

The trainees were loaded up in four two-and-a-half ton trucks and transported to one of the grenade ranges that were mainly used for practice. The deuce-and-a-half was not a vehicle designed with comfort in mind. The twenty-minute ride to the range was extremely bumpy and the bench-like seats attached to both sides of the truck's interior were hard and cold. At one point, the truck Dale rode in hit a crater in the road that sent everyone airborne and into a heap in the middle. A few wondered out loud if they had broken some bones.

Once off the trucks and into formation, the drill sergeants turned the troops over to the range instructors. The three platoons were informed of what they were expected to encounter that day. Before they were issued practice hand grenades, the trainees were divided up into groups of ten and shown three throwing positions, kneeling, standing, and prone-to-kneeling. Quite a few were warned and reprimanded for tossing the weapons like John Wayne or Sandy Koufax.

There were also told about the five stations they would have to pass before getting a Go. Each station involved the engagement of targets, three included silhouette targets, one involved a strike against a machine gun position and the final one required throwing from a bombed outbuilding to a vehicle on the road. They were shown different types of grenades, as well as their capabilities, functions, characteristics and how to grip them. If the weather hadn't been so disagreeable, Dale knew that the training had the possibility of being fun.

Damp, cold and muddy, the troops marched back to Raburn Hall after chow. They attended another morals class by Chaplain Harrison, which was a waste because everyone was too miserable to pay attention and then they returned to the company area where PT was conducted.

The drill sergeants directed PT differently this time. Each senior and junior platoon sergeant led his group of trainees in two exercises and then the trainees rotated to the next two platoon sergeants and repeated the action until all PT was complete. Audi was still on his own leading first platoon since MacArthur's departure but a replacement was expected shortly. The variation on the dreaded exercises helped break up what felt like an extremely long afternoon.

Dale observed the budding alliance between Holmquist and McCoy and was pleased at the way they worked together. They both appeared to be on the same wavelength, unlike McCoy and his dense-as-a-rock former partner, Kathan. Holmquist seemed quite professional but he also, clearly, had fun with his job. He and McCoy were a nice balance, as Dale and Shannon discussed later when they did laundry.

"Holmquist thought he was real cute this afternoon," Dale said and smiled. She looked around to see two first platoon males deeply involved in conversation. They were the only four people in the laundry room.

"Oh, you mean when he gave that command and everybody performed a Right Face and what he really commanded was Fried Fish?" Shannon separated her clothes from Dale's after she removed them from the dryer.

"I thought he said Right Face, too, and I knew the joke." She leaned in close to Shannon. "*I did it once,*" she whispered.

"I think he's going to be okay to have around," Shannon said. "I thought it was funny when he put Hepburn down for pushups and when Hepburn protested with *I didn't say anything!* Holmquist told him that he listens like a smart ass."

"Hepburn is a smart ass. A misogynist smart ass. Holmquist is very perceptive," Dale said.

"He's really attractive, too." She looked pointedly at Dale.

"I didn't notice," Dale told her, smirking. "And neither should you."

"You didn't notice? You are such a liar!" Shannon poked her, playfully. She lowered her voice. "I'm only saying it because I think we need to keep an eye on him."

"Oh, now who's the liar?" Dale laughed. "You'll keep an eye on him, all right."

"Hey, he's more your type than mine," Shannon said.

Dale shrugged. "At one time." She looked at the other two occupants who had just finished their boots and held them up like they were trophies. "Besides, now more than ever, it's look-but-don't-touch, remember?"

"Don't remind me. I doubt there will be any nookie until we're out of here. We're in E's up to our eyeballs," she said, referring to enlisted personnel.

The two male trainees paused before they exited the laundry room. One stopped and spoke to them. "You girls better hurry up. Lights out in five." He smiled and winked at Shannon.

"Okay, thanks, Darrell," Shannon said and returned his smile. After the door closed, her focus returned to Dale who shook her head and laughed.

"Some things just never change. Man…if you were still enlisted, you'd have yourself a smorgasbord here."

"Like you wouldn't?" Shannon challenged.

"Come on, Shan, I didn't the first time. You were the one with all the luck in that area."

"It wasn't luck, it was skill." Shannon's tone was highly amused. "Hey, before I forget, what did Colton have to offer yesterday?" She threw clothes Dale's way.

"Did you get all my underwear this time?" Dale checked the empty dryer. "I don't need a repeat of Audi strolling through the bay with my panties dangling from his pinkie, announcing to the whole world that one of his Joes found them in his laundry."

"When did that happen?"

"Before Christmas exodus."

"I must have missed that," Shannon said.

"I think you were on CQ…or downstairs, flirting with God-knows-who at the time," Dale said. "Probably someone you met in the laundry room while you should have been paying attention to our laundry."

"Whose laundry did they get mixed up with?"

"Drago's." Dale grimaced.

Shannon burst out laughing. "So now everyone thinks you slept with Fat Frank?"

"No. They think you did. I told Audi the panties were yours." Dale smirked.

"You didn't!"

Dale nodded. "I did. I told him I recognized that pair because I had done your laundry that night. I told him you weren't feeling well."

"You puke!" Shannon punched her in the arm, not so playfully this time.

"Oh, I'm the puke? How come it was so funny when it was my underwear?" Dale asked and rubbed her arm.

"Lucky for you nobody really thinks I slept with him. That's a rumor I would have heard long before now."

"Yeah, I don't think anyone bought it. Although I do think Drago was hoping everyone did."

Shannon shuddered. "The horror. Subject change, please."

"Colton. The man needs an exorcist, although not as much as Ritchie. Ritchie is evil and Colton, in all his arrogance, just doesn't have a clue. He thought by doing what he did yesterday that it would cure us from taking another problem into our own hands as a group. I also think he thought he was getting to me. The guy needs some serious therapy."

She put her folded clothes into the laundry basket and sighed. "What really pisses me off, though, is that a lot of the women think Henning betrayed them. Especially after I told Minty that Henning saw right through McKnight."

"I'm sure she's thinking the same thing. Someone will have to get past it and my guess is it won't be Stubby."

"I also hope she doesn't try to discipline me because I didn't go to sick call but I just don't feel I can miss anything right now." Dale checked her watch. "We need to get upstairs. Do you think I should speak to Bishaye about Colton?"

Please concur…any reason to see her again would work but a legitimate one would work so much better.

"Nah. We're big girls, Dale." Shannon held the door open for Dale and they crossed the patio. "We can handle him unless he gets extreme. I'm sure just the thought of the next couple of months with us, depending, is giving him an ulcer."

Dale laughed as they reached the stairs. "Yeah. The same thought is probably giving Bishaye an ulcer, too."

The next morning, after PT, the occupants of all four bays

prepared for a barracks and issue inspection by Colton. The entire morning the soldiers stripped off old wax, mopped, re-waxed and buffed the floors, scrubbed the latrines and showers, straightened lockers to military perfection and arranged field-issued equipment. The gear was uniformly lined up on the top of the bunks so that the inspecting officer could easily review it.

The drill sergeants' behavior drove the women crazy and they wondered aloud if the same thing was going on in the male bays. Every time a different drill sergeant would go through the barracks and check the progress of the preparation, he would contradict the previous drill sergeant. The fourth time this happened, the women congregated and collaboratively wondered if the drill sergeants were doing it on purpose and if they were being set up. The next drill sergeant who entered the bay was confronted by an angry mob. Fortunately for the women, it was the gentle-natured Audi.

"At Ease!" He commanded and the women assumed the position and quieted down. "There is no conspiracy. Each drill sergeant has his own way of doing things and often it conflicts with someone else's way."

"Drill Sergeant, shouldn't there only be one way? The Army way?" Caffrey asked. Caffrey had settled down since her arrival, to the point of being almost non-existent. After a few run-ins with women of stronger personality, she realized she had met more than her match.

"That's an excellent point, Private Caffrey. We do teach you the Army way…as interpreted by each of us. All the ways we show you are the right way."

"Then what do we do, Drill Sergeant?" Minty asked. "Because we just get everything lined up the way Drill Sergeant Robbins says it should be and then Drill Sergeant Holmquist comes up and tells us to do it another way. We just get that done and Drill Sergeant McCoy comes up and says, no, it's this way. Twenty minutes ago, Drill Sergeant Putnam told us we were wrong and it should be his way."

"My advice is that each of you should probably listen to your individual platoon sergeants and do it their way," Audi said.

"But even they contradict each other, Drill Sergeant,"

Caffrey said.

"Well, then, I guess First Platoon lucks out because they only have me," he said to the sound of the women in First Platoon collectively sighing in relief. "I will speak to the other drill sergeants about the confusion."

"Thank you, Drill Sergeant," Minty said.

Fifteen minutes after Audi left the female bay, Putnam walked into the barracks and contradicted everything Audi had told them to do. When they spoke out, he put them all in the front leaning rest position. Five minutes later, after their unanimous compliance, he told them to recover and everything was modified to Putnam's guidelines.

At eleven hundred hours, three drill sergeants entered the female bay and announced they were going to stand by for the company commander. Instead they got the senior drill sergeant. While Colton started on the second-floor barracks with two platoon sergeants in tow, Ritchie inspected the women's side.

He stormed into the bay with a growl and a snarl. It was not a good sign. He began with the first locker of First Platoon, Almstead's, and verbally tore her apart. As he made his way through the first row of lockers, his nature got worse and his comments more vicious. By the time he reached the Second Platoon females' lockers, he had begun to throw things with no regard as to where the items landed, who they hit or if the items broke on impact. He conducted his inspection as far as Mroz, then breezed by everyone else so fast, they almost got windburns.

"Just what exactly did all of you do all morning?" Ritchie bellowed. "Your hair? Your nails? You certainly didn't prepare for this inspection. You're Goddamned lucky I did the inspection instead of the old man. This bay is a disgrace! Your lockers and personal areas are a joke! You women are a joke! You're royally fucking up, as usual!" He pointed to the floor, where he had tracked in mud. "Look at these floors! They're unacceptable. These bunks? I couldn't bounce a quarter off these blankets if it had its own spring. I won't even attempt to go into the latrine because if I do, you will all be sorrier than you already are. Time is running out, ladies! You have to do a hundred and ten percent better than this or none of you will make it to LE School!" Ritchie shook his head in disgust and stomped out of the bay.

The three drill sergeants appeared to be too stunned to follow and the women were shocked into total silence. Dale turned her head and sneaked a glance back at Shannon, whose expression said what Dale felt. *What the fuck was that about?*

The somber silence continued through noon mess, PT, and evening chow. As Shannon reached the ground floor to take a cigarette break before Lights Out, she met Dale on her way back up to the bay. She had just finished her detail of sweeping the patio.

"How are things upstairs?" Dale asked.

"Crazy. Nobody knows what they're supposed to do and they're discouraged. I know the drill sergeants are supposed to be tough but that's to make us better, not make us feel hopeless," Shannon said, clearly still worked up. "What did you think?"

"Bullshit. That's what I thought. Ritchie is a synonym for bullshit. Holmquist called us all into the laundry room afterward and you could tell he was pissed. He told us he wasn't pleased with the way we cleaned up but he wasn't displeased, either. He told us we needed to pay more attention to detail and then he told us he thought we deserved a little more consideration from the senior drill sergeant."

"Putnam wasn't as diplomatic. He told us he was really proud of the way Third Herd looked and, as our drill sergeant, he wanted as much gratification as we did. He told us we done good and he thought the whole thing stank. Now, granted, all that is encouraging but if the senior drill negates it, all the encouragement in the world won't help. You and I know a majority of the women will make it through but it will be a lot less with this kind of incentive."

"You know it's got to be bad when members of the cadre can't hold themselves back from expressing their dismay. I just don't know what to do about it," Dale said.

"I don't think we have much of a choice other than to ride with it. Keep taking your notes. When all of this is over, if he isn't involved in why we are here, we can nail him for all this other crap."

"If he doesn't kill somebody before that," Dale said, pragmatically.

Shannon took a long drag off her cigarette. "I don't know about you but if it comes to him or me, I guarantee you, it won't be me."

"I'm not worried about you or me. He obviously didn't learn his lesson with Kirk. He is doing everything in his power to break these women. Have you looked at some of them? Hewett? DeAmelia? Newcomb? I think he's succeeding."

"Yeah. But, Dale, they're weak to begin with. They're always bringing up the rear in everything. Even if Ritchie was non-existent, I don't hold out much hope for those three."

"Maybe they'd do better if they had more positive reinforcement. Hard to say, I guess. You're on your own tomorrow, by the way. I have CQ with Mroz."

"Well, it's Sunday and everyone's restricted so that's not a problem." As Dale passed her, Shannon grabbed her arm. "Keep an eye on Mroz since you're going to have concentrated time with her."

Dale cocked her head in curiosity. "Mroz? Since when?"

"It's nothing concrete. Just a feeling. She's pushy...and then ingratiating. There's something about her that just isn't right."

"I haven't gotten that from her but I'll definitely watch her," Dale said before returning upstairs.

CHAPTER TWENTY-NINE

Sunday was the easiest day of the week to be assigned to Charge of Quarters duty. The worst problem was how to pass the time. Dale and Mroz switched on and off doing periodic checks of the company area but with no incidents to keep them occupied, they spent most of the eight-hour shift telling jokes and exchanging tales of their pasts. Of course, Dale made up most of her stories and wondered if Mroz did the same but the enthusiasm with which Mroz spoke of her adventures made Dale feel as though she hadn't fabricated her history.

In the afternoon, a couple of hours before Dale's CQ shift ended, Ritchie stopped into the office to visit Robbins, who was the Staff Duty NCO. The senior drill sergeant's presence was not a welcome one and since it was his day off, his appearance was unexpected. Ritchie's demeanor and unpredictability altered the former lighthearted mood and placed both women on edge. They eagerly anticipated his departure and hoped it would be before anything they did set him off on one of his unprovoked tirades.

As if Dale didn't have enough to fill a notebook concerning Ritchie's unprofessional behavior, she and Mroz witnessed another breach of protocol and both experienced different degrees of shock when Robbins played along. The two drill sergeants walked outside the Orderly Room and returned minutes later, laughing. They didn't even try to hide the fact that they were ridiculing Henning and found fault with everything about her. Once they tore her apart as an officer, they moved on to her personal life.

"She doesn't have a boyfriend," Robbins said.

"No guy would want her," Ritchie said. "Maybe she has a girlfriend." Then he laughed. "Nah, women don't want her, either. She's just an all-around loser." There was another round of guffaws.

Dale had to look away from them to disguise the visible anger burning within her. She glanced at Mroz and saw that Mroz could not hide her shock and disappointment, either.

Ritchie caught Mroz's expression and his eyes narrowed. "What are you looking at, Mroz?" he snapped. "Nothing that is said in this room leaves this room! Is that understood, Private?"

Before she could respond, Holmquist entered the Orderly Room through the First Sergeant's office. "Private Mroz, Private LaForest has lost his locker keys. Take the bolt cutters up there and open it for him, please."

"Yes, Drill Sergeant," Mroz answered, quickly, noticeably grateful for an excuse to leave the room. She grabbed the bolt cutters and fled.

"Hey, Jay, we were discussing our company albatross, Henning. What are your thoughts on her majesty?" Ritchie grinned like a mule.

Holmquist studied the senior drill sergeant with a disgusted expression. He remained respectful of Ritchie's rank but clearly, at the moment, he didn't think much of the man behind the stripes. Holmquist looked at Dale who stared back at him as though she expected him to join the party.

"Excuse me," Holmquist said and passed between Ritchie and Robbins when he exited the office.

"Give him another week around her, he'll join right in," Ritchie said, unaffected by Holmquist's maturity. Robbins, however, now seemed embarrassed. He looked at Dale, cleared his throat then bowed his head and left the Orderly Room. Before Ritchie followed Robbins outside, he turned to Dale and pointed at her. "What I said to Mroz goes for you, too, Oakes."

"Yes, Drill Sergeant," Dale said only because she had to.

Shannon practiced G-3 testing with Wachsman. G-3 was a part of the four Gs, short for General Staff. G-1 set personal policies, studied the Army's manpower problems and was responsible for the hiring of civilian employees. G-2 was the intelligence branch. G-3 was the operations branch that covered troop training, troop information and education, special services, maneuvers, field problems and other miscellaneous responsibilities. G-4 was the supply and logistics branch. All trainees were required to go through G-3 testing before they graduated from basic training.

After Wachsman questioned Shannon on military ranks, they were called downstairs for noon chow. On her way out of the bay, Shannon passed Snow who was doing push-ups between her bunk and Steele's. Shannon couldn't resist.

"Hey, Professor, looks to me like you lost your girl."

The remark set Snow's teeth on edge and caused her to lose count. She held the front leaning rest position until she got her bearing. She ignored Shannon and resumed the exercise.

Once inside the Mess Hall, because it was Sunday, the trainees were allowed to talk freely. They all took full advantage of that.

"What do we have here this noon?" Wachsman mused as she scanned everything edible before her. "Ahh...gruel."

"What difference does it make what it is or what it looks like? Trainees don't taste food anyway. It's not allowed," Shannon said. She moved her tray through the chow line. "You know, the old inhale it and get out policy."

Ahead of Shannon in line, Travis was perturbed about an item on her plate. "I didn't want this."

"Take what you want but eat what you take," a drill sergeant from Charlie company recited. He was there to maintain order in the chow line.

"But I didn't want this," Travis complained.

"Then why did you take it, young lady?"

"I didn't, Drill Sergeant, she put it on my plate." Travis nodded toward a server.

"Then I guess you'll just have to buck up and eat it, young lady."

"I hate spinach," Travis mumbled, all the way to her booth.

"Stop bitching and eat the damned spinach," Ryder said, and slid in next to her. "It'll put color in your cheeks."

"Who the fuck wants green cheeks? I'm seventy-five percent green now! I'm going to od on o.d."

By the time Snow got served, most of the trainees were seated and she spied only two available places. One was at a booth with Travis, the other was a booth with Shannon. Snow decided to take what she felt was the lesser of two evils, at least that day.

Travis, especially after the spinach ordeal, was not pleased with Snow's choice. As she studied the last item left on her plate by pushing it around with her fork, Travis sullenly listened to the conversation of the other three women who spoke of past civilian interests.

"I was in a band once," Brewer said as she finished her coffee. "I always wanted to play a musical instrument but I never had the time to learn so I sang instead."

"I bet you were good," Ryder said. "You have a very melodic speaking voice."

"Thank you." Brewer said.

"I played a musical instrument once," Snow said. "But I had to give it up."

"Why?" Travis asked and eyeballed her suspiciously. "Did your monkey die?"

Snow slammed her fork down on her tray. "What? What is it? Do I have a target on my back today or something?"

"No," Travis said and smirked. "But thanks for the idea." She picked up her tray and moved out of the booth.

"Travis, you didn't finish your spinach," Ryder pointed out.

"I know, Mommy. I've decided to take my chances with punishment." She lucked out, though. The drill sergeant was busy with another trainee so Travis dumped her tray and escaped, unscathed.

Later that evening, after the women had returned to the bay from the Dayroom, where they uniformly marked their military clothing according to regulations, Dale still seethed about the CQ incident. She didn't have to tell Shannon. By the time she got upstairs, Mroz had already spread the word regarding what jerks Ritchie and Robbins had been. The fact that Dale didn't refute Mroz's story told Shannon all she needed to know. The only detail Dale added before bed check was that Holmquist didn't participate.

The next morning began the coldest, bitterest day so far. Nothing was able to help the trainees, or cadre for that matter, maintain body heat.

PERMISSION TO RECOVER

The company was issued equipment for Bivouac from the Bravo Company supply room. The instruction on how to pack it all up properly to keep it minimal and dry and how to keep warm in specific weather conditions, had to be given inside. First Platoon conducted their classes in the First Platoon male bay, Second Platoon in the Second Platoon male bay and Third Platoon in the Third Platoon male bay. The major problem in that concept was that the troops couldn't be shown how to pitch a tent, being that it was difficult to dig a trench and to get the wooden tent pegs to stick in linoleum.

The drill sergeants assured the trainees if the cold wave didn't break that Bivouac, scheduled to start the next day, would have to be postponed. A notice was passed around that one training company who had gone out that morning had to be brought back by buses because two GIs landed in the hospital with frostbite on the lungs.

The rest of the day was spent in the individual male barracks. Bunks were pushed aside and classes were held with lectures and diagrams on all the information the trainees would need for field survival. To break the monotony, just before evening chow, the company spent minimal time outside and practiced Drill and Ceremony. Fifteen minutes was all the drill sergeants could stand in the freezing cold before they dismissed their troops.

All the trainees retired early, well before Lights Out, because if Bivouac continued on schedule, they would be awakened excruciatingly early and no one was sure when they would get the next decent shuteye.

CHAPTER THIRTY

The word came up before morning formation that Bivouac was rescheduled due to an unusual weather front that was predicted to hit the area in the next couple of days. This should have been good news but since everyone's duffel bags were already packed and ready to go, the company commander decided to make use of the effort. As cold as it still was, Alpha Company prepared for a hike.

Formation was called after breakfast and the trainees stood in their respective platoons, equipped with seventy-five-pound backpacks, ready to conquer a notoriously steep hill, affectionately referred to as Coronary Climb. It was a five-mile jaunt from the company area to the peak and the same distance back unless the drill sergeants decided to take another route.

The temperature started out in the low teens and that caused the new GIs to quietly grumble, curse, and speculate that anything important, like extremities, just might freeze and fall off. After they marched approximately two hundred feet, dressed in several layers with the duffel bags strapped on, they heated up quickly. When they reached the base of the mountain and began to ascend the incline, that extra seventy-five pounds on their backs felt like a ton. It was suddenly uncomfortably warm and that made the task predictably more difficult.

Dale held back her comments when the gasps of the trainees around her turned victorious about reaching the top. Experience reminded her that the descent took much more control and concentration than the climb. After approximately one hundred fifty pairs of boots traipsed up the same path, it had become quite muddy and slippery, so by the time First Platoon began their descent, some of Second and Third Platoon beat them to the bottom on their backsides. This caused a domino effect for some, which caused a comical looking pile-up at the base of the mountain.

The unintentional human heap amused quite a few but it greatly disturbed the senior drill sergeant. Upon Alpha-10's return to the company area, their successful conquest was rewarded by another degrading lecture from Ritchie. He ranted and raved for several minutes about the trainees' lousy performance on the hike and the climb and then, as the company stood at Attention, sweating profusely, Ritchie ordered the trainees to remove their backpacks, field gear, fatigue jackets and fatigue shirts. He marched them to the parking lot for PT. For Dale and Kotski, who still had not been issued long john tops, their damp bodies nearly froze during the first exercise.

Two hours after a steaming hot shower and safely inside, where it was warm, the bone chill remained. If Dale had been allowed to bring her service weapon into the assignment, Ritchie would have been a dead man by sunset.

Putnam entered the bay ten minutes before Lights Out and announced the dreaded news that a trip to the Gas Chamber was on the agenda for the next day. He advised the women not to take morning showers nor to shave any part of their bodies, as he had also advised the men. Tear gas had a nasty way of seeping into open pores and wreaking havoc under the skin, as though one had just bathed in jalapeño pepper juice.

The Gas Chamber exercise was the most feared by all trainees. Horror stories were passed on from cycle to cycle about incidents that were exaggerated or never occurred that resulted in unnecessary panic. Both Dale and Shannon knew how foul tear gas could be. They had both experienced being gassed without their masks at one time or another and to say it wasn't pleasant was an understatement. They also knew, however, in this particular situation that if a trainee paid full attention to the instructor, contact with any tearing agent would be minimal.

Dale woke up with a slight tightness in her chest and, as usual, her foot ached but she fared better than Kotski, whose cough was continuous and deep. Several others sneezed and sniffled, obviously congested, yet only a couple women requested to go to sick call. None of them wanted to fall under Ritchie's wrath and be accused of being weak.

The company was transported to the range first thing in the morning. They were given instruction and familiarized themselves with the M60 machine gun. The trainees learned how to fire an already zeroed M60 loaded on a bipod, then tripod with an assistant gunner. Only a rare few actually hit what they aimed at but it didn't seem to matter as long as all the rounds were expended.

Next, they marched to the Bivouac site and practiced for G-3 testing, mainly the section that required them to report to an officer outside. When that was completed, they were marched to the Gas Chamber location and given limited instruction on nuclear, biological, and chemical, or NBC, warfare. Instructors demonstrated and then the trainees practiced how to put their gas masks on in nine seconds or less. They were taught how to check for leaks in the masks and if anyone discovered a defective mask, he or she would be temporarily loaned a properly working one for the exercise and the defective one would be replaced back at the company.

When the field instructor believed the trainees had practiced enough, he divided the troops into groups of approximately twenty-five. With masks firmly attached to each individual face, he led them inside the small cabin in the middle of the woods called The Gas Chamber. As the trainees shuffled inside, they felt a slight burn in their eyes, nose, and throat. The air was misty with the gases of chloroacetophenone, CN, and 2-chlorobenzylidene malononitrile, or CS, and the group was ordered to form a circle around a centerpiece that emanated the tearing agents. Two masked field sergeants stood by while the masked instructor spoke of what the next task would be.

"One by one, you will stand in front of me, take a deep breath, remove your mask, and *very clearly* repeat your name, rank, social security number and unit. Then you will walk to the exit and leave the building. Anyone who does not complete the spoken words in a clearly understood voice or runs out of the building will be made to do it again. I suggest you get it right the first time." He scanned the sea of black masks. "Once you get outside, another instructor will meet you. He will direct you to run around in a wide circle with your arms out like airplane wings until your senses clear. This will take about a minute. If you listened to everything I just told you, this should be a painless experience."

That evening, Dale caught Shannon on the south patio, taking a cigarette break. "I think I might have to take up smoking just to get the amount of outside breaks you do," Dale said.

"Why would you want to pick up a nasty habit just so you can freeze your ass off with the rest of us nicotine addicts?" Shannon asked.

"Good point." Dale scanned the area to make sure they were clear. "At least the gas chamber is behind us. That's always a good thing."

"I thought the NBC field sergeants were excellent," Shannon said. "Few in the company messed up."

"And the ones who did, messed up because they couldn't follow instructions. How hard is it to take a deep breath before removing your mask, hold it, then repeat those four items clearly and walk the ten paces to get to fresh air? It takes less than twenty seconds."

"Apparently harder than we thought," Shannon said. "But, let's face it, the ones who messed up today are the same trainees who have been messing up since we got here. I'm pretty shocked that not even one of them bolo'd out before Christmas exodus."

"Yeah, me, too," Dale agreed. "I'm frustrated, too, that here we are, in our fifth week of basic and this case seems to have stalled. No one, male or female, has demonstrated any abnormally questionable behavior toward the cadre. Even Dizzy has kept her distance from the drill sergeants and you can't get much looser than she is."

"So has Snow. I haven't seen anything untoward from her since before exodus, as much as I would love it to be her."

"Down, girl," Dale told Shannon, smiling.

"On the other hand, she knows I know about her sneaking off with Bradbury so she could be laying low, not wanting to keep herself in any kind of a spotlight," Shannon said, with renewed hope in her voice.

Dale patted Shannon on the shoulder. "I don't think it's her."

Shannon sighed. "Yeah, me, either."

"I really hope whoever is doing this hasn't skipped a cycle. I'll be pissed if we're going through all this for naught."

Shannon crushed out what was left of her cigarette and field stripped the butt. "Yep, me, too."

Four hours of the next morning were strictly devoted to map reading. The trainees were provided with topographic maps, papers, pencils, protractors, and straight rulers. They were taught to measure a grid azimuth, measure distance with a certain percentage of the correct interval, orient a map by alignment, utilize marginal map statistics and identify topographical symbols. It was a tremendous amount of information to retain in a half-day.

Chow was served at Raburn Hall and then Alpha company had a surprise exam on what they had learned in NBC training. They were tested on how to identify color symbols, yellow for gas, white for atomic, blue for biological, gas mask deficiencies, correctly putting on the gas mask and symptoms of the side effects of different gasses and chemicals. Their final evaluation was what to do in case of a nuclear, biological, or chemical attack and as Holmquist inserted before anyone else could—bending over and kissing your ass goodbye was not an option. Even though the test was unexpected and no one had a chance to study, everyone scored a Go.

It turned out to be an early day for everyone. Rumors spread that the huge, predicted snowstorm that postponed Bivouac was only hours away and the post literally shut down. All personnel who lived off post were sent home, all on post residents who were not in some kind of mandatory assignment were released from duty and all new GIs were restricted to their company areas. Alpha-10's trainees were snugly back in the barracks by 1515 hours.

Dale decided to make good use of the unexpected free time and checked the availability of the washing machines. The north laundry room was empty so she was back downstairs in minutes armed with a load of her and Shannon's laundry. Moments after Dale had started the washing machine, Kotski entered with her second pair of boots in one hand and shoe polish paraphernalia in the other.

"How are you feeling?" Dale asked her.

"Well, I don't have pneumonia yet." Her voice was shaking.

Dale noticed that Kotski had not been herself last evening and most of the day. She thought it might have been because Kotski was getting sick but now she suspected it was something else. "Are you okay?" The concern in Dale's voice was genuine.

The caring and gentle tone from Dale seemed to trigger the flood valve behind Kotski's eyes and suddenly Kotski could not control her tears. "Sorry," she whispered.

"What? What is it?? What happened?" Dale was alarmed and instinctively protective.

It took a few minutes for Kotski to calm down. "I've just had a bad day," she said, finally.

"I thought you seemed a bit skittish on the way back from Raburn Hall. What's wrong?"

"Well, first off, I'm having a really hard time with my period. It's getting harder and harder to concentrate when my uterus feels as if it's involved in nuclear warfare."

"Is that normal for you?" Kotski shook her head. "Maybe you should have that checked out."

"Oakes..." Kotski hesitated.

"What? Come on, Laurel, you know you can talk to me," Dale said, soothingly. She saw that Kotski was in great emotional, as well as obvious physical, pain.

"It's just that I don't want you to think that everyone is using you as a sounding board, with what happened to Kirk and all."

"I don't think you have to worry about that, people barely spoke to me after Kirk, much less used me as a sounding board. So, come on, tell me what's going on."

Kotski took a deep breath. "Yesterday on the NBC range, when I was standing in the back of the crowd, Ritchie came up behind me and grabbed me by my hair so hard..." She started to cry again. "It felt like he yanked out a fistful of my scalp."

Dale started a slow burn while Kotski continued. "He started to scream at me to pay attention because he knew I had to be too stupid to know any of this on my own. He said that only dummies and morons stood in the back of the class so that they had an excuse for not learning anything. I'm tall, Oakes, I can see above half these guys' heads. I was in back because I didn't feel well. I was going to go to sick call but I didn't want to miss anything. It took everything I had not to slap him into the ground."

Dale took a step toward Kotski and hugged her tightly. "You should have. That son-of-a-bitch." Under different circumstances, Dale might have enjoyed having the beautiful woman in her embrace but the last thing on her mind was her libido. Not only that, the kiss with Bishaye pretty much cured her attraction to anyone else.

Kotski broke the comforting hug and wiped her eyes. She set up her boot polishing equipment. "There's more."

Dale narrowed her eyes. "Worse?"

"Yeah," Kotski said and nodded. "But it has nothing to do with here."

"What is it?" Dale asked as she checked her wash cycle.

"A month before I started basic training, I was beaten up and raped. I was in the hospital for two weeks. They still haven't found the guy. I was walking home from a friend's house and this guy came up behind me, grabbed me by my hair and dragged me into a wooded area a little ways from the sidewalk. I've walked that same route a thousand times and I've never even felt nervous, never expected anything like that to happen. I was pretty messed up with some internal injuries. That's why I think I'm having such a bad time with my period. And all this strenuous exercise isn't helping."

Dale tried not to look horrified. "You're awfully pale. How bad are you bleeding?"

"Bad."

"Why didn't you postpone your enlistment?"

Kotski wiped away more tears. "Because getting away from it all did seem like the best idea. And I honestly thought I was coping just fine. Until yesterday."

Now furious, Dale found it extremely difficult to disguise her anger. "You've got to tell Audi."

"I can't."

"Maybe not about the rape but tell him about Ritchie. He had no right to do that! And he will feed off your fear of retribution not to tell anyone. Audi should know you well enough by now to know you would not gain anything by making this up." Dale observed Kotski's uncertainty. "This is serious shit, Laurel. If he's allowed to get away with it, next time will be worse."

"I know what you're saying, Oakes, but I don't need any more trouble in my life. I think I just want to get through this and get on with everything else."

"Laurel..." Dale was so angry and concerned that she was actually at a loss for words.

"Promise me you won't say anything," Kotski pleaded.

"Okay, okay," Dale agreed, frustrated. "I won't say anything to Audi or any of the other drill sergeants. But I think you're making a mistake. Ritchie is a bully. He's an unsavory, untrustworthy, slippery bastard. A statement from you might help put Ritchie out of commission for a while. Then maybe we wouldn't have incidents like the one you went through and we'd start getting some positive motivation for our training. I won't say anything if you don't want me to but please, please think about this."

CHAPTER THIRTY-ONE

Shannon decided to stroll downstairs for a cigarette break and thank Dale for doing their laundry. Also to make sure her panties weren't accidentally on purpose left in a dryer. When she entered the laundry room, Kotski was gone and Dale had just finished rolling her last item of clothing.

Hey, Dale, thanks for –"

"Perfect timing!" Dale met Shannon halfway and handed her the clothes. "Take this stuff upstairs, will you? I'll explain later."

"Sure, but —"

Dale raced out of the laundry room. She checked her watch. It was four-thirty. She prayed Bishaye was still in her office, even though it was doubtful with the threat of bad weather and the post-wide orders to leave before the snow started to fall. Dale took a huge risk and sneaked away from the company area, around the corner to Battalion and then to the headquarters parking lot, where she was relieved to see Bishaye's car. She made it out of the parking lot, undetected, and across the small street to a wooden barracks-style building that had been converted into a pizza establishment. Ordinarily, she would have been foolish to pick that place, as it was the premiere hangout of roving drill sergeants who stopped in to monitor their troops who had achieved their freedom through passes. Due to the post shut down, however, the business had closed early and it was dark and empty. Dale used the pay phone on the side of the building. She dialed Bishaye's number and was surprised when the colonel personally answered. She usually had minions for that.

"Colonel Bishaye," Anne said.

"Meet me by the pizza place ASAP," Dale demanded.

"Who—? Dale? What are you doing there? If you get caught, you'll—"

"If you get your ass over here, I'll be less likely to get caught." Dale hung up.

Less than five minutes later, the stunning LTC Bishaye walked around to the back of the building. "The way you talk to me sometimes really pisses me off," she snapped. "I'm still your boss." Her attitude appeared to be a mixture of irritation and curiosity. "Why am I here? No, actually, why are *you* here?" She had not brought a coat and crossed her arms in an effort to ward off the chill.

Even as angry and upset as Dale was, Anne Bishaye still caused her to catch her breath. The recollection of the kiss flooded back full-force and Dale was momentarily paralyzed. A loud sigh from Bishaye snapped Dale out of her lustful daydream.

"This better be good to take a chance like this. Now, what's wrong?"

"What's wrong? I'll tell you what's wrong. Number one, most training units get at least a two-hour pass after their fourth week of basic. We are restricted to the barracks for not one, good, valid reason, forcing us to remain together, causing even the closest of new friends to jump down each other's throat. There is no release! The Dayroom is off-limits which confines us to the laundry room, the patios, the bay itself or the Goddamned toilets! We're driven to the max, we put forth five hundred percent and then we are put down for being unfit, incapable, and unmotivated. There isn't one person, at least in the female bay, without some kind of injury or illness, the weather sucks and number two, there's Ritchie!"

"Ritchie?" Bishaye repeated. "I thought your gripe was with Colton."

"I do have a gripe with Colton. He's a lousy, fucking CO. But he's small change compared to Ritchie."

Bishaye gestured helplessly. "What's Ritchie done?"

Dale removed her notebook from her left-side, droopy saddlebag pocket and started to rattle off dates and events to the Battalion Commander. She then finished up with what Kotski had told her earlier. When she looked back at Bishaye, she was encouraged that Anne looked so startled.

"Any witness to the Kotski incident?"

"I have no idea," Dale answered, annoyed. "If there was, she didn't say and, as far as I know, I was the only one she told."

"Could she have made it up?"

"Why would she have gone to all the trouble of making up a story like that and then make me promise not to tell anyone? Anne, do you honestly think I would have risked this if I had any doubt in her story?" Dale asked, exasperated.

Bishaye shrugged. "No, I guess not."

"You *guess* not?"

She circled Dale so that she stood next to her and was out of sight of the main road. "All right! No, you wouldn't. I'm just not sure what I can do about it."

"Well, you'd better think of something because the man is heading for serious trouble and if something isn't done about him, *I* will be heading for serious trouble. It's getting harder and harder to contain myself and what the hell am I supposed to do if you're not willing to cross the man, either?"

"Hey! You watch your tone with me!" Bishaye said, sharply. "I will not tolerate that from you of all people. I never said I wasn't going to do anything, I said I wasn't sure what I could do. Maybe Ritchie's involved."

"Or maybe he's just a psychotic woman-hater. Anne, you aren't right in the thick of this. I am. This assignment is perplexing enough without a Neanderthal like Ritchie complicating it."

I want to kiss you again. Really bad, she thought. She hadn't realized her heart had started pounding when Bishaye lost her temper. She wanted to harness that passion and utilize it for something more intimate. Like Bishaye pushing her back against the wall and hungrily devouring her.

"What about Colton?" Bishaye was either oblivious to Dale's lascivious musings or artfully ignoring her.

"Colton can be controlled. He can be intimidated and he knows we're here so he's only going to go so far before he quits. Trust me, he is scared shitless of you and fear of being disciplined by you is enough to keep him in line. But Ritchie clearly knows no boundaries. He's dangerous, Anne."

Please, God, make her kiss me! Just one more time, that's all I ask.

"Okay, he's dangerous. I will keep my eye on it." Bishaye started to shiver.

Who am I kidding? If she thought about me that way, she would have arranged to get me on Battalion CQ runner duty. This realization made Dale's sour mood worse and she went on the offensive. "Keep your eye on it better than you did the Kirk case, okay?"

"That was uncalled for!" Anne snapped.

"This whole assignment is uncalled for!" Dale snapped back. "But we're all here, anyway, so I am trying to make the best of it. In the meantime, I don't need an obstacle like Ritchie. And neither do you." Dale started to walk away. "I have to get back before I get caught."

Bishaye reached out and snagged Dale's arm. "Dale, why are you so angry with me?"

Dale spun back and glared at her. "Why? You're joking, right? Why do you think?"

"If I had any idea, I would not have asked."

Dale's shoulders slumped in disappointment and she stared at the ground.

"Oh," Bishaye said and nodded in acknowledgement. "The kiss." She removed her hand from Dale's forearm. "I thought we understood each other about that."

"I guess not," Dale said, quietly. "I guess I have deeper feelings than I originally thought." She looked back up into Bishaye's eyes. Dale could not decipher what she saw there. What she didn't see was encouragement of any kind. "Definitely deeper feelings than you."

"Dale …" Bishaye began in a perplexed tone of voice. "You know why I cannot take what happened between us any further."

"I know what your excuse is."

"Dale, please, this isn't doing either one of us any good."

"Let me just ask you one question. If things were different, if there was no Army and no Jack, would there be any hope?"

Bishaye was silent for a moment. "I honestly don't know."

Dale accepted that answer. "Okay."

"This wasn't really what you called me out here for, is it?"

Bishaye asked. "Please don't tell me you took this kind of a chance just —"

"Jesus, no! Did you not listen to my notes? My complaint about Ritchie is not only legitimate, it's urgent. How can someone like him continue to hold a position of power with the attitude he has and the way he treats people? Especially women. He's not making it easier for any of us."

"Including me," Anne said.

"Yes, including you! So why are you so hesitant about stopping him?"

"I have to be cautious. If I go after him right now, he's going to know that there is definitely someone planted in the company. And he's going to quickly figure out that if the spies weren't put in there by the cadre then they were put in there by Battalion to spy on the cadre. Then, if these people, like Private Kotski, don't want to make statements or have the stigma of snitch following them around then I would have to expose you or Walker as my source. Where do you think that will leave this investigation?"

"We don't even know if there is anything to investigate."

"There isn't enough freedom in the company yet for anything too much to happen."

"One of my points." She pressed her notebook into Bishaye's hand. "Read these notes. I'm not kidding. He needs a comeuppance."

"What about reporting him to the IG?"

"Shannon or I can't do that. At least not until this case is solved. We can't keep subtly suggesting this kind of stuff to the other women when we're not supposed to have any knowledge of how things work."

"True. I'll see what I can do. I can't promise anything under the circumstances."

"Hopefully he won't push me into blowing my cover."

"You've had run-ins with tougher guys than Ritchie. I have faith that you won't jeopardize this case because of someone like him."

"You obviously don't know me anymore, Anne." They locked stares, neither openly acknowledging the heat that still radiated between them. "I have to get back," Dale said, reluctantly.

"Listen, if you happen to get caught, tell them I was taking some things to my car, I saw you on the south patio and I asked for your assistance."

"How come you haven't left post yet?"

"I was on my way out when you called. And I would still like to get out of here before the snow flies. I'll wait a few minutes before I walk back. Just signal me if the coast isn't clear."

"I will. But it's pretty deserted, it should be fine." Dale glanced at Bishaye one last time before she took off for the south patio of the Alpha company area. Bishaye nodded silently and managed a half-frozen smile.

Bishaye watched Dale leave. She rubbed the palms of her hands against her face to get some feeling of warmth. If Dale only knew how much emotion was behind Anne's restraint. The fire in Dale's belly really turned Bishaye on. It always had. She really did want to experience something much more intimate with Dale. There was just too much at stake. Once the undercover lieutenant was out of her sight, a sexually frustrated Anne walked back to her office.

Dale made it safely to the barracks unnoticed. No one even missed her. She found Shannon and related some of the conversation with Bishaye to her. Both lieutenants were irked that Ritchie might still be allowed to get away with his antics and they agreed to do everything within their power to make Ritchie's life a living hell any chance they could.

The demeanor of Anne Bishaye once again puzzled Dale. She could not understand how that kiss couldn't be first and foremost on her mind. It was a rude awakening that the adoration she felt for her colonel was not reciprocated. Maybe that was the way things needed to be in the military...but, if so, it was awfully damned cold. Dale suddenly hated the fact that an indiscreet fling with a stranger of the opposite sex would be much more acceptable to Uncle Sam than an affair borne out of love with someone of the same sex.

The predicted snowstorm had not arrived by the next morning but the company remained on an alternate schedule anyway. The troops were left at a drop off point in WacVille at approximately 0730 hours by bus, where they marched to Raburn Hall for more G-3 testing.

The trainees were quizzed in groups of three on Military Courtesy, Saluting The Colors, Guard Duty, and Military Time.

The testing ended about two hours later. Those who didn't get a Go were given a few minutes to regroup and then retested. The second time, everyone passed.

The trainees were then rounded up and transported to one of the muddy, icy grenade ranges. Between the time they had entered and left Raburn Hall, a monsoon had erupted and that made being outdoors extremely unpleasant. As they had the last time, the company practiced throwing fake grenades and then was taken to a different section of the range where they played with real grenades.

As a majority of the group observed, a safe distance behind, three range instructors stood with one trainee each behind three steel barricades that stood four and a half feet high. The range sergeants made sure that the soldiers used a proper and correct throwing stance and that the inexperienced GIs didn't hold onto the weapon any longer than necessary after the pin was pulled. For the most part, the company did well.

The exceptions were, unfortunately, female. Kerrie Hewett had never thrown anything more lethal than a ping-pong ball. When her turn came, her apprehension only hindered the strength of her throw. Upon instruction, she let the grenade go in a wide, upward arc. When it hit the ground, six feet away, it rolled another eight inches into a huge mud puddle.

The only criticism came from the tower. "Oh, shit. Everybody duck," the voice crackled. A second later, the grenade exploded and drenched everyone with mud. Two firing orders later, Dizzy repeated Hewett's performance. Twice.

After chow in the field, the trainees were trucked to a different range where they were to qualify that afternoon. Dizzy and Hewett were herded off to the side and forced to toss big rocks until they acquired some distance in their throw.

Shannon, who was placed in the first group to qualify,

breezed through the course and maxed it out. She had nailed sixty out of sixty targets and that registered her as an expert. Quite a few people had perfect scores. The art of grenade qualification turned out to be a lot easier than the trainees thought it would be.

Since the first group had to wait around for the last group to finish their qualification, the range sergeants had to occupy the trainees' time so that they wouldn't be idle. The sergeant who was in the barricade with Hewett, decided to select six of the bored trainees and made them use their steel pots to remove water that had built up in one mud hole and move it to another mud hole. Every time a helmet full of sludge was loaded into the emptier hole, the muddier hole would fill right back up. The range instructors got a big charge out of the fruitless effort and, after a while, Shannon had to laugh, too, even though she was the first one picked for the detail. There was no way one hole would ever get empty nor would the other hole get filled. It got to a point where everyone involved got silly about it...until they had to put their steel pots back on their heads.

Holmquist was the drill sergeant assigned to accompany Dale's firing order through the course. He was encouraging and supportive and he tried his best to keep morale up even though the weather was miserable. He kidded with Dale and told her that she had a good throwing arm *for a woman*. He also asked if she was sure she had never played pro-ball every time she placed a grenade exactly where it was supposed to land. At the end of the course, Dale had scored sixty out of sixty, as a majority of the group had done. Dizzy and Hewett barely made Marksman and the only reason they did was that some of their grenades rolled downhill into the target area.

The next day was Saturday. After PT, where the wind and sleet froze each trainee to the core, Captain Colton conducted a locker inspection in the barracks. Rumor had been passed around the night before that an inspection might take place so the company was prepared.

As hard as he looked, he could not find fault with Dale or Shannon on anything. He was too anxious and they were too experienced. In fact, in the female barracks, the only things he could find to correct were minor, such as clothing in their personal drawer that should have been hung up as opposed to rolled up. His in-depth inspection prompted Travis to read aloud a letter she wrote to her husband.

"Dear Joey, The captain looked in my drawers today..."

The rest of the morning was spent passing the time, keeping the barracks and the other details up to standard. After noon chow and one o'clock formation, the company was marched to the athletic field where they were told it was mandatory for them to attend a football game between the Averill Police Department and the Fort McCullough MPs.

It started out to be fun with everyone seated on the ice-cold bleachers and they cheered on men no one felt any real allegiance to. The only company drill sergeant in attendance was Putnam who disappeared once the game got underway. Dale and Shannon were both surprised to see Bishaye make an appearance and they may have been the only two to recognize her in civilian clothes.

The minute Dale spotted Anne, she paid no attention to the game. Bishaye stood off to the side of the bleachers, huddled with three people Dale did not recognize. She watched Bishaye casually interact with the two men and one woman, watched her laugh, watched her animatedly converse with her entourage and Dale felt an ache deep her heart. She remembered when Anne used to behave so lightheartedly with her. The dynamics of their friendship had changed so drastically during the past year and suddenly Dale mourned the loss of a more intimate relationship that neither existed in the past nor would ever in the future.

Bishaye stayed approximately thirty minutes, until it started to spit snow again and then she and her friends left. The flurries soon turned into a full-blown storm and within forty-five minutes, several inches of snow had accumulated. The bitterness in the air mixed with the wind and the quickly forming blizzard conditions made it quite unpleasant and uncomfortably wet. Minty and Sherlock announced that they were about to head back to the barracks and they didn't care what kind of trouble that caused. Visibility was limited to the point that no one could see the field anymore, yet no one had the common sense to call the game. They couldn't see the players but they could still hear them.

No drill sergeant was anywhere in sight and Shannon wondered where Putnam could have disappeared to. She exchanged looks with Dale and they stood up with Minty and Sherlock.

"Maybe one of us should search the field first to see if we can find our drill sergeant," one of the guys sitting with them said.

"You go look for him," Minty said and placed her scarf over her nose and mouth.

"You can't even see the field," Sherlock said. "For all we know, they all went home and turned on a recording of a game being played and are broadcasting it on the loudspeakers."

"That wouldn't surprise me," Dale mumbled to Shannon.

"Listen," Shannon spoke up so that everyone could hear her. "I think we're all in agreement that it's too brutal to stay out here in this. I think whoever wants to leave should come with us. If we march back as a group, it will look better for our lynching later."

"Good idea," Minty said.

Once the group of about fifty were in a formation, one of the higher-ranking squad leaders, a male from Third Platoon, took command and marched them back to the Alpha company area. They were disappointed that no one of any importance was around to see the trainees return to the barracks, singing cadence and looking every bit like the soldiers the drill sergeants refused to let them be yet.

Later that evening, Putnam commended the group who had left the game together. A colleague had informed him of their uniform departure. Putnam had been called away for an hour to handle a personal matter. When he returned to the game, he noticed that only about half the troops he originally brought were still there. He advised them that the only thing they should have done differently was to find a phone and call the CQ office and advise whoever was on duty that they were on their way back. Other than that, he was very pleased with their initiative.

Before he left the bay, he informed the women that on the weekend they would have a formal inspection in their Class-A uniforms and the company commander would be conducting it.

When he was asked if basic training was almost over, he actually provided an answer. "According to my sources, in about two weeks."

When he exited, a resounding cheer went through the bay that could have shaken the rafters. Dale and Shannon hooted and hollered the loudest.

CHAPTER THIRTY-TWO

Everyone did exceptionally well in the inspection. That may have been more of a surprise if a two-hour on post pass had not been the incentive. The trainees were not allowed to wear civilian clothes, drink alcohol, or leave McCullough and they were all ordered to be back for an 1800 hours formation.

Dale and Shannon could not be everywhere with everyone, so they split up and each went to one of the two places a majority of the troops congregated — the Pizza Place where Dale had met with Bishaye earlier in the week and the bowling alley which was just down the street from Tenth Battalion. Two hours felt like two minutes and allowed no time for suspicious activity. The pass, as short as it was, put everyone in a good mood and was long overdue. Dale wondered if the order came down from Bishaye because the timing seemed too coincidental.

The next morning was warmer by about ten degrees, which brought the temperature up to thirty. Maybe. The trainees all commented on the heat wave as they stood in line to get clean linen.

Military coaches picked up the trainees at their usual waiting point on the north side of Tenth Battalion after formation. They were then driven to a range they had never been before, deep in the woods where marsh met solid ground and there, they would learn and practice an exercise the military called Live Fire.

In order to qualify for Live Fire, the trainees would be tested with an M16 equipped with a fully loaded magazine of live ammunition, where they would move up a hundred-meter course to an enemy bunker or sniper. The company would be divided into two-person buddy teams and then sent on a predetermined route, using acceptable cover and concealment. They would be taught teamwork in the use of hand gestures and, if discovered, verbal signals to assist one another to overtake the bunker or capture the enemy.

"Your most important objective is to get to that bunker or sniper alive," the range instructor told them. "That means your second most important objective is to know what you need to look for in your search for cover and concealment. You want to do everything possible not to expose you or your partner to enemy fire.

"You will be using a high crawl and a low crawl on this course." He explained what each meant and another range sergeant demonstrated. When that was completed, the range sergeant showed the company how to rush for cover.

"It should take you no longer than three to five seconds to rush for cover. If you should need longer to reach your chosen cover, hit the ground halfway across the open area and roll to either your right or your left, crawl, then get on your feet and complete your rush."

The more the instructor spoke, the more disenchanted the trainees became. The weather had now warmed up to the point where the ground had started to thaw and it had become increasingly slippery. Dale just knew they were going to spend the rest of the day resembling a blooper playback of a football follies parody.

The range instructor pointed out when the use of each particular crawl would be to the trainees' advantage. He then indicated surrounding nature to make another point.

"High grass, weeds, anything along those lines provide only partial concealment. It is not wise to think of these areas as helping you because movement of this sort of vegetation can be easily spotted by anyone trained to look for it. I would suggest using this kind of concealment only when you are rushing." He paced and scanned the faces of the three platoons.

"The best cover and concealment come from using ditches, walls, ravines, gullies, trees, stumps, large rocks, fallen timber, vehicles, folds, or creases in the ground, etc. Thick vegetation and rows of hedges are good for concealment but not cover. The difference is that bullets can penetrate concealment, they should not be able to penetrate cover."

PERMISSION TO RECOVER

One of the male trainees put up his hand. "Sergeant?"

"Yes."

"Why do we have to use live ammunition?" It was a question they all wanted to ask because no one was particularly thrilled with the idea.

"Because in a war, you use real bullets," the sergeant answered, tersely. Then he added, with a hint of a smile. "Besides, how else are we supposed to eliminate the slower trainees?"

No one in the company smiled back at him. They all discreetly eyeballed one another. Each knew how badly the person who stood next to them had shot on the range.

He ended his class with emphasis on the importance of moving as a member of a buddy team. "You *must* stay in communication with one another. Watch your partner, talk to him, listen to him. Use your signals. Cover your buddy's movement by fire. Stay with him. This is imperative."

Before noon chow, the trainees were assigned a buddy and provided with camouflage make-up. Another range sergeant showed the trainees how to apply loam, light colored mud, cork, charcoal, white, light green and sand colors from greasepaint sticks to their faces and necks and how to affix assorted brush to their clothes. When that was completed and they were all in full character, the mess hall trucks arrived for lunch.

Travis sat down with a full tray, next to Dale and Private Tanner, a young man who had possibly gone a bit overboard by the amount of sticks applied to his uniform. Travis studied his inventiveness. "I think that I shall never see a GI as lovely as a tree…" She turned to Dale. "How do I look?"

Dale looked up from her quickly cooling meal. "Great." She looked closer and squinted. "Really great. Who are you?"

"Me. Travis." Even she had stopped referring to herself by her married name.

"Wow. You really did a good job, Travis. I thought you were Muscatello," Dale told her.

"Jeez, I did do a good job," Travis said and beamed. "He's a lot taller." She took her first bite of food. "Maybe that's what I need to do to look statuesque…paint my face four different shades of jungle tones."

"Either that or just keep that tree trunk stuck in your steel pot." Dale referred to the small branch she had secured to her helmet by the band.

"A bit much?"

"It might hinder your movement a bit. But you might be in real trouble if there are any deer hunters in the neighborhood. I know guys back home who would kill to hang a rack like that on their wall."

Travis looked up, startled. "Ooh. That's a good point." She turned to Tanner. "Hey, buddy, can you spare a twig?"

He shook his head. "Nope. But when you're done, I'll show you where I got my supply."

Field chow finished, Dale divided and dumped her edible and non-edible trash. As she walked back to join her platoon, she saw Ritchie with Tierni, out of hearing range. Their interaction seemed very tense. At first, Tierni appeared to hold her own against him but then she burst into tears. They finally left the range together and returned an hour later. Dale wondered what that was about but figured she or Shannon would find out one way or another by the end of the day. If they could not get it straight from the horse's mouth, they would get their information elsewhere. There were too many gossips in the company.

The trainees practiced all afternoon. They ran the course several times and fired blanks and they all enjoyed it much more than they thought they would. Sherlock and Ryder ended up in a playful scuffle that looked much more like a mud wrestling match. They misinterpreted one another's signals and then got into a robust discussion about which buddy was the leader. By the time they slipped and slid to the end of the course and into the bunker, they were already friends again. It was when the range instructor made them shake hands that Ryder lost her footing and took Sherlock down with her, which caused them to slide and tumble to the bottom of the course. By that time, they had collected an appreciative audience, mostly male, who cheered and clapped. The two buddies, now in a tangle of arms and legs, were giggling like schoolgirls, even though they had sustained quite a few bumps, bruises and scratches.

Everyone stopped laughing, though, when the first pair of

muddied combat boots to stand before the two women belonged to the senior drill sergeant. Ritchie glared at them for only a moment, then turned and silently stalked away.

Dale and Shannon weren't the only ones who noticed he was acting peculiar.

***

Shannon sat on the north patio with her platoon while Dale sat on the south patio with hers and First Platoon sat in the open area between the two patios. Two male trainees dragged the trunk that contained the weapon cleaning paraphernalia to a mutually available area so that all platoons could have access.

Dale took what she needed from the trunk and settled in a spot where she would be the most comfortable. After separating the upper and lower receivers from her M16, she removed the bolt carrier group. She fiddled with the cam pin and then looked at Mroz. "Can you help me with this?" Dale could break down an M16 blind if she wanted to but her intention was to draw one of Tierni's friends into a conversation.

"Sure." Mroz reached out to Dale's rifle. "You have to give it a quarter turn and lift it out easy. Like this," she said and demonstrated.

"Oh. Okay. Thanks." Dale smiled at her. A lot of the women had grown to dislike Mroz, including Shannon. Even Dale had to admit that Mroz could be a little overbearing at times but she still liked her. They had shared quite a few laughs since they were at McCullough and if Dale was suddenly going to turn against somebody, there was going to be a good reason for it. Besides, at that stage of the game, alienating anyone was not a good idea.

"What I have trouble with is the hand guards," Mroz said.

"Yeah, me, too," Dale said. She removed the bolt, the extractor pin, then the extractor. "You're pretty close to Tierni, what was going on with her and Ritchie today?" Dale's manner was very off-hand.

"She didn't tell you?"

"No, she didn't have the chance." Dale laid the parts of the bolt carrier group by her foot and started on the buffer assembly and the action spring. "I always break my nails or pinch my fingers on this."

"I got a fucking blood blister from this last week." Mroz dipped a small brush into the bore cleaner. She worked on the dirt and the carbon deposits in the locking lugs of the bolt. "Why is this area discolored on mine and not on yours?"

"Yours was probably used more than mine. My guess is the heat did it." *Come on, Mroz, don't make me ask you about Tierni again.*

"Oh, yeah, about Dee," Mroz said, as if she'd read Dale's mind. "I don't have all the details but it has something to do with her and Silva." Her voice was considerably softer now. "From what I understand, the MP Dispensary called the Orderly Room and left word for Tierni to come back and re-do a pregnancy test. She's been worried for about a week. Anyway, Ritchie intercepted the call."

"So where does Silva come into it?" Dale started to scrub the extractor. "Is he the prospective daddy?"

"She and Silva spent Christmas exodus in Atlanta. Remember when McKnight blasted that around?"

"Yes, we all remember. So what business is it of Ritchie's?"

"None. He's just a dick."

Dale and Mroz discontinued their conversation in the presence of McCoy, who was patrolling the patio. They worked silently, cleaning their bolts. When McCoy moved on to another area, Mroz resumed talking.

"Ritchie spread it around to the cadre like a dirty joke. He humiliated her," she said, angrily, as she slapped a coat of LSA oil on the inner surfaces of the upper receiver.

Dale wondered what possible sick pleasure Ritchie got from mentally torturing an emotionally frail woman like Tierni. Especially when it had nothing to do with training. At least not immediately.

On the other patio, Shannon had just heard the same story from Travis, only Travis went into more detail.

"The fucker made a point to chastise her for what he called morally reprehensible behavior, then he told her it was women like her that promoted the sluttish image Army women were notorious for."

"God, what an asshole!"

"It gets worse. Then Tierni said that Ritchie said that as long as she was being free with her body, she should smarten up and do it with someone who matters."

"Meaning what?" Shannon asked, outraged.

"Meaning either he's looking to get laid or he's pimping for Colton. But, like Tierni said…it's her word against his."

Shannon furiously cleaned the lower receiver components with a toothbrush and bore cleaner. Ritchie was not a nice man and Shannon prayed that he was involved in something, anything related to the case because she wanted a piece of him.

As long as the day was already, it still wasn't over. Before evening PT, Ritchie called only the women into the Dayroom. He gave them a speech on how he was really their friend and he wanted them to come to him with their problems before they escalated into fights like the vicious brawl he had witnessed between Sherlock and Ryder that afternoon. This was news to everyone, especially both mentioned women who were seated together, still filthy.

"These kind of explosive actions will not work in the MOS you have chosen to pursue. If this kind of behavior keeps up, you will be eliminated from the program and recycled to something more suited to…your gender." He never raised his voice but his intent was clear.

Dale knew the Alpha women were getting tired of hearing the same old song from him. Instead of causing them to warm up to him, he turned them colder. They didn't trust him. He had never given them a reason to and after his appalling conduct with regards to Tierni earlier, which had spread company-wide, they had no faith that he'd suddenly changed.

Ritchie gave the women the opportunity to discuss their problems and he seemed genuinely surprised when no one spoke. The women either looked at their folded hands in their laps or the wall. When no one would engage him, especially at is most charming, he returned to his recognizable, nasty self.

"Why you ungrateful bunch of fuck-ups! I am giving you the chance of a lifetime to redeem yourselves and I can't believe you're all too stupid to take it. What do you think? We're just going to give you a free ride to MP School? You have to reach a standard of conduct and none of you, not one, have even come close. Take the telephone call I received this afternoon, for example," He was greeted with the sounds of disgusted sighs and barely audible growls. Remarkably, he did not react to that, he continued as though he had not even heard the dissension. "One of your barracks mates is a slut and a careless one at that." Most of the women knew he referred to Tierni but almost every head swiveled and looked at Dizzy. "You women disgust me. You do nothing to change the stereotype of military women. Those of you not here to entrap a husband are trying your best to be men. But none of you are trying to be soldiers and the ones who are trying are doing a piss-poor job!"

The women clearly seethed and Dale and Shannon sat there, dumbfounded. This tirade was too much. Dale was used to male chauvinism in the military but Ritchie took the description to a new level. With few exceptions, neither Dale nor Shannon had ever seen a group of women try harder, especially when, for the first time, the training the women were successfully completing, was set to the male standard. Dale figured what pissed Ritchie off more than anything was that the women *were* doing so well despite his continued demoralization. In fact, Creed and Michaelson usually did better than their male counterparts and the senior drill sergeant never acknowledged them, or any of the females' accomplishments. There had to be something Dale and Shannon could do to stop this bastard. The question was, could they do it within military regulations? This was a discussion they would have to have later because either Bishaye had not confronted him or it just didn't matter.

"As for the promiscuous behavior of Private Tramp –"

Tierni stood up and left the Dayroom much to Ritchie's shock.

"Private Tierni! Get your disgusting ass back in this room! Tierni!" Ritchie looked as though several blood vessels in his head were about to burst. He slammed his hand against the wall. "You will all pay for Private Tierni's actions! If I can't get to her, maybe you all can. Since you all seem to have excess energy for fighting and sex, I guess we just aren't working you hard enough! Outside for PT! Now!"

Once he had them in formation, he ordered them to remove their field jackets, leave them on the patio and he double-timed them to the PT field. He made them do grass drills, which were exercises performed either lying or rolling around on the grass. Then he commanded them to run until *he* felt like quitting. He kept them out through evening chow and did not dismiss them until it was too late for them to make the last possible serving time.

Later than night, with a majority of women on the verge of tears from sheer frustration, Minty finally mentioned that Ritchie should be turned in to the IG. Sometimes her knowledge from her Army brat days was a blessing.

"What's the IG?" Caffrey asked.

"It stands for Inspector General. The IG is a military officer who investigates complaints of wrongdoing within the system," Minty said.

"Kind of like an Internal Affairs of a police department?" Dale asked, to give more relevance.

"Pretty much."

"Like Ritchie would allow us to go to anyone to complain about him," Sherlock said, and continued to rub her feet.

"That's the good part. He can't deny a request to see the IG. I mean, he can but that would just get him in more trouble. And, who knows? Maybe by requesting to see the IG in front of Colton, he might be instantly cured of whatever is the matter with him."

"Why? Obviously, Ritchie isn't intimidated by much," Almstead said.

"Because the IG has a lot of power and soldiers are afraid of them. My daddy used to say that fear of the IG is almost as powerful as the IG himself."

"Will the IG listen to us? Ritchie's been a fucking prick to us ever since the beginning and no one seems to care," Sherlock said.

"True," Shannon said, and jumped into the conversation. "We just won't make that our major complaint because I think the military thinks we women deserve anything we get for enlisting in the first place. But he denied us chow. He can't do that, right Minty?"

The expression of realization on the majority of the women gathered by Sherlock's bunk was palpable. "Is that true?" Sherlock asked.

"I believe it is. Good thinking, Walker," Minty said.

"Okay, I'm in. How do we do it?" Ryder stepped forward.

"We've got to use the chain of command because they are a real stickler for that. We make sure we tell our squad leaders who will then give us permission to go to our drill sergeants, who will let us see the First Sergeant and then we get to see Colton. We are clearly being victimized and it's the IG's job to protect our rights." Minty looked quite proud of herself.

"I thought we didn't have any rights," Caffrey said.

"The few rights we do have are bound by regulation. And what he did violated one of the most basic ones." Minty looked around. "Girls, I think we might just have him by the short hairs."

The women reacted enthusiastically. Both Dale and Shannon were happy that Minty had thought of the IG. It didn't look natural for either of them to constantly and conveniently have read up on something before they enlisted, every time someone needed to be prodded in the right direction. Asking to see the IG with Ritchie's knowledge may not have remedied the situation but they hoped it would calm him down while, also, letting Bishaye off the hook.

Shannon dried off from her shower and slipped into her sleep attire. She shut her locker and watched Tierni with growing concern. Tierni appeared to be close to the edge, with good reason. Ritchie's vicious personal attacks on her were not building her inner strength or confidence, traits she would need to make an efficient MP. Maybe Tierni wasn't cut out for law enforcement but there were other ways of discovering it without a total destruction of the woman's character. Shannon walked to Tierni's bunk.

"How are you doing?" Shannon asked.

"I'm not going to let that bastard get me down, if that's what you mean," Tierni said with conviction.

"Good. He doesn't deserve the satisfaction."

PERMISSION TO RECOVER

"Can't something be done about him spreading rumors and lies about my personal life?" Tierni started to cry again.

Shannon sat down and pulled Tierni into a hug. "You don't have a personal life, Dee, you're in the Army. And no, I don't think the five-star boys in charge give a rat's ass about your feelings." As she comforted Tierni, Shannon wondered if the treatment of women in the military would ever improve.

Dale watched her colleague's interaction with Tierni as she sought out Kotski, who did not look well.

"You look bad, Laurel," Dale told her, honestly.

"Love you, too, Oakes," Kotski said, weakly.

CHAPTER THIRTY-THREE

Kotski reported to Sick Call at 0730 hours. She actually felt a little better when she first awoke but when she discovered that the company was scheduled to tackle the obstacle course, she quickly changed her mind. She knew there was no way her body could survive that.

The four women designated to make the complaint against the senior drill sergeant to the IG — Minty, Sherlock, Ryder and Tierni — had done so that morning and were returned to the physically stressful course near Range 28 before class started. They were back less than ten minutes before a jeep that carried a representative from the Inspector General's office arrived. The IG aide approached Ritchie, spoke to him briefly and escorted him back to the jeep. Ritchie climbed in after the aide and the jeep drove away. Dale bet the women who witnessed this wanted to cheer loudly enough for the senior drill sergeant to hear them but wisely decided against it. That didn't prevent the word from spreading like fire, though, and morale instantly improved.

When the trainees successfully completed the obstacle course, they were marched to Range 28 where they were taught how to overtake a city. The range was equipped with false-front buildings that resembled an abandoned village, like a movie set. Alpha company went through the motions of learning how to get into a built-up area quickly and effectively, how to climb into second story windows and how to move within the area to surround structure access points. The exercise was played like a competitive game between the platoons, timed and evaluated for efficiency. The trainees enjoyed that.

Third Platoon won but not by much. First Platoon came in second, followed closely by Second Platoon.

They marched back to the company area, ate chow, and then were bused to Raburn Hall for First Aid tests. Everyone got an easy Go. After all tests were completed, they marched halfway and double-timed the rest of the way back to Tenth Battalion. They spent the rest of the afternoon cleaning their weapons on the patios.

***

Word spread quickly and gleefully that Ritchie had been called before the IG and given a stern warning. Henning told Dale, who swept Henning's office that evening, that Ritchie was advised if he pulled another stunt like the denial of chow or continued to badger the females as a group or harass individuals with sexual innuendo, he would lose his drill sergeant status. This would further trigger another investigation whereupon he could actually be busted a stripe. Both Henning and Dale agreed that it couldn't have happened to a nicer guy.

After evening chow, Dale looked around for Kotski to no avail. She finally tracked down Audi. "Drill Sergeant, I was wondering if you know where I might find Kotski."

Audi motioned for Dale to walk with him to the picnic table on the north patio where no one else was around. "Did she tell you what was going on?"

"Yes, Drill Sergeant, she did. Did she tell you?"

"No, but the hospital did. She's going to be there for a while so that they can get the bleeding under control and treat some things that were aggravated by training. I wish she had come to me or gone to Sick Call sooner."

"I think she thought everything would be okay. She's very guarded about herself, understandably, and I don't think she wanted to spend any more time in the hospital or miss any training. Did she happen to mention anything about the senior drill sergeant?" Dale asked, cautiously.

Audi sighed in disgust. "Yes, she did tell me about that. She said you were the only other person she told. I don't think, after today, the senior drill sergeant will be pulling unforgivable stunts like that again."

"Will Kotski be able to finish training, Drill Sergeant?"

"I don't know. We'll see. A lot will depend on how much she misses." He was genuinely upset and his tone was compassionate.

"She's really smart, Drill Sergeant, I know she can make up the work. She wants to be an MP so bad."

"Understood, Oakes, and she might still be able to finish the training but I'm not sure it will be with this company. That decision won't be up to me." He started to walk away from Dale but he turned back. "Don't say anything to the other women about this. I will let them know when the time is right."

"Yes, Drill Sergeant."

Dale returned to the second floor in deep thought. The bay was abuzz regarding Ritchie's visit to the IG but soon that was overshadowed by the news that Kotski was in the hospital. No one knew why so to stop the speculation, Dale suggested that maybe it was bronchitis or pneumonia and reminded everyone how much deep coughing Kotski had done the night before. That seemed to satisfy everyone's curiosity and the conversation drifted back to Ritchie.

The next morning Alpha Company was marched to a different range than the one where they had practiced for their Live Fire test. The trainees ran through the course once and were then paired up to do the exercise for real. Dale went through the course with Private Van Hoesan, a male trainee, who worked so well with her as a partner, the instructor who followed the path with them complimented them on the smoothness of their run.

Shannon did not fare as well. She was assigned Beltran as a partner. The first time Beltran hit the ground, she landed on her M16 and somehow caused it to fire. The bullet grazed the tree branch just above Shannon's head. At that point, the instructor lost his cockiness and fell back a few steps as opposed to running even with Beltran.

"Please, God, let me get through this course alive," Shannon prayed. "I'm surrounded by homicidal maniacs! Dizzy is trying to knock off Drago and now they've put Beltran behind me. I'm only trying to do my job," she mumbled in desperation as she hid behind a boulder.

"Three seconds — go!" the instructor yelled from behind them.

Shannon jumped up and rushed for her next cover.

PERMISSION TO RECOVER

"Pleeeeeze, God."

"Bunker, go!" The instructor hollered at them.

Both Shannon and Beltran ran toward the bunker, twenty-five feet away. "Please, God, I've never been so scared in my life. Please get me to that bunker alive."

Shannon maneuvered through a slender, jagged path, ducked under a protruding branch, leapt two bushes, tripped on a rock and fell face first into a freezing cold stream. She slowly rose and deliberately looked heavenward. "I think I could have made it from here...pushing me was not necessary."

Beltran made it to the bunker, laid against the side of it and threw her cap grenade inside. When it popped, she jumped up and down and yelled. "I did it!"

"Put that weapon on safe!" The instructor screamed at her as he dove to the ground. When he saw her obey, he got up and walked back to Shannon. "Are you okay?"

"That depends...does this mean I have to do it again?"

"Hell, no!" His tone was emphatic. "Anyone who could even get halfway through that course alive with her, deserves a Go."

"Thank you, Sergeant," Shannon said, relieved.

"Are you sure you're okay?"

"Yes, Sergeant. Just frazzled."

"Understandable. Give me your grenade and I'll toss it. Put your weapon on safe and don't forget to have it rodded."

"Yes, Sergeant. Thank you."

The Live Fire test took a majority of the morning and miraculously, no one got shot. When the exercise was over, the trainees were returned to the company area to clean their M16s.

***

A military coach transported the trainees to Raburn Hall after noon chow. Following a filmstrip and lecture on military intelligence, the trainees marched back to Tenth Battalion where they participated in PT.

Finally Dale gave in to the agony and fell out of the run. Henning, who monitored the exercises, approached Dale. "Hi," she said, smugly. "Problem?"

Dale looked up at her and nodded.

"I bet if you'd had that foot taken care of when I told you to, it wouldn't be so bad now."

"Maybe."

"God, you are stubborn. Perhaps a consultation with Colonel Bishaye on this matter might bring about an attitude adjustment."

"You don't have to go that far. I can't put it off any longer. I have to go in tomorrow."

"Why didn't you go in before?"

"I might have missed something. Tomorrow all I'll miss is the seven-mile hike, which I probably won't be able to complete anyway."

"I'm just glad you're doing something about it."

They were joined by Holmquist. "Everything okay here? Oakes? Ma'am?"

"Yes, Drill Sergeant," Dale told him, sitting on the cold ground. "Old injury flaring up."

"Old football injury?" Holmquist said, kidding.

Dale nodded and smiled. She wondered if Holmquist was fishing to see if she were gay. The offhand remarks he had made recently indicated that he was. A year ago it would have bothered her to the point where she would have done something blatant to prove to him that she was not. Shannon was right, Holmquist was definitely her type – if she were still into men. In the last several weeks she had analyzed and reevaluated her feelings for Anne, and women in general, and she came to the definite conclusion that she may have always been attracted to women and buried it.

She knew the military's strict stance on homosexuality but she also knew many women, like Theresa Burke, who respected their military mission yet still managed to keep their sex lives away from unwanted, prying eyes. Theresa had done the smart thing and had married a man right after she was assigned to her first permanent duty station. Her husband, David, was Spec4 in her MP unit and he was also a closeted gay. He occasionally worked dispatch when she did and they became great friends. They married for the cover and the extra money Uncle Sam allotted them for Basic Allowance for Quarters. They used their

PERMISSION TO RECOVER

BAQ to rent a two-bedroom apartment off-post so that they could each live their own lives. Her marital status saved her from scrutiny when others fell under investigation. Both Theresa and David were happy with the arrangement and it worked very well for them.

Maybe Dale needed to find a cover like Theresa and so many other of her colleagues had, even if she weren't going to spend her life serving her country. It certainly seemed a lot easier. *Oh well...thoughts for another time...*

She then wondered if she came on to Holmquist if he would respond or continue to be professional. Dale had the feeling Holmquist was attracted to her. She didn't have those vibes often so when she did, she was usually on target, hence her suspicion of his fishing.

"What were you, Oakes, the QB? I mean I've seen you throw a grenade."

"No, Drill Sergeant, I was a tight end," Dale said before she could stop it from coming out. Both Henning and Holmquist stared at her, eyes wide and blinking. Then Holmquist burst out laughing. Henning looked at her as though she'd lost her mind. Dale smiled at Karen and then said as straight-faced as possible. "But I really always wanted to be a wide receiver."

"Private Oakes," Henning said, blushing uncontrollably. "I think that's inappropriate."

"Oh. I'm sorry, Ma'am. I was talking football positions. What did you think I meant?" Her question was delivered with the utmost innocence.

Holmquist stopped laughing long enough to look at Henning, who was now flustered. "I...you...nothing!" She turned and walked away. "Carry on."

Holmquist extended his hand. "Do you want help up, Private?"

"No, thank you, Drill Sergeant. I'll wait for the jeep." She referred to the vehicle that followed the hikes and runs and picked up the injured or sick along the way. When the jeep got full, the driver deposited the riders at the company area and came back for more.

"Okay, Oakes. Get that injury looked at."

"Yes, Drill Sergeant." She watched him jog away to catch

up with the company.

Dale, you are very bad girl, she told herself.

When Dale awoke the next morning, her foot was swollen and she could barely walk on it. She could not entirely lace her boot and reported to Sick Call at 0730 with her left shoelaces dangling. Ordinarily, this would have been considered out of uniform and she could have been punished for it but Holmquist was in charge that morning and he understood her circumstances.

It was 0845 before anyone saw her and when she entered the office, she found herself face-to-face with an old friend. Lieutenant-Colonel Kathlynn Bell had been her physical therapist at Fort Ord when Dale had developed bilateral tendonitis. The problem reoccurred three times so Dale and Bell got to know one another well.

"Oh my God! What are you doing here? I can't believe it's you." Bell was a pleasant woman with short, salt and pepper hair. "I nearly fainted when I saw that name and, ridiculously thought after I saw the rank, that there must be two of you. Undercover, I take it?" She lowered her voice for the last question.

"Yeah. My turn came up as a cycle spy." There was no reason to lie to her. Bell knew who she was and what she did.

"An officer as a cycle spy?" Bell asked as she sat opposite Dale on a stool. She picked up Dale's left foot and gently pulled the boot off. "That's a bit unusual, isn't it?"

"Yes, but this is the Army. What's considered normal?"

She removed Dale's sock. "There's something you're not telling me."

"Yes. Only because I can't," Dale said.

"I understand. My guess is that it has something to do with the mess that's occurred at Alpha-10 the last couple of cycles."

"You know anything I might not have already heard?"

"I doubt it. Just worthless GI gossip. My only connection to Tenth Battalion is providing whatever guidance is needed for physical therapy." She smiled warmly at Dale. "I'm happy to see you."

"It's nice to see you, too."

PERMISSION TO RECOVER

Bell studied Dale's swollen ankle. "Well...what have we here? A sprain?"

"No. Old injury flaring up." When Bell pushed directly on Dale's heel, the pain almost made Dale black out.

"Was that necessary?" Dale asked, out of breath.

Bell looked directly into Dale's eyes. "Do you want to tell me exactly what kind of old injury we're dealing with here?" She gently kneaded Dale's foot as Dale unraveled the story for her. When she was done, Bell just sat there, her usually warm hands now feeling like ice. She stared at Dale, horrified. "Why the hell weren't you medically discharged?"

"I was on my way but —" Dale shrugged.

"But nothing! This foot is in no condition to endure this kind of stress."

"Come on, Ma'am, we're almost through basic training. I need a profile, that's all."

"You need extensive physical therapy, perhaps even surgery."

"I'll miss too much."

"You know all this stuff!" Bell protested.

"Colonel, I'm not here because I need retraining," Dale said quietly. "I'm here to be a spy. By spending two hours each morning in physical therapy, I could miss a lot that goes on."

"By not spending two hours in physical therapy each morning, you could permanently damage that foot! Who the hell was the idiot that put you on this assignment?"

"Anne Bishaye."

"Oh." Bell let that information settle in. "I thought you and Anne were friends."

"We are."

"What did you do to piss her off?"

Dale laughed. "Nothing. Well, nothing that I'm aware of, anyway."

"Then maybe I can appeal to her better judgment."

"Don't waste your time. She is aware of my injury and how I got it. She is also the one who personally came to Vermont to get me for this assignment. So, please, make it easier on me and forego the therapy. I'll be careful. I promise."

Bell shook her head, defeated. "It may not be good enough.

I can't take responsibility for this if it becomes worse. You'll have to sign a medical waiver that you've absolutely refused physical therapy."

"Just to be put in my CID medical record, right? I can't have that going back to the company."

"If that's what you want."

"It has to be that way."

"I think you're making a grave mistake. You could end up in a wheelchair."

"I know and you may be right. But this is the last thing I have to do for the Army. I just don't want any complications."

Bell shook her head again and left the room to retrieve the medical forms.

Dale returned to the company area with a slip of paper signed by Bell that restricted Dale to seven days of lighter duty, which meant no running, jumping, concentrated marching or strenuous PT. She was also ordered to wear civilian shoes for seven days. At the end of her restricted period, she would return to Bell for an evaluation that would either bring her more restricted duty or a clean bill of health. Dale snickered and wondered if Bell's first phone call, after Dale left her office, was to the Tenth Battalion Commander. She would have loved to have been a fly on the wall for that conversation.

After she turned her Sick Call note in to Fuscha, Dale was told she could go to the PX, by herself, to buy civilian shoes. Instead of telling Silva to drive her there, Ritchie sent word for her to walk, that the company driver could not be spared. Even though he was in the parking lot, leaning on the jeep, intently watching Tierni with the rest of the company, cleaning their M16s.

Dale returned a half-hour later, wearing a pair of black Converse high-top sneakers that felt much better on her feet than the limiting combat boots, which she had in a bag and took upstairs to secure in her locker. When she got back downstairs, the women were congregated on the north patio with the men nowhere in sight. She was surprised to see Henning conducting this relaxed class and wondered when MacArthur's replacement would report and if the successor would be female, also. Right now Henning was the only female liaison A-10 had.

Dale looked at the items spread out on the patio. Barbells, free weights, a huge baseball bat, a box with three steps to the top and a stationary bicycle set up in such a way that when one sat on the seat, her back was parallel to the floor and she pedaled facing the ceiling. Henning observed Dale walk into formation, wearing sneakers. She smiled triumphantly and gave Dale the option to participate or observe. The painkillers were working on her feet, so she participated.

After the women spent two hours with Henning on the patio, the rest of the day was spent cleaning. Shannon was put on a detail that designated her to clean the Battalion offices and Dale was assigned to scrub the north patio laundry room. She finished about two minutes before Lights Out, ran upstairs and changed in the dark. After Robbins performed Bed Check, Dale took a long, hot shower in the one stall left open.

CHAPTER THIRTY-FOUR

The trainees spent the next morning on written make-up tests. Shannon had missed Map Reading by one point and Dale had purposely flunked the Military Intelligence test "The title of the class confused me," Dale had told Holmquist who had expressed his surprise at her failure to get a Go the first time.

Mid-morning, after the make-up tests were completed, Alpha-10 was marched to Ashlin Gym for a PT test. Inclement weather had prevented the trainees from taking the test outdoors, where the scoring would not have been accurate.

Dale did as much as she could do without violating her profile, the restrictions imposed by Bell, so she spent most of the time scoring the others. During the push-up test, Dale was scoring a male trainee who exercised right next to Shannon. Holmquist strolled by and noticed that Shannon continued to struggle with her last sets.

The next thing Dale knew, Holmquist was in the front leaning rest position, face to face with Shannon. He matched her push-up for push-up. "Come on, Walker! Just five more," he urged. "Five more and you'll max out. Come on, I'll do them with you."

She did two more with his encouragement. "Straighten your arms and get your bearings. You can do it! I'm right here with you." She slowly did one and, even more slowly, another one. "Okay. One more and you're done. Come on, Walker, don't you quit on me now, I know you've got one more in you. Take a deep breath, let it out and we'll do it together. Ready?"

Shannon was beet red and her arms were shaking but somehow she managed that last push-up. She dropped to the floor amid cheers and applause. She had never maxed out push-ups before. She received an attaboy pat on the back from Holmquist. She rolled over, sat up and shook his hand.

PERMISSION TO RECOVER

Dale watched Holmquist, admirably. What a nice man he had turned out to be. He didn't have to do what he did with Shannon. She wasn't even in his platoon. A cloud shadowed Dale's thoughts. They were probably going to have to keep a close eye on him. He was too good to be true.

While the company waited for the results of the PT test, Putnam brought out a basketball and tossed it to a group of female trainees. As they began to pass it back and forth and take shots at the basket, Ritchie appeared from nowhere and snatched it away from Ryder. He was not exactly gentle about it.

"You women even want to play men's sports," he snarled at them. "Why don't you all just fly to Sweden and have that operation?" He turned from them and threw the ball to one of the male trainees.

Henning disappeared into the equipment room and returned with another basketball. She handed it to Minty who was the closest to her. She was obviously furious at Ritchie by the expression on her face. The women formed two teams and played half-court on the side the men weren't using.

Dale sidled up to Henning, her voice quiet. "Why is he still saying shit like that? Does the man have a death wish because I am sure we can oblige him. Doesn't an IG warning mean anything to him?"

"Sure," Henning said, her arms folded tightly across her chest. "But the IG who warned him went on a thirty-day leave, starting this morning. Ritchie seems to feel that all incidents will hold little validity after thirty days. Or will be forgotten about."

"Then he'll be in for a surprise, won't he?"

"I surely hope so. Until then, we're stuck with him, aren't we?"

"We'll see."

Prompted by the senior drill sergeant, a quintet of males approached the women playing basketball and challenged them to a game. The women chose who they thought would be their best five players, Verno, Creed, Michaelson, Minty and Ryder, with Sherlock on the side. Sherlock had height but not much experience in anything other than a driveway game of Horse.

Fifteen minutes into a vigorous workout, the game was called because the PT scores were ready. The women didn't win but neither did they roll over and play dead. The score was forty-four to forty. The men made a mistake and thought the women were going to play like girls. Instead, they played like athletes who had all lettered in high school or college basketball. After spending nearly two months with these women, the men should have known better than to underestimate their abilities.

The company froze on the north patio the next morning, after chow. They stood in formation in their dress greens, or Class-A uniform, and waited for the company commander to finish his open ranks inspection. He appeared to revel in the fact that he moved slower than a tortoise. By the time he got to Dale, she was visibly shivering, as was mostly everyone else. Colton told her to stand still. She took a deep breath and tried to stop shaking as the CO examined every inch of her uniform, shoes, and appearance. Dale knew he would be looking to burn her so she worked extra hard on her brass, her dress greens and spit-shining her low-quarters. Even though, no one looked sharper, with the possible exception of Shannon and the prior-service males, all Colton said to McCoy, who was next to him, keeping track, was *average*. He moved to the GI in line next to Dale.

"Average!" Dale spit out later. "It should have been Outstanding. You can see yourself in my shoes. You could get a headache off my brass. How can that be Average?"
"Calm down, Oakes. It's not fair but at least it's not an Unsatisfactory. At least you didn't fail," Ryan said.
The platoons were gathered in the individual male bays again, awaiting the results of the inspection. As Colton had moved through the troops, Henning and the First Sergeant inspected the barracks. McCoy informed the women in his platoon that he felt the results of the inspections were touch and go for them. He said they had done too much bickering and not enough cleaning and preparation.
"Excuse me, Drill Sergeant, but now you sound like the Senior Drill Sergeant," Minty told him, with a respectful grin.
McCoy plainly did not take offense. He paused, thoughtfully. "Sergeant Ritchie has some valid points. He's just not real tactful. Or discreet. Or diplomatic. But this isn't charm

school, ladies, this is the Army. Good training is more important than good manners. We wouldn't be doing our job of toughening you up if we had to walk on eggshells so that we wouldn't hurt anyone's feelings. You wouldn't have gotten as far as you have if we hadn't been tough on you."

"None of us want to argue that point, Drill Sergeant," Mroz said. "I don't think any of us enlisted with the idea that it'd be like pledging a sorority. But Rit – I mean, Drill Sergeant Ritchie's attacking us every time we blink doesn't really do much to keep us inspired."

"That's right, Drill Sergeant," Minty said. "You get your point across without being obnoxious or cutting us down. We do more for you, willingly, than we do for the Senior Drill Sergeant, reluctantly. We thought our visit to the IG would stop his war against us but it didn't even slow him down."

"It will." McCoy wore a curious smirk. Dale had heard that McCoy was tickled that the females had displayed the guts to turn Ritchie in. "These things take time."

Holmquist then entered the Second Platoon bay and announced that everyone had received a six-hour on post pass, except for the trainees who'd had unsatisfactory brass, less than spit-shined shoes, a wrinkled uniform or had done poorly on the PT test. There were eight, all total, who remained behind when the rest of the company raced to their lockers to get ready for six hours away from Tenth Battalion.

Dale went to the Pizza Place with a crowd of fellow trainees and Shannon, with a smaller group, scouted out the PX snack bar and then headed across the street to the bowling alley. Slowly, almost all of those trainees filtered back to the Pizza Place, so after bowling three games, Shannon also made her way back to the more popular hangout, accompanied by Private Zachary, who was clearly smitten with her.

The minute Shannon and Zachary walked inside, they could barely move, the small establishment was so jam-packed. By the time Shannon found Dale, she had run into various people wearing fatigues. They were GIs she did not recognize. A face she did readily identify, though, was Jane Bradbury, drill sergeant from Bravo company.

"Who the hell are these people?" Shannon shouted to Dale to be heard above the jukebox.

"Bravo. They're on a six-hour pass, too. Can you beat it? They're a week behind us in training and they get their pass privileges at the same time."

"I notice a big fan of yours is here." Shannon nodded her head toward Bradbury.

"Aw, hell, she's long lost interest in me. There's too much else to choose from. What's happening at the bowling alley?"

"Nothing. Why do you think I'm here?"

"I see *you* have a fan."

Shannon smiled. She knew Dale referred to Zachary. "Yeah. My protector. Or so he thinks."

"Even though you could probably kick his ass from here to kingdom come."

"Yeah, well, hopefully that won't be necessary. Anything happening here?"

Dale's throat was sore from the need to make herself heard above the music, the yelling and because of all the cigarette smoke in the air. "Let's go outside so we can breathe."

"I just got in here. Besides, we'll never get back in."

"Sure we will," Dale said and moved through the crowd to the door. Once they got outside, Dale took a deep breath of fresh air.

"What's up?"

"Everybody's behaving pretty normally. I counted about thirty-six people drinking beer."

"Thirty-six of our people?"

"Yes. Do you think anyone from Bravo is going to risk drinking beer with the iron maiden standing there, burning holes through them with her eyes?"

"They might. How many females?"

"Eleven," Dale said. "Laraway, Ryan, Troice, Sherlock, Ryder, Lehr, Caffrey, McTague, Travis, York and McKnight."

"God, I'm dying for a beer," Shannon confessed.

"So am I," Dale said. "But you know they're going to get caught." Dale smiled at Shannon. "So…is this serious with *Private* Zachary?"

"Please. He's cute but very young for his age. He's a nice

guy. It's flattering."

"He can't keep his hands off you," Dale teased.

"I know. I'm not encouraging it but his hand on my shoulder or my back feels pretty nice. It's been a while, you know?"

"If you like him then maybe you should go in and stop him. As we were leaving, he was ordering a beer."

Shannon seemed to contemplate this. "No. He'll have to learn like everybody else."

With only an hour remaining on the pass, they went back inside. Neither was able to move very far. Shannon saw Zachary wave to her from the other side of the room. As he braved the crowd to reach her, she felt an enormous hand on her shoulder. She looked up – way up – to see the six-foot, five-inch Jimmy Judd towering above her. Dale just laughed and shook her head. "Some things never change," Dale shouted to Shannon.

Judd, who was apparently under the influence of more than just a couple of sixteen-ounce cups of beer, tried to converse with Shannon above the noise. Apparently, sometime during this ten-minute, non-sexual exchange where every other sentence consisted of *what?*, Judd fell in love.

He excused himself to use the rest room and both Shannon and Dale used that opportunity to escape back to the company area. As they crossed the Battalion parking lot, Zachary caught up with them and escorted them the rest of the way. The company had to be back and signed in by 1800 hours. Attendance would be taken during a formation. Three-quarters of the trainees were signed in by five-thirty while the rest straggled in.

Dale, Shannon, and Zachary checked in and remained on the north patio by the far picnic table and rehashed their afternoon. Mroz, who had signed in an hour earlier, joined them. Neither Dale nor Shannon had seen Judd cross the patio and enter the CQ Office since their backs were to that particular door but they definitely knew when he came out.

He must have seen Zachary's hand rest on Shannon's shoulder in the course of conversation and it made him go wild. It took no time at all for legs that long to carry him to the group of four. He reached out and shoved Zachary and knocked him off balance. This action startled everyone more than it did anything else.

Zachary didn't look too happy to have his good mood ruined by getting into a fight, especially with a wiry giant like Judd. Zachary regained his footing, walked back, and stood between Shannon and Judd. Before Shannon could ask Judd what his problem was, Judd yelled at Zachary to stay away from Walker, that she belonged to Judd.

This came as news to Shannon, who looked at Dale and rolled her eyes. While the two suitors verbally dueled, Dale couldn't hold back a smile. She leaned in close to Shannon. "Just what exactly did you say to him at that passionate setting a half hour ago?"

"I know what I said but it's not obviously what he heard. Although, I'm pretty sure 'I do' didn't come into it at any time."

They returned their attention to the battle and Shannon had had enough. She took a step behind Zachary about to tell them both to grow up when Judd threw a punch. Zachary ducked and the biggest fist Shannon had ever seen connected with her face. The power of it alone sent Shannon reeling backward, across the patio where, after three or four shaky steps, fell flat on her back.

Dale wasted no time and jumped in front of Zachary to hold him back. "Knock it off!" Dale yelled at both men. She had a pretty good grip on Zachary but Mroz had a tougher time with Judd. He wouldn't stay put. When he moved toward his opponent, he took Mroz with him.

"Judd, stop! You guys are going to get us all in trouble," Dale warned, as Zachary tried to break free so he could defend Shannon's honor.

Shannon had just barely made it to a sitting position, shaking the stars from her eyes when Audi, followed closely by Holmquist, charged out of the Orderly Room to see what the commotion was.

"*At ease!*" Audi demanded in a voice that sounded more like the roar of a pissed off grizzly bear. He grabbed Judd away from Mroz with one hand and even though the trainee had a good five inches on the drill sergeant, Audi outweighed him by at least one hundred fifty pounds. With one swift motion, Audi moved Judd from standing on his feet to hugging the patio floor.

"Everybody in the front leaning rest position right now!" Audi ordered.

The five trainees obeyed and assumed the position immediately. They had never seen Audi this mad.

"What the fuck is going on here?" He strolled around the five bodies. No one answered him. "Oakes? What happened here?"

"I'm not exactly sure, Drill Sergeant," Dale answered. She sounded genuinely puzzled. What both she and Shannon did know was that this incident would insure a restriction against the company again.

"You're not sure," he repeated, not pleased with her response. "Mroz! On your feet!" Mroz obeyed and jumped up to the position of Attention. "The rest of you stay exactly where you are. Private Mroz, come with me."

"Yes, Drill Sergeant." Mroz reluctantly followed Audi to the Orderly Room.

While the four others held their position, Holmquist sent a male and female trainee to the bays to round up the company for an emergency muster. Fifteen minutes later, Audi and Mroz came out of the CQ Office. Mroz joined the gathered troops on the south patio and Audi approached the four semi-prone trainees by the picnic table.

"On your feet!" Audi bellowed. Dale, Shannon, Zachary, and Judd rose to Attention. "Get over there with your platoons!"

At the formation, Audi, Holmquist and Putnam went through the ranks and individually asked every trainee, at very close range, if they had been drinking. Once that task was completed, the company was ordered to return to their respective barracks until further notice.

The women who had no idea why the muster was called, found out quickly as the incident spread like wildfire. Dale approached Shannon, who was lying on her bunk with a cold washcloth on her eye. "Are you okay?"

Shannon removed the washcloth to reveal a purplish-red mark on her cheekbone, below her left eye. "How does it look?"

"Like he smacked you with a redwood."

"Why do these things always happen to me?"

Dale let go with an exaggerated sigh. "You've just got to

stop being such a femme fatale."

This made Shannon smile. "Hey, when you got it, it just oozes out, you know?"

"I'm so glad you can make jokes, Walker. While you're enjoying all this attention, our passes for tomorrow will probably be pulled." The crisp voice belonged to none other than Professor Snow.

"Oh, Christ, it figures," Shannon said. She pointed to her face. "Does it look like I'm enjoying this?"

"How should I know? Maybe you're into pain."

Shannon sat up. "Only if it comes out of kicking the shit out of you. Oh, my mistake. That would be classified as pleasure."

"If you ask me, you got exactly what you deserved," Snow said with a sneer.

"Nobody asked you," Dale said. As far as Shannon knew, up until that point, Dale and Snow had gotten along but maybe that was because their paths had so rarely crossed. "Since you brought it up, though, I'm interested, why did she deserve it?"

"Yeah, I'd like to know that, too," Shannon said.

"Oh, please…you let both of them paw you like animals, leading them on and now you play innocent?"

Shannon stood up and headed for the latrine so she could rinse her washcloth under more cold water. She started to pass Snow and then stopped in front of her, unthreateningly. "Now, Professor, how would you know that unless you were watching me the entire five-and-a-half hours? You wouldn't be jealous of the boys, would you?"

It was almost as if Snow could be seen sprouting talons and fangs. She clearly held herself back from lunging at Shannon. "Believe me," Snow hissed through clenched teeth. "You're nothing I'd want."

"Thank God for that," Shannon said as she continued her journey to the sinks. By this time, everyone in the bay was watching.

"You're a cockteaser!" Snow snapped at her.

Shannon stopped and turned to face her. "You're the last person to be calling anyone names like that."

"What do you mean?" Snow asked, indignantly. "I have never gone out anywhere and teased or picked anyone up or let them have free reign of my body just because I felt a little horny. You cannot make the same statement, Walker."

"Yes, I can but not with such a straight face. You're very good, Snow, but…let's not forget about your oh so virtuous behavior the night before Christmas exodus."

There was dead silence in the bay. By the stunned look on Snow's face, the rest of the women wore expressions that seemed to wonder what apparent incriminating secret Shannon held about the Professor. As Shannon stared her opponent down, she sneaked a glance at Dale to see a tiny smirk on Dale's face. She looked back at Snow, whose mouth dropped open but no sound came out.

Dale stepped to Snow and patted her on the back. "You're wise not to say anything. An argument Walker loses is never over. Keep that in mind." As Dale then quickened her pace to catch up with Shannon, a loud noise made everyone jump. It sounded like the barracks door had been kicked open and slammed against the wall.

"*Man on the floor!*"

Shannon and Dale exchanged glances. This was not good. It was McCoy and he sounded furious.

"At ease!" Someone called out, unnecessarily.

"Second Platoon, I want you all in my office immediately!" His voice was more of a bark than a yell.

"Yes, Drill Sergeant," a few females randomly answered. McCoy then exited the bay.

"Oh, boy," Dale mumbled to Shannon. "This is going to be like facing a strict father after coming home an hour after curfew."

"Hey, it could be worse, we could all be standing in front of Ritchie."

"I know but just when we were beginning to form a rapport with McCoy and he was expressing pride and trusting us."

"Sorry, Dale."

"Not your fault, partner. This, too, shall pass."

CHAPTER THIRTY-FIVE

McCoy's small office was near the second-floor landing. When the women entered, he was at his desk. He gestured for them to find a spot and sit. They settled, most of them on the floor and awaited their lecture.

"Okay, I want to know who in this room was drinking tonight."

Laraway, Ryan, McTague, Lehr and Ryder admitted they had imbibed. McCoy nodded and scanned the rest of the women.

"I know six of you were drinking. Only five of you are being honest. I'm going to ask you again. Who was drinking?" He got the same response.

Dale was positive the rest of the women knew McKnight was the sixth one. She wanted to throttle her. McKnight's silence made the situation worse.

"Look, ladies, we honestly expected some of you to drink, even though you were told not to. It happens every time. And it would have gone unmentioned if it hadn't been for the Walker incident."

"Excuse me, Drill Sergeant, may I say something?" Dale asked.

"Certainly, Private Oakes."

"Why will it hereafter be referred to as The Walker Incident when it was Private Judd who caused all the trouble?"

McCoy smiled, unexpectedly. "Are you asking this because you two have become good friends?"

"No, Drill Sergeant, I'm asking because it doesn't seem fair to me that she's getting blamed for this and, as you can ask Mroz, all she did was stand in the wrong place."

"Point well taken. I stand corrected. From now on, in our group anyway, it will be referred to as The Judd Incident."

"Thank you, Drill Sergeant."

PERMISSION TO RECOVER

McCoy kept the women in his office for a gab session that he hoped would help some of them talk out their aggression. While they aired their grievances, Dale wondered if the conversation would get back to McKnight's drinking and if she would finally admit it or if someone would end up reporting her.

Although turning in fellow GIs for minor violations was not a wise thing to do in any close quartered military environment, McKnight had crossed the line too many times, causing grief for her barracks-mates, platoon and company and everyone was sick of it. They were tired of her never owning up to her infractions, never taking responsibility and whatever mess she got into was always someone else's fault. Dale was interested in how this office meeting would end.

"Do you know how I feel when I am called at home by the Staff Duty NCO, a drill sergeant from another company, to tell me about the unmilitary-like goings on in my own company?"

Shannon stood at Attention in the Senior Drill Sergeant's office. She thought she had escaped this fate as Ritchie had allegedly been on a three-day leave. She faced his desk. "No, Drill Sergeant."

"You don't? You have no idea how I must feel?" He stood in front of her.

Didn't I just say No?

"I can only imagine, Drill Sergeant." A massive headache had set in. An audience before the Senior Drill Prick was the last thing she needed.

"No, you can't!" Ritchie yelled at her. He began his usual tirade about how humiliated and ashamed he was and as he ranted, he roamed. His movement, mixed with his nasal bellyaching, did not make it any easier on Shannon's head. He progressed to what amounted to some kind of ritual dance while he hollered about the irresponsibility of the troops. He stopped right in front of her, nose to nose. "And it's your fault!"

"My fault?" Shannon tried hard to maintain focus but there still seemed to be at least two of him. All she really wanted was to go back upstairs and lie down.

"Yes. Your fault. Because you are a female."

"What?" Regardless of her suspected concussion and how

bad her head hurt, she wanted to hear his rationalization.

"Don't you mean What, *Drill Sergeant*?" He was still in her face.

"Yes, Drill Sergeant, that's what I meant," Shannon said. She would have blatantly rolled her eyes if it wouldn't cause her so much pain.

"If you hadn't been a female, Walker, the two men wouldn't have reacted the way they did. In fact, if the females weren't here at all, these men wouldn't have to have been restricted for as long as they have been. They would be through basic training and on to LE School. They would have focused on something other than the female anatomy and what part of them they can stick in what part of you." He walked around her and faced her back. "How did you qualify for the MP training, anyway, Walker? Open your legs for the recruiter?"

Shannon had enough. She moved out of the position of Attention and turned on her heel to face him. "What is it with you, Drill Sergeant, that you can't keep your mind out of the gutter? It seems to me that you have a problem with women and sex."

Ritchie's shock was apparent. He got very quiet and squinted at her. "What did you say to me?"

"You heard me."

"Why you disrespectful little bitch! Get in the front leaning rest position! Now!"

"Are you sure you want me in that position? I would think you would prefer me on my knees in front of you." She watched Ritchie get flustered. His expression told her that his immediate inability to speak was a mixture of not believing her defiance and wondering if she were serious about a possible blow job.

Finally, he spoke. "I told you to get into the front leaning rest position and if you are not in it in three seconds, I will put you there!"

"You lay one hand on me and I swear to God, I will scream rape but not before I kick your balls up through the third-floor bay," Shannon told him through clenched teeth.

"Now you're threatening me? I think your Army experience just ended, young lady. With my report that you propositioned me and then threatened me when I didn't accept your offer, I don't see a simple trainee discharge in your future." He returned to his smug state. His sneer disappeared, however, when Shannon crossed her arms and glared at him. This was not the reaction he expected or intended.

"Drill Sergeant Ritchie, you better think really hard about what you're doing. You are out of line here, not me. Your mistake is that you think all women are stupid, inferior, and subservient and that we're all afraid of you. Understand this, Drill Sergeant, I want to be a good soldier and I want to respect you but it's clear that you do not respect yourself or you wouldn't treat people, females, especially, the way you do. Now, I know I'm in a man's Army but I also know I do have rights. They are limited but I do have them. And I know that you are on the IG's shit list and I think if it comes to who the IG will believe, it won't be you. Not with all the complaints against you in this unit. But I won't wait until the IG gets back from leave. I'll go to the Battalion Commander. As your boss, I don't think she'll appreciate your opinion of how she must have made her rank."

A bead of sweat trickled down his forehead. "Walker, you certainly have some brass ones." He returned to his desk and sat down. "I'm going to overlook this little display of insubordination because you obviously have a head injury. We'll keep what was said in this office between us and I won't give you a counseling statement."

You sleazy son-of-a-bitch.

"Yes, Drill Sergeant. May I please go back upstairs and lie down before I pass out?"

"Do you need to be taken to the hospital? Maybe Silva should drive you just in case." His tone was expressionless and he no longer made eye contact.

"No, please, Drill Sergeant. I'd like some aspirin and to get off my feet."

"See Sergeant Audi for the aspirin. You're dismissed, Walker."

"Thank you, Drill Sergeant." Shannon closed the door behind her and entered the CQ Office.

"Damn, Walker, look at that shiner," the male CQ runner shouted and whistled.

The piercing noise hit her like a shot. If she thought her head could not have hurt any worse, she was just proven wrong. "You do that again and we'll have matching shiners," Shannon threatened.

"You okay, Walker?" The voice belonged to Holmquist, who was seated at Sergeant Fuscha's desk. He at least had a tinge of compassion in his tone.

"I will be, Drill Sergeant, if I can just get some aspirin and go back upstairs to lie down."

Holmquist nodded, reached into one of the desk drawers and handed her a bottle filled with white pills. "New stash. Take a couple for later, too."

She took six pills from the bottle. "Thank you, Drill Sergeant."

"Next time? Duck," Holmquist said, as he returned the aspirin to the desk.

"There won't be a next time, Drill Sergeant," Shannon said. "There never should have been a this time."

"I think I made him suspicious," Shannon said. She and Dale ran into one another on their way back to the bay. Shannon decided to smoke one last cigarette, even though her head was still throbbing, and she pulled Dale back down to the south patio with her.

"I don't think I would have done anything differently," Dale said, soothingly. "You know me, I probably would have been a lot worse. Although…it would have been interesting to see him try to drive home with his desk shoved up his ass. Shan, if I didn't blow it after Kirk's death with as blatant as that insubordination was, I am pretty sure you didn't blow it with Ritchie. In fact, you probably made him stop and think."

"I know but about what?"

Dale folded her hands together as she leaned on the railing and looked out across the parking lot. "Let's just see what happens here before we go packing our bags. If we are found out now, the sooner we can flush Ritchie's toilet and get the fuck out of here."

"Dale…what if it hasn't been a set up? What if those other drill sergeants have really been guys like Ritchie? He's been disguising himself so well that Bishaye certainly hasn't seen anything wrong with him. And Henning thinks he's an asshole but obviously she doesn't suspect him of anything else."

"That's why they have us here."

"But what if it really *is* the drill sergeants?" Shannon looked at Dale. "What if it's been the drill sergeants all along? God, and the way I badgered Willensky…"

"No, I don't believe that. Those girls are hiding something. Your instinct was that Willensky was holding something back."

"Yes, but was that because I had already convinced myself that she was involved?"

"No, I talked to the same people you did. None of those drill sergeants had an attitude anywhere near what Ritchie has," Dale said.

"In front of us."

"Boner said it was out of character for the last two."

"Boner? Now there's a pillar of reliability. And you two were such good friends that you know for a fact she couldn't possibly be mistaken."

Dale arched an eyebrow. "That's my point. Boner is as military as one can get. Do you think that if she really thought Halpin and Fransciosa were guilty that she would have so readily defended them?"

"Unless she was involved with one of them."

"Boner? She's a freaking Army droid!" Dale sounded frustrated.

"Nothing's impossible."

"Okay, what about Carolyn Stuart?"

"I don't know!" Shannon squeezed her eyes shut as the level of her own voice echoed through her still aching head. She took a deep breath. "Maybe that's just coincidence."

Dale was momentarily silent. She chewed on her bottom lip. "Shannon, I know Ritchie really threw you into a tailspin here. I'm your partner. If you want to go to Bishaye first thing in the morning and tell her you think the drill sergeants are guilty and we're wasting our time then I am right beside you. God knows, I would love to get out of here. But I know her. She will grill us until we come up with something solid enough to convince her that it was the drill sergeants and not the women. And we don't have that. We only have speculation. Ritchie's behavior is not enough."

Shannon stabbed out her cigarette, field stripped it and stuck the butt in her pocket. "I wanted to kill him. I wanted to wrap my fingers around his throat and just choke the life right out of him."

"I understand." Dale sat down at the picnic table and Shannon seated herself on the bench.

"God, remember when we got here and we were so worried about Robbins? He's turned out to be no more than a flirt. And harmless, at least to my knowledge."

"Mine, too."

"I'm just stunned that with all the trouble caused by in the last three cycles, Ritchie would so freely say that stuff to me. And to Tierni."

"True. Although he never actually put himself in the situations."

"What?"

"Think about it. What he said to Tierni about doing it with someone who matters or any of the other things, he never actually said doing it with me or sleeping your way to the top, starting with me."

"He implied it."

"Maybe not. Maybe he meant exactly what he said. He seems to hold such contempt for women, maybe he was just thinking out loud. He probably believes the only way women get anywhere in life is by bartering with their bodies. I'm sure it makes him feel superior to voice that opinion any time he gets the opportunity."

"Are you sticking up for him?" Shannon glared at her.

Dale looked down at her. "You know, that would be a much more menacing expression if both eyes were open equally. You kind of look like a pirate." Shannon didn't seem to find that humorous. "I'm not sticking up for him. I'm just trying to reason it out. I can't believe a man like Ritchie, smart enough to retain senior drill sergeant status for several cycles, would be stupid enough to offer up his stud services in light of everything that's been happening, especially after an IG warning," Dale said.

"What about what he did to Kotski?"

"That was wrong. All of it is wrong, Shannon. He's not a boy scout. I just don't think Bishaye will go for it."

"He seemed to calm down when I brought Bishaye into it. She must have a lot of power."

"She does. She knows a lot of people. In fact, she's probably as influential as the IG on this post. You know the post commander, General Oberman? He's an old boyfriend of hers. He's the one who talked her into a military career. And it was through him, she met Jack, her husband."

"I didn't know that."

"Yep." They both got up and crossed the patio toward the staircase. "He and Jack aren't pals anymore but he and Anne are still great friends. Through him and others along the way, she has a lot of acquaintances that carry a lot of weight. And they just *love* to do her favors."

"I'm sure it has nothing to do with her looks," Shannon said, sarcastically. They slowly climbed the stairs. The higher they went, the more her eye ached. "Do you think we'll get that six hour pass tomorrow?"

"No. How about getting a good poker game going?"

"Sounds good. If we're going to get busted anyway, we might as well go out gambling."

"I'll bet you anything we'll still be on the case. How is Ritchie going to justify your insubordination without incriminating himself? He won't tell anyone. And even if he does tell Colton, Colton won't let it get as far as Bishaye. That would make him look too irresponsible."

"Okay, then poker it is...unless Michaelson wants to play. Then you might as well just hand her your money and leave the room."

Dale opened the barracks door. "She's that good? Better than you?"

"She makes me look like an amateur. But I love the competition and so will you." It was the first time that Shannon managed to smile. It hurt.

CHAPTER THIRTY-SIX

Word was sent up early that the doubtful six-hour pass was granted to all but the thirty-six GIs, including McKnight, who'd been drinking beer the day before. Shannon was almost disappointed as she had geared herself up for really good running card game. She was also more than surprised that she was not restricted because of her unwilling involvement in the events of the night before.

Her head finally stopped pounding after Lights Out and she dozed off moments later. Judd's fist was huge and powerful and the bruise it had left covered at least a quarter of Shannon's face. The impact had been startlingly solid and she had no doubt she had a mild concussion. By the time she was ready to sleep, the blurriness was long gone, as was the lightheaded feeling that attacked her every time she stood up. Sleep may have not been safe but it certainly was welcome.

The bruise was purple and blue. Lovely shades of each if she were buying a bouquet of violets. It looked even more conspicuous right out of a hot shower. In a few days, it would look like camouflage and fit right in. *In a week, it will be gone,* Shannon chanted mentally.

In an effort to keep one another out of petty complications, Shannon and Dale made a pact to stay together for the duration of this pass. They hit every trainee hangout, the bowling alley, the PX, the gym, the movie theater, and the Pizza Place. The atmosphere in each was subdued after yesterday. Both Shannon and Dale kept their observations of A-10's GIs as discreet as possible while still participating as much as they could in the activities and not become distracted. Shannon, especially, kept her distance from any prospective amorous advances. No one drank. No one got into any trouble. It was a quiet, boring afternoon but at least they had the semi-freedom to be bored of their own accord.

After the company signed in, it was back to the normal routine of locker maintenance, shined boots, pressed uniforms, and clean laundry.

Dale reassured Shannon that since they had not been sent for, their cover was obviously still good. Neither agent trusted the senior drill sergeant, though, and agreed to be extra vigilant when he was around. Dale had talked Shannon out of advising Bishaye about what had occurred the night before. She said it would be better to wait to see if Anne contacted them. When they got through the day without Henning showing up to pull them away from wherever they were, that was a good sign.

The trainees were brought to a pickup point by bus and marched to a field where they were to be transported by helicopter to the starting site of their ten-mile hike. Each soldier was outfitted with full backpack, war gear and M16, a total of seventy-five extra pounds. They were advised that they would definitely be attacked somewhere along the route so they attached their gas masks to their war gear. It only added another pound but after the first mile, it felt like a lot more.

The troops were instructed on how to approach and exit the helicopter while the tandem rotors of the aircraft were in motion. There were three transport Chinooks that lifted the soldiers in small groups to an area in the woods that led to a dirt road starting point.

Shannon, deathly ill at the thought of riding in a helicopter again, timidly tapped the pilot on the shoulder before her group took off. "Excuse me, Sir," she asked, shakily. "What happens if someone gets sick in your helicopter?"

"We kick them out in mid-air," he said, not even cracking a smile.

"Looking forward to it." She mumbled then moved to her seat, holding her stomach.

Shannon survived the five-minute trip but she rode with her head between her knees the entire time. Once she was back on the ground and out of the Chinook, she was fine. She could never understand why everyone around her, including Dale, seemed to get such a big thrill out of that mode of transportation.

Dale went on the hike even though she knew Colonel Bell would have considered it a violation of her profile. She persuaded Holmquist to let her go since the restriction said nothing about hiking and it technically wasn't marching and definitely wasn't running. She promised him that if it got to be too much, she would drop out, which she did after eight miles.

During the jaunt, the trainees were gassed on three different occasions to test their reaction time. They were besieged by blank gunfire several times and once, Putnam drove a jeep between the two single files of soldiers on both sides of the road. "Bomb!" he yelled.

McCoy, who was in a jeep about an eighth of a mile behind Putnam, evaluated how long it took the trainees to take cover. It wasn't the scramble to get down into prone positions in the ditches on the sides of the road that was difficult, it was getting up out of the ditch with all that extra weight on their backs that was the real task.

At the conclusion of the hike, when the soldiers marched back into Tenth Battalion, five females and only one male had fallen out and needed to be transported back to the company area by vehicle. The women were afraid that Ritchie would land on them again because of it and the collective thought of another episode with him tended to lower everyone's spirits a bit. Fortunately, completion of the ten-mile hike was not mandatory for graduating basic training.

Uncharacteristically, Ritchie said nothing derogatory to anyone, in fact, he was overheard telling McCoy that he thought everyone did a good job. When that got around, the main rumor was that the trip to the IG was finally working. Shannon was worried it was something else.

"What's he up to?" Shannon whispered.

Dale sat at the foot of her bunk frame and removed her boots. She massaged her heel and ankle before she attempted to put her sneakers back on. "I don't know but I'm sure it's not what you think. Henning would have let us know immediately."

Both agents were quiet as several females passed them on their way to their lockers, the latrine or on their way back outside. When the bay traffic became scarce, Shannon watched Dale knead her bad foot. "Why the hell did you go on the march, you dumb shit?"

"It wasn't a march, it was a hike."

"March, hike, doesn't matter. The results were the same and it wasn't that smart, Dale."

"Really, Private Blackeye?" Dale said, good-naturedly.

Shannon pushed Dale off balance and she fell backward onto her bunk. "I didn't choose to get punched in the face."

Dale bounced off the bed and immediately took the wrinkles out of her blanket. "I didn't choose to injure my foot."

"No, you chose to aggravate the injury and maybe do more damage. You're going to end back up at the TMC."

Dale slipped her sneakers on and tied them. "I can't go back to Bell. She'll put me out of commission for sure. She'll override everything and put me right into physical therapy."

"Can she do that?"

"She can."

"Then, Dale, maybe you should go," Shannon said, gently.

"No."

"Then take it easy on yourself, would you? There is nothing about why we are here that requires you to possibly make yourself crippled."

Dale sighed. "You're right. I should have sat this one out."

It was pouring rain when the company attended their 0700 formation after morning chow. That did not deter the drill sergeants who held the scheduled class on digging foxholes, anyway. The trainees were bused out to the field where they spent the morning up to their armpits in muck. The faster and deeper they dug, the more rapidly the water filled the space.

They returned to the company area after noon chow in the field and spread out on the patios, filthy and wet, and cleaned their M16s. With the little time left before 1700 formation, the company practiced Drill and Ceremony.

There were few Alpha trainees in the mess hall for evening chow. A majority of them voluntarily gave up eating to stand in line for a shower. It was a wonder there was any hot water left in Alabama by the time the trainees were through.

Dale returned to the bay after she cornered Audi about Kotski. Had he heard anything? No, he was just as much in the dark as she was. Could it be arranged that Dale could go see her? He would check on it and get back to her.

"Hey, Oakes," Tramonte called out. She, Tierni, and Travis were standing by Shannon's bunk. "Come here a minute."

Dale walked to the happy looking group. "What's up?"

"We're celebrating," Travis said and then pinched Tierni's cheek. "The rabbit lived. So we thought about taking our celebration to the Atlanta Underground on our first weekend pass. We wondered if you wanted to go."

"The Atlanta Underground? I've heard about that place." Dale glanced at Shannon, whose eyes were flashing a rapturous *yes*.

"I liked it. I got my feet wet there during Christmas exodus," Tierni said.

"That wasn't all you got wet there," Travis said, laughing.

"Returning to the scene of the crime," Tramonte chimed in, affectionately, and put her arm around a blushing Tierni's shoulder.

"I'm game," Dale said, smiling. She imagined partying with that little group would be fun.

"Me, too," Shannon said. "I never miss an opportunity to tie a good one on."

"Just be careful what you tie it on *to*," Tierni said. "It can get a little wild."

"Then it's settled," Travis said and rubbed her hands together.

It was far from settled. After the excitement of the idea and all the frivolity that surrounded the initial planning, both cycle spies instinctively knew they would not be able to go. With a handful of women off in Atlanta, there was still the rest of the company to consider and because the Atlanta group would be occupying one another's time, there was no reason either of them would be able to come up with that would convince Bishaye their presence off in another state was essential to the case. To say it would have been essential to their sanity would not have cut any ice with Bishaye. If they were sane, they wouldn't be where they were.

PERMISSION TO RECOVER

Dale was on her way to her locker to get ready for bed when someone called Attention. Everyone stopped what she was doing and assumed the position.

"Good Evening, Ladies." Karen Henning said.

"Good Evening, Ma'am!" the women replied.

"I thought I would stop by to let you all know that you will spend tomorrow afternoon at a GI party, preparing the barracks for the Battalion Commander's Inspection. This is a biggie, Ladies, so this place has to look better than it has ever looked before. Am I understood?"

"Yes, Ma'am!"

"Outstanding. As you were," she commanded. She turned as though she was about to leave and then she stopped. "Oh, I need a volunteer to run a few things to Battalion." No one spoke, as was expected. "Hmmm…Private Oakes, you have running shoes on. How about you help me out?"

"Yes, Ma'am," Dale said, sounding purposely unenthusiastic. She figured something must be going on and she was anxious to find out what it was. Maybe Ritchie was suspicious, after all. She caught Shannon's eye and shrugged before she exited the bay.

***

"What's going on?"

"I have no idea," Henning said. "I just got a call from the colonel who said to send you over."

"Interesting." Dale said, mostly to herself. She looked at Henning. "Have you heard anything, any rumors from the cadre, Ritchie, specifically, about Shannon or me?"

"No. Should I have? Do you think you've been made?"

"It's always a possibility." Dale wasn't sure how much Henning knew about Shannon's last encounter with Ritchie. "I just can't imagine why I'm being called on the carpet, so to speak, if I have nothing to report."

"I don't know. She told me to get you and send you there and that's what I'm doing. She must have a reason." Henning stopped at the edge of the north patio. "Maybe she just wants an update."

"Maybe." *Or maybe she just wants to see me.*

Dale left Henning and walked to Battalion. She climbed the steps while those old, familiar butterflies swirled around in her stomach. She was ambivalent about seeing Anne again, especially not knowing why she was there. That kiss replayed in her head and her inner devil hoped Anne just wanted some alone time with her. Her inner angel laughed uproariously at that idea.

Dale let the door close behind her and she took a deep breath. There was no Battalion CQ in sight. The reception area was empty and dark. The only light came from Anne's office, down the hall.

"Lock the door behind you." Anne's command was clear and it came from within her office.

Dale complied and walked to Anne's office. She leaned in the doorway and watched the colonel write her signature on a few pieces of paper. *Why does she have to be so damned gorgeous?* "Where's your CQ?"

"I sent him on an errand." Anne looked up at her and smiled. She gestured to a chair. "Sit." She appeared relaxed in her crisply starched fatigues and boots, so well spit-shined, they looked like patent leather. Her face didn't seem to have the tense lines it had the last time they had seen one another.

Dale seated herself in a chair against the wall. "Am I in some kind of trouble?"

"Aren't you always?" The tone in Anne's voice was playful. "Is that why you think you've been called in here?"

"I can't think of any other reason." Dale folded her arms. *Unless you've changed your mind and want to fuck my brains out.*

"I didn't know I would be staying so late, so I thought since I had the opportunity, I'd get an update on your progress."

Dale's insides caved in at the disappointment.

Of course. Why do I do this to myself? I should have known better...

"There is no progress. There is nothing to report. If there had been, I would have let you know."

"Are you telling me nothing has happened?" Anne pushed her chair away from her desk and turned to face Dale squarely. Blue eyes temporarily imprisoned Dale's brown ones until Dale looked away.

And what really hurts is that you know what you do to me.

"Plenty has happened. Just nothing that stands out as being attached to the case."

"And what about Colton and Ritchie?"

"Colton has backed off completely. Ritchie is still a prick but I think he's learning to rein in his biases about women. His conduct is unacceptable by anyone's standards except his own."

"Think he's involved?"

Dale shook her head. "It crossed my mind. I doubt it but we're keeping an eye on him, anyway. There are other infractions he can be reported for when we are done here but I honestly don't think he has anything to do with this. I wish he were involved. Nothing would give me greater pleasure than to see him taken down. Especially by you." Dale showed a hint of a smile and got one back from Anne. "The other drill sergeants seem to be going about their business and nothing seems suspicious about the cadre. The prior service males are acting normally, just a little more familiar with the system than the rest. The trainees are adjusting and no one is behaving in any manner that would draw suspicion."

"What happened Saturday night, after you got back from pass?" She folded her hands on her lap.

"People getting their first taste of freedom and alcohol after being incarcerated for too long. I told you we needed a pass sooner. Two months is longer than most basic training lasts. Most trainees would be on to AIT and have more freedom than two fucking hours after eight weeks. We're lucky all we had was a handful of drunks. And no, Walker instigated nothing. She was just in the wrong place at the wrong time."

Anne nodded in comprehension. "I got a call from Kathlynn Bell. She's not happy about your foot."

"I had to tell her why I was here. She would have blown it otherwise."

Anne blew out a small chuckle. "She's not too happy with me, either."

"Like me, she doesn't understand why I'm here."

"You're still angry. I can hear it in your voice, see it in your demeanor."

"What difference does it make?" Dale snapped. "It won't get me out of this assignment and it...won't get me you." She looked back up into Anne's eyes.

"Dale...I...don't know what to tell you."

"That's my point. I don't want you to tell me anything." She stood up and broke eye contact. "It's my problem. I'll deal with it." She turned to leave the office.

Anne also rose out of her chair. "Where are you going? You haven't been dismissed yet."

Dale stopped, did an About Face and stood at Attention. She refused to look at her. "Yes, Ma'am."

The colonel ran her fingers through her hair, exasperated. "What exactly do you want to happen between us?"

"Are you really that blind?" Dale glared at her.

Anne stepped closer to her and pulled Dale back toward her. She shut off the light, took Dale into her arms and pressed her against the wall. Her voice was low and sexy. "You want this again, don't you?" She bent her head and captured Dale's lips with her own.

The softness of Anne's lips and the intensity of the sexual charge that sizzled between them were almost too much for Dale. That and the fact that she was actually kissing her again. The kiss became immediately aggressive and hungry and Dale tasted whiskey on Anne's tongue. When she tried to break contact, Anne became more forceful in her desire to stay physically connected. Dale finally was able to get out a few words. "Need to breathe."

Both women were panting heavily when Anne rested her cheek on Dale's forehead.

"Are you drunk?" Dale asked, her breath still gasps.

"What?" Anne immediately seized Dale's lips again.

Dale's head was spinning. Anne's mouth felt so wonderful moving against her own and she told herself to just enjoy the moment while it lasted. Her common sense screamed something else. She reached up and separated their faces from one another. "You've been drinking."

"I had a couple shots. It isn't important."

"It isn't like you. You're still on duty. In your office. You're acting irresponsibly...with me in your office."

Anne sighed and pushed herself away from the wall and away from Dale. "You're right." She turned away from Dale. "And you should go...before I do something we will both regret."

"I won't regret it." Dale's tone was quiet.

"Yes, you will. And so will I." She returned to her chair and switched on her desk lamp. "I won't leave Jack. Even if I wanted to be with you full-time, Dale, I wouldn't do it. My career would be over. And I can't see you hanging around, feeling okay about being my dirty little secret." She watched Dale wince at that. "Sorry but I won't lie to you."

"May I go now?" Dale felt as though her insides were imploding.

"Make sure your barracks are really squared away. I won't take it easy on them just because of you."

"I know that," Dale snapped. "Did you kiss me because you're drunk?"

"I'm not drunk." Anne went back to studying the paperwork on her desk. "You're dismissed."

Dale shook her head, confused. "Do me a favor. Next time you get the urge to kiss me, just walk away, okay? I can't go through this again with you. It hurts too fucking much."

Anne watched Dale leave her office. "You have no idea," Dale thought she heard Anne mumble before she left the Battalion headquarters offices.

Dale still reeled as she climbed the stairs to the barracks. *What the fuck was Anne doing?* Dale was less concerned about the kissing than she was about Anne drinking while still at work. Although the kisses were once again everything Dale dreamed they'd be and more, she could understand Anne's confusion in regard to the urge for physical contact. After all, Dale was experiencing the same upheaval of emotions. The difference between them was Anne wanted to fight them while Dale wanted to embrace them.

Was it Anne's feelings for Dale that provoked her to drink? Perhaps she wasn't drunk but she was certainly intoxicated to the point of allowing her defenses to be down. Or was it something else? Maybe she should have taken Anne more seriously back in Vermont when Anne told Dale she had no idea what it was like, being a female at her rank, running a battalion. Anne was so strong, Dale just couldn't see it as being too much of a challenge but maybe the good ol' boys, who thought like Ritchie, were actively trying to sabotage her.

The idea of that made Dale angry. Anne was as good as, if not better than, most men her rank and position. She should've been allowed to do her job without interference from a misogynist bunch who still thought women had no place in the military, unless it was nursing, supply, clerical or food service, areas that were considered subservient to the warrior or leadership positions a majority of men held. If that's what was driving Anne close to the edge, Dale could understand Anne's breach of protocol. She didn't necessarily agree with it but she could understand it. If only Anne would talk to her, share with her like she used to.

Now it seemed that nothing between them would ever be the same. Dale cursed herself for allowing her crush to develop into something more and cursed Anne for her own, rare vulnerabilities lending to the confusion.

As much as Dale tried not to, she fell asleep with that kiss on her mind.

CHAPTER THIRTY-SEVEN

The trainees practiced for G-3 testing the next morning and occupied the rest of the day by preparing for the Battalion Commander's inspection.

As Shannon was one of the lucky few who got to strip the floor, Dale helped Ryan scrub the showers, Brasso the drains, clean the tiles with a toothbrush and disinfect the shower curtains. Only two showers, two toilets and two sinks were allowed to be used until after the inspection. It was a long, hard day. The earliest anyone got to bed was 2230.

Alpha Company, Tenth Battalion was informed at 0520 formation that this was their last day of Basic Combat Training. They were advised that there would be no official graduation ceremony and that upset most of the group, especially the two lieutenants. Personally, they didn't need the ritual but they both felt bad for their fellow trainees. This would be the only time these members of A-10 would go through this particular phase of the Army and they should have been rewarded for their accomplishment. Making it to this point had been a difficult road to travel and both Dale and Shannon felt the trainees deserved some emotional compensation for this distinct feat. No one suffered from overstuffed egos and a little positive reinforcement would have been nice and, maybe just the motivation these soldiers needed to go on.

When one of the male trainees questioned Audi about the reasoning behind no graduation, he was told that Captain Colton and Sergeant Ritchie had decided that because the course continued straight through to Law Enforcement School, a ceremony would defeat the purpose of uninterrupted training. Dale and Shannon could not have disagreed more. A group of women were about to complete an endeavor they were constantly told they would not make it through – male combat training. Dale, especially, believed the results should have been recognized.

The first order of business that morning was the PT test. To make it fair, there was no peer scoring. Drill sergeants and instructors from other companies kept track of all the repetitions and exercises and, even despite that, everyone got a Go. The next half of the morning was spent at Raburn Hall for additional G-3 testing.

Dale, for one, got the impression that the testing was just a formality. Her first task was to answer questions on how to clean an M16 and then to demonstrate to the tester whatever he requested concerning the same. This, as Dale discovered, was all hypothetical. He only asked Dale one question: "How much oil do you put in the bore?" Her correct answer was: "Two drops." Then he asked her to show him, to which she put the required amount in and proceeded to get the bore brush jammed tight in the bore. She got a Go.

She moved on, amazed, to her test in Drill and Ceremony, where she performed Inspection Arms. She always had trouble with the precision and sharpness that was needed to make the drill what it was supposed to be. She learned along the way, however, that if she couldn't do it to the exact standards of the inspecting officer or NCO, doing it with a lot of enthusiasm and emphasis seemed to make up for it. That morning, she slapped the butt of that rifle so hard that her palm still stung an hour later. She may not have looked like she knew what she was doing but she sure sounded as though she did. It was enough to get her a Go. A few more tests later, Dale had, once again, graduated from Basic Combat Training.

The company had also passed the Battalion Commander's inspection that was performed while the company was being tested elsewhere. Dale was glad she did not have to face Bishaye so soon.

Regardless of the fact that basic training was officially over, it was handled like another training night. No special treatment was given and no restrictions were lifted. The trainees, as they were still called, remained in the company area, and were only allowed in the normal places that still didn't include the Dayroom. The only laugh Dale and Shannon had that day was the G-3 fiasco. Some of the others agreed that those activities were a bit absurd but they didn't seem to be in the mood to laugh. They voiced a unanimous opinion that it had begun to feel like a sentence as opposed to an enlistment.

The next morning, after chow and before 0700 formation, an argument broke out between Mroz and Saunders that almost came to blows. If Lehr and Mackey hadn't jumped between them, it might have turned into a free-for-all. With all the unreleased tension up in the bay, any provocation would have been enough.

It started when Saunders conveniently disappeared into the latrine, claiming the food was taking its toll on her digestive tract. This was the fourth day that Private Saunders had done this and used the same excuse. She and Mroz were tasked with splitting the detail of policing up the company area, removing and dumping trash from the female bay, laundry rooms, CQ office and washing out the garbage cans. Four days in a row, the entire assignment was left to Mroz.

"Stop shirking your responsibilities, Saunders. I'm fed up with doing the detail for both of us!" Mroz snapped.

Saunders was clearly indignant. "What are you going to do? Report me? Everyone here knows you're the company snitch."

"If I were a snitch," Mroz countered. "You certainly would have been written up by now. And your opinion of me doesn't address your habitual laziness."

"I'm lazy?" Saunders roared as she towered above Mroz by several inches.

"What would you call it? For four days I've barely gotten the detail done on time. I refuse to run the risk of getting a counseling statement just because you don't feel like cleaning garbage cans!"

Saunders pushed Mroz and Mroz pushed back. Saunders shoved harder and accompanied her action with some extremely unflattering name-calling. Mroz was about to take a swing when Mackey and Lehr came to the rescue. No one had any doubt that if the incident had become any more physical, Saunders, who had height and weight in her favor, may have put Mroz in the hospital.

The yelling peaked just when Holmquist walked in and he immediately ordered everyone into the front leaning rest position. The silence in the bay was deafening. Holmquist let first and third platoons return to what they were doing and told the second platoon women to recover and follow him downstairs to the north laundry room.

"I'd really like to know why you ladies can't seem to get along," Holmquist said. The women sat haphazardly in the laundry room, either on the floor or on the concrete bar that was used to fold clothes. "What was the particular problem today?" No one spoke. "Private Mroz?"

MJ Mroz was on the verge of tears, out of anger and frustration more than anything else. She shook her head. "I have no problem, Drill Sergeant."

"Is that so?" He was not convinced. "How about you, Private Saunders?"

"I have no problem either, Drill Sergeant."

Holmquist nodded. "I assumed as much. Mackey? Lehr? I suppose you two suddenly also have amnesia?"

"Yes, Drill Sergeant," they chorused.

He sighed loudly. "I understand what you're doing but it's got to stop. You are adult women, you are not catty high school girls. Now, I don't know what the problem is but if it doesn't straighten itself out, passes are going to start being pulled. I know it has been threatened before but this time we'll do it. If you think you're at each other's throats now, think of what you'll be like when the males get their evenings free and weekend passes and you'll all be stuck right here doing details." He paused to let that sink in. "I want some answers before you leave this laundry room. I want to know why this is a recurring problem that seems to be affecting second platoon females more than any other. You're making Sergeant McCoy and me look bad. If you talk openly and frankly to me right now, I promise you I will work with you to try and straighten it all out right here and now."

None of the women looked persuaded and he expected that.

PERMISSION TO RECOVER

They didn't completely trust him. He was a drill sergeant and still the enemy. He studied them all, looking for which one he would choose to open the discussion. He picked the one female he felt would not be afraid to speak her mind and, at the same time, probably not point a finger at anyone.

"Private Oakes, let's start with you," he said.

Dale stared up at him, startled, and he smiled patiently at her. She looked around at the other women and was greeted by stony, silent glares. She returned her attention to Holmquist. "Do we have to, Drill Sergeant?"

"I'll get to you sooner or later, it might as well be now."

She cleared her throat. "Well, Drill Sergeant, I think half of our problem is that we don't have enough time away from each other. I think this specific platoon has the majority of the explosive personalities and taking into consideration our different backgrounds, upbringings and opinions, we have the biggest unity problem. To me, it seems to be a case of too much individuality and not enough teamwork."

Holmquist looked like he was trying not to smile. "You think that's the brunt of it, Oakes?"

"Yes, Drill Sergeant."

Holmquist scanned the other faces in the room. "Anyone agree with that? Disagree? Have anything to add?"

Minty finally spoke up. "No, Drill Sergeant. Oakes 'bout said it all."

He gave all the women time to make their opinions known. No one did. "If anyone wants to say anything, now is the time to do it. You won't have another chance." There was more silence. "Okay. Consider this fair warning. Next time I walk into a fight, I start handing out restrictions. The more times you get written up, the more it goes on your record. As a possible future MP, you do not want it following you around that you have disciplinary problems. Just being female will be problem enough in this MOS, you really don't need a spotted record to go with it. You need to start readjusting your attitudes because if you don't, I can guarantee Uncle Sam will do it for you and not in a way that will make any of you easier to live with. Believe me when I say that you have made it through the worst end of it and, for the most part, you've all made it through with flying colors. Do not fuck it up now. You've come too far, so if you let me down, I will get even." He checked his watch. "Fall In on the north patio."

***

Dale thought about the way Holmquist had handled the situation with strictness and patience, in a calm but firm manner. She believed he had their best interests at heart and that he was sincere about wanting them to make it the rest of the way. She thought about this as she posed for her graduation photograph that was taken that morning. After Holmquist's lecture, the company was marched to the PT field where they did enough grass drills to make them appear dusty and unkempt and then they were double-timed back to the company area where they were immortalized on film.

One by one the trainees were brought into the Dayroom directly from the field where they were denied the request to return to the bay to towel up their appearance. Before they were perched in front of the camera, they were instructed to go behind a partition and change the top-half only of their uniform.

When the future MPs stepped out from behind the screen, he or she was attired in filthy fatigue bottoms and soiled combat boots but above the waist, the GI sported the crisp, clean dress uniform of a Military Police officer that included the white dress cap. Although with what they had just put themselves through on the PT field, they still looked like what Wachsman phrased as a turd struck with a club. It was with great difficulty that anyone kept a straight face so that they could look like dignified soldiers in the photographs.

After noon chow, they all marched to a wooden building and were seated at desks inside where they were given a brief introduction and then a driving battery test. The first part was a written questionnaire with common sense, multiple-choice answers. The second phase consisted of picking out words that got smaller and smaller until they looked like a speck of dirt on the paper.

The class was then divided into two groups where they were

given eye tests and color blindness tests. The fourth phase found the trainees individually pulling a string to straighten pegs for judgment. Next, the trainees had to step on a specifically designed brake, to test for quickness.

When that class was over, the trainees were marched back to the company area where they were herded to the gym. This class, conducted by four LE School instructors, explained what the trainees would face in AIT. The attitude of the teachers was offensive and condescending and Dale suddenly felt as though the next eight weeks were going to be a harder version of what they had all just spent the past eight weeks doing.

Back in the company area, after mail call, Shannon caught up with Dale. "Wow. We made it through basic training." Then she leaned in close to Dale. "Again."

"Those asshole instructors at the gym today certainly did not make it feel like it."

"Yeah but you know they're all bluster," Shannon said. She gestured for Dale to follow her outside. Once they reached the patio, they strolled to an area where they were alone and Shannon lit a cigarette. "You never got to tell me what Bishaye wanted the other night."

"Oh. That." Dale had tried her best to put that little meeting out of her mind. It hadn't worked. "She was working late and called me to get an update."

"What'd you tell her?"

"Told her we didn't have anything to tell her."

"And she didn't have anything for us?"

Dale shook her head. "Nope."

"Is she as frustrated as we are?"

"Oh yeah, she's frustrated, all right," Dale said and decided not to elaborate.

"How's your foot?"

"Better. Standing down on the tri-daily PT has helped a lot."

"You're not standing down." Shannon said and laughed.

"Well, I'm only doing the exercises that aren't strenuous on my feet."

"Which is none of them."

Dale looked out across the parking lot. "I hope something happens soon, Shan. I can't wait to wrap this up and get the hell out of here."

"No shit. Right now, I'd be grateful for an off-post weekend pass."

"Hell, we won't even get that if second platoon doesn't smarten up."

"Is that what Holmquist threatened you all with?"

"Uh huh."

"Trust me, when it comes to their freedom, I bet you'll find perfect angels until it's granted."

"I'll tell you what, if we don't start getting weekend passes soon, Bishaye won't have to worry because she'll have a mass murder on her hands to fret about. These bitches will kill each other." Dale watched as Shannon field-stripped the rest of her cigarette. "I told Holmquist that we all needed to get away from each other, that being cooped up was what was causing the flare-ups."

Shannon had a silly grin on her face. "And just when did you tell him that?"

"This morning in the laundry room when—" She looked at the expression on Shannon's face. "Don't even start."

"C'mon," Shannon playfully poked Dale in the ribs. "You can't tell me you don't think he's a cutie."

"I think he's adorable but I am not going after him. I'm not even thinking in that direction. Good God, Shan! The last thing I'm going to do. You do remember why we're here, right?"

"Of course I do. I'm just teasing you."

"Why? I mean, why about him?"

"Because I see the way his eyes twinkle when he looks at you."

"You— you what? He does what?" Dale stared at Shannon, incredulously.

"You haven't noticed the way he looks at you?"

"No."

"You should." She smirked and Dale crossed her arms.

"You know I don't fraternize and…why are we even talking about this?"

"I was hoping you weren't as uptight as you used to be."

"Uptight?" Dale was now defensive. "I am not uptight. I'm just…focused. I've had my share of flings and I haven't always been discreet but, God, Shannon, the *last* thing I'm going to do is sleep with a drill sergeant, especially on this case."

"No matter how hung up on you he might be?"

"You're reading an awfully lot into a few looks."

"It's my job."

Dale tapped Shannon's forehead. "You were hit too hard. It's affected your brain."

Shannon nodded. "Probably. I need sleep and I have fireguard tonight, 0300 to 0500."

"I hate that shift. It makes for too long of a day."

"I agree. So after chow, I'm going to try and get some sleep," Shannon said.

"You do that. Maybe it will help calm that overactive imagination of yours."

"It's not overactive. I bet if you were a civilian and you met him in a bar, you'd go for him."

"And you'd be wrong." Dale wished she could confide in Shannon but she wasn't sure she was ready to talk about her sexuality out loud and she was less sure Shannon was ready to hear it.

Shannon looked as though she were about to debate her again when someone announced to line up for chow.

CHAPTER THIRTY-EIGHT

Fireguard duty was usually uneventful. The required two hours was normally spent reading or catching up on personal details.

Shannon knew, not even five minutes into her shift, that boring was going to be the last word to describe it. She was awakened at 0300 by off-going fireguard McKnight, who, by the way she looked, was most likely asleep before she even hit her bunk. Shannon made her way to the latrine and planned to occupy the rest of her shift by polishing her brass. She stopped at the sink to wash her hands and contemplated smoking a cigarette when she heard what sounded like the barracks door close.

When she stepped into the hallway between the latrine and the open bay, she spotted Holmquist. He seemed to be searching for something and when he saw her, he approached her quietly.

"How long have you been on duty?" Holmquist asked her, his voice barely above a whisper.

"I just got on, Drill Sergeant. 0300 hours."

"Let me ask you something...what's one of the first things you do when you take over as fireguard?"

Shannon blinked in bewilderment before she answered him. This certainly was a weird time to give her a spot quiz, even for a drill sergeant. "Well, honestly, the first thing I do is relieve my bladder. And then I walk through the bay to make sure everyone's accounted for and that nothing is plugged in or —"

Holmquist crooked his finger at her. "Come with me."

She followed him through the bay and they stopped at Dizzy's bunk, which was obviously empty. Dizzy hadn't even attempted to disguise the fact that she wasn't there by arranging pillows under the covers to look like a sleeping form. Shannon then shadowed Holmquist back out into the hall. She made a quick, visual search as she walked behind him and made sure everyone else was present.

"Would you have noticed that?" He asked as they stopped in the hallway.

"Yes, of course, Drill Sergeant," Shannon answered.

"Then what would you have done?"

"I would have checked the stalls and the showers to see if she were there."

"And if she wasn't?"

"Then..." *Should I be honest with him?* She decided she would take the chance. "Knowing Private Zelman, I'd go to the other three bay doors and check with the other fireguards to see if she was in any of them."

"And what if she was?"

"Then I'd tell the fireguard I would give her exactly five minutes to get back in her own bunk or I'd report her."

"Would you have actually reported her?"

"Yes, Drill Sergeant, because if I don't, I run the risk of getting in trouble myself if, say, the Staff Duty NCO decides to pop in for a surprise bed check...like, um, now."

"I assure you that is not why I am here. Bear with me for another minute or so, Private Walker. What would you do if you didn't find her in the other three bays?"

"I would check the laundry rooms, the Dayroom and, as a last resort, the CQ office."

"And if she wasn't in any of those places?"

"Then I am awfully afraid I would have to report her missing."

Holmquist nodded. "Who did you relieve?"

"McKnight, Drill Sergeant."

"Get her up and have her wake up whoever she relieved and bring them here."

"Yes, Drill Sergeant." Shannon disappeared into the darkness of the bay. On her way to rouse McKnight and then Creed, she wondered where Dizzy was and what the hell was going on. Within a minute, all three of the night's fireguards stood before Holmquist. After a few quick questions, he sent Creed back to her bunk and he repeated the same questions to McKnight, who gave all the wrong answers. It was apparent to both Shannon and Holmquist that McKnight had fallen asleep during fireguard and that's when Dizzy left the bay.

"Walker, you come with me. McKnight, you'll take the remainder of Walker's shift and I'll deal with you later."

McKnight probably saved her tantrum for after their departure.

Shannon stayed downstairs in the First Sergeant's office and guarded an extremely doped up Dizzy until Colton arrived forty-five minutes later. He seemed even less pleased to see Shannon than he did Zelman, a reaction both women shrugged off without much care. While Shannon waited in the Orderly Room for further instructions from Holmquist, she persuaded Joseph Elnicki, the CQ, by bartering with cigarettes, candy bars, a beer, and a dance at the Pizza Place when they got their next pass, to let her read the CQ log and a copy of Holmquist's report. With Shannon at her most charming, Elnicki was putty in her hands.

By the time first formation was called, rumors of the incident had already spread but Shannon was able to tell Dale what she read herself and overheard in the course of the two hours she had spent downstairs.

"Sometime between 0115 hours, the last time McKnight remembered being awake, and 0225 hours, when the laundry room gang was discovered, Dizzy walked right by a snoring McKnight and continued downstairs to the south patio laundry room, where she had a pre-arranged rendezvous with Tom Court from First Platoon. The deal was, for a gram of cocaine, Dizzy would do anything he wanted."

"Where do you think he got the coke? Ingersol?" Dale asked.

"That's my guess."

"Why the hell wouldn't she go through Ingersol herself?"

"I have no idea. Nowhere in the report did any common sense seem to play into this on the part of the trainees," Shannon said, as she and Dale walked around the parking lot. They both hoped it looked as though they were exercising while members of the company milled about, waiting to be called into formation. Shannon wanted to get all this information out before one of the other trainees tried to join their little stroll.

"We all knew she was sneaking into the guys' bays at night.

PERMISSION TO RECOVER

I just didn't think it would amount to anything," Dale said, shaking her head. "So, what happened next?"

"Court kept his part of the bargain and, I guess, was bound and determined to make Dizzy keep hers. When she got to the laundry room, she was greeted by Court and two other First Platoon males, Canfield, and Brooks, who had cameras. Court and Dizzy evenly split the gram, as the other two declined. I guess the thought of a naked woman was stimulating enough for them. She stripped for the three of them while Canfield and Brooks took pictures of it all.

"Dizzy then lay back on the washing machine and dryer and provided the boys with shots a gynecologist could appreciate."

"Really?"

"Really." Shannon tilted her head in recollection. "I don't know, I never understood the appeal of that particular shot. The close-up kind of looks like open heart surgery to me." She returned to her task of reporting the story to Dale. "Anyway, Court then fed the machines quarters so they would vibrate while going through their normal cycles, then he removed his fatigue pants and climbed on the appliances with her. The performed oral sex on each other and several different positions of intercourse while the two photographers took turns masturbating. It was during a frenzied exercise where Dizzy was accommodating all three men when Holmquist walked in. This shocked everyone, including Holmquist. They thought they had timed their activity around Holmquist's last patrol, which meant he should not have been back for at least two hours."

"Holy shit," Dale said, shaking her head. "Every time you think you just can't hear anything new."

"Right. So, Holmquist, who was Battalion CQ, had forgotten something and was on his way back to the company Orderly Room when he saw light coming from under the door of the laundry room. The light hadn't been on when he had patrolled the area earlier, so he automatically investigated. I overheard him tell McCoy later that he was sure all he would find was the CQ or CQ Runner doing their laundry. Then he opened the door."

"Wow." Dale shook her head. "Did you get a chance to see or speak with Dizzy?"

"Oh, yeah. She provided me with a few other blurry details. Dizzy said that, even for her, the look on Holmquist's face was unbearable. She told me she was used to people being disappointed in her and, until that moment, she thought she had reached a point where it didn't bother her anymore. She said that Holmquist turned away from her, told her to get dressed while he grabbed the cameras from Canfield and Brooks. He ordered them all to Battalion Headquarters, where he took them inside an office a good distance from the hearing range of his runner. He then demanded all the rolls of film and advised them to give them up freely now because he still had not decided whether he would call the MPs. If any additional film was found on them, they'd be in worse trouble than they already were. Dizzy said after Holmquist had the six rolls of film, he put them in his fatigue jacket pocket and told them that he would turn them into the Battalion Commander first thing in the morning. He said he was sure they would be secure and remain unexploited in her possession."

"Oh, she's just going to *love* this," Dale said, caustically.

"So then, Dizzy said Holmquist told them that the least of their problems would begin with all of them receiving Article 15s and then he verbally ripped them apart. He spoke to the males first for their participation in the ordeal and then started in on Dizzy. He scolded her for not only the lack of respect for herself but also disrespecting the oath she took when she was sworn into the Army. She said he then yelled at the men, too, and when Court protested by calling Dizzy an obvious slut, Holmquist countered angrily with, So, what does that make you? You willingly had sex with her and arranged to have it photographed. Stop me when your sterling character starts to shine through here, Court. She said Court kept his mouth shut the rest of the time they were with Holmquist."

Dale sighed. "And the women were doing so well."

"True, but we should accept a small part of the blame because we knew, well, at least I knew, she would get caught. We should have somehow dropped a dime on her way before now."

"I know, I know. I've been so convinced that she's not a part of this that I didn't really consider a different scenario. I thought either one of the guys she wasn't fucking would turn her in or she'd get caught in the act by a roving drill sergeant doing a surprise headcount after bed check. Honestly, she was always in her own bed when I was fireguard."

"Yeah, she was when I had fireguard, too. I was going to do a walk-through after I washed my hands. This, for me, would have been the first time she so obviously wasn't there."

Both women heard the command to Fall In and quickly walked back to the patio. "This is going to suck," Dale muttered before they split up to stand with their respective platoons.

The mood of the cadre decidedly reflected in the merciless PT the company was put through that day. Every member, male and female, was run ragged and drilled unforgivingly. The drill instructors were closed mouthed on the incident but the male trainees weren't as discreet. A majority snarled their discontent at the females whenever a cadre member was out of hearing range.

Dale wanted to scream back at them. "It was *three* men and one woman! Why does it make it our fault?" But she knew that conversation would be futile.

It was reported by the CQ that Holmquist's recommendation was that all of the trainees involved be released from their contracts. In the end, however, as both Shannon and Dale agreed would happen, only Dizzy would be forced to leave. The three males would be written up, verbally reprimanded, given two weeks restriction and recycled to Bravo Company. All that meant was Court, Canfield and Brooks would still be still in MP training, except now they would be a week behind Alpha Company. Dale had to admit, as unfair as it was, maybe Dizzy was the lucky one.

There were jokes passed around the company, such as, Dizzy has a speech impediment. She can't say no and every time she hears the word *drugs*, her legs automatically snap open.

Even though some of the females joined in the frivolity, almost all of them confessed that they felt stung by the incident. It was something else that would reflect on them as a whole, whereas the three males involved were blamed individually for their behavior.

The only good thing that happened that day was that Kotski returned. She was relaxed, refreshed and ready to get back into the swing of things. Several women, including Dale and Shannon, offered to help her make up what she had missed as she was given a week to catch up to everyone else or be recycled to the same MP company the three Alpha guys were to be sent back to.

Kotski was determined to catch up.

That evening, Henning gathered the women together in the Dayroom and gave them a mandatory speech on morals. Henning told them that the talk was just an afterthought and she knew that Private Zelman was not an accurate representation of the rest of them. With weekend passes soon to be upon them, however, she needed them to remember that how they acted when they were away from the company area still reflected on Alpha-10. She said that if she didn't address the subject with them then the Battalion Commander would and it would most likely not be a comfortable lecture to sit through.

"Colonel Bishaye has seen the developed photographs. I have not seen them but I understand they are quite...graphic." Henning looked down at the floor and her face colored at the thought of what those pictures might reveal. She finally refocused on the women in the room. "The colonel takes it personally when a female has such little regard for herself and degrades herself the way Private Zelman has. We all realize and understand the three males are just as culpable but other males don't look at it that way.

"Men are perceived as cool when they do things like this and the only uncool part, for them, is getting caught. Women are always considered the ones at fault and the ones who are tramps and the bad reputations follow the females around for the rest of their lives."

Even though it seemed as though Henning were preaching to the choir, she took a breath and continued. "Colonel Bishaye had hope that the women in this cycle of Alpha-10 would be remembered for all they had accomplished, not for the best French postcards in the south."

The thought of Anne looking at those pictures of Dizzy

brought a smirk to Dale's face. If Henning was the only one to notice it, Dale was sure Henning would have ignored her but at least five other women clearly saw Dale sitting there with a smug smile on her face, which is probably why Henning spoke up.

"You find this situation amusing, Private Oakes?"

Dale snapped out of her preoccupation to find everyone in the room staring at her.

She sat up and cleared her throat as Henning's question registered with her. "No, Ma'am. May I speak freely?"

"Go ahead."

"What I really resent, Ma'am, is the fact that we have to be subjected to a lecture on morals and the men don't. None of us here would go behind the door to say that Dizzy, I mean Private Zelman, was a model soldier. I'm sure I also speak for everyone here by saying none of us want to be categorized with her, either. But one woman fucks up and all of us have to be treated as though we're guilty? Just by association of gender? Three men run circles around her actions and they're upstairs slapping each other on the back. So, no, I find nothing about this situation amusing, Ma'am."

"I fully understand and agree with your complaint, Private, and as I said before, it isn't fair. The fact still remains that this is very much a man's Army and that is going to be thrown up in our faces every chance they get. We have to work twice as hard for half the recognition, especially as an MP. What we don't need to do is give them any reason, no matter how small it may seem to you, to make them feel validated in their claim that women shouldn't be here. Yes, it's wrong but get used to it."

Dale found out later that while Henning was reluctantly meeting with the women in the Dayroom, Bishaye was dealing with Court, Canfield, Brooks, and Dizzy in her office. Dale was sure the term, wrath of God, took on a whole new meeting for them. Especially when she heard they'd been in Bishaye's office a little more than two hours.

Dale wondered if she'd ever know exactly what was said in the Battalion Commander's office but she was pretty sure it was a conversation the four trainees would never forget.

The next morning at 0730, the trainees attended an informal graduation ceremony on the south patio. It was conducted by Colton who, with certificates, gave recognition to Outstanding Overall Trainee, Haviland, Outstanding Trainee in PT, Lasher and Highest M16 Score, Oakes. Dale could see it just about killed Colton to have to recognize Dale in anything. In fact, when he shook her hand, he nearly crushed every bone in it.

The trainees were then rewarded with the traditional company commander speech of what a good job they'd done and how proud he was of all of them for making it through. Not a word was mentioned about the laundry room scandal, which Dale was sure must have been at the insistence of Bishaye. Dale was also sure the women were grateful because they had heard enough.

Before he dismissed them from formation, Colton introduced the company to eight Marines. Five males and three females would join them for LE School.

The Marine Corps sent their trainees to the Army Law Enforcement School because the Marine Corps didn't have one. Any USMC trainee who chose or was chosen to be in police work received their schooling from the Army. The merging of the two military cultures was always interesting. Especially the female culture.

Female marines didn't get the type of extensive combat training Army women did or so the female members of Alpha-10 were immediately informed by the three new inserts. Marine women were taught to be *ladies* not *grunts* and they began their integration by playing that to the hilt. The trio of women in camouflaged fatigues were named Saundra Navarrete, Pamela Chellemi and Denise Endres and they entered the bay with a chip on their respective shoulders and their noses held high in the air.

Dale found this amusing. She couldn't wait to watch these femme fatales when they were ordered to do a little low crawling in the mud, to match prowess on the obstacle course or to accompany everyone else to a week of Bivouac that had been postponed from basic training. They would change their tune extremely fast.

The five marine males seemed to adapt just fine. They were

PERMISSION TO RECOVER

named Briere, Cadwallader, Parisi, Raye and Waylon. Nick Parisi had the chiseled features and well-fit appearance of a movie star and it was a good bet he was going to be very popular among the Alpha women. He stood out from the group of eight not so much due to his looks but that he just exuded confidence. Of the three women, Navarette, who was a mixture of French and Spanish, appeared to capture the attention of the Alpha men. She had the dark, European beauty one expected to find in exotic women overseas. With her particular attitude, however, it was evident she wasn't going to waste her time on common trainees or drill sergeants, either, for that matter.

The locker assignments were once again rearranged to incorporate the marines alphabetically into the platoons. After that annoying task was completed, as the marines were being individually oriented, Alpha-10 was set free for the entire weekend. This came as a shock as they were sure they would all be restricted once more because of the laundry room incident.

Before they were released from formation, the trainees were told a list of seven establishments that were off-limits because they had a reputation as hotbeds of trouble. Three names on the list were on-post and the other four were off-post. The trainees were warned that MPs patrolled all of these places frequently and took names of GIs on the property. If any trainee's name came back on any of the lists or if any of the trainees were picked up for breaking any rules and regulations, they were going to be severely dealt with.

Dale and Shannon showered, changed into civilian clothes and split up the duties, agreeing to meet back at the Pizza Place at 2000 hours. Each agent hit the usual hangouts the trainees had frequented on their other limited passes, only to find very few members of Alpha-10. When they walked into the Pizza Place at eight, it was a different story.

The first area both lieutenants made their way to was the counter to order a beer, which they both knocked back as easily as drawing a breath. Dale and Shannon each ordered another and calmly observed the members of their company closely. By the time forty-five minutes had passed, the two lieutenants had danced with a few different males and Shannon had paid off her debt to Elnicki. Even though they relaxed and had a good time, they took particular notice of who left the Pizza Place and which trainees had paired off. When nine o'clock arrived, Shannon suggested to the female group that was left to hit The Enlisted Club.

"Isn't that place off-limits?" Segore asked as fear registered in her eyes.

"Yes, it is," Travis said. She clearly liked the idea instantly. "Come on, Segore, get real. It's a club for enlisted people like us. We have every right to be there. How's an MP going to tell us apart from anyone else who goes in there?"

"But they'll take our names," Segore protested.

"I hear it's always packed. I hear it's like being in a nightclub. What MP patrol in their right minds is going to spend the night in there taking names?" Dale threw in. She knew if a patrol came in there at all, it was because they were specifically called or just for a visual scare tactic. "Do you really think a drill sergeant is going to go check a list of, maybe, a hundred people or more who might be there when the MPs get there?"

"And that's just one place." Travis picked up Dale's momentum. "If they have to look at the names of hundreds of people..."

"Still, I don't want to take that chance," Segore said.

"Well, I do," Tramonte said.

Tierni stood behind her and nodded her head vigorously.

"We'll walk you back to the barracks," Shannon said to Segore. "I have to get some money out of my locker. And we have to call a taxi."

Shannon didn't really need to get money out of her locker, she just wanted to check on who had returned to the barracks. She discovered, happily, that the women who had not stayed at the Pizza Place were safely back in the bay. If they had been granted Dayroom privileges, she was sure a lot of people would have been in there, watching television, but the Dayroom wasn't available to them until tomorrow.

The Enlisted Club was a nightclub not too far from the A-10 company area, whose patrons consisted of soldiers who held the rank of Specialist 4th class and below. Dale and Shannon knew the club and the location well and they also knew they could have walked there faster than the cab could get them there. The building itself was the size of a normal barracks, big enough to house two bars. One room had a long bar, a huge dance floor, a disc jockey with an immense sound system and about seventy tables. The other room was smaller with a shorter bar, twenty-five tables, two pool tables and a game room.

When the group of five walked in, each paid a two dollar cover charge and were immediately overcome with a sense of freedom they hadn't had since Christmas exodus. There was a loud, pulsating beat that they could feel in the floor and there were people everywhere. Some of the crowd was in fatigues but most were wearing civilian clothes.

Dale and Shannon let the three Ts go in first. They walked through a hallway that led to restrooms, the offices, the smaller bar with their main objective being the source of the music. They stood in the doorway of the big bar and knew they had found a new home. The type of establishment that would have been considered a dive in their native territories now looked like heaven to them. The room was dimly lit, illuminated by colored lights and occasionally a strobe light or a spotlight focused on two suspended silver balls that hung from the ceiling.

As Travis, Tierni and Tramonte went to get drinks, the two lieutenants found a recently vacated table near the dance floor. The DJ played songs by Donna Summer, The Bee Gees, The Ritchie Family, KC and the Sunshine Band, Peter Brown, Foxy, George McCrae, and everyone else who had a hit that was danceable. Disco music was making an overwhelming debut and hearing that beat pound was just the release the Alpha women needed. By the time everybody congregated at the table with the alcohol, Shannon and Dale had already been up to dance twice and the other three were asked several times on the way back to the table with drinks in their hands.

"This place is great!" Tierni shouted above the music. Her smile was wide and enthusiastic. "You two crack me up," she said to Dale and Shannon. "You're so brave."

"I just don't like being told I *can't* do something or I *can't* go somewhere," Dale said and sipped her beer.

"Then what are you doing in the Army, ya bozo," Travis said and laughed. She took a swallow of her rum and cola and looked immediately as though she were breathing fire. She stood up and started to jump up and down.

"What are you doing?" Tramonte asked her.

"Either the bartender forgot to mix my drink or he put the cola in with an eyedropper. Jesus Christ, this is strong," Travis explained as her eyes watered profusely.

"You need to dance," a voice behind Travis said. They all turned to see a very handsome young man in fatigues who gestured her to the dance floor. Amid jokes of adultery, Travis set her drink down and let herself be escorted to a spot near the empty bandstand. That was the last the others saw of her until the club was ready to close, not that the others suffered from being ignored. Once they got up on the dance floor, they rarely sat down. Dale's second beer had exactly two sips taken out of it and Tramonte got only four swallows out of her first drink. They all had such a wonderful time, they ended up closing the place.

"So, who's not here on the right side?" Dale asked her partner. They had returned to the barracks, taken a personal headcount, and wandered down to the patio.

"Swinegar, Sherlock, Troice and McTague. How about the left side?" Shannon said and yawned. She was almost done with her cigarette.

"Laraway and Ferrence. Everyone else is accounted for, including Dizzy and the marines."

"Okay, let's narrow it down. Ferrence is probably in Averill with Brownell, McTague is probably with Halliday and Swinegar is probably with whomever she happens to be engaged to at the moment."

"Unless she's with Robbins," Dale said, arching an eyebrow.

"If she is, we'll find out as soon as she comes back to the barracks. She tells Tierni everything and I'm sure one of us can get the info from Tierni."

"Then it looks like the real guessing is with Troice, Sherlock and Laraway."

Shannon shrugged. "I'd be surprised if we'd have to worry about those three. If one of them is secretive about her whereabouts then one of us will follow her next time. Who's

PERMISSION TO RECOVER

Staff Duty NCO tonight?"

Dale looked skyward as though that would help her remember. "Uh...Lederman from Charlie."

"All our drills are free as the wind tonight, eh?"

"Yeah, scary thought, isn't it? What if one of them had popped into the EC tonight?" Dale asked.

"If I had seen him first, I would have tried to alert you or the others and then I would have sneaked out the back but you know as well as I do that most of the drills hate the enlisted clubs. They don't want to go anywhere near them unless they're called."

"Or you get a particularly sadistic one."

Shannon grinned. "A sadistic drill sergeant? Surely, you jest." She put the extinguished cigarette butt in her pocket. "What do you want to do tomorrow?"

"Hop a plane home but I guess that's out of the question. I'd really like for you and me to be able to take off by ourselves and have some real fun away from watchful eyes. I feel like I'm going nuts here."

"*You're* going nuts? At least you got to get out of here for two weeks," Shannon reminded her.

"Listen, by all rights, we should get weekend passes from now on. Why don't we just take off to Averill next weekend? I'm sure some of the others will be doing the same. If Bishaye asks, we'll just tell her that a majority of the company headed to town. Friday night we can party by ourselves and Saturday, it will be business as usual. What do you think?"

"I think it sounds pretty good as long as nothing major is happening with our three unaccounted for females." Shannon stretched. "You know, I would like to have seen how Bishaye handled the laundry room crew. I bet she's impressive when she's really mad."

"Impressive isn't exactly the word I'd use. I've been on her bad side in the past. I swear to God, it's like she sucks all the oxygen out of the room and you are in pure agony until you can get back outside and breathe again." Dale instantly wanted to get off the topic of Anne Bishaye. She still felt wounded. Not enough time had passed.

"That's what I mean. I would have liked to have been a fly

on the wall in her office. She's not a person to mess with, is she?"

"Nope."

"I saw her briefly in the Orderly Room before she went back to her office to deal with that mess. She looks really frazzled. Don't you think so? You know her better than I do but she doesn't seem as calm and collected as when I first met with her."

So much for dropping the subject of Anne.

"I know. She's changed. She says it's the position. Rank does have its privileges but I guess it can have its headaches, too. We talked about it a little bit after exodus. I'm sure once this fraternization thing is cleared up, she'll relax some. She seems to be under a lot of pressure and this kind of bullshit does reflect on her." It made sense and the more Dale said that the more she felt she understood Anne's behavior.

"I would have liked to have gotten to know her under different circumstances." Shannon stopped Dale by grabbing hold of her sleeve. "Hey, wouldn't it be great if after this, she got charge of her own CID unit and they put us under her command?"

Dale looked at Shannon as though she'd just grown another head. "Are you kidding me?" She removed Shannon's hand. "No, it wouldn't be great." *It would be torture.* "I want to go home when this is over. I'm sick of Uncle Sam."

"Even if it meant working partners with me?"

"Shan, that concept is very inviting but I still have one clear thought in my head. Somebody tried to kill me. That somebody may try it again if they know I didn't die. Now what is going to happen to me is going to happen but I don't want to be responsible for anyone else who, by circumstance, might be with me. Especially not you. Or Anne."

"Dale, if somebody wants you that bad, wouldn't you rather they try and get at you when you're away from your family and loved ones? When you're maybe in a position to defend yourself because you're armed or with other people who are trained to defend themselves and maybe help protect you?"

"No. I think as long as I steer clear of the military, this person or these people will leave me alone. If I just live a quiet life, minding my own business in small town, Vermont, what harm can I possibly do anyone?"

"You? Live a quiet life? Mind your own business? Then why are you here? And don't give me that garbage of Bishaye not giving you another option. I know you. And we both know that's a crock. If you really didn't want to get back into this kind of life, you'd be in Vermont right now, sucking back beer every night and running the pool tables. This is me you're talking to, remember? It's not in your nature to live a quiet life."

There was a brief silence as Dale stared out across the open area between patios. "I guess I really thought she needed me."

Wanted me.

"And?"

"And I was bored, okay? Happy? But I'm not kidding about wanting to get out of *here*."

"That I can believe."

CHAPTER THIRTY-NINE

Dale woke up early and couldn't get back to sleep. She finally gave in to consciousness, got up, got dressed and went downstairs to see who was on CQ.

She passed McNulty on the way out of the Orderly Room and saw Michaelson behind the desk. "Aw, man! You got stuck with CQ on our first weekend pass? That rots," Dale said.

"I don't mind. I didn't plan on doing much anyway. So where'd you end up last night?"

Dale looked around the office to ensure there was no one in authority around. "The Enlisted Club."

"I thought that place was off-limits."

"It is. That's what made it so much fun."

"Weren't you afraid you'd get caught?" Michaelson asked.

"It was a thought. We just figured it was an idle threat. If I were an MP, I wouldn't want to go anywhere near that place on a weekend."

"Good point."

"Did I miss anything exciting last night?" Dale casually asked.

"With all the action everywhere but here?" Michaelson smiled. "It was pretty boring here last night."

"Did you happen to see Laraway, Sherlock or Troice anywhere?"

"No. Why? Are they missing?"

"No, no. I mean, not that I know of. It's just...they were supposed to meet us last night and they didn't show up," Dale lied.

"I didn't see them but honestly, I wasn't paying attention to them, either."

Dale studied Michaelson momentarily. The trainee was so unpretentiously lovely, Dale wondered who, if anyone, was paying attention to her. "So, Deb, have you been beating them away with a stick?"

"Who?" Michaelson looked genuinely perplexed.

"The guys?"

"Oh. Well, there have been a few persistent ones but I'm not really interested in dating while I'm here. There'll be plenty of time for that after I get assigned when I get out of here. How about you?"

"Nah. I've never had to beat anyone away with anything."

"Seriously?" Michaelson seemed truly surprised.

"Never had that kind of appeal, I guess." Dale didn't want the subject to become herself. "Any drill sergeants show interest in you?"

Michaelson laughed. "Other than Bradbury from Bravo?"

"Well, not that I'm trying to insult your appeal but —"

"Yeah, yeah, I know. I've heard she's a good drill sergeant, though."

"I'm sure she is but she's clearly not good with boundaries."

"Other than her few failed attempts at flirting, there's been nothing else. The male drill sergeants have shown no interest."

Both Michaelson and Dale turned to see her relief CQ enter the office. "Finally," she said. She stood up and stretched. "Why do you ask? Is one of them bothering you?"

"No. I just heard a lot of rumors about last cycle and was curious."

Michaelson nodded. "I'm going to brief Renaldi and then go to bed. What do you have going on for today?"

"I don't know yet. Breakfast and then I'll decide."

Michaelson glanced at her watch. "You'd better hurry. Chow stops in about eleven minutes."

"I'll pick up something at the PX."

"Why pay for something you can get here for free?" Michaelson asked.

Dale smirked. "Because I take what little freedom of choice we get seriously."

She waited and walked Michaelson to the second-floor bay to her bunk, then moved on and woke up Shannon.

"Get up," Dale urged.

"Go away," Shannon mumbled.

"Come on, get up. Let's go to the PX for breakfast."

"No. I'm asleep. Can't you see my eyes are closed?"

"If we stay here too long, we're going to get tagged for details, so get up."

"I hate you."

Dale and Shannon chose a booth near the window at the PX restaurant/snack bar and sat down with their trays of coffee, eggs, and toast. They had just begun to eat when they both saw a civilian taxi pull into the parking lot and stop. An MP sedan moved up slowly behind it with its lights revolving. Two military police officers emerged from their vehicle and opened the back doors of the taxi. Out of the back popped Sherlock, Troice and Laraway with Ray Wotek, Jeffrey Souther and Perry Sargent. Dale and Shannon watched with interest.

Sherlock was the first one to enter the snack bar. She spotted Dale and Shannon, grabbed a cup of coffee, and proceeded to their table where she sat down.

"What'd you get stopped for?" Dale asked.

"Something called overloading a civilian taxi. I didn't know there was such a regulation," Sherlock said.

"Are you going to get reported?" Shannon asked.

"No. They didn't take anybody's name. I guess it was just a warning."

"Where'd you go? Anywhere good?" Shannon wondered.

"Not really. We rented a room at a motel in Averill, bought some booze and passed out."

"You mean to tell me the three of you were in a motel room with three of the hottest guys in A-10 and nothing happened?" Dale asked.

"Who are you, my mother?" Sherlock stole a slice of Dale's toast and took a bite. "I guess it's going to take a while for all of us to get back onto the party-hearty habit again. I was the last one to fall asleep and the first one to wake up. Everyone was just as I left them so I can pretty much say nothing happened. Unfortunately. I'm about due, myself."

"Hell, you and the entire company," Shannon said.

"I'll tell you who I'd like a first crack at. That marine," Sherlock said as she finished the toast.

PERMISSION TO RECOVER

"Which one? Navarrete?" Dale said in a kidding tone.

Sherlock glared at her and pretended to stick her finger down her throat. "Here's your toast back."

"Kidding. I knew you meant Parisi. If his ego is as big as his...fan club, there's going to be problems in the bay," Dale said.

"Oh, you mean, like there aren't problems now?" Sherlock asked.

"Not boyfriend problems."

"Shit, I don't want to be his girlfriend. I just want him for an hour or two," Sherlock said and waggled her eyebrows.

"So the height difference doesn't bother you?" Dale asked her.

"Why should it," Shannon said. "A slow dance would put his head right between her boobs."

"In case you haven't noticed — and you better not have — I don't have any boobs so he'll be out of luck there. He'll have to be satisfied with putting his head between something else. And, no, the height difference doesn't bother me as long as he's tall in the right place. If the way he wears his fatigue pants is any indication, he is." Sherlock's expression was quite bawdy.

"Never heard you like this, Toni. You *must* be needy," Dale said.

"I am, you knucklehead," Sherlock said and smacked Dale's forehead playfully with the palm of her hand.

"I thought McTague and Swinegar would've been with you," Dale said, slipping that in. "Did you run into either one of them anywhere?"

"They were in the same motel. McTague was with Halliday. They argued the entire night. Their room was next to ours. He sounds like a little crybaby and she sounds like a real bitch to be in a relationship with. I wouldn't give you two cents for either one of them. Swinegar was a few rooms down with Doug Lasher."

"Lasher? What happened to —?"

"Bachelor number four? Who knows and really, who cares?" Sherlock said. "What did you two do last night?"

"We were at the Pizza Place for a while and then we went to the EC," Shannon said.

"What's the EC?"

"You know, the Enlisted Club."

"So you were brave and went anyway, huh? We talked about going but didn't want to take the chance of getting caught by the MPs," Sherlock said.

"Unlike overloading a civilian taxi," Shannon said, dryly.

Sherlock shrugged. "At least no names got taken. So how was the EC?"

"Fun. Tierni, Tramonte, Travis, Oakes and I went," Shannon said. "We didn't have time to drink because we were too busy dancing."

"Maybe I'll hang out with you guys next weekend. Sounds like it was a lot more fun. I'm exhausted and I didn't even do anything to earn it. I'm going to head back to the barracks and take a hot shower. Maybe it'll perk me up." She stood up and finished her coffee. "See ya."

They watched Sherlock leave. "Toni's a good source of information," Shannon said.

"Yes and she loves to talk. Good combination."

Everyone who left the barracks for some kind of recreation the day and night before now appeared to be quite burnt out and wandered around all day Sunday as though they needed new batteries.

Shannon hit the Pizza Place again toward early afternoon to hang around and see who'd show up.

Dale was at the bowling alley snack bar, drinking a beer, searching for more of her company members to show up. She and Kotski spotted one another at the same time and Kotski, with two third platoon males, VanHoesan and Tetrault, made a beeline right to her. Kotski invited Dale to join them. At first, Dale wasn't sure she should accept and be in one place too long but then she decided an hour of bowling wouldn't hurt. She had planned to stay there a while anyway because the barracks buzz was the trainees were either going to be there or where Shannon was.

Three games went quickly with Kotski scoring the highest every time. She had been on a bowling league back home and her style and form showed it. Dale's score was the lowest and

Tetrault kept saying it was a good thing she didn't bowl like she shot because being beat by one female was enough.

Dale looked around the interior of the building to see how many from Alpha-10 were there. She estimated about twenty and dwindling and wondered if other trainees had drifted back to the Pizza Place.

"Why don't we go roller skating?" Kotski suggested, a little too enthusiastically. "Doesn't that sound like fun?"

"I'm really good on roller skates," Tetrault said, bragging.

Dale finished her last swallow of beer. "I didn't know it was possible on roller skates. Don't you get, like, motion sickness or something?"

Tetrault laughed. "I wasn't talking about sex, Oakes. Unless you want me to."

He picked up his beer glass. "Anyone up for one more?"

Both Dale and Kotski waved him off but VanHoesan also stood up. "I'll get another one with you."

As the two men walked away, Kotski tapped Dale's forearm. "What do you think of Tetrault?"

"I think he'll make a fine MP," Dale said. She knew what Kotski was asking and she hoped the conversation in that area would go no further. No such luck.

"Do you like him?" Kotski asked.

"I don't know him, Laurel. I suppose he's okay."

"What do you think about maybe going out with him?"

"I'm not interested, Yenta, but I appreciate the thought," Dale told her.

Kotski was disappointed. "What's wrong with him?"

"Nothing. I'm sure he's a nice guy but I've got myself a nice guy back home," Dale lied. "I want to be a good girl."

"It's not like I'm asking you to screw his brains out…he just really likes you and I thought, maybe, while you both were here…"

"Thanks but no thanks. Can we close this discussion now?"

"Fine. Are you going to spend the next eight weeks by yourself while everyone else is hooking up with people?"

"No, of course not. I've got you," Dale said, teasing.

"And Walker," Kotski said and grinned. "By the way, where is Walker?"

"No idea." Dale shrugged. "What about you? VanHoesan seems interested."

"No. We're just friends. He's married and he loves his wife. I'm terrified of getting involved with anyone and he doesn't want to be tempted so we talked it out and we're going to be non-sexual companions. He wants to stay faithful to his wife and I want to stay faithful to myself. Constantly being around each other will hopefully keep others at bay."

"You know the rumors will spread that you're sleeping together anyway, right?"

"Yes, but that's okay. We both know we aren't and that's all that matters."

"To thine own self be true," Dale said, quoting Shakespeare. She wondered if she'd have the courage to follow the bard's advice.

When Dale arrived at the Pizza Place, she could barely get through the door. She had taken about three steps inside when she was grabbed for a dance by Gilbert Fanuele. Ordinarily Dale wouldn't have minded because Fanuele was an excellent dancer and the music was a slow song she liked called "Don't Ask My Neighbor" by the Emotions. Dale soon found out that she could have been any warm female body as Fanuele danced exceptionally close to her in a ploy to make Melanie Mackey jealous while Mackey danced in cheek-to-cheek oblivion with Richard Snead. Fanuele's tactic wasn't working and that made Fanuele mad. He then turned and dipped Dale with such force, he almost wrenched every muscle in her back and the sweet nothings one expected to be whispered in one's ear during a slow dance should not have been words like *pendejo* or *maricon*. When the song ended, Fanuele was in the midst of spinning Dale around when Mackey moved away from Snead. Fanuele abruptly let Dale go and she nearly bowled down an entire line of people. When her dizziness went away, Dale focused on Shannon, who was standing next to her. She handed Dale a beer.

"Hey, a dance with Gil, eh?" Shannon asked.

"Is that what that was? I thought I was playing the victim in a self-defense class. Boy, he's really pissed that Mackey is with Snead. I didn't know he and Mackey had anything going."

"They don't. They were here together on our first pass. He considered it a date and she considered it company. She told me he was too possessive, even after just two hours, that he got upset if she even spoke to anyone else."

"Huh." Dale took a sip of beer. "She's smart to get away from him now. So what else have I missed?" Dale rubbed her shoulder. "Any other soap operas going on?"

"Nope. Same old shit as far as I can tell. What about your neck of the woods?"

"Boring. Kotski tried to fix me up with Tetrault."

"Yeah? Tetrault's cute. Are you going to go out with him?"

"Even if I thought that was a good idea to date another trainee to keep my cover, the answer is no. He picks his nose in public. He picks other parts of his anatomy, too. I didn't say anything to Kotski because her intentions seemed so sincere but…yuck."

At eighteen hundred hours, everyone from A-10 was signed back into the company area. Except Dizzy.

"Did you notice Zelman isn't here?" Shannon asked Dale, who had just come out of the latrine.

"Yes. I was going to mention that after I finished my, um, *bizness,*" Dale said.

"Her bed has been stripped and her locker is empty."

"Pretty sure she's been sent home," Dale said.

Sherlock walked toward them, about to enter the shower. "You talking about the A-10 gang banger? Yeah, Laraway was CQ and she said Dizzy was moved to another company to start processing out. She's being discharged for unsuitability or something like that." Sherlock tilted her head. "You know…I'm actually going to miss her."

"If nothing else, at times she certainly was a riveting distraction," Shannon agreed.

"She was fun, too. I mean, she never really took any of this shit seriously, like she knew she wouldn't make it through training, so she just had fun," Sherlock said.

"True. But let's not forget, in the beginning, how many times our platoon got punished because of her not taking it seriously," Shannon reminded Sherlock.

"At least she didn't end up dead, like Kirk," Sherlock said. She then focused on Dale and lowered her head, almost embarrassed. "Sorry, Oakes."

"No, thank God she didn't end up like Kirk," Dale said, taking no offense to Sherlock's statement.

"What about her three accomplices? Are they gone, too?" Shannon asked.

"Yup. Laraway said they've been *resituated* at Bravo, where their restrictions will still carry over," Sherlock said.

"So, no discharge for them," Shannon said, her tone, disgusted.

"You're kidding, right? Uncle Sam will probably give them scholarships to the Army Photography School," Dale said. That was what bugged Dale the most. It wasn't fair or equal punishment and it wasn't enough as far as she was concerned but it did no good to dwell on it. She wondered if, eventually, the partiality would balance out. If it did, she probably wouldn't see it in her lifetime.

PERMISSION TO RECOVER

CHAPTER FORTY

The next morning was the first official day of AIT. The company was marched to Raburn Hall to be taught a lengthy class in Military Law. The afternoon class was the familiarization with the .45 caliber pistol. The class instructed them on the weapon's history, capabilities and how to take it apart for cleaning. As usual, reassembling the weapon was always more difficult. After approximately six attempts at a measured reassembly, the class finally got it done in the required time.

When Dale and the others were settled back into the barracks after evening chow, news came up from the CQ and spread quickly that Ingersol was in trouble. Rumor had it that he attacked a female from Bravo company in the Supply Room. Ingersol's defense was that he was set up, that he walked in on Bradbury and the female going at it. Ingersol complained that Bradbury got the physically sound female on profile and started working her in the supply room so that they could have their liaisons there. Bradbury's story was that she walked in on Ingersol right when he forced himself on the young woman who was clearly trying to fight him off.

And Anne thought her initial headaches ended with Dizzy's departure, Dale thought.

This explained why when Dale was in the mess hall earlier, she saw Bradbury at a table alone with a trainee, having an intense discussion. The female looked rattled and Bradbury looked as solemn as either Dale or Shannon had ever seen her. Of course, Shannon couldn't possibly let the news slip by without saying something derogatory to Snow. Snow's one-time indiscretion with Bradbury was plainly not worth the grief she was still getting from Shannon.

"Personally, I hope Bradbury is innocent this time," Dale

said.

"You do?" Shannon was surprised.

"Sure because if Ingersol isn't guilty of this, he's guilty of plenty of other things, including indirectly causing Kirk's death. I'd like to see him go down before he gets away with anything else."

"But Bradbury is a predator."

"Maybe. It seems to me that the women she hooks up with don't seem to really mind."

"Regardless, she's a drill sergeant and it's wrong."

"Agreed, but she doesn't seem to be hurting anyone, as far as we know. Ingersol is a different story."

An investigation into the charges and accusations Ingersol and Bradbury leveled at one another began the next morning as the trainees marched back to Raburn Hall for another four-hour class on Military Law. Before they were released for noon chow, the company was tested and everyone got a Go.

The weather had warmed up and was milder so the drill sergeants took advantage of that and spent the afternoon conducting PT and Drill and Ceremony outside at a PT field in WacVille.

That afternoon, Alpha-10 caught the first glimpse of the drill sergeant finally sent to replace MacArthur. Her name was Jennifer J. Cassidy, JJ to her friends, and when Dale got a good look at her, she nearly fell over. This was the same Drill Sergeant Cassidy who confronted Dale in the WacVille mess hall her first week there. The same Drill Sergeant Cassidy who nearly took her breath away at first glance. Dale briefly studied her to see she was still as sultry and gorgeous as she was that day and Dale just knew she was in deep trouble. If anyone else remembered her, they didn't mention it.

Dale assumed Cassidy was somewhere around the same age as MacArthur but in much better physical condition. She wore crisply starched fatigues, highly spit-shined boots, hair pulled back into a tight bun and everything about her looked meticulously tidy. She was tall and slender with coal-black hair and dark brown eyes. Her skin tone was tanned and Dale wondered if it was her natural coloring or she'd maybe just returned from a sunny vacation.

Nobody, including Dale, dared to stare at her too long because, at that point, they had no idea what her temperament was. Even though she barely said a word to anyone, her disposition seemed better than MacArthur's but then Dale thought a pit viper had a better disposition than MacArthur. Cassidy maintained a serious, no-nonsense expression as she walked amongst the exercising troops, apparently memorizing everyone's face and matching it with his or her nametag. When Cassidy got to Dale, if she recognized her or her name from their brief encounter, she didn't let on. Dale was a little disappointed by that. She then wondered when Cassidy would be officially introduced to the company.

The next morning the trainees waited two hours at the corner of Bravo Company area for the military buses to pick them up. When the buses didn't show, the drill sergeants marched the troops six miles to the pistol range.

Once at the range, the males were issued .45 automatic pistols and the females were marched down to another range, where they were going to familiarize with a twelve-gauge shotgun. The women were told they would be given a twenty-minute time limit to load and fire five cartridges, reduce stoppage, and completely unload the weapon.

Dale noticed that Cassidy had arrived at the range by jeep after the company had been there a while and now stood behind her, monitoring her procedure. She had not turned to look at Cassidy directly because just the thought of Cassidy's presence made her stomach jittery. Dale was suddenly nervous and wanted to make sure she did everything by the book because she wanted to impress her. Dale mentally chastised herself as she tried to get control of her emotions. This was crazy. She didn't even know this woman, why was she reacting like this? Before she could answer herself, she heard the preparation command come from the tower.

Dale checked the bore and chamber prior to loading the shotgun to ensure it was empty. She pushed the safety lever to the on position and kept the fore end forward while she loaded four cartridges. She released the disconnector assembly and cycled the fore end, which chambered one cartridge. She loaded the fifth cartridge, placed the lever on off and waited for her instruction to fire.

Upon the command, Dale pulled the trigger and shot all five rounds, pumping the slide between each shot to eject all spent rounds and load the next cartridge into the chamber. She repeated the action until the shotgun was empty. It was letter perfect and Dale controlled the kick as opposed to letting the kick control her. She smirked, knowing she had given Cassidy a good show even though Cassidy said nothing and moved on to the next lane to where Ferrence had just completed the same ritual.

Ferrence was flat on her back in Cassidy's path. Ferrence had never fired a shotgun before and had not expected nor anticipated the kick that knocked the slight-of-build Ferrence a good three feet backward, nearly taking her shoulder off in the process. She flew back with such force, Audi and a range commander had to scramble out of her way. It looked as though she were a bowling ball heading for a seven-ten split. Ferrence looked up to see Cassidy staring back down at her.

Cassidy glanced at her watch and then back down at Ferrence. "If I were you, I'd get back up on that firing line, soldier. You have exactly eight minutes to lose those four other rounds." The newest member of the cadre then stepped over Ferrence and moved onto the next in line. Dale stifled a laugh. At the very least, Cassidy was going to be entertaining. Unlike MacArthur, Cassidy did not seem like the type who was easily flustered or unnerved.

Dale moved to the group who had already fired and thought about Drill Sergeant Cassidy. She was not only going to be most delightful to look at every day but Dale had no doubt the staff sergeant would keep them on their toes.

As soon as the women completed their requirement with the shotgun, they were herded down to a muddier range where they were issued .38 caliber revolvers. This really irked Dale. She felt, with the progress of training men and women together, they should have been trained in and allowed to carry on duty the same weapon, the .45. The Army's argument was that the .45 was too heavy for a female's delicate wrist and the accuracy of the shot would be severely hindered. Both lieutenants knew this assumption was untrue. Dale could max out her score every time with a .45. The .38 gave her more difficulty. Dale's wrist may not have been delicate but it was small and her score was preferable with an automatic than a revolver. The .45 was a more accurate weapon, had stronger stopping power and, with a magazine as opposed to a cylinder, it was easier to load and reload. Regardless of her preference to the automatic, Dale still hit forty targets out of fifty whereas Shannon dead-eyed all of them and maxed out at fifty.

They waited for everyone to get through the firing orders and in the meantime, Robbins told Dale to help Ryan, McTague and Laraway clean off oily magazines while Shannon's group was ordered to police up spent rounds. When everyone was done on the range, they were called into formation. They were surprised to see MacArthur standing there, chatting with Cassidy and Robbins.

"Pssst, Snow," Travis whispered as she got into her squad. "Look! Frigid Bardot's back."

"Fuck you, Travis," Snow growled back.

"Not in your wildest dreams, baby," Travis said.

"Travis! Drop and give me ten," Putnam said. He was standing behind her and she hadn't seen him. Reluctantly, Travis dropped and started counting out the ten push-ups. This put a huge smile on Snow's face, which Travis didn't see but Putnam did. "Snow, you drop and give me ten for finding this funny."

Snow's smile quickly disappeared and she assumed the front leaning rest position and copied Travis' movements.

MacArthur received permission to march the women back to the barracks with Cassidy accompanying them. She put them into a separate formation and moved them out. There was something different about MacArthur. She seemed rested and relaxed and she had a smile on her face. She was actually pleasant and some of the females admitted later that they were almost happy to see her.

MacArthur began a cadence the female trainees had never heard before. It was called The Lady Dressed In Red and it was Shannon's favorite. MacArthur started them out singing about three colors, rhyming the activity with the shade and invited the women to participate. The female trainees came up with all kinds of interesting verses.

See the lady dressed in red, she likes to do it in a bed.

See the lady dressed in black, she likes to do it on her back and

See the lady dressed in green, she thinks she is a sex machine.

They had six miles in which to sing it so they got to try different rhymes with repeated colors. The women had a ball and even Cassidy laughed at the way some of the women used their imaginations. It was a nice moment for all of them, including both drill sergeants. Dale thought that if MacArthur had been more like that when she was assigned to Alpha-10, things might have been different.

Shannon qualified Expert on the range the next day with another score of knocking down all fifty targets. Dale made Sharpshooter by qualifying shooting the best she ever had with a .38 at knocking down forty-five targets out of fifty. Almost everyone, with very few exceptions, needed two tries to finally qualify. This was usually the time when the company eliminated trainees to weed out the best of law enforcement training. Helping the recruits through basic training was one thing, assisting them through LE School was quite another. It simply wasn't done. If they couldn't make it on their own ability then they were not fit to be MPs. Fortunately, everyone made it through the first round of elimination.

The afternoon class was called OLEV, which stood for Operating a Law Enforcement Vehicle. The company gathered at Raburn Hall and was shown a film on how to mount and talk on a jeep radio, the AN/VRC-47, a film on the quarter-ton truck or jeep, and the military police sedan. After that, they were given a class on the military ten-series and prowords.

It was a lot of information to retain in just one afternoon and that night everyone studied from the notes they had taken in class and the fact sheets that were passed out. As the women paired off and formed groups to quiz one another, Dale went downstairs for some fresh air. On her way back upstairs, she ran into Holmquist on the landing. She stopped dead in her tracks and moved into the position of Parade Rest.

"At Ease, Oakes," he told her. She relaxed. "Who had garbage detail tonight and during the weekend?"

"I believe it was Minkler, Drill Sergeant."

"Minkler. I see we still can't break her of her slovenly habits. I will have to talk to her in the morning. In the meantime, someone better get that trash emptied of that contraband. If the senior drill sergeant ever walked in and saw those beer cans in there, you'd all be restricted until you left here. Carry on, Oakes."

"Yes, Drill Sergeant. It will be taken care of, thank you, Drill Sergeant."

Dale walked inside and wasted no time finding Ryder, the squad leader, and relayed Holmquist's message. Ryder, in turn, went directly to the culprit and said something in Minkler's ear that put an absolute look of horror on her face and gave her incentive to race that trash right out the door. Whatever those inspiring words were would forever remain a secret between the two of them but it did the trick. Ryder returned to her bunk, smiled, and thanked Dale.

Dale and Shannon discussed Cassidy later on as Shannon finished laundry. Dale neglected to mention her prior encounter with the comely drill sergeant or her instant attraction to the woman who might possibly be the only one to take Dale's mind off Anne Bishaye. Shannon stated that she would refrain from passing judgment until she'd actually seen her in action as a drill sergeant. Dale told her, even from the little bit she'd seen, she was sure Cassidy was much more professional than MacArthur. The subject eventually came around to Holmquist and Dale once again praised him for his militarily ethical conduct. Shannon teased Dale regarding having a crush on him and the more Dale protested, the more Shannon believed it was true. Dale felt she was between a rock and a hard place, so she just stayed silent on the matter.

What Shannon didn't know shouldn't hurt her.

The radio code test was administered the next morning and

couldn't have been any easier if the answers had been sitting right in front of the trainees. Everyone got a Go.

The next class taught the trainees how to direct traffic. The company was divided into groups of approximately twenty-five, instructed in what to do, shown a filmstrip and then taken out to a busy intersection on post to demonstrate what they had just learned. Their demonstration was also their test.

In Dale's group was Shannon, Minty, Snow, Caffrey, Chillemi, Navarrette. The rest were assorted males from each platoon. The practical application of what they had learned was confusing, even with the instructor right there with them. They stood on a box that raised them about eighteen inches off the road and directed traffic to the regulations of their class. Shannon had Audi and Cassidy drive through her test and the usually level-headed Minty got so flustered when a lengthy wedding party drove through her test, she almost caused a four-car pile-up. Minty recovered and everyone got a Go in that class.

The group of traffic directors were brought back to the school and then given a class on how to properly fill out vehicle maintenance forms and how to read a trip ticket. Trip tickets were small sheets of paper issued to the driver of a specific vehicle, along with a set of keys. The information the driver wrote on this document included starting and ending mileage and on whose authority the driver was making the trip. Before they were dismissed from class, the trainees were told that the test which would qualify them for a military driver's license would be on Monday.

It was hard to believe it was Friday already. At 1700 formation, the company was once again set free for the entire weekend. Except for Minkler who was restricted to the barracks by Holmquist for one night only because of the garbage detail, all the females, including the three marines, started their weekend pass at five o'clock.

With everything going on around the company area, Dale and Shannon discussed whether or not they should postpone taking off to Averill by themselves. There was a new drill sergeant to consider, one Dale almost hoped needed to be watched, the Bradbury-Ingersol case and three new females to consider. Shannon suggested they take it a day at a time and monitor as many activities as they could. If something specific bore watching, Dale and Shannon agreed to put their personal time aside and zero in on whatever the issue was. If not, they would strike out on their own and just hang out for a night, as irresponsible as that may have been. They couldn't be everywhere at once and would take advantage of that adage sooner than later.

The Pizza Place was predictably packed, even by the time Dale and Shannon got there. Shannon immediately hooked up with Renaldi, who had been the CQ when the Bradbury-Ingersol incident went down. At first he told her he shouldn't talk about it but after several beers courtesy of Shannon, his memory became vivid and his tongue became loose. By the end of their flirtatious encounter, Shannon had plied Renaldi with enough alcohol that he was not capable of pursuing her any further. He could barely stand up.

"I love to watch you work." Dale leaned closer to Shannon after Renaldi left to use the men's room. "Did you get anything useful?"

"Well, that depends. It's useful for a military tabloid but has nothing at all to do with why we're here. He talks as if Bradbury is guilty but from everything we already know and the few facts he added, my money's on Ingersol."

"Couldn't happen to a nicer guy," Dale said. She felt no remorse for Ingersol, or Bradbury for that matter. She watched Rutledge, another male in Shannon's platoon, watch Shannon. He was obviously a man in lust. When Rutledge saw Renaldi stagger out of the Pizza Place, he made his move. "You're going to have company," Dale said to Shannon.

"Who?"

"Rutledge." She saw Shannon's face light up.

"Really? I can't see him. He's coming here?"

"Yup. You interested?"

"A little. He's a pretty nice guy."

"Hell, he's just plain pretty." Dale smiled. "He hasn't been able to take his eyes off you since we got here."

"Hi, Walker," Rutledge said. He barely looked at Dale. "Oakes," He said, politely acknowledging Dale.

"Rutledge," Dale answered. Rutledge asked Shannon to dance and Dale stopped her before she joined him on the dance floor. "If you want to keep an eye on things here, I'll head to the EC."

Shannon nodded. "No problem." She and Rutledge moved closer to the jukebox as Dale finished her beer and headed toward the exit.

Dale was stopped by Fanuele, who was on his way out, also with two other males from Second Platoon. "Oakes, where are you off to?"

"The EC."

"Us, too. Why don't we all go together?"

"Sure. Uh...are you going to dance with me?"

"Probably."

"Are you going to pin me to the mat this time or body slam me up against the ropes?" Dale asked.

Fanuele bowed his head, embarrassed. "Sorry about that."

"I guess I've recovered." Dale playfully pushed him.

"Are we going to walk or take a cab?" One of the other males asked.

"Walk," Dale said. "We can beat a cab by at least thirty minutes."

They left the Pizza Place and nearly knocked down Cassidy, who was about to enter to check on things.

"Hey, slow down gang," she said, good-naturedly. She looked the four of them over, then directed her attention to Dale. "Y'all sure look different out of uniform." Her voice was low, smoky, sexy as hell. Cassidy's observation returned to the males. "Where is this happy crew off to?"

"We're going somewhere else for a change of scenery, Drill Sergeant," Fanuele offered.

"Yeah, um, somewhere not so loud and crowded," Kreiger, one of the other males, lied.

The men fidgeted while Dale seemed totally relaxed. In reality, she was the most nervous of the quartet. Something about Cassidy heated up Dale's blood to an intensity that nearly boiled her brain and rendered her stupid. Especially when Cassidy looked into her eyes, like she was doing right now. "Three escorts, Private Oakes?"

PERMISSION TO RECOVER

Huh. She remembered my name. Interesting.

Dale bet Cassidy couldn't recall the names of her three escorts. "Yes, Drill Sergeant, I figured these boys needed a female influence."

"Really? Female influence for what? Y'all wouldn't be heading to the Enlisted Club, now would you?" Cassidy's tone was less accusing and more mock warning.

While her cohorts looked guilty and immediately found interest in the Alabama clay, Dale met Cassidy's penetrating gaze. "Why, Drill Sergeant, that place is off-limits. Why would we go there?" She punctuated her question with an indulgent smile.

Cassidy visibly swallowed, evidently not quite prepared for Dale's coy response and then, smirked. She was tough to read and Dale wondered if she'd overstepped her trainee boundaries. That smirk, as sultry as it came across, could have given way to a nasty *are you flirting with me, soldier?* tirade. Instead, Cassidy broke into a grin. "Go on, get out of here."

"Yes, Drill Sergeant," Kreiger answered for the rest. They watched Cassidy disappear inside the noisy Pizza Place and close the door. "Oh, man, she is such a fox!" The other two men enthusiastically agreed. "Hey, Oakes, how come she knows your name already?"

"I'm going to guess it's because she monitored me at the shotgun class and stared at the name on the back of my steel pot for fifteen minutes." As the words left her mouth, Dale realized that was exactly how Cassidy could have remembered her name and she felt her stomach flip in disappointment.

Careful. Don't be pulling another Bishaye on yourself. You don't even know if Cassidy is gay and even if she is, she is totally off-limits.

Dale sighed in chagrin but her companions did not hear her amidst their fantasies of what they would do if they got Cassidy alone. It was all boastful man talk and Dale finally joined in, teasing them that they wouldn't know how to handle such a woman. That started a whole new round of boasting.

Dale and the three males she went with closed the EC. Quite a few company members were there but once again no drill sergeant or MPs showed up to take any names. She took turns and danced almost every dance with her companions. They seemed like really good guys, serious about being MPs and not at all bothered, anymore, by taking training with females. Dale enjoyed the small group.

On their way back to the barracks, Fanuele whispered suggestively in Dale's ear that they go somewhere else, just the two of them.

"Gil," she said, softly. "I'm very flattered but let's be honest here. I'm not who you want and I'm not going to pretend I am just because we're both a little intoxicated and horny. I love being around you guys, we have a good time, but that's where it starts and ends. If it's Mackey you want then go get her."

"She doesn't want me, she wants Snead."

"Then stop wasting your time on her, Gil," she said, gently. "You're handsome, you're nice, you can dance…a lot of women would be thrilled with that."

"But I want Melanie."

"Then why do you want to get me alone?" Dale asked him. When he was silent, Dale answered for him. "Because you're horny."

"Aren't you?"

"Yes, but I have someone at home," Dale lied. "And he deserves my loyalty."

"You sure he's being loyal to you?"

"I can only have faith that he is."

"Hey! No secrets back there," Kreiger called to them.

"It's nothing like that," Fanuele said. "She's playing *Dear Abby*."

Kreiger and the other male with them looked at one another. "Mackey attack," they chorused.

They parted ways once back in the company area. Dale walked by the Orderly Room door to see who was on CQ duty. It was Minkler and Freddie Swan. Dale rolled her eyes. As one of her first drill sergeants used to say, those two were, individually, about as useless as tits on a bull. Together, on CQ, they were a disaster waiting to happen. When she got to the female bay, she did a quiet walk-through and noticed several more empty bunks than last weekend, including Shannon's. Dale smiled, wondering if her partner got lucky or just decided to follow the rest into town. She would find out tomorrow and went to bed.

CHAPTER FORTY-ONE

Shannon paid Dale back for last week and roused her out of a sound sleep early Saturday morning. Dale showered, dressed, packed a few things for overnight and they took a taxi into Averill to the room Shannon had booked the night before at the Journey Inn.

The Journey Inn was the most popular motel in Averill for the members of Alpha-10, the same as it had been six years earlier when Dale and Shannon had gone through training the first time. It was clean, inexpensive, with a military discount and centrally located in town.

Dale bounced down on the unused bed in the room. She was happy to recline on something larger and comfier than a regulation cot.

"So, how was your night?" Dale asked Shannon. They had discussed Dale's evening and who was on CQ on their taxi ride. Dale wanted to get into the dirt.

"It was interesting. I can honestly say I slept with Rutledge and I mean slept. What's with these GIs? They have a few drops of alcohol and they're dead to the world. I don't get it."

"They're just not old, grizzled veterans like us. What about this morning? No action either?"

"He was sick this morning. That's why we came back early. He went to the barracks and I got you."

Dale sat up, scooted down to the edge of the bed, and turned on the TV to cartoons. "Thanks. You make me feel so wanted." She switched channels. "Cool, Rocky and Bullwinkle." She looked at Shannon, who sat down on the other bed. "It's probably good nothing happened between you two. No ethics to think about."

"True but I could really use a good fuck," Shannon admitted. "This is the longest I've gone without any action since, well, the first time we were here."

Dale laughed. "It's so tough being you."

Shannon threw a slipper at her, which Dale deflected with her forearm. "I'm hungry. Let's see what's still in town for restaurants."

They found a breakfast café, ate heartily and leisurely, and then wandered around downtown Averill to see if they could find any of their fellow A-10 trainees. They were surprised they didn't spot anyone after a few hours of circling the usually popular and frequented section of the city. Finally, they gave up and decided to see a matinee of a new movie release called *Saturday Night Fever*.

"I am in the mood to dance, aren't you?" Shannon asked Dale as they split a pizza and a pitcher of beer at The Stop, the same bar where Shannon met Henning a month before training started. It was Happy Hour and people had begun to file in, which caused the noise level to rise and the jukebox to be turned up. Most of the music was country-western until Dale fed money into the machine and the new sound of disco filled the room.

"Now I definitely have to dance," Shannon said and smiled.

Two men, civilians by the length of their hair, approached Dale and Shannon's table. They introduced themselves as Matt and Warren and sat down with another large, full pitcher of beer. The men seemed pleasant enough but acted as though their company was automatically wanted and welcome. That irked Dale because, in her opinion, it said a lot about their attitude toward women. That women couldn't be out just to have a good time by themselves. That the only way their night would be complete is by adding the company of men.

Dale considered their presence an intrusion and then wondered why. It wasn't as though she had feelings for Shannon other than friendship or that they had anything specifically planned but Dale had hoped for some uninterrupted time between them for one evening so they could just be themselves. She also felt if they'd wanted company, they would have invited some.

On the other hand, Shannon appeared fine with it and embraced the male companionship. Dale had to admit that Matt seemed adorable. He had shaggy blond hair, bright green eyes and a dazzling smile and he clearly only had eyes for the other green-eyed, shaggy blonde with the dazzling smile. If either of them were to pair off, it would be Shannon and Matt.

Warren wasn't unattractive by any physical means but the chip on his shoulder about military women was, particularly military women from the north. It was also quite clear that he would rather have had the first crack at Shannon. What bothered Dale was not that fact because she wasn't interested in him anyway but his conduct toward her as second choice.

After their third shared pitcher of beer, Shannon suggested they leave The Stop and find a disco so they could dance. Dale was the first one to jump up from the table. Maybe they could salvage what was left of the night.

The group of four piled into Matt's car, drove a couple of miles north to the outskirts of town to another bar, one on the off-limits list. This club had three different dance floors and an uncanny resemblance to the disco John Travolta was king of in the movie they had seen that afternoon. The foursome found one of the last available tables in the bar, sat, ordered drinks, then instantly got up to dance. They caught the tail end of a fast song and stayed on the floor for the next song, which was a ballad. Shannon and Matt, who seemed completely infatuated with one another, easily moved into one another's arms and swayed to the music. Warren stepped toward Dale, expectantly, and Dale excused herself to use the ladies' room.

On her way to the bathroom, Dale's path was blocked by an outstretched arm. The owner of the arm belonged to Sherlock who was seated at a table with, surprise of all surprises, that handsome new marine, Parisi. Also at the table were Troice and McTague with their respective dates from Alpha-10.

"Figured if we ran into anybody from the company here, it would be you two," Sherlock shouted above the music and grinned. She nodded toward the dance floor. "Nabbed yourselves some civilians, I see."

Dale shrugged. She leaned down, close to Sherlock's ear. "I see you might just get to find out if he's tall in the right place."

Sherlock broke out in a huge smile. "I found that out last night. He is more than tall enough for me," she said in Dale's ear.

Dale patted Sherlock's shoulder and put up her hand in greeting to the rest as she continued on her way. When she returned from the bathroom, she wanted to warn Sherlock and the rest that MPs really did check this place, or they used to when it was a live band bar called Hayseed's. Then she remembered she wasn't supposed to have that information, so she walked by the table unnoticed by McTague who was kissing Halliday with her hand stuck under a coat draped on his lap. From the motion underneath, it wasn't hard to figure out what she was doing. Dale ignored them, though it annoyed her that neither cared enough about their self-respect to indulge one another in public as opposed to private.

You're one to talk, she then chided herself. *If Anne had pushed you just enough, you would have let her do you right in her office and you know it.* That thought sent a wave of butterflies fluttering around her insides as the fantasy picked up where the reality left off. *Well...at least it wouldn't have been in front of others.*

Sherlock, Troice and their dates were up dancing, as were Shannon and Matt. Before Dale could even reach her seat, Warren grabbed her hand and pushed his way onto the dance floor so that they were close to Shannon and Matt. Dale got Shannon's attention and mouthed the words *MP check soon* to which Shannon reluctantly nodded. She held up her hand, indicating she wanted to wait out at least five more songs. Dale was fine with that as long as they were out of there before ten. Staying longer was just taunting fate.

Warren became more obnoxious as the evening progressed. Before they had arrived at the disco, all he could talk about was what an exceptional dancer he was and that he could do anything Travolta did in that stupid movie. His wild, uncontrollable movements, however, undeniably contradicted his boasting. What he made up for in rhythm, he lacked in coordination and he danced as though there were no other bodies on the crowded dance floor. When he would knock into people, he would glare angrily at them, ready for a fight if anyone looked like they wanted to make something of it. Finally, gratefully, Shannon suggested they leave.

On the way back to the motel, Matt drove with Shannon in the front seat and Warren in the back with Dale. Intoxicated, pompous and rambunctious, Warren finally lunged at Dale just before the car pulled into the parking lot. He kissed Dale so forcefully, she thought he'd given her a fat lip. His big, rough hands grabbed at her breasts, painfully squeezing them and when the car slowed, Dale wrestled her way out from under him, opened the door and escaped before the car even came to a complete stop.

Shannon sprang out of the car right behind her. "What happened? Are you all right?"

Dale took a deep breath and turned to face her friend. "I am now. Look, Shan, that asshole is clearly fucked up. You and Matt can do what you want but Warren's not coming anywhere near me. Literally and figuratively."

"Okay. No problem. How do you want to handle it?" Shannon asked, as both Matt and Warren got out of the car.

Dale studied Shannon. "You really like this guy, don't you?"

"So far. He's not a trainee or a drill sergeant or even in the Army so no ethics to consider." She glanced at him. "And he's cute as hell." She looked back at Dale. "Sorry Warren turned out to be such a prick."

Dale wasn't sorry. It saved her from an awkward situation if he'd turned out to be as nice as Matt. She fleetingly wondered if she had subconsciously provoked Warren's behavior so that she wouldn't have to be confronted with the decision to sleep with him to save face with Shannon and then she dismissed that thought. No, Warren was a jerk from the beginning with no pretext from her.

"Who's here who isn't with somebody?" Dale said, as the boys started toward them.

"Uh...Travis, I believe, is in two-oh-four," Shannon said.

"I wonder if she'd mind company?" Dale asked.

"Probably not yours," Shannon said, smiling. "You two get along well. Thanks, pal, I owe you one." She walked back to Matt, spoke quietly with them and they headed upstairs. Dale walked toward the opposite stairway.

"Hey! Where are you going?" Warren yelled at Dale.

"Away from you."

"But they're going up to the room," he protested.

"No shit. I'm sure they'll have a great time."

Warren reached the bottom step. "What about us?"

Dale whirled on the stair and glared at him. "This was not a package deal, Warren. You don't like me and I don't like you. Nowhere in my book does that add up to sex. You want a piece of ass? Go get one. Unless it's changed, the corner of Fourth and Knight is a good place to start." She pointed to herself. "This is one piece of ass you won't be getting."

"Hey, what about all that beer I bought for you and we took you to the disco? Don't I get anything for that?"

"You got my company and you didn't even deserve that and, excuse me, but I paid for my own cover charge and you got free beer because Matt's brother-in-law is the bartender."

"You bitch!" he spat.

He started up the stairs after her. Dale braced her hands on the railings, bent her knees and kicked out, connecting both feet with Warren's shoulders, which propelled him backward. He fell on his behind at the bottom of the stairs, the impact knocking the wind out of him.

"Fucking dyke," Warren sputtered as he regained his breath.

Now you've got the picture. "That's the nicest thing you've said to me all evening."

"How the hell am I supposed to get home?"

"That's not my problem. Take that up with Matt. He was your date before Shannon came along."

"Fuck you!"

"You'll never get that lucky." She reached the second landing and knocked on the door of Travis's room. Dale looked at her watch. It was almost eleven, hopefully Travis wasn't asleep yet. She knocked again and a very tousled-looking Travis opened the door.

"Hi, Travis. Can I come in?"

"Sure." She yawned and closed the door after Dale was inside. "Where's Walker?"

"She's busy," Dale said and then related the whole story to her. "So that's why I'm here. Why'd you want a room by yourself?"

"I didn't. I got it with Tierni."

"Like I said, why'd you want a room by yourself?" Dale smiled. "Why didn't you invite your husband down, Mrs. Novak?"

"We can't afford it just yet. He usually works weekends anyway and, besides, the way Ritchie is, I wasn't sure we'd get this weekend off."

"That's true. Nothing ever seems definite in the Army, does it?" Dale sighed.

"Watch TV or whatever you want. I'm going back to bed," Travis said.

"Nah, I'm beat," Dale admitted. She stripped down to her shirt and panties and crawled into the other bed that was supposed to be Tierni's. "I hope she doesn't pop back during the night," Dale said.

Travis turned out the light. "Well, if she does, it will be interesting to hear just how lucky someone is to fuck you."

"Heard that, did you?" Dale said and chuckled.

"I'm sure half of Averill heard it."

"Well, my secret will still be safe because Tierni just isn't my type," Dale said and meant it. Tierni wasn't tall, dark, and brooding.

Shannon knocked on Travis' door early. She gave Dale the room keys and told her to check out whenever she was ready because she wouldn't be going back to the room. While Travis was in the shower, Dale stepped outside in the cold, crisp morning air.

"Where are you going?" She kept her voice low.

"Matt and I are going to spend the day together."

"Wait a minute, I think it's great you found Matt and I'm thrilled you got laid but I'm not going to track everyone down by myself."

"I don't want you to," Shannon said, defensively. "Look, Dale, this guy is a great cover for me. One day away from all this is not going to hurt anything. Give me a break, would you? You got two weeks away from this crap. Six hours, in the grand scheme of things, isn't that long."

Dale looked down at the keys in her hand and nodded.

"You're right. Go. Have a good time. Just don't get used to him. We've got work to do."

"Yes, mother." Shannon grinned. She walked downstairs to Matt's idling car, got in and they left.

Dale and Travis returned to McCullough early. Travis went back to the barracks and Dale walked to the Pizza Place, accompanied by Caffrey, Creed and Almstead.

Caffrey had toned down a lot since that first week and other than being a little klutzy, she was turning out to be a pretty good soldier. Her test scores were average but her enthusiasm was high and she tried extremely hard. Harder than a lot of her male platoon members.

Creed was just naturally good at everything she attempted and Almstead put forth the extra effort because she was so enamored with Creed and wanted to impress her. Dale had seen a close relationship develop between the two women, very supportive and, although discreet, physically affectionate, too, wisely stopping short of openly behaving like lovers. Almstead came across as being convinced they were fooling their barracks mates and seemed to think everyone believed she and Creed had become great friends, like Shannon and Dale or the three T's. Creed knew that Dale, among a few others, was onto them and appeared happy and surprised that Dale regarded them no differently. Dale felt no need. She was all about duty and mission and, so far, their actions weren't interfering with hers. She wasn't there to be barracks snitch, especially not a hypocritical one.

After the four of them split a pizza, Dale table-hopped to get the lowdown on everybody's weekend. No one mentioned a drill sergeant, no one alluded to anything covert, no one suggested that anything remotely suspicious went on during the weekend. Dale finished her beer and went outside to find an isolated pay phone. It was time to check in with her boss.

She let the phone ring eight or nine times and was about to hang up when the Bishaye picked up.

"I didn't think you were home," Dale said. As usual, Dale's heart pounded.

"I just walked in," Bishaye said, out of breath. "I was in

town, buying groceries. What's going on?"

"Nothing you don't already know," Dale said, nonchalantly. All she could think about was that she wanted to be the cause of Bishaye being breathless. She could hear full paper sacks being set down on the counter. She pictured Bishaye in her kitchen, balancing the phone between her ear and shoulder, putting groceries away.

"Really? Are you sure?" Bishaye asked.

Dale almost stared at the phone. "Am I —? Of course I'm sure." She didn't disguise her annoyance.

"Okay, then where is your partner?"

"Shannon?" Dale asked, surprised, then felt an impending dread.

"You have another partner I don't know about?" Anne's tone was still neutral.

"Funny," Dale said, humorlessly. "Shannon's back at the Pizza Place. Eating pizza."

"Better check your nose, my friend, I do believe it may be growing."

Fuck. Now Dale's heart was pounding for a different reason. She hated getting caught in a lie with Bishaye.

"I just saw her in Averill."

Fuck, fuck, fuck! "Did she see you?"

"No. I made sure of that. Why'd you lie to me? What's going on?" Now Bishaye's tone was a little stronger.

"Nothing is going on."

"If nothing is going on, why did you lie to me?"

"Anne, come on, it's not a big deal," Dale said and sighed.

"Nothing's going on, it's not a big deal…" Bishaye repeated. "Fine. Who's the young man she was with?"

"Matt," Dale said.

"Dale. Do *not* make me drag it out of you!"

"He's some guy we met last night. He's a civilian, which I'm sure you guessed already by the length of his hair. She was just spending the day with him. It didn't hurt anything. Nothing is happening. She needed it. Nothing's been happening."

"Yes, you said that."

"It hasn't been. You saw him. He's an attractive guy. Everybody's been pairing off. We can't pair off with male trainees and we really need to pair off with somebody other than each other. Wouldn't you prefer she be with someone totally removed from all this?"

"Are you sure he is removed? How did you meet him? Did he approach you or did you two approach him? What do you know about him?"

"Uh…well…he can dance," Dale said, knowing that's not what Bishaye wanted to hear. "I'm sure at this point Shannon knows much more about him, but —"

"He can dance. I'm so glad for him. Look, Dale, I don't care if Shannon picked someone up as long as she's cautious. I'm sure she's in need of company other than you and a hundred forty-four of her closest friends. It's none of my business who she goes to bed with as long as she *knows* who is slipping between the sheets with her. Now the fact that nothing is happening within the company, that you know about, is making me a little nervous and then this guy, Matt, just pops in out of nowhere."

"Well, not really nowhere. Matt and a friend picked us up at The Stop. They were there first so they didn't follow us in. I dumped the friend. He was a dick."

"Did you sleep with him first?"

The question surprised Dale. "I thought you said it was none of your business who we went to bed with."

"No, I said it was none of my business who Shannon went to bed with."

Dale took a deep breath and blew it out. "Since you so obviously don't want to join the club and would rather play cruel head games with me, it's none of your business who I go to bed with, either."

Her words had evidently stunned Bishaye as there was dead silence on the other end of the line. Bishaye cleared her throat. "Just because I won't betray my husband doesn't mean I don't have feelings…in that area…for you. I just can't act on them any more than I already have." Her tone was sincerely apologetic. "Is it just me or do you have feelings for other women, too?"

Dale closed her eyes and took another deep breath, letting it out slowly. This was torture. "It's none of your business who I go to bed with."

"You're right," Bishaye said finally, with resignation, There was another awkward silence. "What were two doing at The Stop? You know that place is off-limits."

Dale was both relieved and sad the subject had changed. "I know. I went to off-limits places when I actually was a trainee. I figured if I did it then, some of my barracks mates might do it now."

"Why did you both have to go in there together? What happened to splitting up and covering twice the territory?"

"We have been splitting up but sometimes it also looks better when we travel together."

"Not so sure I buy that one, Dale," Anne said, wisely. "Anything unusual about the people pairing off?"

"No. Not so far. Some of the coupling is a little odd but not a damned thing suspicious."

"And the cadre?"

"The drill sergeants have backed off completely. Ritchie hardly comes around at all anymore which is no great loss."

"I told Colton to back him off, to involve him in other projects that would occupy his time and keep him away from the company area. I felt he was working you two into such a frenzy, one of you would blow your cover with him."

"That's not so far-fetched. He was out of control. Everything else is running like a normal cycle, considering, although, the laundry room incident broke the monotony."

"I don't want to talk about that," Bishaye said, flatly. "Except...you two had no idea any of this was going on?"

"We thought she might be slipping out at night, we were pretty sure she was buying drugs from Ingersol but we couldn't prove it and after our daily rituals, neither of us had the energy to sit up all night, spying on her."

"You suspected all this and you never said anything?"

"I believe I did say something to Henning. Not my fault if she didn't pass it on to you. And Dizzy, I mean, Zelman had already been caught once with drugs in a surprise inspection. We knew she wasn't going to be here for the duration and it was just a matter of time before she got kicked out. I'm curious, though. Why did she get booted out and not the guys? They were all caught doing the same thing. I thought you, of all people, would be fairer than that."

"I wanted to release them all from their contracts but Sergeant Audi begged me not to. So they got recycled and a second chance but that's all they get. Private Zelman requested a discharge. She admitted she is not now and would probably never be military-oriented and the last thing she wanted to be was a cop. We agreed it would be better all-around for everyone if she left. I have to tell you, I felt no guilt in accommodating her as quickly as possible."

"She never should have been recruited or enlisted, in the first place. Just like Kirk."

"Water under the bridge, Dale, we are not going there."

"Fine. What's happening with Ingersol and Bradbury?"

"Ingersol's luck ran out. Two of the men he works with signed sworn statements saying Ingersol told them both he was planning on nailing that female that afternoon and for both of them to stay out of the Supply Room. One went on an errand and the other felt guilty and went to Sergeant Bradbury. You know the rest. Ingersol will be gone within the week."

"Discharged or to confinement?"

"Discharged."

"Anne! It was attempted rape!" Dale was outraged.

"The victim requested it not go any further. She prefers her military career not be stigmatized by the incident. Dale, you know as well as I do, she's right. She would be tainted forever in this good ol' boys club."

"I don't know how you've stuck it out for so long," Dale admitted. It suddenly hit her like a blow to the stomach that Bishaye was right. Life as a woman of rank in the military was not a picnic and that it wasn't the system that had changed her, it was the politics. She was clearly a David up against many Goliaths. It really *wasn't* personal. Dale felt chagrined and it came through in her voice. "Shit. Anne, I —"

"Forget it. Listen, you probably need to get back to sign in."

Dale checked her watch. "I do, yeah."

"Get back to me when you can."

"I will." Dale paused. "Are we okay?"

"We have to be."

"Yeah," Dale said, quietly and hung up.

Dale waited outside the CQ Office for Shannon. She showed up exactly three minutes before the sign in time of 1800 hours.

"How was your day?" Dale asked, after Shannon exited the Orderly Room.

"Nice. We had a great time. He's a really sweet guy."

"Think you'll see him again?"

"Probably." Shannon noticed Dale's odd demeanor. "What?"

"Bishaye saw you in town this afternoon with Matt."

"When did you talk to her?"

"About thirty minutes ago."

"Christ. Is she really upset?"

"No. She was just concerned because she didn't think we were doing our job and because neither of us know him and he might not be who he says he is or, at the very least, he will interfere with your work. Other than that, she was fine."

Shannon was visibly relieved. "I think he's who he says he is. Before we went for a drive, he stopped at his house to change his clothes. I met his family. He's just a local guy out to have some fun and that's all I want. He won't cloud the investigation, I promise."

"That's good enough for me." Dale told Shannon the other pertinent parts of her conversation with Bishaye. They also talked about Dale's day, Shannon's day, the previous week, and the week to come. They speculated about Cassidy, suspecting she was going to be tough but fair and would hopefully work with the females and not against them like MacArthur.

"I've got to say, she is one of the most attractive female drill sergeants I have ever seen," Shannon said, stubbing out, then field stripping her cigarette.

"You think so?" Dale asked her with perfect nonchalance.

"You haven't noticed?"

Dale shrugged. "I guess I haven't."

"Come on, I'm bushed. I didn't get a lot of sleep last night," Shannon said and headed toward the stairway.

"Braggart," Dale said, kidding.

"Hey, you could've had some loving, too, it's just that

you're too picky."

You have no idea. If only she could confess to Shannon just how dangerously attractive she thought Drill Sergeant Cassidy was.

CHAPTER FORTY-TWO

The first order of the morning was a test on jeep upkeep. The procedure was called First Echelon Maintenance and it was a necessary policy for all military personnel before they operated any military vehicle.

An instructor had a checklist and marked off all the required series of steps as the trainee explained it to him. The trainee was not expected to actually perform the maintenance, just to indicate what he or she was describing on the parked jeep.

"Okay, Private Oakes, tell me how you'd check your vehicle," the instructor asked.

Dale pointed to all the applicable areas as she spoke. "I'd check the cooling system and oil levels. I'd check the engine compartment for any leaks or foreign objects. I'd look for bent fan blades, faulty wiring connections or loose fan belts. I'd make sure, if the canvas top wasn't intact, that it was stored under the seat, that all the correct tools were in the driver's compartment and that the map compartment contained all the proper forms and logs. I'd check tire pressure. I'd check the safety belts for serviceability or damage..."

The instructor scribbled notes and crossed off numbers on his list as Dale eliminated each one without hesitation. "Excellent, Oakes."

"Thank you, Sergeant."

"I take it you've owned a jeep."

"No, Sergeant, I just studied really hard this weekend." Dale blinked at him, innocently.

The instructor chuckled. "If that's true, you'd be the first. Okay, Mr. Peabody, just one last question. If you discover any deficiencies, what form would you fill out?"

"That would be DA Form 2404, Sergeant."

The instructor wrote another note on the sheet of paper at

the top of his folder. "Very good, Oakes. You've got yourself a Go."

That afternoon, the company was transported to the vehicle driving range to learn how to operate a four-speed jeep. The trainees who already knew how to drive a standard shift, like Dale and Shannon, could easily breeze through the course but, for others, it was not that simple.

The company had to pass four stations. Station Number One was where they had to successfully start the jeep and put it into first gear. Station Number Two was learning how to shift from first to second gear and then second to third gear. Station Number Three was learning how to stop and start on a hill without rolling backward and Station Number Four was learning how to shift into reverse and how to parallel park.

Shannon was paired off with Travis who effortlessly taught Shannon how to manage Station One while Dale was peer instructed by Tramonte. The two partners agreed to feign ignorance on stick shift driving. Neither wanted to get stuck with instructing their peers, a thankless, hair-raising task, especially on the driving range. Whiplash and an excessive case of nerves were usually the result of that particular class, which was probably the reason there were peer instructors in the first place. No military person in his or her right mind would ever be able to handle a permanent position on educating a GI to drive a vehicle. Not without hazard pay, anyway.

Before the afternoon ended, Dale and Shannon had successfully completed Station Two with two males as peer instructors. The next morning, after 0700 formation, they would immediately report to Station Three.

Dale was fortunate enough to hook up with Tramonte again and they practiced how to stop and start correctly on a hill. They repeated this exercise for at least forty minutes before Tramonte reluctantly gave Dale a Go and sent her on to Station Four. Tramonte informed Dale that her reluctance was not reflective of Dale's ability for Dale had accomplished the task perfectly the second time around. Tramonte just didn't want to risk getting the likes of Freddie Swan and Kramer, who she'd had at Station

One. They were both so afraid of the damned vehicle, they could barely start it and when they did, the violent jerking motion of trying to put the stick shift into first gear without stalling nearly put poor Tramonte in the hospital. She kept Dale as long as possible until Audi came and told her that Station Three was backed up.

Shannon ended up at Station Three with Private Pamela Chillemi, USMC, as a peer instructor. This was the first time Shannon had a solo chance to speak with the marine insert, so she greeted her pleasantly with the thought if there was no rapport between them, at least there would be civility. She grossly misjudged the marine.

"Hi, I'm Walker," The lieutenant barely said before Chillemi snapped.

"Just get in the jeep, will you? I haven't got all day!"

Shannon swallowed her smile and the urge to remind her that she really did have all day. She also resisted the initial desire to slap the facial features off her new peer instructor. In the one lap around the course that it took Chillemi to show Shannon what to do, the marine berated her for not knowing how to drive a standard vehicle before she enlisted. She let Shannon know, in no uncertain terms, that she detested and felt above Army women and made it obvious she'd only wanted to get male GIs to instruct.

They switched places and Shannon demonstrated the perfect stop and start on an incline. She didn't want to spend any more time with Chillemi than she absolutely had to. Chillemi told her it was wrong and made her do it again. They went around the course several times and the marine yelled at Shannon for her so-called inability to grasp instruction, each verbal attack worse with every lap. When Shannon did interrupt her to mention that she couldn't correct a mistake if she didn't know what the mistake was, Chillemi's tantrum became worse.

Shannon finally had enough of the obnoxious marine's tirade and picked up speed around the sharp curve just before the hill and purposely hit a pothole in the middle of the road, which caused the jeep to go airborne. When the tires touched the ground again, the jolt bounced Chillemi completely out of the jeep and onto her behind on the dirt track. Shannon's butt slammed backed into the driver's seat with such force, she nearly cracked every tooth in her head. It was worth it. Chillemi seemed more shocked than bruised.

Shannon jumped on the brakes and kicked up a tornado of dirt, gravel, and dust into her new enemy's face. Shannon let the jeep roll back a meter or two, making Chillemi scramble to her feet and get out of the vehicle's path. Shannon looked at her. "What are you doing out there? Get in, I don't have all day!"

This provoked Chillemi into an even more vicious attack on Shannon. Chillemi stomped toward the driver's side of the vehicle and demanded that Shannon get out.

Shannon shrugged, put the jeep in neutral and obliged. She exited before Chillemi was even close to being in the driver's seat. "I can't deal with you, man, you've got problems." Shannon walked to the Station Three pick up point to wait for another peer instructor.

A horrified Chillemi watched helplessly as the jeep rolled backward down the hill, unoccupied, passed Drill Sergeant Cassidy and into a ditch. Chillemi chased after it screaming.

"The brake! You forgot to set the emergency brake!" she cried out.

Cassidy had showed up at morning formation and stood in front of First Platoon with Audi. She had slowly begun her integration into Tenth Battalion. She had still not made any special effort to speak to the women as a group yet nor had her appointment to the company been made official by any announcement or introduction. Still, as short as the encounters with her were, it was quite obvious Cassidy was not going to be a clone of MacArthur.

She made herself quite visible on the driving range and made her strong but silent presence known by mostly observing. She had heard most of the one-sided conversation and witnessed Chillemi's unplanned departure from the vehicle. She looked at Shannon who was getting into a jeep driven by Wachsman.

"I'll deal with you later, Walker," Cassidy said, authoritatively, for the benefit of the others. Since no one got hurt and many a jeep ended up in that particular ditch and survived, Cassidy probably wouldn't discipline Shannon. In fact, after seeing that Chillemi's ego was the only thing damaged, Cassidy looked as though she had to bite back a smirk at how Shannon had handled the situation.

The new Alpha-10 drill sergeant approached the still-sputtering marine and commanded her into the position of Attention. Then she tore into Chillemi about abandoning her post. When Chillemi tried to respectfully argue the point, Cassidy reminded her that, as a peer instructor, whoever was in her vehicle was her responsibility and if she couldn't handle the situation, she should not have accepted the assignment. Chillemi steamed and continued to point at Shannon and affix all the blame on her. Cassidy then put Chillemi down for twenty-five push-ups for breaking the position of Attention while the jeep continued to idle. Finally Cassidy told Chillemi to recover and to get the jeep out of the ditch and back on course. This was a task only the most experienced of drivers could do and Cassidy knew it. Thirty minutes later, as Chillemi triumphantly maneuvered the two front tires to the top of the ditch, the jeep ran out of gas.

Both Dale and Shannon reached the parallel parking phase of Station Four at the same time and Dale noticed there was no one waiting at Station One. She shouted that observation to Shannon as their jeeps passed. They both flew through the steps of Station Four, got their respective Go's and hurried back to Station One to be peer instructors, each receiving their own jeeps. As was common practice with the two friends, they did not have to confer on the idea they both got at the same time. They just gave one another a look and drove their jeeps to the deserted figure eight course in the back and began to drag race.

The sight of plumes of dust and dirt and the sound of grinding and groaning gears caught Cassidy's attention. She jogged toward the commotion in disbelief that a trainee would actually have the guts to race on an unauthorized course. If nothing else, she certainly seemed to have gotten involved in a *spirited* cycle.

These two females were clearly having the time of their lives and were oblivious to her presence. She looked back to see that almost everyone else had finished the learning stations and had moved on to the driving obstacle course. She returned her attention to the drag racers and recognized Walker, then Oakes. Once again, she had to bite back a smile.

Cassidy was amused, impressed and pissed off at the same time by the actions of the two trainees. It was a pretty ballsy thing to do and she might have even written them up if she hadn't been caught doing the exact same thing herself when she was a trainee. She knew well the urge that pushed them over the edge. Still…

Oakes, who trailed Walker by less than a couple feet, spotted Cassidy first. It only took Walker another second to notice her, too, since Cassidy stood directly in their path. Cassidy saw Walker mouth. "Oh, shit."

They brought their jeeps to a halt by downshifting instead of braking to stir up as little dust as possible. That action made Cassidy conclude they knew more about driving a standard vehicle than they let on. She certainly couldn't discipline them for that. Peer instruction never felt like a privilege and if she knew then what she knew now, she wouldn't have zipped through the course, either.

When the jeeps came to a complete halt, Cassidy watched the two trainees frozen in place, not even breathing. Cassidy just stood there and glared at them, her arms folded across her chest. She never said a word to either one of them, she just stared at them and pointed to the direction of the driving obstacle course. They both nodded.

"Thank you, Drill Sergeant," Oakes said.

Walker responded as well. "Yes, Drill Sergeant."

She spun and followed them, shaking her head in mild amusement the rest of the way. She was going to have to keep her eye on those two.

Dale didn't dare to breathe at the sight of Drill Sergeant Cassidy. Cassidy just stood there, silently inviting them to make a move and, of course, they would have been idiots to take her up on it. Shannon had already pushed her luck once with the new cadre member and had been given a reprieve. To think there was a second one in there was too much to hope for. When she let them go without a word, both Dale and Shannon looked upward and wondered how they could have possibly earned such a favor.

Cassidy's denial of discipline didn't mean she was weak or intimidated, on the contrary, her expression and bearing told them that they were, indeed, lucky she elected to send them on their way. Dale had seen how Cassidy handled Chillemi so there was no doubt that the new drill sergeant had the grit needed to deal with the trainees.

No, it wasn't that. It was her posture that oozed with attitude, that *look* that caused Dale's heart to jump to the vicinity of her throat. Cassidy's dark eyes blazed with an intelligence and intensity she had only previously seen in Anne Bishaye. To say she neglected to see and feel the aura of raw, smoldering sensuality that surrounded Cassidy would have been a lie.

No, Shannon, she's not one of the most attractive drill sergeants I've ever seen, she is one of the most attractive women I've ever seen, period.

Dale knew she had to pull the reins back. Cassidy was going to be a pleasure to look at and an alluring diversion to the monotony of this case but it had to be hands off, even if Cassidy did prefer women, which Dale doubted. The last thing Dale needed in her life right now was another complication that involved the military.

Just concentrate on the case and get the fuck away from Uncle Sam.

Sure, like she was going to be able to concentrate on much of anything with Cassidy around.

Dale and Shannon felt sorry for the trained instructors at the driving obstacle course. Dale feared only a lobotomy could have reduced the stress of that job.

It looked so easy when they maneuvered around the pylons on *Charlie's Angels*, a majority of the trainees voiced later. It turned out to be not as easy viewing it from behind the wheel.

There were two stations in the class, the second being the more difficult of the two. The first station had the pylons set up in a circle that became smaller as it wound around in a spiral. At the end of that track, the driver was required to stop the jeep by a tree, where he or she would get a score and either move on to the second station or have to repeat the test.

The second station had the pylons set up in half-circles and at intervals where the driver would maneuver the vehicle in and out and between the orange cones. The objective was to leave all the pylons standing. If more than three cones were knocked down during the test, the driver was required to do it again, three tries being the limit.

Dale knew the secret of this little game was to forget how Mario Andretti would do it and not to build up any speed. The second someone got cocky and decided to accelerate was the second they would fail.

There were always trainees who, like Vanessa McKnight, panicked and plowed down every single pylon in the circle. McKnight hit a few at such an angle that the little plastic cones turned into missiles and beaned two instructors who stood off to the side and then she ran the jeep into the tree at the end of the track. McKnight did no damage to herself but put a substantial dent in the fender of the jeep. She was assigned another jeep and instructed to try the station again. Although she missed the tree on her second attempt, she still left no cone untouched. The instructors gave her a third and final chance and even after patiently being talked through the track, McKnight still failed to leave most of the cones upright.

She was the first female to receive a No Go. She was driven back to the company area to await instruction on reclassification. It was a sad moment for McKnight and a disappointment for the females, as a whole. After the loss of Kirk to death, Barbara Kramer to AWOL and Zelman to discharge, the Alpha women were hoping the rest of them could make it through to the end of LE School, a goal Dale knew was unrealistic.

McKnight joined eight males who had also bolo'd out. Reclassification was a long process and the people who didn't make it through MP training would most likely be still hanging around when most of the others had graduated and been sent to their first permanent duty stations.

Later, after Chillemi confronted Shannon in the barracks and the two had it out regarding the former's attitude, they reached an agreement to stay out of one another's way. Shannon and Dale stopped by McKnight's bunk and offered their condolences. None of the females really cared for McKnight but that didn't mean they wished her to fail.

McKnight tried to shrug it off in the pretense of not caring. She told Dale and Shannon she was glad it happened because now she could hopefully get an MOS that was a little more nine-to-five. When Dale got up in the middle of the night to use the latrine, she heard crying from the direction of the showers. She knew McKnight had fireguard duty but she was nowhere in sight. If it had been anyone else, Dale might have gone in and tried to comfort her, someone else who may have been more suited to continue the training and become an asset to the job. She decided to leave McKnight alone. Dale had told her earlier that she was sorry, and she meant it, but she didn't think the elimination of McKnight as an MP was unfair. McKnight was given every chance to succeed, more actually than she deserved, and her not making it was her own responsibility. Dale would've had more sympathy for someone else but not McKnight.

The next morning after formation, PT, chow and details, the trainees were marched to the vehicle driving range and divided into two groups. While one group took their road test in the jeeps, the other half took their road test in the sedans. Those who finished the test with the jeep would then cycle around to take their test in the sedan and vice versa. Everybody got a Go and qualified for their military driver's license, a process that took most of the morning.

The company was marched back to Tenth Battalion for noon chow and then remained in the company area the rest of the afternoon. Dale, Shannon and Wachsman were sent to the Supply Room for details but when they got there, there was nothing for them to do. No one had advised the GIs assigned to the Supply Room that trainees were being sent to do their cleaning for them so after hearing rumors of an inspection, they had already squared the room away that morning.

Dale, Shannon and Wachsman now looked at a couple of boring hours ahead of them. PFC Singleterry, a GI who was temporarily assigned to run the Supply Room since Ingersol's impending discharge, told them to just hang out and if they needed anything, he'd be in the front room with the M16s.

PERMISSION TO RECOVER

After he was gone, the three women looked at one another blankly. "Anyone for poker?" Shannon said, finally.

The two others readily agreed and Shannon rummaged around what used to be Ingersol's desk, not surprised when she found a deck of cards. By this time, pretty much everyone in the company knew Shannon loved poker and that she was good at it, something Dale also knew from past experience. She'd almost lost her life savings to her partner the very first time they'd played.

Two and a half hours later, Wachsman barely held her own and Dale was just twenty dollars in the hole. The game was really one-sided but it was passing the time and even though Shannon was thirty dollars up, they were all having fun.

Shannon was bluffing her way to winning another hand when the cry of Attention came from a fearful sounding Singleterry in the other room. Cards flew everywhere, chairs were thrown back and the women were on their feet.

"Well, well, well, what have we here?"

Even though they didn't show it, Dale and Shannon were relieved to hear Henning's twang. Dale hoped Henning was alone.

"Do I see money and cards on this table? Whose money is this?"

"Not mine, Ma'am," three voices chorused.

"Does anyone know the regulations about gambling on post?" She stepped up, nose to chin with Dale. "Private Oakes?"

"No, Ma'am."

"I didn't think so or you wouldn't be this stupid. At least not openly," Henning added, rather dryly. She gathered up the money left lying on the table, one hundred fifty dollars, all totaled. "Looks like you were winning, Private Walker." Shannon said nothing as the training officer paced in front of them. She stopped in front of Dale again. "Were you gambling, Private Oakes?"

"No, Ma'am," Dale said. She looked straight ahead.

Her response should have angered Henning but the training officer appeared to take it in stride. She stepped in front of Shannon. "Were you gambling, Private Walker?"

"No, Ma'am," Shannon said.

"Hmmm." Henning then stepped in front of the rookie, certain the petrified private would be the undoing of the dynamic duo. "Were you gambling, Private Wachsman?"

Without hesitation, Wachsman looked Henning squarely in the eye. "With whom, Ma'am?"

Both Dale and Shannon, unable to contain themselves, revealed tiny smiles of triumph that didn't go unnoticed by Henning. The training officer smiled, too, as she folded up the money and stuck it into her right breast pocket. "If no one was gambling then I guess there was never any money on the table, was there?"

"No, Ma'am," the three women answered, not quite as enthusiastically.

"I didn't think so." She circled them once more, slowly, then walked toward the door. "As you were."

"Yes, Ma'am. Thank you, Ma'am." Dale spoke for the group as Henning left the Supply Room.

Singleterry bounded around the corner. "I'm sorry. She was just suddenly here and I didn't have time to warn you. Of course, I had no idea you were playing poker, either. How come you didn't invite me to play?"

"You're lucky we didn't," Wachsman said. "She confiscated all the money. Fifty bucks apiece. That was my spending money for the weekend."

"Listen, it's better than an Article 15," Singleterry said. "At least this way it's done and nobody's the wiser. If she'd done it by the book, you three would be in some really deep shit."

"I wonder why she didn't rake us over the coals," Wachsman asked, curiously.

Dale and Shannon just looked at one another.

CHAPTER FORTY-THREE

The women of Alpha-10 officially met Drill Sergeant J.J. Cassidy the next day in the middle of a GI party.

The floors had been completely scraped of old wax with razor blades, tile by tile on hands and knees when she walked in. At Ease was commanded by Tramonte, who was closest to the door, and they stayed in that position until Cassidy was finished walking up and down both aisles.

"Carry on. Everybody come to the left side of the bay," she said, finally. When they all gathered, she told them to sit on the floor so she could see them better.

"I apologize for taking so long to introduce myself to you. I wanted to get used to the way training was being done in Tenth Battalion first. The entire post is talking about OSUT and especially about the, so far, successful progress of the women.

"So, if you've heard anything to the contrary, let me reassure you, this is a pretty remarkable battalion and you are the pioneers." This observation from Cassidy made the women smile. It made Cassidy smile in return, a dazzling action that caused Dale to secretly lose her breath momentarily. It should have been against the law for a drill sergeant to be that beautiful. Dale wondered how long Cassidy's hair really was when it wasn't up in a bun under her Australian Bush Hat and if it was really as black and silky as it looked. Cassidy had dark eyelashes that were so naturally long, they were a fashion model's envy and eyes so dark brown, they were almost ebony. She also had dimples when she smiled that just added to her physical appeal.

"You've already experienced that there is no special treatment for you because you are women," Cassidy continued. "I am an advocate of that. I have one standard of training a soldier and it does not get altered. If you want to survive in a man's Army, you need to learn to concentrate on job performance and credibility. You will go nowhere if you put your gender first. If I can instill the thought into your brain that you are a soldier first and a woman second, I'll have accomplished my personal mission. Take a little advice from someone who has been there.

"A majority of military men still hate the fact that you are here. They don't think you're up to it, especially not law enforcement. They believe you'll drag them down, even though you're out there every day, side by side with them, doing the exact same things. They don't care. And they will burn you just as soon as look at you just to eliminate you. It doesn't matter that some of you are more instinctively talented and can run and shoot circles around them, you are a female and that is automatically two strikes against you.

"We are, unfortunately, victims of stereotyping. The men think that we operate as a whole. It's too difficult to believe that we are individuals to be taken on our own merit. In their eyes, we are only as good as the last female who screwed up and we are stupid. According to them, the military is no place for us because we don't have the ability to adapt to the environment. That we are temperamental, incompetent, emotional, immature, self-centered cry-babies and we can't handle stress and responsibility. How many of us here think that's the ol' pot calling the kettle black?" A majority of hands rose. "That couldn't be further from the truth, could it, ladies?" she asked them.

"No, Drill Sergeant," they responded in unison, clearly mesmerized by her allure and her frankness.

"I don't know Sergeant MacArthur that well. I don't have any idea what training here was like with her as your drill sergeant or as your liaison. How she ran things up here is of no concern to me. As of right now, this day, this very second, your conduct, your progress, your reputation all reflects on me. If you think this has been fun and games so far, you can knock that thought right out of your head."

She looked at Shannon when she said that. "If anyone does anything to piss me off, you're going to wish you could go face to face with the devil as opposed to dealing with me."

She then looked at Chillemi, who immediately found

interest in the floor. "Don't cross me and don't lie to me and you will find that I am very easy to get along with. Any questions?"

Minty put her hand up.

"What's your question, Private Minty?"

"Why are we still being treated like recruits, Drill Sergeant? We graduated basic training."

"Private Minty, I know you understand that the program you are all involved in is experimental. That, mixed with the fact that training you to be a cop is no trip to the candy store, means you need to be more disciplined and in better condition than the normal trainee. You need to be driven and you need to earn the privilege of liberty. Besides, you're getting weekend passes and soon you'll be set free during week evenings from 1700 to 2130 to do whatever you like. That's as much as any other AIT student gets. You really don't need more freedom than that. All you'll do with it is get drunk and be wild and promiscuous."

"Yeah!" That came from someone in the back.

Cassidy smiled in spite of herself. "Sex. That's all trainees think about. It's a bad habit while in training, ladies. All that energy you use up needs to be concentrated on much more serious things right now. Besides," she added. "Did you know scientists have proven that people who think about sex too often lose one or two of their senses?"

"What'd she say?" Shannon said to Dale.

"What'd *who* say?" Dale said and looked around the room, squinting.

After the chuckling died down, Cassidy stared at them both, amused. "Ah, yes," she said, finally. "You two. A.J. Foyt and Al Unser. You owe me. How about coming up here and knocking out twenty apiece for me." Her tone indicated there would be no debate.

Yep, real easy to get along with, Dale snickered to herself as she stood up and moved to the front.

"Oh, and Private McTague? Why don't you come up here and join them for yelling out 'yeah' a minute ago."

While the three women started their discipline, Cassidy strolled around their bodies and continued to speak. "Ladies, despite the opposition, the military is a good opportunity for you right now. With the right determination, you can be limitless.

Look at General Mary Clarke. She's been in the Army since nineteen forty-five, where she started out, enlisted, as a finance clerk. Today she's a major general who works out of the Pentagon. Your own Battalion Commander is an excellent role model." Cassidy looked down at Dale, Shannon and McTague who had finished their push-ups, and were in the front leaning rest position, waiting to ask for permission to recover. Cassidy ignored them.

"Last year, General William Westmoreland was quoted as saying that he didn't believe women could carry a backpack, live in a foxhole or go a week without taking a bath," Cassidy continued. "We're slowly but surely proving him wrong. I didn't enlist to be a homecoming queen and if any of you joined up just to find a husband...or a wife...or to be Miss Popularity, then do me a favor and get out. If you don't, you'll just make it that much harder on the women who come in after you. But...I've been observing you for a while and it looks like we have a pretty sincere group here." She looked at the three women, still waiting, their arms shaking as though they were experiencing an earthquake. "Recover."

Dale, Shannon and McTague moved to a standing position and Cassidy pointed at them, then toward the group, indicating she wanted them to sit back down.

"Now, we are all quite aware we cannot match the men in upper body strength, that we are physically weaker, that we weigh less and are smaller. But research also proves that we are better educated, instinctively smarter and we score higher on aptitude tests. We usually remain on active duty longer and we lose less time than the males due to drug or alcohol abuse, disciplinary problems, and AWOL. The only thing they won't let us do yet is combat but I firmly believe that will change in the future for those who want it to."

"Drill Sergeant, if we're not allowed in combat or even a combat-related MOS, why are we expected to take basic *combat* training?" Ryder asked.

"Essentially to give you a better understanding of your jobs in non-combat roles. You will be much more effective soldiers having familiarized and qualified with combat weapons, for having low-crawled in the mud, survived live-fire courses and for having slithered under barbed and razor wire. Especially as military police officers, you may be assigned to a unit that specializes in the processing and confinement of POWs, which usually includes rear-area protection."

"What does that mean, Drill Sergeant?" Sherlock asked.

"It means you're not in the front lines of a battlefield, you're in the rear area of it." She looked at her watch. "I'm going to let you get back to your GI party now. I want you to know I have an open-door policy. Anytime you need to talk, you can come to me." As Cassidy was leaving, Michaelson came running in. Michaelson had just been relieved of CQ and looked as though she desperately had to go to the bathroom. At the sight of Staff Sergeant stripes, Michaelson stopped and stood at the position of Parade Rest. Cassidy also stopped and looked at the private who stared straight ahead. "Private Michaelson?"

"Yes, Drill Sergeant?"

"Drop and give me twenty."

Michaelson looked at Cassidy, puzzled. "Drill Sergeant?"

Cassidy stepped closer to Michaelson and put her face approximately an inch away from the pretty blonde's. "Do it," she told her quietly but firmly.

"Yes, Drill Sergeant." Michaelson dropped.

The other women regarded the dropping of Michaelson, of all people, as curious. Michaelson pushed up and counted off as Cassidy watched. When Michaelson reached the count of fourteen, Cassidy told her to recover. Before Michaelson could even get to her feet, Cassidy was out the door and gone.

"Why the hell did she drop me?" Michaelson wondered out loud, staring at the door.

"Because she can?" Dale responded, also wondering what drove the gorgeous drill sergeant.

Another Friday had rolled around and the company was itching for the final Fall Out call at 1700 hours.

The members of Alpha-10 were given their graduation pictures that morning. None of them turned out well. A majority of the women immediately hid them away in their lockers, in hopes that the photographs would self-destruct over the weekend.

That afternoon they were surprised by an unscheduled PT test. Cassidy kept a strict, watchful eye on the females to make sure they didn't make it easy on themselves. At one point, when Shannon attempted push-ups, she didn't get her chest close enough to the ground for Cassidy's liking so the drill sergeant rested her boot on Shannon's back, which made it impossible for Shannon to push up at all. Shannon had already been counted as completing six repetitions.

"You've done nothing," Cassidy told Shannon. She then looked at Mark Morse, the GI who scored Shannon. "Start her at one."

Cassidy proceeded to explain what Shannon did wrong. She took her foot off Shannon's back, got down on the ground with her to demonstrate and then stood back up. "Now, you try it." Shannon obeyed. "Again," Cassidy commanded. "Better." Cassidy turned to Morse. "Start scoring her now. If I see you cheating for her, I'll remove from your score the amount of times she incorrectly completes a repetition."

"Yes, Drill Sergeant," Morse answered her.

When Cassidy moved on, Morse was in the process of making a very un-gentlemanly gesture with his finger.

"I'd think twice about doing that if I were you, Private Morse," she said, without looking at him. "It's supposed to be a hell of a nice weekend. You don't want to be stuck in the barracks during it."

"Yes, Drill Sergeant," he responded, meekly.

If it wasn't for a certain sparkle in her eyes that Shannon recognized and found so familiar, she would have somehow ensured that Cassidy would not have enjoyed her upcoming weekend, either. Shannon discerned that Cassidy hadn't singled her out though because Cassidy was right. Shannon did a horrible, below-regulation push-up and, as much as she hated to admit it, Cassidy was only doing her job. Five minutes later, Cassidy did the same thing to Caffrey and Segore, who were messing up their squat thrusts. Shannon figured that once Cassidy got through establishing her authority, she'd relax and hopefully be more pleasant.

"I'm seeing Matt again," Shannon told Dale, reluctantly.

She looked as though she expected a reaction and Dale didn't disappoint her.

"Tonight?" Dale's eyes narrowed before Shannon could answer.

"Yeah. We arranged it last week. He's picking me up in thirty minutes." Shannon finished putting on her make-up and closed her locker.

"Shannon, this isn't fair," Dale sing-songed, annoyed. "I'm not going to go through this with you again."

"Nor am I with you." Shannon looked directly at her partner. "I like him, Dale." She lowered her voice, almost to a whisper. "I can see him and still do my job. You and I are both inventive and capable enough to work around this."

"Would you feel the same way if it were me who was occupied by an outside interest?" Dale asked. She also kept her voice level down. She didn't want to give the impression they were having a jealous lover's quarrel. Dale drew a deep breath and then exhaled it. "All right, let's flip a coin."

"Another coin? I don't think so."

"Fair is fair. The least you could've done is told me last week so I could've made other plans."

"I knew you'd be like this."

"I don't think I'm being unreasonable," Dale said.

"Okay," Shannon said, testily. "Flip the goddamn coin!"

"Heads, you go with Matt," Dale said and took the quarter out of her pocket. "Tails, you stay with me."

"And make your entire weekend a living hell. Just keep that in mind," Shannon said, solemnly, as Dale flipped the coin.

"Heads," Dale announced, disappointed. "That's it. I want a divorce."

Shannon smiled, triumphantly. "I planned on compromising, anyway. I'll bring him to the EC with me. I can bring a guest, right? Then we'll just follow the crowd wherever they happen to go. I'll just be with him instead of you."

"Great," Dale said, tepidly. "What happens if we see or hear something?"

"Don't make it more difficult than it is. We can't do anything without Bishaye's approval, anyway, so we'd have to sit on it until we got hold of her or Stubby."

After Shannon had left with Matt, Dale took off with Travis, Tramonte and Lehr. They headed into Averill to stay at the Journey Inn. They got a room between McTague, Halliday, as well as Charlene Keival and her date, a trainee from another company. They left all the room doors open and a small party became uncontrollable within the hour with what felt like half of Fort McCullough showing up.

Other than the Averill police responding to the motel twice to quiet them down, nothing out of the ordinary happened. At approximately four in the morning, Dale finally fell asleep on one of the beds next to God only knew who.

The knock on the door around seven thirty came much too early for anyone. When no one bothered to get up to answer it, the door opened and Shannon and Tierni peeked in.

"Jesus...who are all these people?" Shannon asked Tierni, who shrugged. Shannon stepped between and over several bodies before she even made it halfway across the room. Tierni maneuvered her way into the bathroom as Shannon recognized Dale's clothes. She made her way through the remaining sea of GIs and prodded Dale awake.

"Go away," the raspy voice told her.

Shannon lifted the pillow off the person's head just to make sure it was, indeed, Dale. "Get up. Come on, I've got something to tell you."

Dale opened a crusted eye and looked up at her. "What? Did you elope?"

Shannon chuckled. "File that under a big, fat no. Let's go get some coffee. I think you need it."

It took Dale a few minutes to swim into consciousness. "Okay." She sat up and looked around at the five people, other than her, squeezed every which way on the bed and then saw all the people passed out on the floor. "Who are these people?"

"You tell me," Shannon said. She carefully navigated her way back to the door. They both heard the shower start.

"Who's in the bathroom?"

"Tierni," Shannon told her.

Dale nodded and followed Shannon outside to fresh air. They walked toward the coffee shop. "Got any gum or mints or anything? I've got a rotten taste in my mouth."

"Why? What have you been doing?" Shannon rummaged around in her jacket pocket.

Dale glared at her. "Trust me, there was no protein in *my* diet last night."

Shannon laughed. "Despite that, I don't have a bad taste in *my* mouth."

"It's just plain morning mouth, okay? Complicated by what now feels like a half-keg of beer that I must've drank all by myself. I hope a bathroom is close."

Shannon handed Dale a stick of gum.

"I know what it looks like in there but if an orgy happened, it happened without my knowledge or, hopefully, my participation. The word must've spread that there was a party and people just kept showing up. I crashed sometime before five, I think."

"Five what?"

"O'clock, smart ass."

"I'm surprised no one called the police."

"They did. Twice."

"No arrests?"

"Are you kidding? One of the cops came back after he got off duty. He took a real liking to Travis."

"How did she feel about that?" Shannon asked.

"She completely ignored him. She's beginning to worry me a little bit, Shan. She's not her normal, sarcastic, razor-sharp, witty self. She really misses her husband. She was telling me last night that, according to her paycheck, the Department of the Army hasn't recognized him as her husband or as a dependent."

"They will. You know how paperwork is around here."

"*I* know that but she's getting really discouraged." Dale suddenly looked around. "Where's Matt?"

"I dumped him," Shannon said, simply.

"For the day?"

"No. For good."

Dale stopped. "You did? How come?"

Shannon pulled on Dale's sleeve to get her moving again. "He started talking to me and treating me like property in front of people he knew. Last night, at the hotel, he told me that if I was going to be his girlfriend while I was here, I would have to start watching my language and my behavior."

"What? You're kidding!"

"I wish. So we argued about that a while and went to bed, where he then started in on me about my obvious experience in the sack, insinuating I was a slut."

"But...you are a slut." Dale saw the shocked look on Shannon's face. "So am I...by our parents' standards, anyway."

"By our parents' standards, we would marry the first man we ever slept with and be trapped in a miserable, dead-end marriage with men we didn't love. It's okay that men satisfy their sexual whims but I guess women aren't supposed to. I mean, I get just as horny as any man does and I have every right to scratch that itch just like they do."

"Shan," Dale stopped her. "You're preaching to the choir."

"Right. Sorry. He really pissed me off. So I gave him his half of what he paid for the room and told him to hit the road."

"And he just went?"

"I don't think he wanted to get me any more riled. He said he'd never seen a woman get as mad as I got."

Dale smiled. "Poor, sheltered boy."

"So here I am, at your disposal."

"Well, good." Dale put her arm around Shannon's shoulder. "And just for the record, you can talk and act any way you want around me and I promise I'll never call you a slut."

"You already did."

" In bed."

"Oh. Well, I never had to worry about that in the first place. Anyway, I have something to tell you."

"That wasn't it?"

"No. Last night, we were driving near Mobile and we stopped at a liquor store to get some wine for the room. I sat in the car like a dutiful, subservient girlfriend and guess who I saw sitting together in the window of the cozy, romantic Italian restaurant next to the liquor store?"

"Who?" Dale got excited and skipped ahead of Shannon then danced in front of her. "Bishaye and Colton? Henning and Ritchie? A trainee and a drill sergeant? Joe and Frank Hardy? Rocky and Bullwinkle?"

Shannon glared at her. "Rocky and Bullwinkle?"

"You told me to guess, I was guessing."

"No. Robbins and Cassidy."

"Robbins and…Cassidy," Dale repeated, deflating.

Shannon noted but didn't mention Dale's sudden dismay "Yes. It probably has nothing to do with the case but it surprised me. It's good information to have possession of, don't you think?"

"Sure, but I agree the information is probably useless to our case. That might explain her behavior toward Michaelson, though."

"How?"

"Everyone knows Robbins would nail Michaelson in a heartbeat if she gave him any indication she wanted to be nailed. I'm sure Cassidy knows it, too. She was probably establishing her territory."

"What territory? Robbins is married. And unless Michaelson knew something we didn't until last night, what would be the purpose of it? Michaelson wouldn't understand. Michaelson *didn't* understand."

"Maybe it's something Cassidy had to do for herself."

"I think they're kind of an odd couple. Cassidy and Holmquist would've made a cuter pair," Dale said, subdued.

An odd expression was on Dale's face but Shannon didn't question it. They arrived at the coffee shop. Dale opened the door for both of them. "Okay, Yenta, let's eat. I'm starving."

"Are you sure you didn't do anything last night?"

Dale drank copious amounts of coffee but didn't do much damage to the breakfast she ordered. She wasn't sure why the idea of Cassidy with Robbins bothered her so much. First, it didn't seem to make sense. Cassidy was a beautiful woman, she should have been able to find a man who wasn't married and second, she couldn't decipher why she was so disturbed that Cassidy was with anyone. It's not like there was ever going to be any promise for her, regardless of what Cassidy's circumstances or orientation were. Dale tried to put it out of her mind.

After breakfast and helping to clear everyone out of the hotel rooms, Dale and Shannon split up again until the next day.

Dale stayed in town again, participating in another multi-room party, this time on the first floor. That started early in the afternoon and raged on until the wee hours of the morning. This made Dale's job easy because three-quarters of Alpha-10 were there. The regular couples paired off and nothing happened to raise Dale's suspicions.

Shannon hit all of the on-post hot spots and found everything running normally, smoothly. At one point, when she went back to the barracks to get a heavier sweater out of her locker to help fight off the sudden dampness of the evening, she talked Michaelson into accompanying her to the Pizza Place for a while. Michaelson drank only soft drinks and split a portion of a pizza with Shannon and Segore. When they were done, Michaelson and Segore walked back to the barracks and were in their respective bunks by ten o'clock.

Shannon had never met anyone as dedicated as Michaelson. The attractive blonde trainee had the opportunity to be the busiest, most popular woman on post and she shunned it all in favor of her studying and her fitness. She was never rude or unfriendly to anyone, she just didn't ever seem to break down and become quite human. Half the female population of A-10 quietly spoke of being relieved she had removed herself from the competition but it made Shannon curious.

Michaelson had done nothing to draw attention to herself, other than being in a different frame of mind from the rest, nor did she fall into any of the other patterns of preceding women who had pressed charges or had been victims, whatever the previous situations turned out to be. Shannon decided that Michaelson's behavior was nothing to worry about. Michaelson probably wanted to be the best soldier possible and Shannon figured she was channeling all her energy into training. Some people just weren't the partying type. Shannon chuckled to herself. She just couldn't understand those people.

Shannon spent the night in the barracks and Dale, with most of the others who shared the rooms, came back to the post early. Quite a few passed out on their bunks but a small group still felt the need to squeeze every ounce of life into their time, so they went to the bowling alley. Dale hated bowling but she was not ready to take a nap, either.

Once at the alley and after they had secured a lane, Dale decided to watch the group as opposed to participating. They had to share a seating area with another group that, in the beginning, no one seemed to object to. Dale noticed that one male member was intently interested in her group's conversation. When he heard one of the males mention something about training, it was as though an alarm went off that cued him to behave like an obnoxious jerk.

On her way back from picking up her spare, Lehr made the mistake of sitting down on *their* side. The interested young man stepped in front of her and leaned his face into hers. "What do you think you're doing, Private?" his voice boomed at her. His friends looked on, amused.

Lehr looked up at him, confused. "I'm sitting."

"You mean, I'm sitting, Sergeant," he corrected.

Lehr hesitated. She wasn't sure of the protocol when both parties were out of uniform, so she continued to stare at him.

"What are you? Stupid? You're in my seat!" he told her.

Lehr looked at the empty chairs next to her. "There are three empty chairs right there."

"I said," he emphasized, raising his voice. "You are in *my* seat!"

Dale rolled her eyes. She was tired and not in the mood for the rank game garbage. She was about three days away from her period and there was rising irritability in her system so the timing of this man's folly was unfortunate.

Lehr started to get up but Dale was next to her before she could actually stand. "Stay put," Dale told her.

She looked at the man who claimed to be a sergeant. "Let me see some ID."

He blinked at Dale, clearly surprised by her boldness. "I don't have to show you shit."

"I didn't ask to see shit, I asked to see your military ID. Before any of us do anything for you, we just want to make sure you're not some asshole bully private fucking around with us." Dale's voice or stance didn't waver.

He pulled his identification out of his wallet and shoved it in her face. The rank said Specialist 5th Class, pay grade status of E5. He just as quickly put it away. "Now, let me see yours," he spat out.

Dale retrieved her ID and let him take a good long look at it. "In case you can't read, I'll pronounce it and spell it for you," she said, unruffled. She was in one of those moods where she wished she could whip out her real ID and make him do what he was trying to make them do.

I outrank you, you pompous shit, Dale thought. *You have no idea who you're fucking with and I wish I could tell you how dangerous it is to assume.*

At the sound of sneers from his companions, the man got into Dale's face. "You're in serious trouble, Private Oakes. I'm going to have to report you to your CO."

"For what?" Dale almost laughed at him.

"Insubordination. In the meantime, all of you get down and knock me out some push-ups," he pointed to Dale's group.

"Fuck you," Dale said, wearily. "We're in civvies and so are you. This is not a training environment and none of us are doing squat for you, got it?"

The Spec5 was now beet red. "You are in so much trouble, young lady. I want your service number and I want it right now!" Someone handed him a pen.

"Do you have a Privacy Act Statement on you?" Dale asked.

"No," he said. Obviously, he had not expected a challenge and he began to sound not as confident.

"Then I don't think so, Specialist, but I'll make it *real* easy on you." Dale looked at Lehr. "Go call the company and see if you can get one of our drill sergeants down here. I'm sure he or she will be more than willing to straighten this out."

Lehr left for the phone and the young specialist blanched just as Dale figured he would. Chances were that no one had ever called his bluff before. They stared each other down and Dale didn't budge. He finally spoke to her. "Well, Private Oakes, you've lucked out. I remember how it was when I was a trainee and I wouldn't want to have to deal with the kind of punishment you're going to face by getting your drill sergeant all the way down here to find out that you and your friends have a disciplinary problem. So we're just going to drop it. How about that?"

Dale's eyes tightened. "How about not? I think you're a bully and a coward and out of line and I think I'll wait here to see if my drill sergeant agrees with me."

Lehr returned. "It's Cassidy. She's on her way down."

Dale did not change her determined expression but, inwardly, was not reassured. If Holmquist had been on duty, his loyalty to them would have backed Dale up, especially if, as Shannon suspected, he was interested in her. Cassidy, on the other hand, was still a mystery and her allegiance could go either way.

Anger and embarrassment burned in the specialist's face. He turned to his friends. "Come on, let's get out of here. Now."

"We just started our second game, Gabe," one of his friends protested.

"I said we're out of here. Now, let's move," he snapped to his friends. Within a minute they were packed up and gone.

"All right, Oakes!" Lehr, Mackey, Snead, and another GI chorused. Two other males in their party, Wolfe and Stillwell, weren't as impressed.

"You're nuts, Oakes. What happens if he reports us?" Wolfe asked.

"Do you really think if that asshole was in the right that he would have taken off so damned fast? He was exercising authority that he really didn't have, showing off to his friends. If you wanted to play along with him and look like an idiot, then you should have spoken up and said so." Dale also wasn't in the mood for skeptical hindsight, either.

"Hey!" Wolfe, insulted, took a threatening step toward her. "Unlike them, I'm not blown away by your big mouth!"

"Then you should have kissed his ass and got down and knocked out the push-ups," Dale countered.

"What's going on here?"

They all turned to see Cassidy standing there. Her eyes looked like they could have blazed a hole into Dale and she did not look pleased to be there. Everyone but Dale surrounded Cassidy and started talking at once. Dale sat in a chair and watched her fellow trainees.

Cassidy calmed them all down and listened to their stories

one by one. When she had all the facts, confirmed by what the gentleman behind the shoe counter told her, she looked at Dale. "All right. Everybody go back to your game. Private Oakes? Outside."

As Dale stood up to follow Cassidy outside, she heard Stillwell say to Wolfe. "I knew she came on too strong."

Outside, Cassidy stopped and turned to Dale. "What happened in there, Private Oakes?"

Dale looked at the ground. It was difficult to look at Cassidy and not picture her in bed, having sex with Drill Sergeant Robbins. Then, of course, that visual segued into just Cassidy in bed. It was unnerving. "What happened was everything you heard in there, Drill Sergeant. The guy was trying to ruin our afternoon by being obnoxious. I asked to see his ID because, for all we knew, he could have been a slick-sleeve screwing around with us. We were just minding our own business, Drill Sergeant."

Cassidy folded her arms. Her tone was reasonable, not upset. "This is the Army, Private Oakes. There is no such thing as your own business to mind anymore. I'd prefer you look at me when I talk to you."

Dale's eyes met hers.

"I'm not going to tell you what you did was bad. You had every right to ask to see his ID and you were also correct to request one of us from the company. However...he was still an NCO and you shouldn't have said fuck you to him." Cassidy smirked. "Whether he deserved it or not."

Dale nodded. "Yes, Drill Sergeant."

"Do you remember this young man's name?"

"Chase, Drill Sergeant. Specialist Fifth Class Gabriel Thomas Chase."

"I will track him down and we will have a chat. In the meantime, I'm going to highly suggest you watch your temper. This is not the streets, you can't just say whatever you want to whomever you want anymore."

"I understand that, Drill Sergeant." Dale was bewildered by the current of excitement she still felt around this woman that seemed...mutual. It had to be her imagination. Wishful thinking wouldn't make it so. It had to be the rush of hormones.

PERMISSION TO RECOVER

"I hope so. From everything I've heard, you're a good soldier. Don't get caught up on a technicality you can control."

"Yes, Drill Sergeant. Thank you."

"You're welcome."

She held the door open for Dale, who entered the bowling alley to join her friends. Cassidy turned and walked back toward Tenth Battalion. Dale was suddenly depressed. She wasn't sure if she'd made points with Cassidy or disappointed her and then was more pissed off at herself because she wasted the energy caring.

CHAPTER FORTY-FOUR

At 1930 hours that evening, the company was called into a formation on the south patio and advised that they would start a week of combat orientation the next morning. All platoon drill sergeants were present as the trainees packed up their duffel bags and prepared to move out to the Bivouac site tomorrow at 0700. The unpredictable and unusually cold weather was only a part of the reason the company faced the week with dread. Simulated wargames mixed with outdoor training for the next one hundred eight hours was not anyone's idea of a swell time and Dale knew that included the drill sergeants.

The company was awakened at 0330 to draw their weapons from the Arms Room and provisions from the Supply Room. After chow, the trainees were loaded up in three deuce-and-a-halfs and, with Lieutenant Henning waving them off, were transported to their Bivouac site deep into the woods of Fort McCullough.

The women were separated from the men and allocated a small area to set up their tiny, two-person tents. Dale was assigned a tent with Mroz and Shannon with Wachsman. Even though the air was bitterly cold and the ground was frozen solid, the women erected their tents to standard regulations without the help from the males or any of the drill sergeants. The exercise of digging the surrounding trenches and the pounding of the stakes kept the women warm, at least for a while.

After the tents had all been raised, the trainees were marched to an area where bleachers were set up. Once the company was seated, they had a class in D-TOC, which stood for Divisional Tactical Operation Center. In a combat situation, it would be one of the MPs' many duties to not only operate from this integral area but to protect it as well.

At lunch, Alpha Company experienced their first taste of c-rations. C-rations were usually issued during field maneuvers or actual combat when fresh foods, or A-rations, and food prepared in mess halls and transported to the fields, or B-rations, were not available. C-rations consisted of individual cans of pre-cooked meat, tuna, if you lucked out, fruit, cookies, crackers, coffee, jam, and sugar. The meal neither satisfied nor filled anyone up. It was going to be a long week where food was concerned.

The afternoon began harmlessly enough with a class on Processing Prisoners of War. The trainees were told and shown how to search prisoners for weapons, military equipment, or documents and to search the vicinity of the capture point for the same. They were told that the POWs must be segregated into groups by nationality, rank, and sex and to maintain control of the prisoners by keeping them silent. The POWs then had to be sped to the rear area for confinement and the prisoners had to be safeguarded against harm or escape.

The field instructors also told the trainees that any personal property or documents taken from a prisoner had to be itemized and described in detail on a receipt, with a copy given to that prisoner. POWs or stragglers, a military person in a combat zone or maneuvers who is away from his/her unit without proper authority, were processed and kept at a Circulation Control Post, or CCP. If one paid attention, it was pretty easy, common-sense information.

It was during this class that the Battalion Commander made a surprise visit to the Bivouac site. The field instructors and the cadre expected her to show up at some point during the weeklong exercise to monitor the company's progress because she usually did but no one thought it would be the first day.

Bishaye, accompanied by her driver and Command Sergeant Major Soledad, who was also from Battalion Headquarters, stood and observed a class on Processing a Prisoner of War. When it was finished and the next group assembled, she stood and discussed the classes with three instructors, Putnam, and McCoy.

Dale never ceased to be amazed by the way men acted around Bishaye. Even the hardcore bunch seemed to melt in her presence. Her looks had a lot to do with it but Dale knew they also respected her mind. Bishaye's reputation as a commander was solid but only because her character as a soldier, dedicated to the military was beyond reproach. Dale watched her laugh at something one of the instructors had said and suddenly she became melancholy. She longed for the time when her relationship with the colonel was less complicated and prickly, when she could easily make Bishaye laugh like that.

As Bishaye continued her conversation, the next class was divided into two groups, those who processed and those who were prisoners. The trainees were told to practice while the field NCOs and the drill sergeants flirted with the brass. The instructors' big mistake was turning their backs on the group.

The five sergeants stood in front of Bishaye and Soledad and outlined the schedule of the next five days as the colonel and the CSM faced the class, listening to the plan. If Bishaye had seen the tiny smile that curled the corner of Dale's upper lip, she might have been more prepared but then, with Dale, she always should have been prepared for *something*.

As the instructors chatted away, Dale helped the POWs to their feet, individually. Silently, she put them all in a straight line as her comrades looked at her, confused. The standing order of twelve soldiers caught Bishaye's attention but when Dale backed away, tightened her M16 firing adapter, aim her rifle, flip the lever to auto and shoot off a magazine of blanks at the prisoners, she clearly had all she could do to keep herself from busting out laughing. Everyone jumped and turned to see who was shooting and why.

"Is this what you're teaching your troops, gentlemen?" Bishaye asked.

"No, Ma'am," one of the field instructors said, embarrassed. It took them but a second to figure out what Dale had done.

McCoy excused himself and stalked to Dale. "Oakes! Come with me!"

"Yes, Drill Sergeant." Dale followed him into the woods.

In a lecture that was conducted about an inch away from Dale's face, one of McCoy's specific inquiries was if Dale had a death wish. He spoke at her in calm intervals of five minutes and then he put her down for twenty-five push-ups between each reprimand. When they emerged from the woods thirty minutes later, Dale was as pale as a ghost and her arms hung limply at her sides. Bishaye and Soledad had since gone and the next class had replaced Dale's. As Dale passed Shannon, she mouthed the words, *I am dying*. She told Shannon later that she never thought she'd recover the use of her arms.

Dale and McCoy caught up to her group just in time to make the beginning of the class on the duties of a perimeter guard.

"A perimeter guard must completely patrol his assigned area and, in the process, avoid setting regular patterns in his patrol. He must know and comply with all general and special orders and make required and frequent radio checks with headquarters. Any defects in barriers must be reported immediately to a supervisor. Barriers include windows, walls, and doors. Once potential breaches of security are reported, immediate action must be taken to prevent all unauthorized entries and exits from the secured area." As the instructor explained what other specific breaches of security were, such as holes in or ditches under the fence line, suspicious persons, washed out areas, poles, ladders, inoperative lights, the group was gassed.

Four people failed to don their masks in time. Dale was one of them and there were two reasons for that. One was her arms weren't functioning properly yet so her speed and coordination were off and the second reason was that the canister of CS had landed right between her feet, not allowing her the nine second time limit before the torturous vapor rose and hit her full in the face.

That action was not an accident. One of the instructors Dale had embarrassed by her POW antics, had searched the different classes until he found her. He then popped a canister of tear gas and rolled it directly at her so that when the gas escaped, she wouldn't have any time to react.

The amount of fumes she took in with one breath was enough to knock down an ox. Dale coughed from the depths of her lungs, gasped for air and teared uncontrollably. Her skin burned as though her pores were in flames and she felt on the verge of throwing up like she'd never thrown up before. The only thing she cared about at that moment was getting away from the gas and regaining control of at least one of her normal, involuntary functions…such as breathing. The three others who didn't get their masks on in time were responding as helplessly but Dale was hit the worst.

When she felt as though she were going to live, she looked up to see the culprit standing above her, screaming at her because she'd left her M16 at the spot where she was gassed. Dale stood upright, the effects still causing tears to stream down her face and her lungs to feel on fire. She was about to tell the instructor that if he was so worried about her M16, she'd bring it back to him and shove it up his ass and that way they could be inseparable but she looked beyond him to see McCoy and Holmquist approaching. McCoy ushered the field instructor away and Holmquist handed Dale her rifle.

"Are you okay, Oakes?" Holmquist asked.

"I am now." Dale still coughed hard, the burn in her throat, intense. "I'm just mad, Drill Sergeant. He went after me on purpose."

"I'm sure I don't need to remind you that almost an hour ago, it was him being humiliated by you."

"And I was disciplined for it, Drill Sergeant." Dale accepted Holmquist's canteen and poured water in her eyes to help flush the residual chemical remnants. She wanted to rub her eyes but knew that would only make it worse and wiping them with her jacket sleeve would just distribute more of the gas that lingered on her clothing.

"Yes, but not by him." Holmquist accepted his now empty canteen back. Before Dale could protest, Holmquist continued. "Remember, Oakes, this is a simulated time of war. In combat, you can't choose when and under what circumstances you get gassed."

"I understand that, Drill Sergeant, but—"

"No buts. Stay on your toes, Oakes. If this sergeant is targeting you, don't let him. Stay ahead of him. You're a smart soldier. Don't let him do it."

Dale was finally able to focus. She studied Holmquist, impressed. He hadn't called her a smart girl, he'd called her a smart soldier, no reference to gender at all. She wondered if he understood how much that meant. She tried to manage a smile. "Yes, Drill Sergeant."

Shannon was right, Holmquist would definitely be her type if she were still into men.

Twenty minutes later that same field instructor attempted to get the best of Dale again. She had taken Holmquist's advice and kept her eye out for him, not that it mattered this time. Seconds after he jumped out of the brush, popped another canister and threw it right at her, the wind shifted and, unprepared without a mask, the brunt of the small, white cloud blew right back in his face. The troops wasted no time expressing their delight. They applauded.

Dale looked at Holmquist and he gave her a subtle, thumbs up sign. He nodded and winked at her before turning away. Maybe Bivouac wasn't going to be half bad.

After evening chow, which was brought out to them, semi-hot, by the consolidated mess hall, the trainees were again divided and transported to different areas to practice their D-TOC training. Dale's group of twenty-four was supposedly guarding a general's tent. Twelve soldiers were assigned to work two hours on while the other twelve slept or rested and then they switched and worked in that pattern until morning, when they were relieved.

Inside the general's tent was enough room for everyone's sleeping bag and a working wood stove to at least cut some of the chill. Unless one stood right next to the stove, however, it was still too damp inside and everyone was worried that the drill sergeants or field instructors were going to pull a surprise attack, using CS, so no one slept.

Shannon's group got to guard the Command Post that, in actuality, was the warm wooden dwelling where the drill sergeants slept and was right next to the warmer wooden accommodations where the latrines were. Anyone who was stationed anywhere outside, spent most of their guard time shivering from the cold. If they had put on any more clothing, they would have been immobile. But the two sets of long johns plus regular underwear beneath the fatigues, wool shirt, heavy male fatigue jacket, wool scarf, cold weather cap, not to mention the two pairs of socks, wool glove liners and the leather glove shells, just were not warm enough. The women vowed that if the next night was as cold as this one, orders be damned, they would all sleep in the five by twelve latrine. Shannon agreed with the sentiment. They wouldn't be much use frozen solid so if they were going to be yelled at and disciplined, they were bound and determined to be thawed out during the process.

The effects of practically no sleep showed in the early morning Physical Security class. The company sat on the bleachers again and tried to concentrate on what was being taught but the majority tried harder to stay awake. Private Swan, who had been the instructors' favorite target during classroom time in basic training, became the first casualty out in the field. He nodded off while seated on the top tier of the bleachers, fell to the side, off the structure and onto the frozen ground, knocking himself unconscious and breaking his shoulder in the process. After the excitement was over, the class continued as though nothing had happened.

The next class consisted of the particulars of working a gate, or as it was technically called, *Security Procedures for Entrance To and Exit From a Controlled Area*. The company was instructed on what was considered proper authority for entering and exiting personnel for movement control. This included personnel on foot or in a vehicle and how to check property, package, and material movement to make sure it was authorized for entry or exit. The trainees were shown how to check the registration log for all vehicles to ensure the vehicle met with all requirements on the document. They were also told to report any and every violation as a breach of security.

That afternoon, while half the company learned about Route Reconnaissance and how to conduct a hasty one, the other half of the company learned how to install and operate a field telephone and to prevent and reduce jamming. The instructors also spoke about how to authenticate transmissions and understand and recognize the duress codes system.

At nightfall, Dale and a large group participated in a long, exhausting blackout convoy, which turned into night number two without sleep. Half the night was spent getting to their destination and the other half was spent getting back. Dale shared a jeep with two of the males who had accompanied her to the Enlisted Club and although it broke some of the monotony, it did not help the dragging hours become shorter.

Shannon was selected to stand guard at a missile site, which was nothing more than a bunch of trash barrels clumped together. It was pitch black where they had assigned her and she finally chose an area between two trees as her best lookout point when she wasn't patrolling the perimeter.

Earlier in the evening, she had assumed correctly that one of the drill sergeants would try to sneak up on her. At the beginning of her fifth perimeter patrol, she heard a noise that sounded like shoes rustling through the grass. She turned around, suddenly alert, and dove toward the two trees and rolled behind one for cover. She focused and saw a figure low crawling toward her about twenty meters away. Shannon, armed with a magazine of blank rounds, flipped her selector level to semi and aimed her M16 at the shadow. She yelled an obligatory, "Halt, who goes there?" with a lot of authority behind it, the only quiver in her voice caused by the cold. When she tried to locate the shadow again, it was gone.

"Shit," she whispered. She stuck her head up above the rifle for a better view. She switched the lever to safe and stood up, carefully shielding herself behind one of the trees. She stood very still and listened for movement but heard nothing. How could she have lost him like that? After a few minutes, she decided to proceed with her perimeter patrol and as she turned around, she came nose-to-barrel with Robbins' M16.

"Bang, you're dead," he said.

He must have based his own movements on every time she took a step so that she couldn't hear him. Although he had scared her half-to-death, she was as impressed by his performance as he clearly was with hers.

"You did very well, Walker," he told her, proudly. "You came the closest of anyone to shooting me, actually even pinpointing my position." He ordered her to continue her mission until she was relieved or rounded up and then he was off to kill another trainee.

Ten minutes later, she heard two shots and a scream of

fright that could have woke the dead in three states. Shannon found out the next morning that Robbins had sneaked up on Chillemi, whose post was closest to hers, startled Chillemi senseless, which provoked her to fire off two rounds randomly just because her finger was welded to the trigger while she ran in place, screaming in fright, peeing her pants. Despite the animosity, Shannon felt bad for Chillemi, having to remain out in the freezing cold, in wet pants for the rest of the shift.

Welcome to the Army, Jarine, Shannon thought.

After breakfast, the company spent the morning being tested on what they had learned so far. Dale's high point was, after she had passed all her hands-on testing, she got to participate in a little role-playing to help out with others being quizzed. Cassidy picked her to be the undercover enemy soldier trying to get through a controlled area. Out of ten trainees she tried to get by, only two of them did everything correctly. The other eight she blew up with a bomb in an envelope she was hand carrying. Most of the trainees were too tired at that point to care about specifics. Cassidy assured them that was what the enemy would be counting on.

The company dined on C-rations for lunch then pulled up stakes, packed everything up and marched to another Bivouac site. They had been set up for about fifteen minutes when they were allowed another C-ration break, called supper. After chow, the troops were divided into smaller groups and given their assignments for the night.

Dale had started her period the day they had hit the field. That was never fun when opportunities to change her tampon were few and super tampons were not always comfortable nor absorbent enough to leave in for the length of time she needed to until she had the chance to remedy the situation. If there was one disadvantage to women in the field, it was that. It was more of an annoyance than a deterrent and Dale was determined not to let it interfere with her training. Her cramping and her lack of sleep, however, contributed to her foul mood. She agreed that Bivouac was essential to Army training but sleep was more essential to the human body. She wasn't sure about any of the other groups but all of the soldiers she'd had classes with so far had not been allotted any real sleep time. She was at the point where she could barely stand upright.

Dale was on perimeter guard and patrolled the tent area where the females were encamped. She was in charge of that post for two hours while whoever was still at the site was allowed two hours to rest. Shannon was scheduled to relieve her and then Dale would hopefully be able to catch some sleep, even if it was only two hours. Two hours out of two days was better than nothing.

She circled the small, outer edge of the camp for the umpteenth time and stopped when she heard the rustling of leaves close by. She listened for a moment and then she was glad she waited to challenge. She would have felt very foolish asking the two, little black, beady eyes that glared at her, *who goes there?*

At first she thought it was a raccoon until she shined her flashlight on the furry little beast with the white streak and discovered it to be a skunk.

"Go away. Go on, get out of here," Dale coaxed, firmly, in a harsh whisper. The animal scurried closer. She didn't want to upset the temperamental creature but she couldn't let him freely wander around their camp, either. She made noise with sticks and leaves in hopes that the skunk would scamper away on its own. He didn't budge. Someone was invading his backyard and obviously, he was going to stay there until these trespassers moved. Dale tossed a few small rocks on the ground near him but not at him. Instead of sending out a warning to him, he stamped both front paws on the ground and sent a warning to her. They were at an impasse.

"Please go away," Dale pleaded. She wondered if he was a she and there were babies around she now couldn't get to. If that were the case, Dale knew the skunk would be committed to getting to them and nothing short of death would stop her. Since Dale wasn't about to kill her, she needed to come up with another solution. While she was thinking about what to do, the skunk inched closer. "Maybe you're hungry. Are you hungry? If I gave you some crackers, would you go away?"

The skunk appeared to listen intently. Dale reached into her pocket and removed a cracker she had saved from supper. She held it out to the skunk and, to her surprise, the skunk approached and gently took the cracker from her and started to chew on it. Dale was about to feed her a second cracker when three gunshots were fired off close by. The noise startled both Dale and the skunk and, as the skunk felt threatened, it sprayed.

Fortunately, the spray was away from Dale and the wind blew in the opposite direction so no spray got directly on her. Unfortunately, the skunk's back end faced the campsite. The skunk was now gone but not with the outcome Dale had hoped for.

It didn't take long for the less-than-fresh odor to assault her nostrils. "Son-of-a-bitch," she gagged as she tried to cover her mouth and nose. She felt her dinner rise in her throat as the smell seemed to freeze and hang suspended around each tent.

It was only a matter of seconds before the perimeter Dale was guarding was alive with activity. The occupants of the tent were now awake and commenting.

"Oh, goddamn! What's that smell?"

"Jesus, Mary and Joseph, what died?"

"Put your gas masks on!"

"What the fuck is that smell?"

"Hey, Beltran, did you fart again?"

"Holy Christ, I'm going to cack!"

Everyone was up and out of their tents, searching for the source. Shannon, who was dressed and ready to relieve Dale, sidled up to her. "What'd you do? Challenge Ritchie and lose? You didn't have to shoot him in the stinkpot, you know. You could have tried wrestling him to the ground first."

Dale smiled, sheepishly. "I don't know any more about this than you do."

Shannon studied her partner. "Of course not. Go get your two hours."

Dale crawled into the tent but there was no sleep to be had. The lingering odor was too strong to allow anyone any peace. She hoped no one ever found out she allowed a skunk to get that close. She'd never hear the end of it.

First order of the day was *How To Lead An Ambush Against An Enemy Convoy*. The troops learned about Point Ambush,

PERMISSION TO RECOVER

Area Ambush, and Hasty Ambush. They learned how to select an ambush site, determine the kill zone, or the area where the shooting was concentrated, to isolate, trap and destroy the target. They also learned to determine equipment needs, establish communication procedures, coordination fire, execute the ambush, control the ambush element, and withdraw. This took all morning. The tired and cranky troops were somewhat pacified by a hot meal brought out to them by the mess hall.

The trainees were called into another formation after chow and ordered to pack up and march to another Bivouac site, two miles away. They spent the rest of the afternoon setting up their tents and digging trenches and foxholes around them. At dusk, the company guarded their areas and because they were told there would be an all-out attack, they were warned to stay alert.

Dale and Shannon managed to finagle a tent together on their fourth night in the field. They hadn't had but four hours of sleep between them since Bivouac began and, ready to collapse from exhaustion, they built a dummy so it looked like someone was sitting in the foxhole in the dark. They created a cross-figure with two sticks, outfitted it with a dirty set of fatigues and stuffed it with whatever they could find to fill it out. Shannon rolled up and packed her sleeping bag cover so that it looked round enough to be a head and Dale stuck her steel pot on it. Satisfied with the appearance of the soldier standing perimeter guard, Shannon and Dale crawled into their sleeping bags and were in deep slumber in a matter of minutes.

About an hour later, in the pitch-blackness, Lieutenant Henning, who had come to the field later that afternoon, decided to make the rounds. She was accompanied by Silva, the company driver. Henning was tired of challenging and being challenged, so when she got to the next soldier, she decided she would like to sit with him or her for a while and discuss the reasons for such security precautions. She had conversed with this unusually quiet and cautious GI for a few minutes before she realized something was not right. She placed her mouth right next to what she decided was an ear. "What's the matter, soldier? Can't you hear me?" she shouted.

As she was about to shine her flashlight on the trainee's

steel pot to read the name, Silva tapped the soldier's helmet with the butt of his rifle. The head jerked spasmodically off the shoulders and thudded heavily to the bottom of the trench. Henning let out a yelp and then covered her mouth to stifle it. She recovered and lit up the dummy with her flashlight and began to burn. "Whose foxhole is this?"

"Well, Ma'am, I don't know," Silva answered as he reached down to pick up the steel pot.

"Private Walker! Private Oakes!" Henning yelled. She stormed toward the tent and ripped the flap open. She got down on her hands and knees and crawled inside.

"Get up!" she hissed at one lump. She slapped the other lump. "I said get up! I want you both outside immediately, if not sooner!"

Henning paced outside with her arms folded tightly across her chest. She stopped, kicked Shannon's side of the tent and heard an ouch from within.

"Get up!" Henning ordered again.

While they awaited the women's emergence, Silva chuckled, apparently thinking about the audacious cleverness of the two trainees. Henning cut him off with a sharp glare and shined her flashlight directly in his eyes.

"I find this incident neither amusing nor ingenious, Private Silva," she snapped. "Furthermore, you will mention this to no one. *No one.* Is that clear?"

"Yes, Ma'am." Silva stood at rigid Attention.

"Not to any of the other platoon members, not to the drill sergeants, not to Sergeant Ritchie and especially not to Captain Colton. Is that understood, mister?"

"Yes, Ma'am." He stared straight ahead, trying to avoid the glare of her flashlight.

Dale and Shannon squirmed out of the tent and stood up. Henning turned around to face them. "You two, stand at Attention." Her voice wasn't loud but it was firm. "Silva, get out of here. Take the jeep, go to the command post and bring me back a cup of coffee. And remember, you say nothing." She looked at him before he walked to the jeep. "I'd better not hear this story coming back to me from any source other than this little group right here."

"Yes, Ma'am."

Henning waited until Silva drove away. She knew she had to keep her voice down because sound traveled during the quiet of the night. She took several deep breaths before she got right in their faces. "Walk with me."

The two undercover lieutenants obeyed and they followed Henning to an isolated area about a quarter mile away.

"I know you two outrank me by time in-service but I feel, in this case, overstepping my boundaries is necessary. What if I had been Ritchie, McCoy, or Robbins ? They might have caused you some real trouble, trouble resulting in restrictions and discipline that prevented you from doing your jobs! Fun is fun and I enjoy a good laugh as much as the next person but this was downright foolish and arrogantly careless!"

"I thought it was rather brilliant, myself," Dale mumbled.

"Oakes!" Henning thrust her finger forward, nearly stabbing Dale with it.

"Hey, come on, Karen, we're running on empty here," Shannon jumped in.

"That's right. We're as used to this kind of bullshit as anyone," Dale protested. "We've been through it all before, remember? They don't have to drive us this hard."

"For some of us, this is our fourth night without sleep," Shannon said. "The human body cannot function without proper rest and that's a scientific fact! Everybody is run down, they're sick and they are not learning a fucking thing. Now I don't think expecting a few hours' sleep is asking too much, do you?" Shannon was suddenly wide awake.

"You've been given sleep time," Henning said.

"Two hours on and off?" Dale asked. "Who can sleep on a schedule like that? You just get settled down and it's time to get up. And if you do manage to fall asleep, there is too much commotion going on, outside the tent with testing, to allow you to stay asleep. Plus you constantly have to be semi-alert for a tear gas attack."

Henning drew a deep breath and blew it out. "I understand your need for sleep and I'm not angry about that. I'm angry at the way you went about it."

"Agreed. We took a chance," Shannon said. "We just reached a point where we got desperate and we didn't care whether we got caught or not. I bet we aren't the only ones you caught sleeping."

"You're not," Henning confessed.

"Then could you please talk to Colton and tell that little SOB that most of the people you challenged were dying from exhaustion? Tell him to alter the schedule to that we're not doing something every goddamned minute of the day. Because if he doesn't? We won't bother with complaining to Bishaye, we'll go right to the post commander."

"Yes, I believe that will make the statement you wish to," Henning reassured them.

The next morning after chow, Colton announced during formation that there would be a schedule change. He told the company that they would be given four hours of commander's time, during which he would allow a rest period for those who *felt they needed it.* With few exceptions, everyone went to sleep, voluntarily giving up their c-rationed cuisine for noon chow.

One o'clock formation completed, the trainees reviewed everything they had learned that week. They packed up and policed their Bivouac site and then were transported back to the company in three deuce-and-a-halfs. They turned in their equipment to supply and cleaned their rifles for the rest of the afternoon. At five o'clock formation, they were set free for another weekend.

It seemed to be when they were in line for a shower that it hit the females that they had just survived five days out in a cold, filthy, nasty environment, playing wargames. The women were pretty proud of themselves and what they had accomplished.

The hot showers revitalized almost everyone and gave them a surge of energy they thought would be impossible to achieve after their week. While standing behind Michaelson, Shannon asked her to go with them for the evening.

"No, I don't think so," she said, politely. "I'm really very tired but thank you anyway."

"Come on, Michaelson," Dale said. "You've been through a very hard week. You need to wind down. Relax. Have a little fun. You know as well as we do that if you stayed here and tried to sleep, you'd be too hyper. You don't have to stay out with us. Just for a while. What do you say?"

"Well..." she hesitated. "I could use a beer. Are you sure you don't mind?"

"We'd love to have you join us," Shannon confirmed.

"Okay, I'll go," she said, looking pleased with herself. "I'm not sure how long I'll stay, but —"

"That's fine," Shannon said. "Stay as little or as long as you like."

Two hours later, the trio walked into the Pizza Place looking and feeling like entirely different human beings. As packed as it was, as soon as a group of young men from another company took one look at the three women, especially Michaelson, she, Dale, and Shannon had themselves a table. After they split a pizza and a pitcher of beer, while Shannon debated a woman's role in the military with the men who gave them the table, Dale and Michaelson partnered off against Buckman and Kulick, two male A-10 members for a pool game of eight-ball.

Dale and the two guys watched, amazed, as Michaelson expertly cleared the table, calling every shot precisely. After four, easily won games, Dale turned to Michaelson. "Is there anything you don't do well?"

Michaelson gave her a noncommittal shrug. "I just like to try and be the best at everything I attempt."

Buckman smiled adorably at her. "Oh yeah? Had any sex lately?"

Michaelson didn't answer him but she laughed. She and Dale turned the pool table back to Buckman and Kulick, grabbed Shannon and left for the EC.

The trio walked up to the door and paid their two-dollar cover charge. "How do you feel about overwhelming attention?" Shannon asked Michaelson.

"I hate it. Why?" Michaelson looked at Shannon and Dale, confused.

"Then this probably wasn't the best place to bring you," Dale said, as they walked through the doorway to the main dance area. Both Dale and Shannon instinctively knew what was going to happen and they were right. It actually felt like time stood still for a minute. Every eye, male and female, seemed to be on Deborah Michaelson and she noticeably felt it.

"Oh shit," Michaelson whispered. In the next moment, six men at once asked her to dance.

At first, Dale thought that Michaelson might panic but then it seemed as though she blinked herself into feeling somewhat comfortable with the situation. She displayed a confident smile and declined all six offers in a charming but firm manner. She told Dale and Shannon she liked to dance but she'd rather have a drink first. By the time the women made their way to the bar, all six men were already there, climbing on top of one another to buy the three Alpha women drinks.

Michaelson grinned at the blatant display of testosterone overload in the men as more joined the crowd.

"Seriously," one of the men said to the bartender. "Give these women any drink they want for the rest of the evening. It's on me." Four others held up money and chimed in with *me, too!*

"That's really nice of you all to be so generous but let's get one thing straight up front," Michaelson said. Her smile was still in place but her intent was clear. "I can't speak for my friends but if you're buying me a drink because you think it buys me for the evening then don't waste your money. I'm here to dance and have a good time. If you want to buy me a drink just to be nice guys with no strings attached, then have at it. Any other reason? I can afford my own drinks."

The men, along with Dale and Shannon, were stunned into silence. The men weren't used to hearing no, thank you, and Dale and Shannon weren't used to hearing Michaelson assert herself. The confidence with which she expressed herself only made her more attractive and the men held firm to their offers. Dale was positive that at least one of the males thought plying Michaelson with enough alcohol might change her mind. She was positive Michaelson wouldn't.

The three women drank free for the rest of the evening. Even though Michaelson relaxed quite a bit, she never came close to losing control. She was fascinating to watch as she handled all the adoration thrust upon her and, at one time or another, it appeared as though the entire capacity of the room was either sitting at or gathered around their table. Dale and Shannon might have been insulted if they hadn't agreed that Michaelson deserved the attention. It was obviously good for her quiet ego.

By the end of the evening, both lieutenants deemed the night a rousing success. Michaelson, the belle of the ball, could not stop thanking them and telling them what a great time she'd had. Dale and Shannon invited her to go into town with them, to spend the weekend away from the post but she politely declined. She told them she'd be the rest of the week working off the pizza and beer she'd already consumed. Dale and Shannon told her they understood, packed some necessities, and left the barracks.

CHAPTER FORTY-FIVE

"This is really odd, Dale," Shannon said, the next afternoon at lunch. "We're practically in our fourteenth week. Nothing out of the ordinary has happened."

"Maybe it won't. Maybe whoever is behind all this is skipping a cycle to let things cool down or to throw us off. It shouldn't take a genius to figure out they'd put spies in the cycle to look for it. If I were behind it, I'd skip a cycle."

"So this has been a waste of time," Shannon said.

"I guess we won't know until it's over," Dale said.

Shannon sighed. "At least the worst part is over. LE School is a piece of cake. Hopefully, I'll be able to take some leave after this TDY is through and I have to go back to Texas."

"How is old Texas these days?"

"The same. Hot, dusty, dry." Shannon finished her coffee. "It could be worse, though. They could've sent me back to Korea. Or to Bayonne."

"No!" Dale said. She mocked being horrified. "Not the armpit of New Jersey!"

"Where do you think you'll go?"

"Vermont, remember? I was promised my freedom."

"You won't stay there," Shannon said. "I know the gypsy that possesses your soul."

"I will for a while, until I get bored. Then maybe I'll move where you are," Dale said and grinned. "Just to haunt you."

"You'll have to find me first." Shannon laughed.

"I'll have Bishaye track you down. She found you this time."

"That she did. Just exactly how much does our battalion commander know about me anyway?"

"Probably a hell of a lot more than you want her to," Dale said.

Shannon picked up the bill and looked it over. "Great. Maybe I will find my ass in Bayonne."

Dale and Shannon found themselves back on post while tracking down their fellow trainees. For some reason, a majority of the company decided to party closer to home this weekend. Even the individuals and couples who had rented hotel rooms in town were either found at the Pizza Place or the EC.

At the Pizza Place, Dale was ready to leave within the hour. Buckman had been there all day, was completely intoxicated and wouldn't leave Dale alone. He constantly touched her with some part of his body and came on to her overbearingly in the process. Diplomatically, Dale continued to refuse him but her patience wore thin. She waited until her suitor went to the men's room and she left the building, only to run into Shannon, who was on her way in.

Shannon couldn't help but notice Dale's annoyed expression. "What's wrong?"

"Buckman."

"Ah. He finally got the balls to go after you, huh?" Shannon chuckled. She lit a cigarette. She had just come from the bowling alley via the barracks.

"Walk with me to the EC," Dale said. "What do you mean, he finally got the balls?"

"Jesus, Dale, Stevie Wonder could see how Buckman lusts after you. I don't know how you can miss it. He's a cute guy. Maybe you should get together with him, Dale. You're getting out after this, it's not like you can get punished for fraternizing and God knows, you need to get laid."

"I *need* to get laid? Are you saying I'm being bitchy?"

"If the broom fits…"

Dale glared at her partner, then softened. "You're right. I do need to get laid. But it won't be Buckman."

"Why not?"

She wanted to tell Shannon, to just blurt out *because he's the wrong sex* but she couldn't. She wasn't sure how to say it or, if she did, how her partner would react. As far as Dale knew, Shannon was okay with homosexuality in general but would she feel the same about a close friend? Dale decided not to test her friendship in that direction. She and Shannon needed one another until the end of the assignment and Dale didn't want anything to come between them. Until Dale established herself in a relationship, she didn't feel the need for Shannon to know. After all, Shannon got married and divorced and Dale wasn't aware until after, so maybe the subject of Dale's sexuality should be kept personal until there was a reason to tell Shannon.

"Don't tell me you're waiting for someone special..." Shannon said.

"I don't even know what that means, Shan, much less waiting for it. Buckman is a cute guy, no doubt, but he doesn't do anything for me. Am I horny? Yes. But if I'm going to sleep with someone, I want to be a little more attracted to them than I am to Buckman."

"More like Holmquist?" Shannon teased.

"Would you drop the Holmquist thing? I'm not going to sleep with a drill sergeant, either. I would never hear the end of that if it ever got back to Bishaye."

"And how could it get back to Bishaye?"

Dale stared at Shannon. "Have you learned nothing about her since you've been here? I know she doesn't look green and ugly but I wouldn't be surprised to find out she has a crystal ball and flying monkeys."

Shannon opened her mouth but shut it just as quickly. "I don't think I have a comeback for that," she said, finally.

The moment they entered the Enlisted Club, they were on the outskirts of a fight. The two lieutenants inched their way around the mob and made it to the main dance floor, unharmed. They surveyed the room first to absorb the atmosphere. Everything appeared routine until they were at the bar, ordering drinks, and Shannon saw a couple of men giving Hewett and Segore a hard time. Another male GI approached their table, pulled a reluctant Hewett to her feet, and dragged her to the dance floor. Shannon poked Dale and made her aware of the situation. They set down their beers.

"I hate this kind of bullshit," Dale said, steamed. "I'll take Fred Astaire."

Shannon nodded and headed toward the table.

PERMISSION TO RECOVER

Dale approached her target on the dance floor and stood next to a frightened Hewett. She assumed that Hewett wasn't so much intimidated as trying to figure out how to get out of the situation without her name ending up on a list. Dale ignored the GI and shouted to Hewett above the music. "Hi, Kerrie. I thought you said you'd never come here."

A look of relief crossed the young Mormon's face. "I should have stuck to my guns."

"Hey! What are you? Stupid? We're dancing here! Get lost!" The GI's tone was nasty and his words were slurred.

Dale looked at him, blandly. "Is that what you call this? A dance? Looked more like an abduction to me." She quickly reached over, snatched his wrist and used a pressure point to cause him to release his death grip on Hewett. "Kerrie, go help Walker and Segore, okay?" Dale didn't take her gaze off the soldier. "Gene Kelly and I need to have a chat."

As Hewett left them, the young man tried to break away from Dale's grasp but Dale turned his hand in a direction it wasn't physically designed to go, causing him agony. It was a simple, discreet move that brought little or no attention to either of them on the crowded dance floor.

"Pretend you're dancing with me or I'll break your hand," Dale said in his ear. "If you make any sudden moves, I'll drop you and then I'll break your balls." Dale looked him in the eye so that he could make certain she wasn't kidding.

He danced.

In the meantime, Shannon had stepped between Marilyn Segore, who was still seated and the young man who was giving her a hard time. "The lady obviously doesn't want your company," Shannon told him. "Go find someone who does."

He took a step toward Shannon that would have been more threatening if he hadn't been off-balance. "Fuck you! I'll be with who I want when I want."

"You boys are drunk," Shannon said to the two male GIs at the table. "Don't get yourself into more trouble than it's worth. Let's all just say goodnight and you guys walk away."

"Let's say we don't," the seated soldier told her.

"So you just want to ruin everybody's good time," Shannon

said.

"You go away and we'll continue our good time," he said.

"Yeah. I just bet you will. Okay," Shannon said. "I wanted to make this easy but we'll play it your way." She gestured to Dale on the dance floor. "See my friend out there with your friend? As soon as you two leave our friends here alone, your friend will be released. But if you stay here and insist on being foolish, your friend will not be very happy. He will be very sore and walking and talking very funny. Understand?" It was difficult sounding like she meant business when she had to practically scream to make herself heard above the music and noise.

The two men looked at their friend, dancing, and then back at Shannon. "He doesn't look like he's in any trouble to me," the one who was sitting said. He stood up and took a step toward Shannon. "Now unless you want to join the fun, get —"

Shannon stuck two fingers directly under his left collarbone, hitting a pressure point, and easily pushed him back a few steps. "Kerrie, go get Sergeant Bascomb at the door and tell him what's happening."

Hewett wasted no time following Shannon's orders and knowing where she was going made the two GIs stand still. Shannon nodded her head toward Dale and the men looked at their buddy. His expression registered discomfort. Dale took a step back, enough to let them see that she had a handful of their buddy's genitals. It wasn't a loving grasp and the sight of it made them wince. Dale stepped back into the dance and continued to guide the GI around the floor.

The young man who was pushed away by Shannon looked at his friend. "Come on, Dave, let's get out of here."

Dave, the other belligerent male, shook his head, defiantly. "No! Fuck these chicks, man. Don't be a pussy!" He then looked at Shannon. "Get lost, bitch, or you'll have a face-full of my fist."

"Oooh, tough guy, huh? That make you feel like a real man? Intimidating women and threatening to beat them up? Is that the only way you can prove your manhood? Huh? Tough guy?"

The man called Dave grabbed both of Shannon's wrists and held on so tightly, it cut off her circulation. "Now what are you going to do? Huh? Tough girl? Maybe you and I should get on the dance floor with your friend," he said to her, triumphantly.

From the corner of her eye, Shannon saw Bascomb in the doorway with Hewett, who pointed in their direction. Bascomb signaled to someone and started toward them but Shannon wanted this one all to herself.

She tried to force her wrists together, knowing that would automatically make Dave try to keep her wrists apart. Then she changed momentum and tried to force her wrists further apart, which instinctively made him resist her effort by forcing her wrists together. At that point, Shannon gave in to the impetus and when her wrists swung together, she passed her right wrist just below her left and, suddenly, without Dave even realizing it, he was holding on to her left wrist with both hands. With her free arm, she used the heel of her right hand to smash him in the nose, sending him sprawling backward, blood everywhere.

Bascomb and the other bouncer reached the group and, witnessing what just happened, the doorman burst out laughing. He looked at Shannon. "Next time, we'll call you," he said.

Dale saw that everything now seemed to be under control and assisted her dance partner to the group. She let him go and he immediately fell to his knees in front of Bascomb. Bascomb and the bouncer then escorted the three young men out of the disco area and to the EC main office.

Segore and Hewett gaped at Dale and Shannon, who wiped blood off her hand with a napkin dipped in beer. "Where did you learn to do that?" Segore asked Shannon.

Shannon looked at her hand. She gestured to the napkin and the glass of draft beer. "It made sense. We didn't have any water and —"

"No, all that karate stuff you just did," Segore said.

"I don't know karate," Shannon said, laughing it off.

"Well, what was that you just did?" Segore asked.

"She picked that up on TV," Dale said, quickly. "Kerrie, what are you doing here? In a bar? Drinking?"

Kerrie Hewett looked at both of them as though they had lost their minds. "Don't you think I've earned it?"

Dale and Shannon exchanged glances. "True. If anything

can drive you to drink, the Army can."

Dale and Shannon sat down with Hewett and Segore and spent the rest of the evening there. Predictably, there was no more trouble from overanxious GIs but there was this one sergeant on the EC staff who kept propositioning Dale. *I'm going to have to change my perfume*, Dale thought. It attracted all the wrong attention.

At first, the NCO seemed very nice but after several of Dale's polite but firm refusals, he became persistently obnoxious. Dale hated to be pressured for sex and being that this was the second incident in four hours, she was ready to call it a night. She knew she had to be careful how she dealt with him and his ego. She didn't want to offend him to the point of banning her from the EC. When the sergeant left her to take care of a problem in a different section of the bar, Dale turned to Shannon. "I have to get out of here. This guy is driving me nuts."

"Hell, Dale, he's a handsome guy. Why not go for a little ride with him? All he wants is sex and you agreed you need to get laid."

"I don't know him. And, last time I went for a ride with someone I didn't know, I almost didn't come back."

Shannon looked at her partner pragmatically. "I forgot about that. I'm sorry."

"Unfortunately, I can't forget it." She patted Shannon's shoulder as the NCO returned and sat down next to Dale.

"All right, good looking. Does your passion match your anger? I saw you with those guys. I *love* strong women," the NCO said smoothly.

Dale was all too familiar with guys like this NCO. They loved strong women until they got into relationships with them. Then they wanted everything to change so that they could corral that strong woman and make her a possession.

"I know you can take care of yourself. Do you want to get out of here?" the NCO continued, His attitude was such that he believed sex with Dale was inevitable. Dale guessed he didn't give up easily or get turned down too often.

Dale got an idea, then looked at him flirtatiously and said in a suggestive tone of voice. "Hmm, looks like you've worn me down. Where do you want to go?"

The NCO's grin was triumphant. He leaned in close. "There's a little room off to the side of the EC office. We can go there now, if you like."

Dale knew all about that room. It was legendary on Fort McCullough as the Enlisted Fuck Room. The only people with access were the NCOs on duty. She had never personally visited that room but Shannon had. There was nothing cozy or romantic about it. It was a place where the door locked, the participants got naked and engaged in sex, the wilder, the better. "Okay. Why not?"

The NCO stood up and held out his hand. Dale remained seated and crooked her finger at him. He leaned down and she put her lips to his ear. "Speak to me, baby," he said.

"I just have a question, if you know. Does penicillin go into effect on the first day of the shot?"

"What are you on penicillin for?" He moved his face away so that he could look in Dale's eyes.

Dale gave him a saucy wink. "Do you think you're the first or only guy that's invited me to that room?"

The NCO straightened up and slowly backed away. "Um...I just remembered, I have to, uh, help Sergeant Bascomb..."

"Aw, come on. It's only a little rash," Dale said to the quickly retreating man. She then sat back with a satisfied smirk.

Damn, I should have thought of that sooner.

Segore and Hewett left the EC an hour later. Dale and Shannon decided to leave when the lights came on and the bar closed down. On the way out, they thanked Bascomb again for his rapid response.

Once Dale and Shannon were outside and around the corner of the entrance with no one in sight, they were confronted by the two GIs they had hurt earlier.

"You broke my fucking nose, bitch!" Dave, the bandaged soldier, yelled at Shannon.

Shannon took a deep breath, tried to stay calm and act unfazed. "Your fucking nose? Jeez, I'm sorry. If I'd known that's what you used it for, I would have aimed much lower."

"You two think you're real cute, don't you?" the other man asked.

"Look, anything you got, you deserved," Shannon said.

A balisong knife materialized in Dave's hand. He held it toward them and flicked it open in a display of well-practiced ease. "Just like anything that happens to you, you're going to deserve."

Both women blanched at the sight of the butterfly knife, realizing this incident had just taken a much more serious turn. They both knew, in purposes of defending themselves, that a knife was just an extension of the attacker's arm but if miscalculated, it could be a lethal extension.

"That's going to get you real far," Shannon said. "You use that thing on us or to force us to do something, you know we're going to turn you in."

"If we don't kill you first," Dale said. This was a tone of voice Shannon had never heard in Dale before. It was even, controlled, and cold. It made the little hairs on the back of her neck rise. Shannon was well aware of Dale's fighting skills and she instinctively knew that Dale was not going to be taken down alive again.

"Look, guys, too many people saw what happened in there tonight. Even if we didn't say anything or you thought it was important enough to kill us, you'd be the first people to be questioned. Is a little hurt pride worth the rest of your lives in Leavenworth?" Shannon said.

"We just wanted to have a good time tonight and you took that away from us," the soldier Dale had injured said. "So we think you owe us some fun."

"Your idea of fun is not our idea of fun and you'll still get thrown in jail," Shannon said.

"If we don't kill you first," Dale repeated. She hadn't lost the edge to her voice.

"You said that already," Shannon reminded her.

"I mean it," Dale said, quietly. She hadn't taken her gaze off the knife since Dave pulled it out.

"Come on, Dave," the other man said, finally. "Let's go. They're right. I don't need any more trouble."

"No! They humiliated us!" Dave yelled at him.

PERMISSION TO RECOVER

"How does it feel?" Dale asked. "What do you think you did to those girls in there? What do you think you'd be doing to us if we didn't resist your plan to have some fun? And we only made you feel a little bit of the way you make women feel when you don't take no for an answer. Let me tell you something, if you boys get thrown in Leavenworth, you'll know exactly how it feels when someone doesn't take no for an answer."

Bascomb and another club worker walked around the corner and stopped at the sight of Dale and Shannon with the two offenders. "Don't tell me these yay-hoos are bothering you again."

"No, Sergeant," Shannon said, as Dave quietly closed the knife and put it in his pocket. "These gentlemen were just apologizing for their behavior tonight and promised not to do it again. Right, fellas?"

Dave and his friend looked from Dale and Shannon to the two massive NCOs standing next to them. "That's right, Sergeant. We were just leaving."

"I think that's wise," Bascomb said. He and the other sergeant waited until the two lower enlisted GIs were out of their sight. "So, what really happened?"

"Nothing," Dale said. "Although, next time they come in, you might want to check them for weapons."

"Why?" Bascomb's eyes narrowed.

"Just a hunch."

"They won't be allowed back in for a month and I'll have to send the report to their CO, so I assume they'll be cooling their heels on company restriction for that thirty days," Bascomb said.

"Why thirty days? I thought the norm was fourteen days," Shannon said.

"It's at my discretion and I figured in thirty days, you two will be out of here and I don't want guys behaving like that to chase away two of my best customers." Bascomb grinned. "Come on, we'll give you a lift back to your barracks."

After they were dropped at the corner of Bravo-10, Dale and Shannon jogged to the company area. Dale did a walk-through of the bay as Shannon visited with Wachsman, who was fireguard.

"Where is everybody?" she asked Wachsman, who had just finished polishing her boots.

"Some went to Atlanta. The rest? Who knows? Hey, Segore and Hewett couldn't stop talking about what happened at the EC and how you and Oakes really cooked in taking control of the situation."

Shannon sighed. "I wish they hadn't. If it gets back to the drill sergeants, Oakes and I could get into a lot of trouble for fighting and the four of us could get into hot water just for admitting to having been at the EC. How many people did they tell?"

"Oh, just the twenty of us who were hanging around the barracks."

"Great. I guess we're just going to have to put ourselves at the mercy of everyone's better judgment and beg for secrecy in the morning."

"I don't think anyone will say anything. Good thing Snow wasn't here, though."

Shannon smirked. With the dirt she had on Snow, she was pretty sure Snow wouldn't have said a word. If that was her biggest threat then she was positive they were home free on the EC incident.

Dale had CQ starting at noon the next day. While she performed eight hours of that tedious task with an extremely hungover and apologetic Buckman as her runner, Shannon traveled around the post and in town trying to find out who went where and with whom.

At nine that evening, when Dale was off duty, she showered and changed into civilian clothes and headed to Averill to join Shannon. The Journey Inn had no vacancies but that didn't stop them. They partied with a majority of the company and sacked out on the floor in a room with Mackey, Lehr, Wachsman and Laraway.

This was the first three-day weekend the trainees had. Washington's birthday was a celebrated holiday on military reservations, also. If Dale had not had to work that day between, she and Shannon would have certainly tried to devise a valid excuse to have landed somewhere in the vicinity of the Atlanta

PERMISSION TO RECOVER

Underground. Unfortunately, it just wasn't meant to be.

After they wandered around Averill for a while, most of the company members who were in town, returned to McCullough and wound up down at the Pizza Place. Tramonte, Jaffe, Minkler and Newcomb, who had used the three days to take a trip home, also returned. Travis had planned to either head home or have her husband fly down to Averill but she was scheduled for CQ right after Dale, so that ruined her plans. Dale had mentioned to Shannon that she thought Travis having been scheduled for duty was done purposely. It didn't help Travis appreciate the system anymore and her attitude continued to sour.

Dale used the pay phone close to the Pizza Place to try and call Bishaye. After three attempts with no answer, she gave up.

CHAPTER FORTY-SIX

Shannon reported to sick call the next morning with an upset stomach. While she was gone, Ritchie held a surprise locker and gear inspection. With few exceptions, no one passed. The senior drill sergeant was unreasonably angry at the condition of the barracks, the lockers and the equipment and ranted at the 0700 formation. He informed the troops that there would be another inspection Wednesday morning. Anyone who didn't pass that would be restricted for the weekend.

Shannon was given something to help settle her stomach and she rejoined her company at the gym. Today was the start of their twenty hours of self-defense training.

Before the class began, Travis approached Robbins and asked him to please tell her by Wednesday evening if she was going to be restricted for anything because her husband planned to fly down for the weekend. If she was not going to be available, she did not want her husband to waste the time or the money. Robbins assured her that he would let her know.

The first hour of self-defense training was mostly an introduction to the type of moves they would be learning and the three hours after that would be spent applying and rehearsing defense against choke holds. The instructors told the class that action against those particular holds must be swift and successful. Speed was imperative, the instructors emphasized, because within five-to-seven seconds, one could be unconscious and within ten-to-fourteen seconds, one could be dead.

The trainees were made to partner off with someone of the same gender and practiced moves called the Throat Push-Off, Front Windmill Defense and the Rear Windmill Defense. They were told to not hold back or be gentle with their attacker so when everyone was released for noon chow, it was suddenly difficult to do tiny little things like stand, sit or swallow because everyone's body was rebelling against the abuse it had just received.

Shannon still felt out of it by the afternoon and wondered if she had picked up stomach flu or something. She forged ahead even though she felt like running to find Ritchie, stick her finger down her throat and throw up on his carefully spit-shined boots.

Staff Sergeant Duane Halloren and two others conducted the Interpersonal Communications class while one or two company drill sergeants monitored it. The purpose of the instructors in this class, among other things, was to attempt to provoke the trainees into a negative response, in spite of themselves. This was to see how much stress they could take because, after all, as they were repeatedly told, as MPs they would sometimes be under an unbearable amount.

For some unknown reason, Halloren selected Shannon to pick on. She reasoned later that it was most likely because if she looked as sick as she felt, she probably came across as weak. She was not in the mood for verbal sparring and she let Halloren get away with humiliating her with his situational attacks at least three times. She stayed calm but after the third time she started to do a slow burn that rivaled her fever.

Ten minutes after his last strike against Shannon, Halloren must have incorrectly sensed he had a prime target in her and pounced again. He presented her with another nonsensical, hypothetical situation and, although she answered correctly, he was less than pleased with her response.

"Private Walker, can you produce evidence to support that statement?" Halloren asked.

"I'm sure I could produce it," Shannon answered him.

"Do you have it on you this very minute?"

"Well, no, but —"

"Then again, I have no choice but to label you an incompetent liar," Halloren said, triumphantly. He got smiles from his fellow instructors and he had the demeanor of someone who had finally proved his point to an irritating female student.

"Sergeant Halloren," Shannon said. She did not look up at him.

"Yes, Private Walker?"

"Can you produce your parent's marriage certificate?"

"Yes," he answered, cautiously. "I can produce it…"

"Do you have it on you this very minute?"

"No."

"Then I guess I have no other choice than to label you a bastard," Shannon said, with more strength and confidence in your voice than she'd had all day. They locked stares as her classmates hooted, hollered, and clapped their approval for her comeback.

"Private Walker, I'd like to see you outside," Halloren said. His tone of voice wasn't as triumphant as it had been. He headed toward the door.

As Shannon rose from her desk, she heard someone familiar.

"That's okay, Sergeant Halloren, I'll deal with Private Walker later. I'd rather not have her miss any of this class, if that's okay with you." The voice belonged to Cassidy, who was monitoring in the back.

Halloren stopped and glared at Cassidy but she didn't back down. He gave in and continued the class but he wisely chose to leave Shannon alone for the duration of the course.

Outside the classroom, during a cigarette break, Cassidy ushered Shannon away from everyone. Shannon looked at her defiantly, expecting a lecture on insubordination. Cassidy couldn't help but break into a grin. "At ease, Walker, I'm not going to yell at you. As a woman, as an individual, I agree with what you did. As a sergeant…" She took a deep breath and released it. "Oh, what the hell, I still agree with what you did but that doesn't give you free rein to do it again. Is that clear?"

"Yes, Drill Sergeant," Shannon said. She suddenly felt better. Cassidy was turning out to be not half-bad.

"Good. Consider yourself chastised, should anyone ask."

"Yes, Drill Sergeant," Shannon said.

"I want to know if he says or does anything to you out of line when I'm not around. Now, get back with your group."

"Yes, Drill Sergeant."

Halloren ignored her the rest of the day.

The locker inspection was held after 0700 formation. The

word spread quickly that everyone had passed.

At 0800, the company was put back into formation and marched to the gym for more self-defense training. Again, the instructors partnered the females with each other and the males with other males. Dale and Shannon were not the only two to think this was unfair but, being it was only the second day, they remained silent and did what they were told.

They practiced what they had learned the day before and then there were shown what were known as come-along holds. The come-along holds were used to control movement of an unwilling individual from point A to point B without injuring that person. The holds produced pain and discomfort because pressure was applied to sensitive areas and temporarily bent specific joints into positions nature did not intend them to go. Those particular moves also had to be executed with speed for success.

That morning, they practiced holds called the Fingers Come-Along, the Reverse Grip Come-Along and the Front Hammerlock. Once more, the trainees left the class unable to perform simple tasks such as being able to button fatigue shirts.

The IPC class went a lot smoother that afternoon. The instructors left Shannon alone and because they no longer seemed to be on the warpath, the class was borderline enjoyable. The instructors released the trainees earlier than scheduled but it wasn't a blessing. The drill sergeants took advantage of the opportunity to give their troops extra PT. To the aching self-defense students, added exercise time was not welcome.

When PT was over, Alpha-10 had to put it into overdrive to keep pace on their march back to Tenth Battalion.

Dale recognized O'Brien right away and took that opportunity to hop off to the side of the road into the trees. Dale watched as O'Brien and Delta-12 passed and nearly jumped out of her boots to catch a branch several feet above her head when she heard a voice behind her. "What are you doing?"

She almost broke her neck turning around and sighed in relief when she saw it was Henning. "Thank God. Listen, we might have a slight problem."

"We?" Henning put her hands on her hips.

"Yes. That company we just marched by? I know one of those drill sergeants. We were once drinking buddies at Fort Jackson. Staff Sergeant Cindy O'Brien."

"Drinking buddies? Fraternizing, were you?" Henning arched an eyebrow.

"No. Well, yes, but that's not important. Look, if she recognizes me, she might unintentionally blow everything. I can't keep diving into the bushes every time she marches by. I think Cassidy thinks either I have a serious bladder problem or an uncontrollable tree fetish because she always sees my butt dashing into the forest."

Dale made sure O'Brien was out of sight before she and Henning walked back to the road. They jogged to catch up with the rest of the company. "What would you like me to do?" Henning asked.

"I don't know. Talk to Bishaye. Get her transferred. Send her TDY somewhere. Convince her it's time for a vacation."

"Why don't you just make an appointment to see her and beg for her discretion? You said you were drinking buddies, shouldn't that count for some honor between you?"

"I could but I'd rather not. That might open up a whole new can of worms."

"Whatever you decide, do it fast because if she makes you or your name comes up in casual conversation at the NCO club for most annoyingly amusing trainee of the cycle, it'll be too late."

"I hope O'Brien would recognize it for what it was and keep her mouth shut."

"Unless she thought you might be out to burn a friend. Or a lover," Henning said.

Dale immediately thought of Holmquist. It wouldn't have been impossible for Cindy to have been involved with him. They were a lot alike and he was her type, too...when her bisexuality swung her toward men. "I guess I'll give her a call. Maybe I can sneak out and see her tonight."

"Think about it. If you make the wrong decision and it backfires —"

"I know, I know." Bishaye would have her ass and not in a way Dale would enjoy.

"Just let me know," Henning told her.

"Fuck it. Call Anne and have her take care of it," Dale said.

"Oakes, you better not be bugging the lieutenant!" Holmquist yelled as they caught up with the rest of the company.

The morning did not start off well. In self-defense class, Dale raised her hand and finally questioned the pairing of women with women and men with men for the third day in a row.

"What's your reasoning, Private?" the NCOIC asked.

Dale tried not to gawk at him but she knew he couldn't be that dense. "My reasoning, Sergeant, is that the women should be given more of a challenge. We're not always going to have the choice of subduing just females. We should know what it feels like to go up against someone bigger and stronger than us."

Sergeant First Class Greenlaw openly laughed at her. "I suppose you have a good point. But we are trying to get you through this course as quickly and effectively as possible. We don't have a lot of time to do this in, Private, and it will eat up valuable time we don't have if we have to stop every other minute to make sure you ladies aren't getting too badly hurt. Do you understand *my* point?"

"No, Sergeant, I don't. If we don't learn in class the difference between using these moves on men and women then we're going to be in serious trouble once we're on the road."

"Private, if we make it fair for you," he argued, as though annoyed. "We make it unfair for the males. I mean, they'll have no contest with you females."

"Sergeant, don't you agree," Dale countered, patiently. "That it would be good practice for the males to work with the females at least one day? They might come up against a female criminal in the line of duty. Not subduing her properly out of gallantry or an old-fashioned attitude about touching her might get them killed. From what I understand, female MPs are few and far between and there won't always be a female available to come out and assist. And," Dale added, knowing she shouldn't but she was already too worked up. "If this class is being taught correctly, if the instructors are ensuring that the women are learning the same way the men are, these moves should be effective on either sex. There should be no challenge."

There was dead silence as Dale and Greenlaw stared the other down. Dale thought if this were a cartoon, she would see steam come from his nostrils and ears. "Private, come up here."

Oh, shit, Dale thought, *me and my big, fat mouth.* Some of the women started to clap.

"At ease!" Greenlaw commanded everyone. Dale stood before him. "Well, Private Oakes, do you think you, personally, could be a challenge to me?"

"Sergeant, I was just trying to make a point," Dale said, wearily.

"And now *I* will make a point. You and I will show the class why we'd be wasting everybody's time right now. I want you to defend yourself against me, utilizing positions you've already learned."

"I don't want to fight you, Sergeant," Dale said, firmly. She knew he would try to embarrass her and put her in her place by using moves they had not been shown defensive tactics for. Dale also knew her enormous, stubborn pride wouldn't let that happen, which is precisely why she didn't want to fight him. As it stood, the trainees had only been shown enough to get them killed.

"What's the matter with you, Oakes, are you a coward?" he boomed at her. He really seemed to enjoy the moment. "I gave you an order. Now, let's go."

"Yes, Sergeant." Dale felt she tried her best to get out of it. She got close enough to take a few obligatory swings and made sure the fake punches didn't reveal anything about any former training she'd had.

Greenlaw grabbed her arm in an attempt to put her in a hold. Dale was prepared for it and she counter grabbed him and the next thing he knew, he was on his back with Dale's foot a hair away from his Adam's apple. He couldn't move. If he did, she could not have only dislocated his arm out from his shoulder but crushed his windpipe, as well.

"Let me up, Oakes," he told her, among the cheers of the entire company. On his feet again, he rubbed his shoulder and faced her as the drill sergeants calmed the troops. "You got lucky," he muttered. The look on his face, though, was still somewhat startled and she could tell he wondered just how lucky she did get.

Dale turned away from him on purpose and pretended to acknowledge her peer praise but was peripherally alerted on Greenlaw. Greenlaw took the opportunity to get his revenge. Two breaths later, he was flat on his stomach, Dale's foot resting on the back of his neck, her hands holding his other arm in such a position that, if he moved, she could have easily broken it. She allowed him to get up without asking this time.

"You're not using what you've been taught," Greenlaw spit out.

"That's not really what this is about, is it, Sergeant?" Dale asked, her volume as low as his. She glanced up to search out the faces of her drill sergeants. She wanted to see their expressions before she continued. The four she saw looked impressed and amused. McCoy clearly got a huge charge out of it.

"Why don't we just stick to the stuff we've learned?" Greenlaw asked. "I will be the aggressor."

"Yes, Sergeant," Dale said, respectfully.

Greenlaw lunged toward her, not unexpectedly, just sooner than she thought he was going to. He got a decent grip on her neck but she easily broke the grasp by counter grabbing him and enforcing a binding hold he could not free himself from. He then threw four consecutive punches that Dale effortlessly blocked. By that time, the company was going nuts.

Finally, Greenlaw put his hand up, signaling a truce. Dale was guarded until he broke into a smile. "Okay, ladies," he announced. "Private Oakes has proved her point. Find yourselves a male partner." Before she returned to her group, he gestured her to him. "Just out of curiosity, what is your experience?"

"Shaolin Kenpo, second degree black belt. As you could probably tell, I'm not as strict with the style as I should be. Now, I just use it as a base for whatever works," Dale said.

"You're very good. I'll definitely remember this for the next company I teach," he told her. His tone was now more sincere than conciliatory.

"I didn't hurt you, did I?" Dale asked, as she watched him rub his shoulder.

Greenlaw laughed. "Just my ego and my reputation for being a hardass but, no, you didn't hurt me. You just surprised the hell out of me."

The rest of the class went well, except nobody willingly wanted to pair off with Dale. She finally talked Bigfoot into it but because of the eleven-inch height difference between them, some of the choking techniques were a struggle for her.

The trainees also learned moves with their nightsticks that morning. They were shown exercises called the Figure Four Strangle Hold, with and without the club, a Groin Lift, a Wrist Takedown, Elbow Smash Defense, Reverse Wrist Grip, Cross Bar Stranglehold, two come-along holds, with the club, and the Bar Hammerlock Takedown.

The thing the trainees learned the most about the multi-step Bar Hammerlock Takedown was that if they were the attackers, their faces were smashed to the mat as many times as it took whoever was playing the cop to get it right. That's if the aggressor didn't get out of the hold first. The smart-alecky trainees who were able to escape the first steps usually decided to not struggle as hard after the fifth or sixth time their faces were slammed against a cushion that was harder than the floor, in most cases. This particular maneuver was the most difficult to learn correctly and was usually one of the moves they were tested on, a Go or No Go situation that would take place the next morning.

The IPC class that afternoon was much more mellow than it was most of the week. Since the incident with Shannon, Halloren seemed to go out of his way to be pleasant to her and the rest of the class. She returned the courtesy but she still didn't like him.

After the class, the company went on a two-mile run, then marched back to the company area for 1700 formation. A majority of the trainees were preoccupied with visions of the upcoming self-defense test. The fact that the IPC test was scheduled for the next afternoon didn't lift anyone's spirits, either.

Once back in the bay after evening chow, quite a few women approached Dale about helping them with their self-defense moves. She provided a few pointers but declined to physically demonstrate anything. When Segore and Hewett reminded the group that Shannon was pretty good, too, the throng moved to Shannon, who also declined.

Dale had just finished folding her laundry when Michaelson entered the laundry room with an armful of fatigues. As she loaded them into an empty machine, she turned to Dale. "You really surprised a lot of people this morning in class."

"Yeah, surprised me, too," Dale said, laughing. "I'm surprised I still had it in me."

"Think you could show me a few things?"

Dale shook her head. "I'm not a qualified teacher, Michaelson, it's not a good idea. I'm not a clean fighter anymore. I've betrayed the style and I don't feel right passing on what I know."

"I suppose you're right. I've had some training myself." She blushed at the admittance.

"Really?" Dale smiled and playfully shook Michaelson's shoulder. "Don't be embarrassed. I think that's great. I think all women should be taught how to fight."

"My parents always thought fighting wasn't lady-like."

"Neither is getting raped and murdered," Dale said, sighing. She hated that misogynist argument. "What style have you been trained in?"

"Just some Hapkido." She started the machine and faced Dale squarely. "I really respect what you did for the women in class today. It took a lot of guts to stand up to Greenlaw in front of everyone like you did."

Dale shrugged, now feeling a little awkward, herself. "Thanks. I just felt it needed to be done. He wasn't doing us a favor by being so chivalrous."

"Do you think you'd feel like sparring sometime in the gym?" Michaelson asked.

Wrestling, maybe, Dale thought as she discreetly studied the pretty blonde. *Damn, Dale, when did you turn into such a hound?*

She blinked the R-rated thought out of her mind. Michaelson was indeed hot but she was a trainee and one who showed no particular interest in either gender, much less fooling around with one. "As I told some of the girls upstairs, that's probably not such a great idea. How much training have you had?"

"Well...I'm a fifth-degree black belt." Again, she hesitated, coming across as very shy about it.

Dale's eyes snapped open in surprise. "No kidding? I was right. Sparring with you wouldn't be a good idea. You'd kick my ass."

Michaelson laughed. "I'm sure not. I may have the belt but you have practical experience. For me, it started as a way to stay in shape, you know? And I just paid attention, passed classes and tested for the next degree of belt. Anyway, that's what I've been doing with my Wednesday nights, when you guys are all out partying. I head to the WAC recreation center and work out."

"They hold Hapkido classes there?"

"No, but they have equipment that is vital to my training."

"I'm sure our gym has newer equipment," Dale said.

Michaelson smiled. "But nobody from here goes to the rec, so nobody bugs me."

"Why are you so...unsociable isn't the word I want to use...I guess aloof is better..."

"I like being alone. I know that sounds strange but I really enjoy being by myself. I have never liked the feeling of having to rely on someone else."

"You might be in the wrong MOS for that, you know. And sometimes forming those bonds require a lot of time spent with your partners."

"You almost sound like you're talking from experience."

Dale had to think fast. "I sort of am. My family is peppered with military and civilian cops."

Michaelson nodded in comprehension. "I feel like if I open up, I lose a little of myself and we have such little privacy here as it is. Once I get assigned to my permanent duty station, I know it'll be different."

You have no idea, Dale thought.

"Any hint on where you might get assigned?"

"Yeah, if all goes well, I was promised Hawaii. My career counselor said my GT scores were high enough, so I got to pick my MOS and my first permanent duty station."

"Hawaii? Cool."

"What about you?"

"I'm going to go wherever Uncle Sam needs me the most, I guess. I didn't score high enough to get a choice. My dream sheet says California or Southern Europe. Who knows?" Dale said, doing her best to sound unconcerned. She gathered her folded laundry, ready to leave the laundry room.

"Sure you don't want to spar?" Michaelson asked one last time.

"I'm sure." Dale grinned. "I really don't think I'm up for you. Your training is recent and probably by the book. Mine is not. I didn't do what I did this morning to show off, I just wanted the women to get an even break."

"I kind of figured that. You didn't look too thrilled to be out there."

"I wasn't."

"Can you, um, you know, keep my fighting experience just between us?" Michaelson asked.

"Sure. I wish I hadn't had to expose mine."

She foolishly believed she had made her point to Greenlaw and the others, which was why Dale was furious when she was told she No Go'd twice on the self-defense test. Everyone else seemed shocked, too, and Dale immediately saw it for the head game it was. Even though Dale was positive her drill sergeants were confident of her experience and training and would not let the instructors fail her a third time. She also knew, as a last resort, that Henning would say something to Bishaye and the battalion commander would step in and not let her get eliminated from training, especially in a situation where it was obvious to everyone Dale could take care of herself. Regardless, Dale was livid to the point where Cassidy had to order her outside.

"Calm down, Oakes," Cassidy said. "You've got to get a grip on that temper."

Dale paced and sputtered and was so irate, she didn't stop and take the time to enjoy being in the sensual drill sergeant's company. On any normal day, when the opportunity presented itself, Dale took full advantage of being able to admire Drill

Sergeant Cassidy from afar. Now she had a private audience with the beautiful woman and was wasting the moment.

"Oakes! At Ease!"

Dale stopped and assumed the commanded position. Her body still seemed to vibrate.

"Let's try something. Relax, first." Cassidy watched as Dale let her arms drop to her side. "Good. Close your eyes and stretch your arms out to the sides." Cassidy couldn't help but smirk at the look Dale gave her. "Just do it, Oakes."

"Yes, Drill Sergeant." Dale obeyed, closed her eyes, and made her body look like a T.

"Now, take a good, deep breath. Exhale. Take another. Now try and touch your index fingers together."

Dale opened an eye and looked at her.

"Close it!" Cassidy waited until Dale did. "Go ahead, touch your index fingers together."

Dale attempted it and missed the first time. The second time she was successful. "Okay…what does this do?"

"Nothing. I just get a kick out of watching it."

Dale opened her eyes to see Cassidy laughing. Dale's breath caught at the sight. Cassidy's dark eyes sparkled and her dazzling white smile revealed those dimples that humanized her movie star beauty. Cassidy's chuckling was contagious and Dale smiled.

"Feel better?" Cassidy asked.

At being face-to-face with you? Absolutely. What a waste, you being with Robbins, Dale thought.

"A little."

"You had to know when you humiliated him that he would retaliate somehow," Cassidy told her. "What did you No Go for, did he say?"

"Lack of motivation and improper execution. If he really wants to see a proper execution, I'll be happy to accommodate him."

"Oakes…" Cassidy's tone was warning.

"Yes, Drill Sergeant." She smirked.

"We all know what you're capable of, Oakes. You don't have to worry about passing this course. He's just getting even."

"I will never understand the male ego, Drill Sergeant."

PERMISSION TO RECOVER

Cassidy nodded. "I don't think most men understand it either." She extended her arm toward the entrance and gestured Dale back inside the gym.

The IPC test that afternoon was a breeze to get through and no one failed. After that, Dale and a few others were returned to the company area to receive counseling statements for having two consecutive failures in one class. This got Dale angry again but since Cassidy was present, she controlled it. When Dale found out she was restricted Saturday for her two No Go's, she hit the roof. A warning glare from Cassidy calmed her down.

Later, after a little digging, Dale discovered the real reason she'd been restricted. The company needed forty people for a detail and, miraculously, forty people were restricted for one reason or another. The restriction started at 1800 hours that evening and, unfortunately, one of the trainees confined to the company area was Tracy Travis Novak.

Travis had not heard anything negative from Robbins regarding her earlier request and had made definite plans for her husband to fly down and spend the weekend. By the time they gave Travis the news, Mr. Novak was already on his way, his plane scheduled to land at 1400, his bus getting to the post at approximately 1830.

Travis furiously confronted Robbins, who told her that she shouldn't have gotten married and that would have solved the whole problem. Robbins had also been in possession of the knowledge that their free time wouldn't have been a full weekend, anyway, but he didn't inform Travis of that little detail, either.

That evening, everyone met Travis' incredibly handsome husband. The cadre had graciously allowed him to spend time with her in the Dayroom...along with at least thirty other trainees. Once again, military head games resulted in pushing someone who would have most likely turned out to be a dedicated soldier over the edge. At 2300 hours that evening, Tracy Travis Novak went AWOL.

Though they tried to talk her out of it, neither Dale nor Shannon blamed her. Dale almost wished they could have gone with her.

AWOL or Absent Without Official Leave had such a stigma about it, Travis probably felt like she would have to hide the rest of her life. Dale wanted to tell her that because of budgetary problems, the FBI had started to steadily phase out the standing search for military deserters and concerned themselves with AWOL cases that involved violent crimes. The Department of Defense had put less money and effort into the search for deserters since nineteen seventy-three because the congress' General Accounting Office had been accusing them of spending too much money in that area. Dale wanted to tell her that even though her decision to desert had not been a wise one, she would not be hunted down and shot for it, as officials liked to infer, but Dale couldn't tell her. That, indeed, would have been knowing too much.

CHAPTER FORTY-SEVEN

Travis' absence was reported at 0700 hours the next morning when the thirty-nine restricted trainees showed up for formation. Another company's M16s were passed out to be cleaned and, still angry about her restriction, Dale took all day to clean just one.

Ritchie was in the area all morning because of Travis' unscheduled departure but he left the troops alone. Dale monitored his comings and goings along with Cassidy's and Putnam's. Their expressions were rather grim and Dale wondered if they had contacted Bishaye yet and, if they had, what reason they had given for Travis' motivation for departure. If, in future conversation, the subject was brought up, Dale wouldn't hesitate to tell her about what a good soldier Travis had been up to that point and what provoked her to take off.

At 1700 hours, after Dale turned in her spotless weapon and stood in formation, the restriction was lifted for the evening. Before they were dismissed, they were reminded that the entire company had to be back in the barracks for a 2300-hour bed check that night.

Shannon was at the WAC recreation center swimming most of the day with Wachsman and Kotski and wasn't back yet so Dale hit the Pizza Place. She drank a beer and waited for Shannon, who came in alone. They agreed to go to the EC for an hour before the mandatory bed check and stopped by the office first, to call a taxi.

"Hey, Oakes," Jerry Renaldi, the CQ, said, as she and Shannon entered the Orderly Room. "Some chick has been trying to get you on the AUTOVON line all night."

"Really? This chick have a name?" Dale tried to act confused but she was concerned. She couldn't think of who would call her on the AUTOVON line as only a privileged few military people knew she was there. She looked at Shannon, who flirted madly with Rutledge, the CQ runner, then returned her attention to Renaldi.

"Yeah, here." He tore a sheet off the message pad. "A Sergeant Burke. Said it was real important."

Dale snatched the note from him and read the name and number. "It must've been. I have to use the phone."

"To call her back?"

Dale nodded and Renaldi shook his head. "Come on, Renaldi, you know it's a matter of urgency or I wouldn't ask you. Please."

"No way. Use the pay phone outside."

"There's a line six deep waiting for that pay phone," Dale protested.

"What if the Staff Duty NCO walks in while you're on the phone?" he asked.

"Who is it?"

"Bradbury."

Dale rolled her eyes. "Great."

Shannon burst out laughing. "I wouldn't worry about it, then. If anyone can handle Bradbury, Oakes can."

Renaldi looked at her with a sly smile. "Oh, yeah? Why is that?"

"No reason," Dale said to him and glared at Shannon. She returned her attention to Renaldi. "If she comes in, I'll pretend I'm calling a cab. We were told we could use the phone for that."

"I don't know, Oakes. It's late. Where would you be calling a cab to? Everyone has to be back in an hour and a half."

"I'll tell her we're going to WacVille. She doesn't have to know we've already been there. Come on, Jerry," Dale pleaded. "You won't get caught."

"That reassures me coming from one of the great escape artists of the company," Renaldi said, dryly. "All right. Go ahead but make it quick." He watched Dale pick up the phone and start to dial. "Hey, how do you know how to use the AUTOVON line?"

"It's in the CQ instruction book. Besides, I had to call a number for Henning once. She showed me how," Dale lied. She listened for the connection.

"Who's Sergeant Burke?" Renaldi asked.

"She's...uh...my recruiter."

"A female recruiter? I didn't know there were any female recruiters."

"You probably didn't know there were any female MPs, either." Dale smiled. She heard the phone being picked up at the other end and Theresa Burke identified herself. "Hey, it's Dale," she said, uncomfortably, as Renaldi watched and listened.

"Thank God," Theresa's tired voice said. "I thought I was going to have to sit here all night. Where've you been?"

"It's Saturday night, where do you think I've been?"

"Well, then, to what do I owe the honor of your return? Or should I say, to whom?"

"I forgot my birth control pills." Dale smiled and winked at Renaldi.

"I'm so happy I don't have to worry about silly little things like that," Burke said.

"Careful. This is not a secure line."

"Really? Who's listening at your end?"

"The entire 7th Division. Why do you think I need my pills?"

"That must have been some trip. Last I knew they were still in California."

"Not after they heard me and my buddy, Walker, were in town." *Okay, enough banter.* "So what's up, anyway? I know you didn't call me just to chat."

"Hurry up, Oakes," Renaldi prompted. Dale waved him off with her hand.

Burke chuckled. "Someone hassling you about staying on the line?"

"I'm still a trainee, I don't have any privileges yet."

"And if they really knew you, they'd keep it that way," Burke kidded. "Okay, I know it took a while but I may have some information for you."

"May have?" Dale grabbed Renaldi's pen and the notepad off the desk.

"Yes. It's not much but it's from Carolyn Stuart's lover, so..."

"What was it?"

"She doesn't know a lot, and I'm not just talking about the case, either. She must be one hell of a lay because there's no way Stuart was with her for the intellectual stimulation. She said she never saw Stuart actually *with* anyone but she recalled her on the phone one night and Stuart was very, very upset. She said she overheard Stuart tell that person that she knew she was going to get caught and, if she did, she wasn't sure she could keep quiet. Now, this is the interesting part. She said Stuart then asked if the revenge against that lieutenant was really worth all this bullshit. What do you think?"

"What lieutenant?" Dale looked at Shannon who suddenly alerted to their conversation.

"I have no idea," Burke said. "What rank is your CO?"

"Captain, oh three."

"Then I would say it's your training officer."

"That would surprise me but it's not impossible. What else?"

"That's it. That's all she remembers, other than she asked Stuart and Stuart refused to discuss it with her," Burke said.

"Why didn't she say anything about this before?"

"I don't know. I told you she was helium from the shoulders up."

"Just how did you come by this information, anyway?" Dale asked.

"How do you think?"

"You're amazing. You'll do anything for your country, you incurable patriot, you." Dale said.

"Actually, it wasn't me. It was somebody else's pillow talk. Can you use it? It's got to be related to why you are there."

"I would say you are right. Certainly gives us a little meat to chew on. Hey, thanks, Theresa, I owe you."

The door swung open and hit the wall with a bang and Bradbury popped in as if she had been sprung from a toaster. Dale quickly crumpled her note into her palm.

"This is not a congregation area, people! If you don't have business in here, get out!" Bradbury bellowed and looked directly at Shannon. She then turned toward Dale. "CQ, what's this trainee doing on the phone?"

"Calling a taxi, Drill Sergeant." Renaldi looked at Dale like he wanted to kill her. "Fuckin' A," he swore under his breath.

Bradbury strolled to Dale, who was still on the phone.

"Yes, that's right. Building eighteen oh one south and could you get here sometime within the next millennium? This is the third time I've called," Dale said, for Bradbury's benefit.

"What?" Burke asked, confused.

"Ten minutes? That's what you said the last time and that was an hour ago."

"Oh, I get it. The Staff Duty NCO walked in, right?" Burke asked.

"That's right," Dale said and hung up the phone. "Sorry, Drill Sergeant but we've been waiting a long time."

Bradbury compared the time on her watch to the time on the wall clock. "Doesn't your company have a mandatory bed check at 2300?"

"Yes, Drill Sergeant," Dale answered.

"We've been trying to get out of here, Drill Sergeant, but the taxi hasn't come yet," Shannon said.

Bradbury whirled to face Shannon. "Are you still here?"

Shannon nodded.

"Well, get out." She turned back to Dale. "Hot date?"

Dale maneuvered around her, attempting to diplomatically avoid her closeness. "No, nothing like that, Drill Sergeant. It's just that I've been cleaning weapons all day and I wanted a drink before mandatory bed check."

"Then I suggest you be satisfied with a beer at the Pizza Place and leave it at that. You obviously can't depend on these taxis to get you to wherever you want to go and back in time and, with one female in your company being AWOL today, I would think you wouldn't want to run the risk of not being there for headcount."

It was good advice but considering she wasn't talking to a cab dispatcher, they didn't need it. "Yes, Drill Sergeant. The Pizza Place it is."

"Good girl," Bradbury said. "Now, get out of here."

Renaldi breathed an audible sigh of relief as Dale and Shannon left. They heard Bradbury ask him if he had a problem as she went to check the CQ log.

Once outside, Dale checked the time. They had less than an hour before they had to return to the barracks so they headed back to the Pizza Place for another couple of beers.

"What was the phone call about?" Shannon asked. Dale explained to her who Theresa Burke was and then related the phone conversation to her. "She said that? Revenge on a lieutenant? That's what Stuart said?"

"That's what allegedly was said with what seems like selective recall. Stuart's supposed exact words were, is revenge against that lieutenant really worth all this bullshit?"

"Is the source reliable?"

"Burke is reliable, I don't know about her source. I trust Burke's judgment so I don't think she would risk calling me here if she didn't feel there was something to it."

"How'd the source get this information?"

"Pillow talk."

Shannon rolled her eyes. "Right and that method is always fail-safe."

"Look, why don't we just accept the information as is. At least it's *something* after all these months of nothing. And if it's about this company, it's got to be Henning. She's been the only lieutenant here since just before this bullshit started," Dale said.

"So what do you think? Stubby did something to piss somebody off to do all this?"

Dale shrugged. "Or maybe someone developed an obsessive crush on her and she shot down his, or her, –feelings."

"But why go to all this elaborate and dangerous game playing? Why not just set her up, personally, and burn her?"

"I don't know that, either, Shan. I guess this is where we start earning those big bucks Uncle Sam has been secretly sending to those Swiss bank accounts we don't have access to. I will try to call Anne right now and tell her this piece of news if you will go in and please get me a beer." Dale handed her the money and walked to the pay phone. She looked around to make sure no one was lurking or in hearing range.

Dale wasn't surprised to find Anne not home on a Saturday night. Jack Bishaye told Dale that his wife was out to dinner with his sister who was in town for a visit. Dale flirted with him a little and then told him she and Shannon needed to meet with

PERMISSION TO RECOVER

Anne in a secure location. Jack suggested the Cloud Club on Monday night. The Cloud Club was an exclusive, downtown Averill men's bar that one needed a pilot's license to qualify and a membership card to get into. Anne had access to the wives' lounge, which was an annex to the building the club occupied.

Dale knew the Cloud Club was a sexist organization that Anne despised, so much so that even though Jack kept his membership current, he only attended for emergency member meetings. Anne had only been in the wives' lounge once, for her welcoming ceremony. Jack was right, though, it was a safe area and no one from the battalion would be in there or anywhere near there. Even on the off chance that one of the drill sergeants possessed a pilot's license, it was a pretty sure bet he wouldn't be milling around the wives' lounge, which had its own separate entrance.

Shannon returned outside with a red plastic cup of draft beer for Dale just as Dale hung up the phone.

"What'd she say?"

Dale took the cup from Shannon and took a couple sips. "She wasn't there. I talked to her husband."

"What'd he say?"

"I didn't tell him what's going on because it doesn't involve him and he probably wouldn't want to hear it anyway. I just told him we needed to see his wife and it was really important."

"So, she'll call us? How?"

Dale took a few more swallows. "No, he'll arrange for her to meet up with us at the Cloud Club Monday night at 1830 hours."

"Where's the Cloud Club?"

"It's a side building off the second entrance to the Coral County Country Club."

Shannon glared at her. "Great. Where are we going to get the clothes to go there by Monday?"

"Jack said The Cloud Club is casual. He also said absolutely no one we know would be there so it's a safe place to meet and relax."

"Think we should bring any of this to Henning?" Shannon asked. "Maybe it would jar her memory toward remembering any incident that could motivate someone wanting to get revenge on her."

"Let's wait and see what the boss says first," Dale said. She and Shannon drank from their plastic cups of beer. "She may have other ideas. We're too close to getting out of here. I don't want to think about doing anything without her approval first. If I mess this up, I will never hear the end of it."

Shannon finished her beer in four gulps. "True. Well, partner, let's hit the bunks for bed check. I wonder what's in store for us tomorrow."

"Nothing that I think we really have to be doing on a Sunday unless they're going to force us into worship services."

Guard duty training was what the company was roused out of bed for at 0500 hours. They stood at a 0520 formation and then they were all herded into the Dayroom, only allowed to leave for a morning chow line at 0600. They were then dismissed to hang around either the Dayroom, the north and south patios, or the laundry rooms. They were not allowed back in their bays, the parking lot or anywhere else out of the Tenth Battalion area.

It wasn't until 1100 hours that the company was divided up and sent to specific guard posts. Shannon was sent to guard the Twelfth Battalion Arms Room while Dale, Bigfoot, Judd and Zachary went sent out to patrol and guard Fort McCullough's Ammo Dump.

The Ammunition Dump was a widespread, hilly, woodsy, dark, and desolate area and the four GIs were literally locked in behind a cyclone and razor-wire fence after they were dropped off by a deuce-and-a-half. The quartet was speechless when they were told that their duty would be two hours on and two hours off until 0500 the next morning. The four would interchange their shifts with four Delta-12 trainees. The off time was spent in a guard shack, outside the fence line that contained absolutely no heat.

Bigfoot and Judd went off in one direction while Dale and Zachary headed in the other. They explored and familiarized themselves with the territory so that after it got dark, they'd have an idea of where they were going. Dale had never been to the ammo dump before but she had heard ghost stories of strange noises and sights emanating from the deep cluster of woods that had to be traversed in order to complete a full perimeter check.

Dale and Zachary, for reasons they could only attribute to fatigue, dwelled on conversations about the real Bigfoot, UFOs, ax-murderers, evil spirits, and alien abduction. By the fourth cycle at 0030 hours, they had understandably scared themselves into what felt like eternal consciousness.

Dale eagerly shared the small flask of brandy Zachary had hidden in his jacket. Regardless, it did help fight off the bitter cold that was battering them internally as well as externally. At 0500 hours, when the deuce-and-a-half returned to pick them up, Dale had never been so happy to hear those familiar grinding gears in her life. She felt like she'd never be warm again.

The trainees were returned to the company area and given enough time to turn in their M16s to the Arms Room and eat breakfast. Their next formation was at 0700 hours. Dale would then head to the gym for her make-up test in Self-Defense. She was tired, cranky and a mite hungover from imbibing with Zachary. At least she got to wash her face and brush her teeth. That helped a little.

Dale was partnered off with Tierni and Dale opted for testing second. Dale did her best to help Tierni along without making it obvious that Tierni just didn't have the momentum. When they gave Tierni the news she had bolo'd out, she fled the gym in tears. This left Dale with no partner and, as she was the last one to test, a slight dilemma.

"Private Oakes, it looks like one of us is going to have to be your partner," one of the scoring instructors told her.

"Then I want Sergeant Greenlaw," Dale said. She watched as the two scorers exchanged glances. "Look, Sergeant, I've been up all night on guard duty. I'm exhausted and, as you can see, very crabby. Not that either of you care a hang about that, I know. I shouldn't be here today and both of you know it. If I have to pick one of you as a partner then I think it's only fair I have the guy who failed me so I can show him, and you, how well I can handle myself in this area. I was promised a nap after this test and I really need it. So, can I get this over with?"

Unmistakably, neither instructor wanted to go up against

Dale. They'd seen what she could do and instinctively knew whoever went up against her would be put through their physical paces. One instructor excused himself and approached Greenlaw at the other side of the gym.

Dale heard him sputter and swear and yell he didn't give a shit what Private Oakes wanted, that she was just a fucking trainee. "You're the one who No Go'd her, and since neither Sergeant Boyer nor I want to feel her wrath, you should be the one to do this…or are you too chickenshit?" the instructor said.

Dale tried to hide a smirk as she knew that would be bait he wouldn't refuse. Greenlaw reluctantly agreed to Dale's request and, as anticipated from all, she showed him no mercy. All of her moves were precise and strong and when she finished her last required takedown, Greenlaw could barely get off the mat. It wasn't completely one-sided, though. Dale also left the gym looking as if she had been put through a wringer.

On her way out, Dale heard Greenlaw speak. "There's one future MP I would not want to fuck with."

When Dale got back to the Orderly Room to present her Go paper, she found out that particular round of testing had not only eliminated Tierni but also Sager, Beltran, Ferrence and Hewett. Dale found Hewett crying outside the CQ office and gave her a hug.

"I really wanted this, Oakes," Hewett said and sniffed back her tears. "I needed to prove something to myself. Now I have to go home with my tail between my legs."

Dale sat with her at the picnic table. "Nah. It doesn't mean you're out of the Army, Kerrie. It just means you can't be an MP. Being a cop probably wasn't for you anyway because it can be a violent job sometimes and you're not a violent person. I'm sure they'll find something suitable for you to do. They could recycle you into confinement or maybe even tag you for interrogation or something interesting like that. What would your choice be if not this?"

"Well…I like languages and I've studied Mandarin Chinese."

"There you go. Make sure they know that. They might send you to the Defense Language Institute and turn you into a translator."

"Ugh. More school?"

"It wouldn't be like here, like you'd still be treated like a trainee. My recruiter told me the DLI is at the Presidio in Monterey, California, right on the ocean."

"That wouldn't be too hard to take." She wiped her tears on her sleeve.

Dale patted her shoulder. "Something tells me you'll be happier wherever they put you." Dale hated to tell Hewett that she just wasn't MP material. She shied away from confrontation and shrunk back when someone displayed an attitude. She didn't have to be aggressive but she did have to be assertive and she had a hard time with that. She excelled in studying and keeping the barracks clean but Dale assumed, from Hewett's Mormon upbringing, that she felt completely out of place performing any task that wasn't officially a woman's place. Dale felt Hewett would do well in an MOS where the job itself wasn't male oriented or male dominated but where her superiors were men. It fit in with her conditioning.

Back up in the barracks, Dale and a few others were allowed to sleep until eleven-thirty. Dale also chose to skip noon chow and set her alarm for 1230 hours. She felt refreshed and alert and ready for the rest of the day.

At one o'clock formation, the company was informed that they were now all on what was called self-pace. Self-pace allowed the trainees to study the remaining required classes to complete LE School by taking as much time as they needed to grasp the individual subjects. It also meant that they were permitted to make their own decisions as to which class they would take in which order, as long as all mandatory classes were completed.

Dale's choice was to attend an Apprehension class and could barely keep her eyes open through it. Her alertness after waking had given in to not really feeling rested enough to concentrate. Her body had begun to tighten after the morning's workout and everything suddenly felt exhausted. She wanted to go back to the barracks at five o'clock and go right to bed after she was dismissed from formation but she knew that was impossible.

Self-pace had its advantages but for Dale and Shannon, it had more disadvantages. Alpha Company was not the only company to have students in these classes as there were other LE School trainees from other battalions also on self-pace. This made it increasingly more difficult for Dale and Shannon to keep track of their fellow trainees because self-pace made it easy for Alpha trainees to hook up with members of other companies, which only complicated the art of spying.

CHAPTER FORTY-EIGHT

Anne Bishaye showed her husband's membership card with her military identification at the door of the Cloud Club wives' lounge. Bishaye led the way as Dale and Shannon followed. They sat at a corner table and ordered a round of drinks.

"Would you mind telling me what's with all the cloak-and-dagger stuff? You know I hate this place," Bishaye said. She removed her leather bomber jacket and placed it on the back of her chair.

"It wasn't my idea, it was Jack's," Dale said. "And it was a good one. No one will see us here." Her hands were ice-cold and her heart was hammering in her chest. She hated that being in Bishaye's presence could do that to her. She tried her best to act blasé.

Bishaye smiled warmly at Shannon. "And how are you?"

"Ready for this to be done." Shannon smiled back.

"Cheer up. It shouldn't be too much longer. One way or the other you should be out of here in another month." She looked at Dale. "Okay, now you've got me here, what's going on?"

There was still heat between them. Dale didn't know how Shannon could not see it or feel it. "I got a call from Theresa Burke on Saturday."

"*She* called *you?* How?" Bishaye was suddenly alert as they accepted their drinks at the table.

"You're buying," Dale said. She held her beer bottle out to Bishaye in a salutatory motion before she took a sip.

"I'd planned to. How did she call you? The AUTOVON line?" Bishaye moved her stirrer around her drink.

"Yes. I passed her off as my recruiter to the CQ. He believed me."

Bishaye looked at Shannon for confirmation.

"Yeah, he bought it a hundred percent," Shannon agreed.

"I hate it when you do that," Dale mumbled.

"It's habit, it's not personal," Bishaye said and sipped her drink.

No, it never is with you. Dale bit her tongue.

"She had a message for us allegedly from Carolyn Stuart's lover. The lover recalled that she overheard Stuart on the phone and Stuart talked about knowing she was going to get caught, not being sure she could keep her mouth shut and then supposedly asked the other party if getting revenge on that lieutenant was worth all this bullshit. What do you think that means?"

Bishaye's expression had changed. She bristled and was all business. "How reliable is Burke's source?"

"I don't think she would have called me if she hadn't been reliable. You know her, Anne, she's not going to disturb me here unless she thinks it's serious. So what do you think?"

"What do *you* think?" Bishaye asked and looked at them both.

"It's got to mean Henning, right?" Shannon said. "A-10 hasn't had another lieutenant since this started, right?"

"We were wondering if we should tell her what we know and —"

Bishaye shook her head. "No. Let me think about this. The way things have been going, for all we know, she could be tied into it all."

"Henning? You handpicked her to liaison this," Dale said.

"I know this will shock you, Dale but I can and do make mistakes," Bishaye admitted.

Dale opened her mouth to comment. Her eyes flashed to Shannon and she snapped her mouth shut.

"Henning may not know she might be someone's target. If she is, I don't want her to panic. Until we get more information, if there is more information to be had, and I figure it out for what it is, then I will call a meeting with her and tell her myself."

"Will you share it with us?" Dale asked.

"Of course I will share it with you," Bishaye snapped. "What good would it do to keep it to myself?"

Shannon looked back and forth between Dale and Bishaye at the visible tension. "Do you want us to watch her more closely?" Shannon's question broke the sudden uncomfortable silence.

"Yes, when she's around," Bishaye said to Shannon. "But she's gone back to school."

"Is that why she hasn't been in the company much lately?" Dale wondered out loud.

"You mean you haven't noticed?" Bishaye asked.

Dale and Shannon exchanged guilty looks that did not go unnoticed by Bishaye.

"Sure we did," Dale said, unconvincingly.

"You did not," Bishaye argued, mildly aghast. "And I hired you two as spies?"

"Okay," Dale said. "We've been concentrating on other things. We're not supposed to be watching her, anyway. She's in on all this, remember?"

"Yes, I remember," Bishaye said. "You're supposed to stay in communication with her."

"We have been...well, when we've had a chance to inconspicuously talk with her. She's been around enough," Dale said.

"When was the last time you saw her?" Bishaye asked.

"Last week. Didn't she call you about that drill sergeant I recognized from another battalion?" Dale's tone had a hint of victory to it.

"And before that?" Bishaye looked at both of them.

"Bivouac," Shannon said.

Bishaye immediately focused on Dale. "About that POW massacre I witnessed..."

"Don't worry your pretty little head about that. I was disciplined. I'm still surprised I can use my arms," Dale said.

"And the perimeter guard business?" Bishaye asked with an arched eyebrow.

Dale grinned. "Oh, come on, that was classic. If Henning hadn't come along, we would have gotten away with it." Both lieutenants beamed so proudly that even Bishaye broke into a reluctant smile.

"I knew you were incorrigible, Dale, but I'm honestly surprised at you, Lieutenant Walker. I brought you in on this to hopefully be her anchor."

"That was your first mistake," Dale said, smugly.

Bishaye nodded. "Evidently." She drained her glass and waited for the two lieutenants to finish theirs. "Is the Burke call all you wanted or is there something else we need to discuss?"

"No, just that. Don't you think that's an interesting lead?" Dale asked.

"Yes, especially if it turns out to be something that can result in solving this. I'll let you know where we go from here. I'll need to make a few phone calls first and do a little more research on Stuart and her training cycle. I still don't see why you couldn't have called me on the phone about it because you know how much I hate this place." The last sentence was directed at Dale.

"I did call you. You weren't home. Again, this was your husband's idea. And this place is not as bad as I thought it would be," Dale said, looking around. "I kind of expected something like, I don't know, Rue Morgue after the murders."

"Or the way Dale explained it to me, the Stepford Wives Club," Shannon said. She looked at Dale. "Did you want to bring up Travis?"

Dale yawned. "You just did."

"Travis? The female who went AWOL? What about her?" Bishaye asked.

"She was a good soldier," Dale said. "I don't know what you've been told or what you may have heard on your own but the cadre fucked her over too much. I would tell you if I thought she was useless or irresponsible but she wasn't. Prime example of how Uncle Sam can turn a good attitude bad."

"It happens," Bishaye said and shrugged.

"It shouldn't," Dale countered. "Just like —"

Bishaye shot her a sharp look. "Don't start."

"What was going on with you two tonight?" Shannon asked Dale after Bishaye dropped them off a block away from the PX. "I thought I was going to have to move to another table."

"Nothing, really," Dale said, as they jogged toward the company area. "She gets pissed off when I remind her about Kirk."

"No, it started before that. You both were on the edge."

"My best guess is that we're getting on each other's nerves. This case hasn't seemed to work out the way either one of us wanted."

"And what way was that?"

"Quickly."

They picked up their pace to get back to the barracks faster. It was cold and they were both tired. Dale hoped that answer satisfied Shannon's curiosity as the atmosphere this time did seem to feel more choleric than sexually charged. Dale and Bishaye were clearly exasperated with one another. The big question on Dale's mind was why? She understood her own frustration with Bishaye but was confused at Bishaye's obvious aggravation with her. It couldn't have been the stunts in Bivouac because Bishaye certainly knew her well enough by now to expect some playing around and Dale couldn't fathom that they shared the same frustration. Even if they did, Bishaye had made it very clear that she could and would control hers.

Maybe she was just in a bad mood, Dale thought, *and it has nothing to do with me.*

Her resignation on the subject still didn't stop her from dreaming erotically of Bishaye that night. It was most difficult to have arousing dreams in room with forty other people. Dale had never been known to talk in her sleep. She hoped she didn't start now.

Dale breezed through the Apprehension test not so much because of her precision in the powers of arrest but more for the NCOIC's attraction to Saunders. The sergeant waved Dale and McTague through the procedure so fast, they practically got windburns. Dale hung around to see what developed and was glad to see that Saunders played it smart. After she received a Go, Saunders thanked him and left the classroom before he could get any more persistent with his flirting.

Dale moved on to the Report Writing class that, for her, wasn't hard. She knew the basic requirements of a majority of the paperwork was to record the who, what, where, when, why and how and the other forms were pretty much self-explanatory. The first section was on how to fill out DA Form 3975, a pre-printed document containing questions that requested all the pertinent information concerning a committed or allegedly committed crime. The report had to include a brief summary of what occurred, who was involved and the circumstances surrounding that occurrence. Dale had completed so many of those forms, she could have written one up blindfolded.

The next section was on How To Write A Sworn Statement, DA Form 2823. This form was a little trickier because what was written on it was the testimony of a victim, witness, accused, etc. Again, this statement dealt with the who, what, where, when, why and how of a crime so questions and answers had to be specific. The form was also used as an integral part of an investigation, so it was imperative that it was as accurate as possible.

The third section dealt with preparing a Rights Warning Procedure/Waiver Certificate , DA Form 3881. This form was a military equivalent of the civilian police department's Miranda warning. The document stated in print that the accused did not have to answer questions or say anything and that anything the suspect did say could be used as evidence in a criminal trial. For others, the class was interesting, for Dale, it was monotonous.

That evening, as Dale hung around the Pizza Place, Shannon finished up eight hours of guard duty at the Charlie-12 Arms Room. It was uneventful, except for witnessing one of the C-12 male drill sergeants openly fraternize with a female she recognized from Bravo-10. After she watched about thirty minutes of their heavy making out, she decided to ignore it. If they were both that obvious and stupid, she didn't need to turn them in. They'd be caught on their own.

The next morning, Dale was selected for eight hours of guard duty at the Charlie-11 Arms Room. That company was in their fourth week of basic combat training and as Dale observed them, she was relieved her company was well beyond that stage.

Shannon caught up on the classes she had missed the day before. She flew through the Search and Seizure instruction and decided to take her time on Protecting A Crime Scene class with Sergeant Montemurro.

Montemurro was a mild-mannered, jovial, by-the-book, but reasonable NCO. He admitted he didn't like failing anyone but

PERMISSION TO RECOVER

he wouldn't give away answers, either. This was a class one had to take their time with because with his method of teaching, there was no other way to do it. Montemurro made sure that when someone left his class, they *knew* how to protect a crime scene and preserve evidence.

"Private Walker, in your opinion, what is evidence?" Montemurro asked.

"Evidence is anything – photographs, weapons, fingerprints, footprints, clothing, broken glass, tire tracks – any physical objects that directly or individually establish the facts relative to the particular incident under investigation. Written or oral statements can also be considered evidence."

"Very good. Tell me about protecting a crime scene."

Shannon looked around at the five other people in class with her and wondered why Montemurro was making her the star of the show. Montemurro must have mistaken her blank expression to mean she misunderstood the question.

"What I mean is, how do you protect a crime scene?"

"By not touching anything." Shannon purposely answered vaguely. She could have given him the correct response of by preventing the destruction, removal, rearrangement, or concealment of any physical evidence and to prevent the departure of suspects, victims, and witnesses until the responsibility of the investigation of the scene is taken by proper authority but she wanted him to call on someone else.

"Can you be a little more specific?" Montemurro gently probed. "Refer to your notes, if you must."

Another trainee raised his hand and Montemurro acknowledged him. "Protecting evidence from people and weather," the student answered.

"Anything else?" Again, he directed his question at Shannon.

"Clearing the crime scene of anyone not related to the incident," Shannon said.

"What else?" Montemurro finally looked at another trainee.

"Keeping witnesses separated and guarding the scene to prevent entry of unauthorized personnel," the trainee answered.

Another ninety minutes of Montemurro's class was the equivalent of ingesting a couple valium. Shannon was glad the day was done because by the time he released the class to return to their company areas for their five o'clock formations, she was ready for bed. She was scheduled for the class test first thing in the morning.

As Shannon cruised through her exam, Dale easily passed the Report Writing class and the partners hooked up for their personal favorite, Questioning A Witness class.

The instructors explained that questioning a witness was a very delicate and important procedure. It had to be done correctly or serious problems could result. To insure the trainees understood the exact process, the instructors handed out several situations written in dramatic script form for the class to study and practice among themselves in groups of two, three or four. After that they met with an instructor individually, who went over the program with them and then they returned to their group to study again and prepare for their test in which they pretended to be the interrogator while the instructor played the witness.

Dale and Shannon paired off and were lucky to get an area partitioned off away from the two sergeants overseeing the course. They had done this routine before and waited for an unsuspecting victim. Their patience was rewarded ten minutes later when a marine insert from Bravo was added to their group by one of the instructors.

Dale flipped a coin to see who played the questioner and who played the witness and when fate had decided, Dale and Shannon began their little act...a little too realistically for the jarhead.

"I want that man placed under apprehension!" Dale said, desperately. She brought actual tears to her eyes. She pointed accusingly at the unsuspecting marine. "He...he...oh, it was just awful!"

"Now, now, calm down," Shannon said, soothingly. "Let's get the whole story in chronological order. What did he do first?" Shannon glared daggers at the AIT insert who looked at both of them a little stunned but stayed silent. Dale and Shannon knew he had been told that he was only supposed to be an observer and not a participant at this point. The sergeant had placed him with them because they were in advanced stages of the class and he was supposed to learn the technique of questioning a witness from them. LE School was big on peer instruction.

"First he locked the door," Dale said and looked at the floor.

"Kidnapping! Ten years!" Shannon shouted joyously and wrote it down. "Then what did he do?"

"He pulled down his pants," Dale spat and moved her chair away from him while she gave him dirty looks.

"Indecent exposure, one more year." Shannon wrote it down. "Then what?"

The marine frantically searched his limited text for this conversational exchange which wasn't in any of the ten examples they were supposed to practice.

"He put his hand on my...on my..." Dale pretended to be upset.

Shannon nodded. "It's all right. I understand. Attempted battery. Five years. Then what?"

"He threw me on the bed!"

Shannon glowered at him. "Pervert." She scribbled quickly on her note pad. "Mayhem and felonious constraint. Ten to fifteen years." She looked at Dale, who played this role to the hilt. "Then what?"

"Then he...he...did it to me." Dale looked away.

"That does it! Twenty years, easy. Maybe we can even get him the chair." Shannon looked at Dale. "And all the while you were struggling and screaming..."

"Well, no, not exactly," Dale said, coyly. "I mean, it was kind of late and I didn't want to wake anyone up and —"

"Well, shit," Shannon said. She sat back and let out an annoyed sigh. She threw her pen down. "That's just a plain, old, ordinary fuck."

The three occupants of the cubicle heard the clearing of a throat and turned to see one of the male instructors leaning against the doorway. He clearly swallowed a smile. He seemed to enjoy their performance as much as anyone, with the exception of the marine, but as he explained, this was a serious class. "Playtime is over, ladies. How about you just stick to the examples?"

"That one wasn't in the booklet?" Dale asked, blinking

innocently.

The sergeant smiled and shook his head. "Just do it as it is written. You do have to pass a test tomorrow."

"Yes, Sergeant," Dale and Shannon said together.

At 1500 hours, class was through and the two lieutenants joined the entire company to take their final PT test for G-3 testing. No one failed. As soon as they were dismissed, Dale and Shannon changed into civilian clothes and made a beeline to the EC.

The disco/dance area was closed for minor repairs, so Dale and Shannon sat at the other bar, drank mixed drinks for a change and relaxed. When they left the bay, most of the remaining Alpha women were studying their notes for their particular classes the next day. The only other females who planned to do anything else were Kotski, who was going to the movies with Van Hoesan and Michaelson, who was going to the WAC recreation center for her usual Wednesday night routine.

Dale and Shannon rehashed their day, the phone call from Burke and what it might mean and speculated about Henning possibly being the target. The bar was empty except for them, two GIs playing pool and the bartender. He engaged in trivial conversation with them to help pass the time.

As with the other two nights that week, Dale and Shannon barely made it back to the barracks before bed check and had to slide into their bunks fully clothed. They covered up to their necks and pretended to be fast asleep as the drill sergeant took a mental headcount and never knew, or didn't acknowledge, that the two women had raced in the door just minutes earlier. Dale and Shannon were getting the reputation of being the party daredevil duo of Alpha-10. Old habits really did die hard.

CHAPTER FORTY-NINE

Time now seemed to be going so fast, it was hard to believe it was Friday again, although since Shannon faced eight hours of CQ at 2200, she wasn't really in a TGIF sort of mood.

The partners passed Questioning A Witness on the first try. Shannon moved on to the Apprehension class and spent the rest of the day covering the definition of procedures, requirements, resistance of and persons not subject to the UCMJ. Dale ended up in Approaching the Scene of an Incident class that entailed learning how to properly approach an individual, group, and building. The class also dealt with how to estimate a situation and how to execute a plan of operation. Those classes lasted until the end of the day.

Dale and Shannon showered and changed and walked to the EC so that Shannon could relax a bit before she had to report to her post in the Orderly Room. Her CQ runner was scheduled to be Chillemi, so she said that was a good enough reason to knock back a few alcoholic beverages before her shift began.

"Are you sure you don't want me to go back with you and hang around, just in case?" Dale asked, sincerely.

"No, it'll be okay. I have no problem with Chillemi, really. She's got the problem with me. If she can't get past it then there are plenty of ways it can be taken care of. I still don't get it. The other marine women haven't been like her. Navarrette and I worked together in Apprehension this afternoon and she was very pleasant. The others have integrated very well. Chillemi just doesn't want to blend in, especially not with me, even after she's been witness to and victim of my resourcefulness. I can handle her, though. Besides, if you're here babysitting me, who's babysitting the other women?"

"The bad part about that is they are now too spread out for even both of us to cover the territory. I can try to stay centrally located but, at this point, what does that mean?"

"I know. I wonder when Bishaye is going to get back to us about Stubby."

"I'll call her sometime this weekend and see what she wants us to do. If there's a chance on us being able to get out of here, then I want to do what we have to do and be gone ASAP."

"Roger that. So...what are your plans for the rest of the night?"

Dale shrugged. "I'm not sure. I guess I'll know when I get there. Tomorrow night, though, I will meet you at the Journey Inn. The room is already reserved in our names."

"Why not go there tonight?"

"I couldn't get a room, they were booked."

"You could crash with somebody, I'm sure," Shannon assured her.

"I haven't made up my mind yet."

Shannon looked at the bar wall clock. "Shit. My time is up." She stood up and put a five-dollar bill on the counter. "I'll see you tomorrow in Averill, if not sooner."

Dale ended up in Franklin, which was the next county past Averill. It was an expensive cab ride but something compelled her to make the trip. She wasn't up to the usual trainee chasing and after the EC started to fill up, she didn't want to stay there and deal with the type of crap she and Shannon had confronted the week earlier. The Pizza Place held no interest for her, nor did hanging around the barracks or partying in Averill.

Without Shannon and on her own for the night, she'd had just enough to drink to give her the courage to check out a place called Gems. She knew about the club because of Theresa Burke and that it was a place with a reputation for being rowdy, fun, and discreet. It was a casual drinking and dancing establishment that catered to lesbians and even though one took her chances going there, Gems was not usually frequented by military women since it was so far away from the post. There were two other such women's bars much closer but for her purposes of just being curious, Dale thought the pricey taxi fare was worth the distance.

She entered the crowded bar cautiously, as was her nature

PERMISSION TO RECOVER

regardless of the circumstances. She pretended to search for someone and observed the women as unobtrusively as possible. Much to her relief she saw no familiar faces. Her biggest concern would have been Bradbury, but she knew the Bravo drill sergeant was stuck at her company area for all-night guard duty training.

As Dale made her way to the bar to order a beer, she noticed that a majority of the women did look like what she understood to be the stereotypical dyke — short mannish hairstyles, heavyset, men's jeans or slacks, flannel shirts, biker boots. It was as though a certain group had their own uniformed code. The women who didn't suit the cliché looked like they could have fit in anywhere from a farm to the Cloud Club's wives' lounge.

It was a busy night, as Fridays were in most bars so she wasn't surprised she couldn't get the cute, boyish bartender's attention right away. She had almost succeeded when she heard a familiar voice behind her, say her last name. She quickly turned to face Jennifer Jane Cassidy.

Oh, fuck, I'm busted, Dale thought. *She's a fucking shoofly! I should have known better. Goddamn it all to hell!*

A shoofly, in this case, was an operative working on behalf of Uncle Sam to try and weed out lesbian and gay soldiers by going to places patronized by homosexuals and catching them there. If the GIs couldn't legitimately explain their way out of it, they could kiss their military days goodbye. Usually after basic training, unless the soldier had a particularly accommodating company commander, there was no cover discharge like Unsuitability. It was a Bad Conduct offense that usually went hand-in-hand with a General or Dishonorable Discharge. Anything other than an Honorable Discharge meant the individual would have a nearly impossible time securing a decent job in the future.

Dale knew she didn't have to worry because a phone call to Bishaye would, hopefully, sweep this under the rug but it was just her luck, her very first time doing this, she would get caught.

Well, Dale thought, *I can always play ignorant and see if she buys I had no idea I just walked into a lesbian bar and gee, drill sergeant, while we're on the subject, what are* you *doing in here?*

"It took me a few minutes to recognize you out of uniform," Cassidy said, in that sexy, husky voice that zipped right to Dale's libido. Cassidy smiled, as her gaze seemed to appreciatively roam across Dale's attire before it rested on her face. "What in the world are you doing in here?"

Dale's mouth lost moisture at the blatant scrutiny. It was most unexpected. Cassidy was really playing up her part. She wanted to be as openly brazen and appraise the drill sergeant in the same manner but that would not help her case any. "I was hoping to get a drink but..." Dale looked at the bartender wistfully as she sped by her once more.

Cassidy studied Dale and tilted her head in a confused manner. "This...um, isn't exactly a usual trainee hangout." She gestured around the interior as if that should have prompted Dale to look and realization to strike.

Dale looked at Cassidy just as confused. "I don't understand. Is it a drill sergeant hang out?"

Jesus Christ, she is too damned gorgeous for words.

"No, no, I think you're missing my point. Look around, Oakes," she said, slowly. "Notice anything missing in this place?"

Dale made a point of scanning the club. She then looked at Cassidy as innocently as she could. "Style? A disco ball? Sacrificial virgins?"

"No. Men! There are no men here," Cassidy said, exasperated. Then she looked around again. "And don't be too sure about the sacrificial virgin part," she muttered, almost too quietly for Dale to hear.

"Oh! Men. Is that a bad thing?"

Cassidy peered at her. "You knew that before you came in here, didn't you?"

There was something about the eagerness in Cassidy's eyes that told Dale she was not trying to warn her as much as she was trying to gauge her. This epiphany sent a flood of relief through the lieutenant and then she was immediately presented with another hard fact. Cassidy was a lesbian. Her brain repeated that piece of information until it finally clicked.

Oh my God! Cassidy is a lesbian! Or at least bisexual.

Dale wanted to dance around the room singing "Oh Happy Day." Her horrible luck had just changed. Or had it? *Now, what?*

"Uh, yes, actually I did." Dale admitted, with a mischievous little grin.

Cassidy shook her head and glanced at the floor before she returned her attention to Dale, matching her grin. "Is this your first time here?"

"Yes. Is this yours?" Dale asked.

Cassidy squeezed in and leaned her forearms on the bar. "No. Actually, this is a regular hang out for me. And if you repeat that to anyone, just remember, you'll have to explain just how you know."

Dale put her right hand up. "They'll have to torture it out of me."

"Really? And just how difficult would that be?" Cassidy looked at her in a way that Dale could only describe as smoldering.

"Um…uh…listen, if my being here is going to ruin your evening, I can leave." Dale was prepared for Cassidy to agree to her going and, in a way, almost hoped she would tell her to leave. The proximity of Cassidy affected her greatly. Dale really needed a drink and searched for the bartender again, who was at the other end of the bar.

Cassidy ignored Dale's suggestion. "How did you know about this place?"

"I'm afraid you'll have to torture that out of me," Dale said.

When Cassidy laughed, Dale lost her breath. She again attempted to get the bartender's attention as she moved by but it seemed like a lost cause.

I'm thinking a shot or two of Jack Daniels would be good right about now.

Dale cleared her throat. "Can I ask you something, since we're kind of off the record here, anyway?"

"Sure."

"I thought you were with Drill Sergeant Robbins."

"Ted?" Cassidy laughed again. "Boy, that barracks really is rumor central, isn't it?"

"Aren't they all?" Dale asked.

"I suppose. Where did you get that idea?"

"One of the A-10 women saw you in Mobile at a romantic Italian restaurant with him."

"Ah. I knew we shouldn't have sat at that window table. Ted and I are good friends. We went to drill sergeant school together. We were celebrating my transference into Alpha-10. There is nothing romantic happening between Ted and me, I assure you."

She assures me. Why is she assuring me? Does she want me to know this for a reason or just to establish her allegiance to the fairer sex? Oh, good Lord, forget the Jack, I need a bottle of Cuervo, all to myself, with the worm.

Then Dale's practical, skeptical side kicked in.

Don't kid yourself, Dale, she is way out of your league. The only reason she is paying attention to you like this is because she's still trying to evaluate whether or not you can be trusted. Just relax and have a nice time with the stunning sergeant as long as she allows you to.

"I still can't believe someone from my company found this place," Cassidy repeated and finished the contents of her beer bottle.

"Is that why you hang out here? Because it's so far from McCullough?" Dale asked.

"That and it's convenient to where I live. My apartment is a half-block away. It's a long drive to work but it's worth it to keep my privacy intact." She looked at Dale. "It will remain intact, right?"

"Like you said, I can't tell on you without telling on myself." Dale was finally able to ogle Cassidy when the drill sergeant momentarily turned away. Cassidy's hair was long, silky, and flowing loosely. The length was halfway down her back, which was a bit longer than Dale had initially guessed. She was dressed in a plaid, cotton shirt tucked into tight blue jeans that accented her perfectly shaped behind and slender waist. She wore cowboy boots as opposed to the motorcycle footwear or earth shoes the other women in the bar seemed to be fond of. When Dale moved her way back up to Cassidy's face, she was met with two seductive, dark brown eyes that, at the very least, were amused. There was something else there, too…a hint of interest, maybe? Was the sizzling attraction Dale felt for this woman next to her reciprocated? Or was that too much to hope for?

Cassidy held Dale's gaze then released it with a shadow of a smile on her face. "You are the last person I expected to see here."

"And may I say ditto?"

"How's that temper of yours?"

"Fine as long as you don't make me mad," Dale joked. "Actually being on self-pace helps a lot."

The bartender flew by Dale again, ignoring her. "What do you have to do to get a drink around here? Dance naked on the bar?"

Cassidy leaned closer to Dale's ear. "Why don't you try it and we'll see if it works." Her tone was suggestive and her breath was hot on Dale's skin.

While Dale remained speechless and glued to her spot, Cassidy signaled the bartender who stopped and approached her instantly.

"How did you do that?" Dale asked, not sure whether to be impressed or annoyed.

"Hey, Cokie, another one for me and a..." Cassidy looked at Dale.

"Bud draft," Dale said.

"A Bud draft for my friend."

"You got it, J.J.," the bartender said and left to fill the order.

"I certainly hope you call her Cokie because of her love for cola," Dale said.

"It's a nickname. Her last name is Cocatello. Where's Walker?" Cassidy asked, suddenly.

"She's on CQ. And she doesn't know this about me."

Cassidy nodded in comprehension. "So, she's straight."

"Yes, but hopefully not narrow."

"Guess you'll never know until you tell her."

"If I tell her."

Cokie set a bottle in front of Cassidy and a huge, frosty, beer mug in front of Dale. "On your tab, J.J.?"

"Yes, that and whatever else my friend wants to drink," Cassidy told her.

"No, you don't have to do that," Dale said quickly.

"I know that but I'd like to. Relax, will you?" Cassidy

looked back at Cokie. "Give us both a shot. Maker's Mark."

Cokie glanced at Dale for the first time, then smiled at Cassidy. "Coming right up."

"How do you know I even do shots?" Dale asked.

"Someone as full of hell as you are? It would shock me much more if you didn't." The bartender handed Cassidy the two shot glasses filled with dark amber liquid. Cassidy gave Dale her glass. "To keeping secrets," Cassidy toasted. She downed the shot and chased it with her beer.

"Amen," Dale responded and swallowed her small tumbler of Kentucky bourbon. She expected bitter and strong and, instead got sweet and smooth, with an aftertaste of caramel and molasses. "That's not bad." She took a sip of her beer. "Thank you."

"You're welcome. Do you want to try and find a table?"

Dale hesitated. "Do you think that's a good idea? I mean, I appreciate the drinks but doesn't sitting at a table kind of insinuate that we are in here together, as opposed to standing at a bar next to each other?" Dale saw a trace of hurt in Cassidy's expression. "No, I don't mean it like that. I have no problem being in here with you, I'm just concerned that someone from the company might walk in here, see us together, get the wrong impression and get you in trouble."

Cassidy appeared relieved by Dale's explanation. "I appreciate your concern but, again, anyone who came in here would be here for the same reason we are and —"

"If they tell on us, they tell on themselves," Dale finished with her.

"Right. Except for you, tonight, I have never run into a female from any of my training companies. Occasionally, I'll see a permanent party soldier that I recognize in here but nothing has ever come of it. Now," she pointed to a table whose patrons were just leaving. "I've been on my feet all day and I'm tired. Would you like to join me?"

"Yes, I would, thank you." Dale followed Cassidy to the empty table.

Don't do this. Do go any further. Thank her for the beer and the shot and get the hell out of here. Forget how sultry and desirable she is! She's a sergeant, you're a lieutenant and she thinks you're a trainee. All the way around it is fraternizing and you're just asking for trouble. On the other hand, despite what you think you might have felt from her, it befits her to be nice to you. She's got nothing to lose by being friendly and making you feel like a friend so you won't betray her trust on this issue. So why don't you just enjoy her company while she's willing to treat you like a peer?

Dale sighed. It was just silly of her to wish for anything more. She couldn't see Cassidy being that careless.

Every individual, couple or group they passed said hi to Cassidy and then eyeballed Dale. The scrutiny made Dale wonder if they thought she was Cassidy's new girlfriend and immediately was envious of any woman who was in the past, currently or would be her girlfriend in the future. As she sat at the table, she took a long drink of her beer and speculated on Cassidy's current status. Did she have a lover? Of course, she had to have at least one. A woman as luscious as Cassidy? There was no way she was alone.

While Dale was momentarily staring into space and hypothesizing, Cassidy had gone back to the bar and returned with two more shots. "Thank you again, Drill Sergeant," Dale said as Cassidy sat down.

They clinked glasses together and drank the shots.

"You're welcome under one condition. You drop the drill sergeant and call me J.J. Just for tonight." She saw the skeptical expression on Dale's face. "We can keep this formal if you'd like but I would think having to say drill sergeant after every sentence has to be uncomfortable. Especially this far away from post. The other, more important reason is I don't know everybody in here and you calling me drill sergeant might alert a shoofly."

"Understood. It's just awkward."

"It's awkward for me, too, but I'm pretty sure you're not going to run back to the company and spread it around that I bought you a drink or two in a gay bar."

"You're right." Dale looked at her. "J.J."

Cassidy grinned. "Much better, Dale."

"You know my first name?" Dale was genuinely surprised. Cassidy winked at her, a gesture that caused Dale's heart to race and a blush to crawl up her cheeks. She cleared her throat. "Do you have a girlfriend?" Dale blurted out, then realized how it sounded. "Never mind, it's none of my business."

"No, that's fine. I've been single about seven months. Do you?" Cassidy asked, her eyebrow arched in curiosity.

"No. Seriously? Someone who looks like you is single?"

"I was thinking the same thing about you."

Dale flushed again and Cassidy smiled radiantly at her.

I am in so much trouble...

Several drafts and a few shots later, both women were quite comfortable and relaxed with one another. They got past the small talk, discussed many varied subjects and, one time, even got into a heated debate about Viet Nam. Just before the bartender announced last call, Cassidy took Dale's hand and lightly tugged her to the dance floor. It was a slow song and suddenly Dale was terrified. She knew she should have said goodnight before the evening got to this point but the second she was in Cassidy's arms, she felt helpless to do anything but stay there and savor the sensation.

The song, "Close The Door" by Teddy Pendergrass, did nothing to deter the intensity of feelings that surged like a current through and between the women. Cassidy pulled Dale as close to her as two bodies could get and moved against her so sensually that Dale thought she might pass out from the sheer bliss of it. She rested her head on Cassidy's shoulder and Cassidy's cheek touched Dale's forehead. When the music stopped and Dale slowly let go of Cassidy, the drill sergeant tightened her grip on the lieutenant. Dale looked up at her in question and was answered with a kiss.

Cassidy bent her head gradually toward Dale's, giving Dale a chance to stop the action but Dale was paralyzed and couldn't have prevented the kiss at that point, under any condition. The drill sergeant didn't close her eyes until her lips were touching Dale's.

There was nothing timid about the action. It was long and tender and thorough. When the women parted, both knew they had definitely been kissed. Without another word, Cassidy grabbed Dale's hand, led her to the rear hallway, backed her against the wall and kissed her again. Dale's body was trapped between Cassidy and the wooden divider that separated the storage room from the bathrooms. Cassidy captured Dale's lips in another passionate kiss. As their mouths and tongues pressed and played, Dale prayed the kiss would never end. She had never melted into a kiss before, not even with Bishaye, and the feeling was devilishly divine.

Making out with Cassidy was different than kissing Bishaye. It was hard to explain, other than to say the incidents with Bishaye felt almost desperate as opposed to the pure desire she detected from Cassidy. Time seemed to stand still as they stood in that one place, explored one another's mouths, and stirred themselves into a sexual frenzy. It wasn't until the sound of a throat clearing behind them brought them to consciousness of the real world around them.

"J.J.? Sorry to disturb you but we have to lock the doors." It was Cokie, the bartender, who spoke.

"Right." Cassidy rested her chin against Dale's forehead. "Can you give me a minute?"

"Sure." Cokie walked to the storage room and left them alone.

"Come home with me." It wasn't a question, it wasn't a demand, it was a simple statement that Cassidy put to Dale. She stroked the side of Dale's face and lifted Dale's chin to give her a quick kiss.

Every ounce of inherent prudence that should have screamed *no* fled Dale's system the second Dale looked into Cassidy's revealing eyes. Dale's body's needs double-crossed her common sense. "Yes. I'd like that," she whispered to the enticing beauty in her embrace.

Dale did not remember much about the short walk to Cassidy's apartment, other than the binding handholding. Neither spoke and Dale thought that was for fear that the other might change her mind before they reached their destination.

Once Cassidy shut and locked the door, she reached for the light switch but Dale's hand grabbed her wrist and stopped her.

"Please, don't." She brought Cassidy's fingers to her lips and kissed them.

"You like the dark. I don't blame you, so do I." She tugged Dale to her and kissed her deeply.

Dale reluctantly broke the kiss and sighed. "I shouldn't be here," she said, quietly.

Cassidy locked her hands together behind Dale's back. They swayed, like they were moving to a leisurely waltz. "I know. It's not too late. Do you want to call a cab?"

"I should. Do I want to?" She looked up into Cassidy's eyes, making sure Cassidy understood her intent. "No, I don't."

Cassidy smiled, relieved. "Good." She kissed Dale again and guided her backward, into the bedroom. "How is it you haven't picked yourself a little playmate while you've been here?"

"How do you know I haven't?" Dale asked, slyly.

Cassidy let her go while she removed her jacket. "Do you think this is the first time I've seen you? I watch you constantly. I know you haven't paired off with anyone except Walker and I know she has." She helped Dale off with her coat.

"You watch me?"

Cassidy smiled. "Ever since I laid eyes on you that first time in the WAC dining hall."

"You remember that?" Dale was shocked.

"I couldn't get you out of my mind. When Ted told me about the vacancy in the company, I knew I had to try for it."

"You...you came to A-10 because of me?" Dale blinked in astonishment.

Cassidy steered Dale onto the bed and joined her. She propped herself up on her elbow. "Yeah. You affected me that much."

"What made you think I was a lesbian?"

"I didn't think you were but I'm a hopeless romantic and wanted to look at you every day."

"Wow." Dale sincerely couldn't believe someone as tantalizing and physically perfect as Cassidy would be that attracted to her. But here she was, lying on Cassidy's bed with her, soon to be engaging in the most intimate act two people could share. Then a different thought crossed her mind. She was careful in the tone she used to ask the next question. "You sure you didn't secretly have a crush on Lieutenant Henning and that's why you wanted Alpha-10?"

"Henning?" Cassidy appeared sincerely surprised by the question and not evasive in the least. "No. She's cute, don't get me wrong, but she's very straight and even if she wasn't, she's not my type."

"I'm your type?" Dale was flattered.

Cassidy gestured to the bed they were face to face on. "Evidently." She placed herself on top of Dale and kissed her breathless again.

"J.J., I have to tell you something…"

Cassidy was busy kissing Dale's neck and nipping at her earlobe. "What's that?"

"It's a little embarrassing…but…I've never, well…tonight was my first time, ever, at a lesbian bar."

Cassidy stopped nuzzling the nape of her neck, raised her head and studied Dale's expression. "Ooookay? And, what are you saying? First time ever with a lesbian, too?"

Dale nodded, not sure what to say next. She was sure she could have made moves up as she went along but she was scared shitless and felt that Cassidy should know the truth.

"Am I just an experiment?"

"No. I've suspected for a long time that this is my true inclination and you can prove to me what I've known all along."

"And what's that?"

"That being with you will finally make me feel connected and be so much more satisfying than anything I've previously experienced." Dale smiled at her. She already felt more intimacy from Cassidy's touch than she'd ever felt with any of the men she'd been with, including Keith.

Cassidy smiled back. "That's a lot to live up to."

"Somehow I don't think you'll have a problem." Dale gasped as Cassidy dipped her tongue into the hollow of Dale's neck. "So, I guess you're okay with this?"

Cassidy unbuttoned Dale's denim shirt, kissing bare skin as she proceeded. "What do you think?"

The pleasure in the undivided attention Cassidy gave Dale's body as she undressed her had already demonstrated Dale's theory. No man had ever produced the exciting, shivering effect in her Cassidy had and this was only foreplay.

Dale's experience with men was always more of a means to an end. She was hot and bothered, she needed release and she wanted to share it with another person. The men she knew were always willing, including Keith. Sex with him was always comfortable and fun but never fulfilling. Dale had been with more men than she wanted to admit to and not one of them had ever, truly satisfied her sexually. There were a few clumsy attempts at oral sex but none of her partners were ever really fond of the act and always stopped before Dale could get there. She had always believed that she was the one with the problem and that maybe her expectations were too high in the sexual arena.

Yet here she was with J.J. Cassidy who had done nothing more than kiss her bare skin, no probing, no stroking, no sucking, just the touching of lips to flesh, and Dale was so aroused, she could barely breathe. She had never felt as though every nerve ending in her body was electrified.

Cassidy undressed herself as Dale watched. Her eyes had adjusted to the darkness of the bedroom so seeing her was not an issue. The differences in their bodies were noticeable. Dale was light complexioned and Cassidy's tone was olive, Dale's build was trim and athletic and Cassidy's was slender but sinewy. Dale's breasts were full, with pink nipples, while Cassidy's were small, with brown nipples. It wasn't as though Dale had never seen another woman's body before, after all, in the military, you forfeited modesty for the sake of expediency, but never one that elicited a fire in her like Cassidy's did.

When Cassidy joined her on the bed again, she covered them in a puffy quilt. It felt nice for a moment but Dale knew within minutes, they'd be kicking it away as they'd generate their own heat. She kissed Dale passionately and moved down her body to concentrate on Dale's breasts. The solicitous reverence with which she treated them was unlike anything Dale had ever been exposed to. Who knew that kind of attention would cause her nipples to be so sensitive? Cassidy sucked, licked, and nipped her to the edge of madness and then moved south. If Dale had been able to speak, she would have cursed soundly at her.

PERMISSION TO RECOVER

Dale was more than ready for Cassidy to settle between her legs. She anticipated this experience with not only someone who knew what she was doing but enjoyed it, as well. Dale was not disappointed.

The fact that her body was already in a heightened state of arousal didn't hurt and the sensations that blazed through her, especially in her lower anatomy led to the most powerful orgasm she could ever remember having. Cassidy didn't stop her ministrations while Dale climaxed and whatever magic Cassidy created with her tongue brought Dale right back to gratification. Dale looked down at the woman pleasuring her and watched her as she proceeded to feast. When Cassidy opened her eyes and looked up at Dale while she continued to indulge her, Cassidy's wanton expression propelled Dale over the edge again.

Dale was convinced at that point that Cassidy must have sucked all the moisture out of her body. She felt weak and spent. Yet when Cassidy inserted a finger and then added another, there was no dryness anywhere, if the sound from her southern regions was any indication. Cassidy slid back up Dale's body and maintained a rhythmic thrust with her hand. She straddled Dale's thigh and rocked her body to match the cadence of her fingers. Cassidy's center was hot and she was dripping wet, a realization that turned Dale on even more than she thought was humanly possible.

The friction was deliciously feverish and less than a minute after Dale crested, Cassidy matched her fervor and cried out Dale's name in ecstasy. She took Dale's hand and brought it to her center while panting. "Go inside me, Dale. Go inside." She guided Dale's fingers to her need and slid them in. Cassidy rode them to a second climax then collapsed on top of Dale, trying to catch her breath.

Dale didn't mention that her fingers were still inside Cassidy as she was sure she knew. With her free hand, Dale stroked Cassidy's back as Cassidy's respiration returned to normal.

"That was amazing," Dale said, the awe clear in her voice.

Cassidy chuckled. "Yeah? Which part?"

"All of it but especially being inside you."

Cassidy brushed her sweaty hair away from their faces and

she answered Dale with a searing kiss. "You're amazing," Cassidy told her.

Dale suddenly felt like weeping. First, Cassidy had given Dale her first honest-to-God vaginal orgasm and second, Cassidy made love to her like she meant it. That was also an alien experience for her and it touched her. She pulled Cassidy down for another torrid kiss. "Tell me what to do to please you."

The heavy drapes kept the bedroom dark but when Dale awoke, she instinctively knew it was light outside. She guessed it to be late afternoon. She felt disoriented and hungover. Her eyes finally focused and she tried to get her bearings and recall where she was. She was naked and her entire body ached. She started to turn to see whose tanned arm was draped across her middle and in mid-turn, she remembered.

She looked at the sleeping form to confirm her fears. Her head fell back on the pillow and she stared at the ceiling.

Bishaye is going to have my head. And well she should. I've probably just blown the impartiality of the whole case.

Dale gently took Cassidy's hand and tried to move her arm without disturbing her. She had almost accomplished that feat when Cassidy stirred, then awoke. Cassidy seemed to be a lot more enthusiastic about Dale's presence in her bed than Dale was.

If only I could have met you under different circumstances, Dale thought.

"Morning," Cassidy said. Her voice was raspier and definitely sexier than usual. She interlaced her fingers with Dale's and placed a lingering kiss on Dale's temple.

"Afternoon, you mean," Dale said, quietly. "Do you mind if I take a shower?"

Dale's demeanor was restrained, much different than the night before and Cassidy noticed. "No, of course I don't mind. Clean towels are in the closet by the bathroom door."

"Thanks." Dale attempted to rise but Cassidy pulled her back down.

"Do you have to run into the shower right now?"

Dale glanced at the alarm clock on Cassidy's nightstand, then looked at the disheveled woman whose bed she shared.

PERMISSION TO RECOVER

Even first thing in the morning Cassidy was a stunner. "Um...yeah...I have to meet some friends in a couple hours."

"Can't you call them and cancel?" Cassidy sounded calm but almost wounded.

Dale avoided her gaze. "No. I wish I could."

"Dale? Hey, please look at me," Cassidy requested, softly. "What's going on? Are you okay?"

Dale's eyes met hers. "I'm not sure."

Cassidy sat up and leaned her back against the headboard. "Is it our situation?"

Dale nodded and looked down at the sheet that covered her, avoiding Cassidy's bare breasts.

"I would love to tell you that we didn't do anything wrong but, according to regulations, we did. On more levels than just the glaring one. Do I feel like we did anything wrong? No. In fact, nothing has felt this right for me in a very long time. It's not that I needed you, Dale, I wanted you and have from about the very first time I saw you. I don't want you to leave here thinking it was anything else, okay? If all I'd wanted was a quick fuck, I could have gotten that anywhere. I wanted to be with *you*, Dale, and it was fate that you walked into Gems last night. And I know you felt something for me. It was obvious." She studied Dale and her voice broke into a whisper. "Please tell me you felt something for me."

"I did. I do...but...it never should have happened, J.J. You're my drill sergeant! It's unethical."

"You're almost out of here," Cassidy protested. "You're already on self-pace. I really have no influence on you anymore, other than you possibly having to corral you in occasionally. You're not in my platoon. Dale—"

"If we get found out, no one else will look at this like you are."

"If we don't say anything to anyone, no one will know."

"We'll know." Dale wanted to kick herself for this. There was no way she would be able to keep this to herself. If Cassidy only knew...

Cassidy watched Dale trying to process the events of last night. "Come on, I wasn't that bad was I?" Cassidy's smile was impish and it coaxed a smile out of Dale.

"Bad? Uh, no. I can think of a lot of words to describe last night and bad isn't anywhere close to any of them."

Cassidy arched an eyebrow. "Want to do it again?"

Dale looked at her for any hint of kidding in her expression. "Jesus, no! I'm raw as a radish now. We had continuous sex for nearly four hours, how can you even sit comfortably?" She laughed as she spoke.

"Because I can't get enough of you and if this is the only time I'm going to have with you, I want to make the most of it." Cassidy reached out and took hold of Dale's wrist. To her amazement and delight, Dale didn't resist.

She pulled Dale on top of her and brought their lips together for a kiss.

"Sorry for the morning mouth," Dale said.

"Me, too. In a moment, it will be the least of our worries." The promise in Cassidy's voice mixed with the touch of her hands on Dale's body riled up Dale's libido again.

"You're hard to resist, you know that?" Dale said. "Could you be a little gentle on the lower half? I'm going to have trouble walking as it is."

"Good. If I had my way, you'd be on permanent bed rest."

Dale allowed Cassidy to start stroking her, feeling a radiating tingle in her groin almost immediately. She was sober now and she knew the consequences of her actions. She couldn't use intoxication as an excuse. As Cassidy sealed her lips around an overly responsive nipple, Dale cast aside any lingering guilt.

Over coffee, Cassidy offered to drive Dale into town but Dale declined. She needed time alone, to think. She called a cab and while she waited at the window, Cassidy silently stood behind her with her arms wrapped around Dale's waist. When the taxi pulled up, Cassidy turned Dale around and kissed her goodbye with such feeling behind it, Dale's knees nearly buckled. It was not just hormones and physical attraction that drew her to Cassidy, it was her emotions, as well. It was as if all her senses were trying to tell her something about this woman, something she really wanted and needed to hear but protocol wouldn't allow her to accept.

She sat in the back of the cab and silently wept, tears

PERMISSION TO RECOVER

streaming down her face all the way into Averill. On her instruction, the driver dropped her at the first bar they came to, where she stayed for the next hour and drank straight shots of Southern Comfort.

CHAPTER FIFTY

Shannon was a pacer by nature. She had always been one and chances were she would never stop being one. When Dale hadn't shown up and was now three hours late, Shannon was wearing a path in the carpeted room and with good reason. Dale was many things, but inconsiderate was not one of them. If she were all right, she would have phoned to let Shannon know. Something was wrong. Dale hadn't returned to the barracks last night and no one at the Inn had seen her. She wasn't at the EC, the Pizza Place or any of the other frequented hangouts. None of the people she socialized with, when Shannon wasn't available, had been with her. It was as though she'd disappeared off the face of the earth. If she didn't show up in another hour, Shannon was going to have to contact Bishaye, which was not an appealing thought, especially if Dale hadn't shown up at Bishaye's, either. Still concerned, Shannon called McCullough yet again.

"Alpha Company, Tenth Battalion, Private Almstead speaking, may I help you, Sir?"

"Lex, it's Walker."

"No, Walker, she hasn't showed up yet."

"Has anybody seen her?"

"York and that guy she's dating said they thought they saw her getting out of a cab in town a couple of hours ago but they weren't positive," Almstead said.

There was a knock on the door. "Somebody's here. I'll talk to you later," Shannon said.

"Maybe it's her. If not and she shows up here, I'll have her call you."

"Thanks." Shannon hung up the phone and walked to the door. "Who is it?"

"Shhmeee," a familiar sounding voice slurred.

Shannon slammed the locks back, opened the door and Dale stumbled into the room. Shannon gaped at her. "You're drunk!"

"So am I," Dale said, barely coherent. She tried to focus on Shannon. "Stand still, would ya?" After three attempts, Dale's hand found the wall. "God help me, I've got the whirlies."

Shannon slammed the door shut. She was so angry she could hardly speak. "I'll kill you, Dale, you shit!" She grabbed her inebriated buddy and walked her into the bathroom where she pushed Dale into the shower and turned on the cold water full blast. Shannon shut the shower door and leaned against it, holding Dale hostage. Shannon folded her arms angrily across her chest and started to yell unflattering things at Dale in a very un-ladylike tone of voice. "We're supposed to be doing a job here. While I'm frantic thinking you've been butchered and used to fertilize someone's garden, you're out getting totally shitfaced! How could you do this to me? This was so irresponsible. You could have at least called me to let me know you were okay or invited me to join you. God, I am so *pissed* at you right now! I'm your partner, for Christ's sake! If I can't depend on you to even let me know where you are —"

"All right, all right already! Okay! You're right," Dale managed to say, intelligibly. "I was wrong, okay? I apologize from the bottom of my heart. I am sorry. Now would you please let me in? It's raining like a bitch out here!"

Shannon shook her head and walked away from the shower door. "Wrinkle, Goddamn it."

"Shan? Shannon? Well, shit."

Dale slowly became aware of her surroundings and realized she was not outside during one of Alabama's monsoons and, instead, in a shower where she could control the water temperature. She finally was able to adjust the hot water and regulated the degree of heat.

When the water warmed up, Dale removed her wet clothes and slowly sobered up as the hot water opened her pores and helped steam the toxins out. Thirty minutes later, she stepped out of the stall, stuck her finger down her throat and threw up whatever liquor was still rolling around in her stomach. She scooped some cold faucet water into her mouth, rinsed it around and spit it out, the taste of vomit going with it. She dried off with a towel and cracked the door open. "Do you have a robe I can borrow?"

Shannon brought Dale a terrycloth robe. Dale accepted it and put it on. She mumbled a thank you to Shannon and bowed her head.

"Are you okay?" Shannon's voice was minus the agitation it held an hour before.

"No. Not really." Dale sounded vulnerable and that made Shannon appear all the more curious.

Shannon reached out, took hold of Dale's sleeve and led her out of the bathroom. They both sat on one of the double beds. "What happened?"

It took Dale a minute to speak. She was at a loss but she knew she had to tell Shannon because Shannon had to help her figure out what to do. "I think I may have done something damaging to the objectivity of our case."

"What do you think you did?" Shannon literally held her breath.

"I...uh...I spent last night with a drill sergeant."

Shannon tried to cover her obvious shock but her tone was panicky. "Spent, as in slept with?"

"Yes."

"One of our drill sergeants?" Dale nodded. "Which one? Who was it? It was Holmquist, wasn't it? I knew it! I knew —"

"No. It wasn't Holmquist."

Shannon was stumped. "It wasn't Holmquist? Then which one?"

"Cassidy."

Shannon looked at Dale as if she hadn't heard her correctly. "What? Cassidy? *Cassidy?* Dale, Cassidy is a *woman.*"

"Yeah...I noticed that...when I was in bed, naked, with her."

Shannon still looked perplexed. "You had sex with a woman?"

"Not just any woman. Drill Sergeant Cassidy." Dale reiterated. The visual of their lovemaking replayed in Dale's head.

Shannon saw the smirk form on Dale's face and swatted her.

PERMISSION TO RECOVER

"Cassidy is supposed to be having an affair with Robbins."

"Don't believe everything you hear. They're just friends."

"I can't believe this. You're a dyke?"

Dale immediately wanted to recoil and protest Shannon's words and get defensive like she would have six months ago but six months ago, Dale wasn't as conscious of her inclinations as she was now. "I have discovered that I am more attracted to women than I am to men," Dale confessed.

"Since when?"

"Since a while, I think."

"You *think?*" Shannon was still gaping at her. "Dale, how could you not tell me?"

"I wasn't positive about myself. I didn't want to say anything until I was sure. If nothing else, last night made me sure."

"Last night was your first time with a woman?"

"Yes."

"So you've never…you know…with Bishaye?"

"No. I have a wild crush on her but that's all. Why would you ask that?"

"That would explain the strange tension between you two. Does she know you have a crush on her?"

"Unfortunately, she does."

"So, she knows you're a dyke?"

"No. And could you stop calling me that?" Dale asked.

"Why? It's not like this will be the only time you'll be called that. You'd better get used to it. Now, you want to tell me how you ended up with Cassidy last night?"

Dale couldn't read Shannon's expression. Her tone still sounded somewhat appalled but Dale couldn't tell if it was from shock that Dale may have blown the case or disgust from Dale sleeping with Cassidy. "I took a taxi to Franklin last night after I left the EC."

"Franklin! Jesus, Dale, that's more than an hour away. And then what? You just happen to walk into a bar in Franklin and coincidently met up with Cassidy?"

"Actually? Yes. I knew of a particular bar and since I was alone, I went in. I didn't see Cassidy but she saw me."

"A particular bar? You mean a gay bar?"

"Yes."

"How did you know about this place?"

"I'd heard about it a long time ago. I looked in the phone book to see if it was still listed and when I saw it was, I decided to go. Just to see what it was like. I never thought I'd run into anyone I knew…especially not Cassidy."

"Yeah, I bet that was a shock. So, let's get to the bigger shock. How did you end up in bed with her?"

"Like I said, she recognized me. We chatted a little bit. I think we both wanted to make sure that neither of us was there to cause any problems."

"Like what?"

"You know what a shoofly is, don't you Shan?"

"An undercover snitch."

"Pretty much. So when we both decided the other was safe, she bought me a beer and a shot."

"Why didn't you refuse her?"

"It was just a beer and a shot," Dale repeated. "I asked her if I should leave and she ignored the question. Then we just started to talk. We sat at a table and we drank more and talked more and…we danced. Then…we kissed. She asked me to come home with her and…I did."

"Jesus, Dale! You should have left if you were feeling that vulnerable."

Dale shot a sharp look at Shannon. "Why? Because she's a woman? Would you be saying the same thing if it was Holmquist? Weren't you the one who just said to me I needed to get laid?"

"Yes, but —"

"But what? It doesn't count if it's another woman?" Dale was livid. "It counted for me! I felt – I *feel* something for her. It wasn't just sex. I apologize if she wasn't who you would have picked for me."

Shannon stood up and started to pace. "This is not about that. Dale, I have nothing against gays and don't turn this into that. I am upset because you didn't tell me about you and what you were feeling. I thought you trusted me!"

"I do trust you, Shan."

"Really? Then why, when I teased you about Buckman and

that sergeant at the EC, didn't you say something then? When you said you were waiting for something special, you weren't kidding."

"So, what you're saying is that you're not mad that I'm a lesbian, you're mad because you think I didn't trust you enough to tell you."

"Exactly!"

"It doesn't bother you in the least that I had sex with a woman?"

Shannon hesitated. "I just need to get used to the idea." She stopped and looked at Dale. "Your first time, huh? You could have done a hell of a lot worse than Cassidy. I'm not gay and even I might've straddled the fence for her." She sat down next to Dale again. "I never, not in a million years, guessed that she's a lesbian."

"You didn't suspect me, either."

"True but I know you. Well, I thought I did."

"You still know me, Shannon. I'm the exact same person I was before you found out. The only thing that has changed is your knowledge of my orientation. It wasn't about not trusting you...I just needed to be sure about myself before I said anything and, frankly, I wasn't exactly sure about where you stood on the issue. We've never really talked about it and I needed to make sure that you weren't one of those people who would suddenly hate me because of who I choose to share my bed with."

Shannon nodded. "Too bad she has to be one of our drill sergeants, huh?"

"Yes."

"Did you get drunk today because you slept with a woman?"

"No. I got drunk because I slept with a drill sergeant and I realize all the negative consequences that could come with that. With that one decision, I could have blown this whole case."

"Further proof that guys aren't the only ones guilty of thinking with their lower anatomy. Do you think she has anything to do with the case?" Shannon asked.

"I don't think so."

"You don't think so or you don't want her to be?"

"Both. My gut tells me that she's not involved but she told me that she transferred to the company because of me."

"What does that mean?"

"When we first got here in November, I got nailed in WacVille for staring at a drill sergeant. It was her. She was such a fox that I was mesmerized. She said she couldn't stop thinking about me and when Robbins told her there would be a vacancy at A-10, she put in for it. With Robbins' help, she got it."

"You know what that sounds like to me? Like she targeted you."

"It does but Shannon, that would mean that she knew who I was and what I was doing here and I didn't get that from her. The only way she would have access to that information would be through Bishaye or Henning."

"Or Colton."

"Well, we know Bishaye wouldn't plant Cassidy and Colton is too much of an idiot and we're not sure what is going on with Henning. I'm sorry but I really don't believe she is a part of it."

"Dale..."

"Shannon, I know how it looks but I did ask questions and I didn't get any feeling whatsoever from her that she had anything to do with this."

"You interrogated her in bed?"

"No, of course not. I joked around about her really wanting to be in our company because Henning was such a cutie and she said Henning was straight and not her type. I brought up the fact that according to rumors, other A-10 drill sergeants had fried for what we were doing. She told me she wasn't sure what was going on with that. I asked her why she would take the risk with me and she told me that I was different and that she had wanted me from the moment she saw me."

"Dale, I can see how that would flatter the hell right out of you but if you listen to that from my point of view? It doesn't sound good."

"We talked for a long time at the bar. It wasn't like *can I buy you a drink, let's fuck*, okay? And remember, she had no idea I was going to walk into that bar because I didn't even know I was going there until I got into the cab. Neither of us could have known the other one was a lesbian before last night.

PERMISSION TO RECOVER

It wasn't planned and if she was one of the bad guys and knew who I was, she would have certainly targeted me in a different setting before last night."

"That does make sense." Shannon took a deep breath and blew it out. "What about the case now?"

"I think we stay on track and just play it through to the end. We're looking for set-up trainees, not drill sergeants, yes?"

"That was before we knew that it has something to do with Henning. And how do you know you're not the only trainee Cassidy has messed around with?"

"I...I don't know that." The sudden thought devastated her. "I don't think a drill sergeant with a reputation would have been allowed anywhere near Alpha-10."

"Doesn't it make you wonder now about the other women who pressed charges? Maybe they were as vulnerable as you and couldn't handle it when it was through and pressed charges."

"The reports didn't read that way and Stuart wouldn't have willingly slept with a male drill sergeant and...Jesus, Shannon, don't do this."

"I have to play the devil's advocate, Dale, because you're a little biased now. This is our sixteenth week here and nothing has happened except for the bite with Henning and now your situation. As far as we know, no one has gone anywhere near Henning and we've kept a pretty good eye on the other women. The only one we haven't felt the need to check on is Michaelson because we know where she is and why and she doesn't go near the drill sergeants or Henning."

"I know. Honestly, Shan, if I had any suspicious feelings about Cassidy, I would have gotten out of the bar. I think she's okay and I think what happened last night was sincere."

Shannon sat in silence for a moment. "I hope so, for your sake." She put her hand on Dale's shoulder. "And now for the most important question of all. How was she?"

Dale's blush and smile told Shannon a lot. "Why is that so important?"

"I want to make sure all the anguish you're going through is worth it."

"She was incredible. The best sex I've ever had."

"Really? The best?"

"Without question. And no, I am not going into details."

"Don't want them. Are you going to see her again? Romantically, I mean?"

"No. I made that very clear."

"Does she want to see you again?"

"Yes."

"Man, that's going to be difficult, isn't it? You still have to see her professionally and take orders from her and —"

"We agreed it won't be a problem."

"Let's hope not. Now, what are you going to tell Bishaye?"

"Why do I have to tell her anything?" Dale immediately became flustered.

"Dale, you have to tell her. If Cassidy is involved in this in any way —"

"She's not."

"*If* she is, Bishaye needs to look into her background. You *have* to give her a head's up. I don't want us to have gone through this to be screwed in the end by someone who screwed you, all puns intended."

"But that will kill Cassidy's military career, either way."

"Look, Bishaye's your friend, right? If Cassidy is innocent, talk Bishaye out of bringing her up on charges. It's not like Cassidy is the only lesbo drill sergeant. If Bradbury can get away with her shenanigans, Cassidy should be fine." Shannon noticed the funny look on Dale's face. "What's the matter?"

Dale took off for the bathroom. "I'm going to throw up again."

"Is that because of the liquor or because you have to tell this to Bishaye?" Shannon yelled after her.

Dale rested the remainder of the evening while Shannon partied with some other members of the company in town. Dale couldn't get Cassidy out of her mind, the way she looked, her expressions while they made love, her scent, her feel, her taste. When she woke up the next morning, she wasn't sure if it had really happened or if she'd dreamt it. When Shannon reminded her that she needed to tell Bishaye, Dale remembered sex with the impressive Staff-Sergeant Cassidy was a reality. And so would be getting her ass chewed out by Bishaye.

After the two lieutenants went out for coffee, they returned to the room and Shannon bugged Dale about calling their boss. Dale procrastinated and watched cartoons until Shannon turned off the television. "*Now*, Dale, before we check out of the motel and leave Averill."

Dale picked up the phone, hovered her finger above the zero and then hung up. She picked up the receiver again and dialed the motel operator. She was connected to an outside line and dialed Bishaye's number. As the phone began to ring on the other end, Dale looked up at Shannon. "I don't think she's home."

"How many times has it rung?"

"Three times."

Shannon gave Dale a tolerant stare. "You want to give her a chance to get to the phone?"

"It's a small house, she would have answered by— hi, Anne, it's Dale." Dale's expression was, at the least, apprehensive. The look on Shannon's face told Dale that Shannon wouldn't have traded places with Dale for any amount of money.

"Well, it if isn't my favorite lieutenant," Bishaye said, jovially.

"You're in a good mood," Dale commented and knew that would be short-lived.

"Any reason I shouldn't be?" Bishaye asked.

Dale ignored the question. "Anything on the lead we gave you?"

"Not yet. Why? Do you have something more for me?"

"No, we've been waiting on you."

After an awkward momentary silence, Bishaye spoke again. "What's going on, Dale?"

Dale swallowed hard and closed her eyes. "I need to see you. Today, if possible. I need to talk to you about something."

"Sounds urgent. Is it the case?"

"It's personal."

Anne sighed. "Is it you and me personal or something else? Because if it's about you and me —"

"It's something else."

"Can you give me a hint?"

"I'd rather not. I really should discuss this with you in person."

"Can you come here?"

"No. Can we find a halfway point?" Dale definitely wanted a public and neutral location where she knew Bishaye would be hesitant to cause a scene.

"We are not going to the Cloud Club again."

"It's the perfect place, Anne. We won't run into anyone we know there."

"All right. But it better be worth it this time and not something you could have told me on the phone in one sentence." The annoyance left Bishaye's tone. "Are you okay?"

"I am right now," Dale said, unconvincingly.

"This doesn't sound good."

"Cloud Club in thirty minutes?" Dale asked.

"I'll be there before then. I'll leave your name at the door."

Dale hung up the phone and looked at Shannon. "Please kill me now."

"Hey, they're supposed to have excellent food there. The least she can do is buy you a last meal," Shannon said.

Dale entered the lounge after being cleared at the front door. She spotted Bishaye at a booth in the corner and made her way there. The women seated at the bar and at the tables paid no attention to the undercover lieutenant as she passed and as expected, she recognized no one.

"Hi," Bishaye said, the interest clear in her voice. She waited until Dale sat down opposite her before speaking. "I took the liberty of ordering you a beer. I got the impression you needed one."

"Thanks." Dale folded her hands on the table like an obedient schoolgirl.

Bishaye scanned Dale's face before speaking again. She sounded concerned. "Okay. Out with it. What happened?"

The waitress appeared at the table with a bottle of beer for Dale and a half-carafe of wine and a glass for Bishaye. "Menus?" she asked.

Bishaye looked at Dale in question. When Dale shook her head, Bishaye smiled at their server. "Not yet. Can you check back when the drinks are gone?"

"Sure."

When she was gone, Bishaye reached out and patted Dale's hand. "You know you can tell me anything, right?"

"Can I? Really?" Dale looked into her eyes, knowing how easy this was *not* going to be.

"I certainly hope so." Bishaye poured the wine in her glass and set the carafe back down.

Dale took a long swallow of her beer. "I think you're most likely going to make more of this than you should. I want you to keep in mind that I'm telling you this as a friend, okay? Not as employee to boss or as it relates to the case. This is a private conversation between Dale and Anne."

"I don't like the way this is beginning. Why do I feel this *is* about the case?" Bishaye's eyes didn't leave Dale's face. When Dale finished her beer in four more swallows, Bishaye chuckled. "My, aren't we thirsty? I should have known better than to have had anything less than a keg at the table waiting for you. Do you want another one?" She didn't wait for Dale to answer. "What am I saying? Of course you want another one." She signaled the waitress and pointed to Dale's bottle. Soon Dale had a fresh beer in her hand and they were alone again. "Don't drink too much of that. I want you to remember the reason you dragged me here. Again."

"Don't worry. I won't get drunk. That's what got me into this mess in the first place."

"And what...mess...is that?" Bishaye asked cautiously.

When Dale finally spoke, she looked everywhere but at Bishaye. "I...uh...I...slept...with one of my drill sergeants two nights ago."

"What did you say?" Bishaye clipped off every word. There was no doubt she'd heard what Dale said. She leaned on the table.

"You heard me." Dale still could not look at her.

"Who was it?" Bishaye didn't yell. Her voice was controlled and that made Dale happy she had insisted on meeting in a public place.

"Let's wait on that," Dale said. "Until you calm down."

"Calm down? This is the calmest I've ever been when I've

been furious. Why should I be upset? I call in several favors to arrange to get you and your partner on this case, promising everyone I asked that you are the best person for the job, that you can be implicitly trusted. I get you here to specifically find out why these women are lying about having sex with their drill sergeants and what do you do, of all things? Have sex with your drill sergeant! When I said undercover work, I didn't think you'd take me quite so literally. Just what, exactly, is your fucking problem, Dale?"

When Bishaye said the word *fuck*, it was never a good sign. In fact, in recent years, most of the situations where Bishaye resorted to a salty vocabulary were usually incidents where Dale was directly involved.

"I don't have a problem," Dale said, quietly. "And would you please calm down?"

Bishaye took several deep breaths and tried to collect herself. "Just tell me what happened, how about if we start with that?" When Dale didn't answer right away, Bishaye took a sip of wine. "What aren't you telling me? Did something...bad happen?"

Dale finally made eye contact with her. "Yes. I slept with a drill sergeant."

"Yes, you said that. What I meant was, did he force you?"

Dale noticed the word *he* in her question. Of course Bishaye would think *he*. She was sure as far as Bishaye was concerned, she was the only female Dale had feelings for. This confession just got harder. "No, I was not forced. It was mutual."

"Mutual?" The alarm in her voice disappeared and the irritation returned. "How the hell did this happen?"

"It wasn't planned. It just, you know, happened." Dale took a drink of her beer.

"No, I don't know. Explain it to me."

"Do you believe in fate?" Dale asked.

"I believe yours will be negative in about another minute if you don't stop leading me in circles and tell me what happened. Circumstances and the conditions shouldn't have been *that* right, despite the alcohol I'm assuming was involved. You're supposed to be a trainee. He's a drill sergeant. Does anything about that scenario ring a bell?"

"If we had not met up, *by accident,*" Dale emphasized. "At this particular place, it never would have happened. That's the truth. This person is a good drill sergeant, Anne. Strictly a professional when on duty. No flirting, no innuendo, no unmilitary-like behavior."

"You're protecting him." It was an accusation.

"Only from you. And it's not a him." Dale held her breath.

Bishaye sat back and looked as though she'd just been slapped. "You had sex with Sergeant Cassidy?"

"Yes." She was surprised that Bishaye looked almost...*hurt*.

"Did you do it just to get back at me?" Bishaye's voice was barely audible.

"I know it will shock you to discover that my world doesn't revolve around you. This had nothing to do with you."

"Oh, but it does, Dale. Do you realize the position you have just put me in?"

"I hope the position of having knowledge of my indiscretion and keeping it to yourself. Other than the obvious, we didn't do anything wrong."

"It's the obvious that worries me. You've just served me up her career on two counts. I have her for fraternization and homosexuality."

"Then you'll have to get rid of me, too. On the same two counts." Dale glared at her defiantly.

Bishaye studied Dale. "Do you...are you saying that you have feelings for her?"

"Yes."

"When did this little affair start anyway?" Bishaye folded her arms across her chest.

"It's not an affair. It happened once and it won't happen again. It's because I do have feelings for her that I won't let it happen again. She's not a sleaze, Anne. This is the first time she's ever been involved with a trainee."

"To your knowledge."

"I believe her."

"That's your libido talking," Bishaye snapped.

"No. What happened between us has nothing to do with the case. We were just two people who got caught up in a moment."

"Two *women.*"

"Two consenting adults."

"How many other women have you been caught up in a moment with?"

"Just her." Dale sat back and played with the label on her beer bottle. "Are you upset because you weren't my first?"

Bishaye appeared to consider Dale's question carefully. "You enjoyed asking me that question, didn't you?"

"Anne, I am not enjoying any part of this conversation. If Shannon hadn't pushed me into it, you wouldn't have known."

"And what does your partner say about all this?"

"She says you needed to know and that you should probably look into Cassidy's background just to be sure and that she trusts my judgment. As should you." Dale glared at her.

Bishaye refilled her wine glass. Her voice was calm again. "In light of the reason why I brought you here, couldn't you have put her off? At least until training was finished or the case was solved?"

"Obviously not." Dale sipped her beer. "I know it's hard for you to understand but it wasn't a case of a drill sergeant seducing a trainee. I'm not a trainee, I know what I'm doing. It was me, the woman, not the trainee or the CID agent with her, the woman, not the drill sergeant. Don't pursue this, Anne, please. She is an excellent drill sergeant and she loves it. She's highly motivated and she gets great results from the troops."

"Especially the women."

"Absolutely but not in *that* way. She did not go after me. And if she was a hound, we would have heard about it."

"That isn't the point. Your actions may have compromised the entire case! Why don't you understand that? And what do I do if it has been compromised? Bring you both up on violations of the UCMJ? I think you both could qualify under a few of the many sections of Article 134. But threatening you with charges doesn't change the fact that if she is involved in any of this, when she finds out you're CID and were sent here specifically to see who is setting up the company by accusing drill sergeants of forcing sex on the trainees? She can come right back on you and me, as your superior, for Article 133, Conduct Unbecoming, not to mention entrapment, maybe even conspiracy."

"She won't do that. She's not involved in this crap."

"You don't know that! You don't know her, Dale. I don't even know her other than she transferred to A-10 on the recommendation of Sergeant Robbins and she has a clean jacket. She could be anybody. She could even know who you are."

"She doesn't."

"You *think* she doesn't. You don't know for sure because the *only* way you'll know for sure is when the case is solved. I am so damned pissed off at you right now, I can barely think," Anne said through clenched teeth.

"I think the reason you're that pissed off is because I had sex with a woman other than you. You refused to let things go further between us but you wanted to be the only woman I had those kinds of feelings for. You wanted to continue to be the one in my life."

Bishaye looked away from her. "Dale...you're obsessed. It's not healthy."

"What wasn't healthy was waiting for you to come around." She paused for Bishaye to focus on her again. "If you weren't with Jack, things might have been different."

"So...did you sleep with Cassidy just to prove a point to me?"

"No. I slept with her to prove a point to me. That's why I'm positive she has nothing to do with this case."

"You used her?"

"Not intentionally."

"Would you sleep with her again if you weren't bound by regulations or in the middle of a case?"

"In a heartbeat," Dale said, without hesitation.

She watched Bishaye closely. The colonel appeared lost for a brief moment.

"Anne, I've been wrong to push you toward something you may never be ready for. I know we have feelings for each other. I wanted more from you than you wanted or were able to give me. I understand that now. Until Cassidy, I thought you would be the only woman I'd wait for. You see, that would give me an excuse not to act on my sexuality and come to grips with who I really am. I thought the only woman I'd ever feel safe with was you. J.J. made me realize that wasn't true. You should be grateful," Dale smiled.

Bishaye nodded slowly as she took in Dale's words. "Tell me, if Walker has CQ and you're off screwing Cassidy's brains out, who is keeping an eye on the women in the company?"

"They're too scattered to watch them all now. They've got too much freedom, too much knowledge of the post and Averill and too much access to get anywhere they want to go."

"What does that mean? Unless something obvious happens that your spying days for this cycle has been completed? It's all fun and games now? Don't you think you should be concentrating on which women are the most suspect and don't let them out of your sight? We're at a very crucial point here."

"I thought everything was on hold because of Henning."

"I never told you to put anything on hold, Dale, and last I remember, Henning was not a suspect, she was a possible target. I didn't want you planting that idea in her head because she might panic. I want you and your partner to keep an eye on what goes on around her when she is in the company area. While she's at school and at home, I have other agents on her." Bishaye signaled the waitress for another round. "Now where did this meeting between you and Sergeant Cassidy take place? In a bar, out in the open, where anyone could have seen you?"

"You don't need to know that."

Bishaye turned an ear toward Dale as though she hadn't heard her correctly. "I...what?"

"We ran into each other at a bar. I'm not going to tell you which one or where to find it. If you're mad at me, take it out on me, not on Cassidy or other gay military women who need a place to go."

"It's against regulations."

"Then catch them doing something in the line of duty that violates the UCMJ and shows them to be bad soldiers. Don't go after them for what they do in the privacy of their own bedrooms. If Cassidy is a bad drill sergeant or uses a sexual quid pro quo to help females get through, then yes, you should do what's appropriate. You know if you kicked all the gay and lesbian soldiers out of the military, Uncle Sam would be shit out of luck if a war suddenly came along. And you'd also be hypocritical because you know you're as sexually attracted to me as I am to you. In different circumstances, it would have been your bed I was in Friday night, not Cassidy's."

Their drinks were brought to the table and the server left them alone again. "You have a one-track mind, you know that?"

Dale smiled. "You didn't deny it."

Bishaye sipped her wine. "I will pull Cassidy's jacket and find out everything I can about her. Of course, I can't do anything overt without drawing attention to you so I guess I have to leave her alone. However, if I find out she baited you and is setting you up. You both are in shit so deep, the stain and smell will never wash away."

"I understand."

They sat in silence, both digesting the conversation. "How about some supper? I hear they have good food here," Bishaye said, finally.

"Only if we've stopped fighting. I want to enjoy my meal."

"Fine. I officially call a truce."

Dale laughed. "You *know* that won't last."

CHAPTER FIFTY-ONE

The trainees were awakened an hour earlier than usual the next morning for a surprise locker inspection, conducted by Ritchie. On Friday, the drill sergeants had dropped a hint to their individual platoons so everyone made sure their personal areas were squared away before lights out on Sunday night.

Dale was nervous about her first encounter with Cassidy after Friday night but when she saw her accompany Ritchie through First Platoon's lockers, Cassidy made no attempt to look at or treat Dale any differently than she had before their night together. Between the two of them, Dale was sure she was more obvious about the oddity of pretending nothing happened, as she watched Cassidy, transfixed, during her participation in the inspection. Dale still couldn't believe the passion and intensity she experienced with the drill sergeant that was unlike anything she had ever known before and now had to feign indifference when the breathtaking woman was anywhere near her. She wanted to be with Cassidy again as much as she wanted anything and knew she had to stay away from her when out of the company area because it would be just too tempting, otherwise.

Dale knew that there would be no future for them when she left Alpha-10 because once Cassidy found out who Dale really was, the deception would most likely be unforgivable. Sometimes life just wasn't fair. Dale had found a reality to take the place of the Bishaye fantasy and couldn't do anything about it. Suddenly she was acutely depressed.

After the trainees were told they had passed the locker inspection, they were marched to the motor pool, where they were assigned MP sedans to drive to specific spots on post. One location was the parking lot of the LE School.

The vehicles in the parking lot of the school were made available for trainees for graduation. The sedans were used for role-playing. Once a week, usually a Friday night, all the MP trainees who had successfully completed all the required courses, dressed up like real military police officers and went out on patrol. Everything was handled just as if the group were really MPs. Roll call was taken, guard mount was conducted, a briefing was given and patrol assignments were decided by the patrol supervisor, usually a designated drill sergeant from another company. There were two partners assigned to one vehicle and the future MPs treated the rest of the evening as though this was their first actual time on patrol.

They were evaluated on everything— radio procedures, correctly filling out all logs and reports, vehicle maintenance, attitude, appearance, the handling of any given situation, etc. Unfortunately, a majority of the trainees knew what they were in for because a week before this test, they had been assigned to play the perpetrators. The speeders, the brawlers, the domestically disturbed, the potential cop killers, the drug addicts, etc. There were NCOs provided as supervisors to make sure that the trainees played their roles to the hilt because they were being evaluated on that, too.

At the motor pool, while she handed out trip tickets that designated destinations, Cassidy stopped when she handed Dale and Shannon her last two. A tiny smirk played on the edge of Cassidy's lips as she barely looked at Dale. It was enough to cause Dale's stomach to flutter and for her to blush. If Shannon noticed, she did not let on.

"Which one of you is the better driver?" Cassidy asked them.

Dale and Shannon instantly pointed at one another. "She is." Then they looked at one another, surprised. Again their responses were simultaneous. "She is. No, you are!"

Cassidy took a step back and observed them. "Do you two rehearse this stuff?"

"Really, Drill Sergeant, neither one of us are good drivers, we're just fast," Shannon said.

Dale wanted to slap her partner as Cassidy turned away to wipe the smile off her face. She turned back to Dale. "Where are you taking your vehicle?"

"OLEV," Dale said, reading it off her trip ticket.

"You're going to the driving range." Cassidy turned to Shannon. "What does yours say?"

"CPO," Shannon answered.

"You're going to the LE School parking lot. I'll ride with you," Cassidy said.

Shannon pasted a smile on her face. "Yes, Drill Sergeant." As Cassidy walked around to the passenger side of the sedan, Shannon looked at Dale with the same stupid grin still adorning her face. "Great."

Dale watched them drive away, not sure whether to be relieved or upset.

On the short trip to the LE School, Shannon and Cassidy had a nice, pleasant chat. Shannon was concerned that Cassidy might try to fish for what Dale may have said about them and was pleasantly surprised when the drill sergeant never went near that subject.

Cassidy related a story to Shannon about a morning when she was in training and misread her trip ticket. "I thought I was supposed to go to the parking lot, so I did. When I got there, I discovered that I was supposed to have gone to the driving range. I thought no one would notice until everyone began looking for my car because a vehicle was missing. When my drill sergeant checked all our trip tickets and saw what mine said…well, let's put it this way, it was already zero eight hundred, my first class had started and we were on a flying trip to the driving range. I ended up being AWOL from two places at the same time." Cassidy chuckled at the memory.

"Did you get disciplined for it?" Shannon asked. She could certainly understand Dale's attraction. Cassidy was enchanting without trying to be.

"No paperwork but because I was already late for class, my drill sergeant made me double-time all the way back to LE School from the driving range while he drove behind me the whole way, beeping his horn."

Shannon later told Dale the story. "The entire time I was with her, all I could picture was you and her in bed."

"Pervert," Dale said.

"It wasn't because I was enjoying it, I just couldn't get it out of my head," Shannon said and glared at her.

Dale grinned. "I can't get it out of my head, either."

Dale passed her Approaching The Scene of an Incident test and Shannon passed her Apprehension class. They teamed up again for Collecting and Processing Evidence.

Later that evening, as they sat in the smaller bar of the EC, Dale and Shannon discussed Dale's meeting with Bishaye the night before.

"What about Cassidy? Is Bishaye going to do anything to her?" Shannon asked.

"She can't, not without blowing our cover."

"That much I know. I meant after we're out of here."

"I have no idea. As long as she's not involved in any of this, I doubt Bishaye will do anything."

Shannon reached out and patted Dale's shoulder. "And…how are you doing?"

Dale smiled, warmly. "I'll be fine." She took a swallow of her beer. "I've got to tell you, though, one night with that woman and I'm still going through withdrawal."

"You sure you're not going to see her again?"

"As much as I'd like to, no. It'll just make things more confusing."

"What about after we're done here?"

"Do you think she's going to want anything to do with me after she finds out who I am and that I deceived her?"

"Dale, you haven't intentionally deceived her."

"I don't think that'll count." Dale glanced at Shannon. "Look at you, all Yenta. I didn't think you were going to be this into my choice of lover."

"I like Cassidy. Like I said, you could have done a hell of a lot worse."

"For the record? I liked Matt. I'm sorry he turned out to be such a dick."

"Me, too. I wonder if women are any better…"

Dale shrugged. "Can't say for sure. I have no doubt that women can be dicks, too." She finished her beer. "I forgot my watch in my locker. What time is it?"

Shannon looked around for a wall clock, to no avail. "Well,

it can't be nine-thirty because we're supposed to be back in the barracks at nine-thirty and we aren't there yet." She stopped the bartender and looked at his wrist. "Ha! I was right. It's only nine-fifteen."

They stared at one another. "Oh, shit, we did it again," Dale said. She followed Shannon at a dead-run out the door. If they were lucky and moved at a sprinter's speed, they'd just beat whoever was going to conduct bed check.

The morning started at the motor pool again, Dale and Shannon each driving their individual sedans to the OLEV training area. They were returned to LE School in the back of a deuce-and-a-half.

Dale stayed in Collecting and Processing Evidence as Shannon took her test, passed it, and went on to MP Reports. By noon chow, both lieutenants were done with those classes.

The afternoon found them together in TAI, Traffic Accident Investigation. This was a long class, three days being the average time spent learning the complexities. Most of the class took place outside, on pre-set up accident scenes and put to use what had been taught inside in the classroom by a Marine Gunnery Sergeant with an abominable attitude.

One of the most confusing requirements of TAI was an animal called triangulation. The trainees had to figure out what were considered and acceptable as permanent structures, such as fire hydrants, lampposts, manhole covers, etc., and measure distances between the damaged vehicle point of impact and these immovable objects. Once calculated, a diagram was drawn depicting the position of the vehicles at the time of impact and where the structures were in reference to that. A key and a legend also had to be provided in the lower right-hand corner of the sketch and the completed drawing, when handed in, had to be exact or the trainee was given an entirely different scenario to illustrate. If the trainee didn't do that diagram correctly, that trainee would be given one more chance to do it right or bolo out of the school.

The reason for the precision diagrams were simple. In an actual accident, if the case didn't get settled right away, lawyers could return to the scene years later and perfectly recreate the incident by using the sketches and measurements made by the original investigating MP. It was a valuable class despite it being so annoyingly drawn out.

Both lieutenants hit the major hangout spots frequented by their barracks mates. The weeknights were not as popular with the future MPs as they were with the two CID agents. Of course, Dale and Shannon didn't need to study and only a rare few opted for partying, instead, and usually those were prior service. An hour at the Pizza Place or the bowling alley was usually the most time an average trainee would spend away from the barracks during the week.

Dale and Shannon did a mental headcount in the bay and then moved on to other locations to find who was missing. There were certain people they knew they didn't have to check up on. Creed and Almstead always played basketball at the gym and Michaelson worked out at the WacVille recreation center. Since LE School put a lot of pressure on the trainees to get high marks, to have as few No Go's on their records as possible and to complete the courses swiftly, there was little frivolity during those precious studying hours of 1700 to 2130. Dale and Shannon were always the last two to make it back to the barracks and usually barely before bed check.

Some of the women expressed that Dale and Shannon's bravery of pushing their luck to the limit was amusing. Others seemed to see their narrow escapes as irresponsible. The possibility of being discovered in bed, with all their clothes on, ran the risk of group punishment and another black eye for the women and no one wanted any more of that.

The climate lately was quite accommodating. With spring less than a month away, the winter chill and dampness were being replaced by much more pleasant weather. Since the trainees were stuck in classrooms the entire day, excluding food breaks, cigarette breaks and PT, no one was able to appreciate it. Unfortunately, the day the partners were outside, measuring and sketching, they were drenched by torrential rains.

Toni Sherlock, who was selected to work with Shannon on her TIA problem, had that same silly look on her face she had worn since the weekend. Shannon wondered why Sherlock behaved so oddly and was beaming.

"What's the matter with you, Toni? I barely need the flashlight when you're around," Shannon said.

Sherlock moved closer to Shannon. "Can you keep a secret?"

Shannon wrote down a measurement. "Of course not. You should know that by now. There are no secrets in the Alpha-10 women's bay. So...what is it?"

They stood, huddled together, and hoped they looked like they were discussing the vehicle damage of the car in front of them. The rain came down so hard, it was difficult to keep their eyes open. If they stood in one spot too long, the pounding rain created small ditches in the dirt around their feet.

"I got married," Toni announced, proudly.

"To whom?" Shannon asked. She was more than mildly surprised.

"Doug Mancini."

"Well, uh, congratulations," Shannon said. She wanted to reach and shake some sense into her. Mancini was a GI Sherlock met two weeks earlier at the EC. He was assigned to Headquarters Company and was hanging around, waiting on orders that would send him elsewhere. "That must've been some two weeks you two spent together."

"It was," Sherlock answered, wistfully.

Shannon pulled Sherlock with her as she walked around to the front of the vehicle. "Does Robbins or Putnam know?"

"I'm going to tell them tomorrow when he's gone. He came down on orders today. He's going to Korea."

She was still glowing as Shannon looked at her, sympathetically. "Where in Korea? Did he say?"

"Um...South Korea?"

"I certainly hope so."

"Something that begins with a U?" Sherlock looked puzzled.

"*Uijeongbu?*"

Sherlock stared at Shannon, still perplexed. "That's not the way Doug pronounced it."

I'm sure not. It won't take him long to pronounce it

correctly. "Second Division?"

"I think that's what he said."

Shannon shivered due to more than just the chilly rain. "By the DMZ?"

Sherlock shrugged. "I don't know but wherever he's going, I'm going with him."

Don't count on it, Shannon thought.

She wanted to tell Sherlock to be grateful if they laughed her out of the office. She wanted to sit Sherlock right down in the pouring rain and tell her to get used to the mud and muck if they allowed her to join her husband in Korea. She wanted to also tell her to get used to the feeling of having to close her eyes because she was going to have to ignore things that would morally outrage her because they morally outraged Shannon. Incidents the public didn't know about because the western press wasn't aware of them. Shannon wanted to slap that silly smile right off Sherlock's face and replace the obvious lust that had gotten her in this situation with no common sense. As disappointed in Sherlock as Shannon was, she still felt the need to try and prepare her for a crash landing.

"Audi and McCoy were talking about assignments in Korea the last time I was on CQ. They said something about spouses not being allowed to go, that the military won't pay the transportation costs of moving the husband or the wife because it's usually only a twelve-month assignment."

The smile faded from Sherlock's face. "But I'm not just a spouse, I'm a female MP who came in with no definite assignment. I heard the units in Korea are begging for female MPs."

"That may be true but, Sherlock, you haven't graduated yet. You still may bolo out."

"I have two weeks left, tops. I've been doing great in all my classes. I only have one No Go to my name. I'll make it," Sherlock said, indignantly.

"That brings up another point. They may have already cut your orders and they may be sitting there, waiting for you to complete LE School. You may already be assigned somewhere."

"You're just a bundle of encouragement," Sherlock said, dryly.

"I just don't want you to be caught unaware when you go in tomorrow and talk to Robbins or Putnam. They're going to be plenty pissed because you got married while still in training and didn't go through the proper procedure. Look at the shit they put Travis through and the stuff they're still doing to Minkler and Jaffe. They won't be very happy. I wouldn't go in there counting on any favors."

Sherlock's bubble had definitely burst. "Well, that was a short honeymoon."

Shannon handed Sherlock the end of the measuring tape. "Are you going to see Doug later?" She watched Sherlock nod as Sherlock walked away from her to a fire hydrant fifteen feet from them. "Hold it tight," Shannon reminded Sherlock. Shannon read the measurement and wrote it on a dry spot on her hand. The sheet of paper she'd had originally was saturated and unreadable.

"Then, if I were you," Shannon said as Sherlock walked back to her. "I'd make the most of it."

Dale and Shannon discussed Sherlock's surprising news later that evening at the EC. They had been witness to these abrupt marriages before and held out little hope for the union to be successful. They didn't know much about Sherlock's new husband but knowing Sherlock the way they did, they gave her two months at a new installation, away from Mancini, before she found someone else.

As usual, lately, there weren't that many people to follow around. Most of the company was inside their respective bays either studying or just staying warm and dry. In fact, both Dale and Shannon returned to the bay earlier than normal, so much so that Tierni led several others in making a big deal out of it.

"I don't believe it! You can usually set your watches by you two. When you come busting through the doors, it's time for bed check. What happened tonight?"

"The EC closed early," Dale said.

The partners played along with the good-natured ribbing until it got old and the participants got bored and dropped it. It was actually nice for Dale and Shannon to be able to change into their sleep attire with the rest of women for a change.

PERMISSION TO RECOVER

The last one in the barracks that night was Toni Sherlock.

CHAPTER FIFTY-TWO

The rain persisted and poured down as it had the day before, making the ground even messier and being outdoors more wretched. The drill sergeants still made their troops do PT once before breakfast, once before lunch and once before dinner. The thought of venturing out anywhere that evening was not an inviting one to anybody, even though it was Friday night.

Sherlock, who had told her secret to more than just Shannon, had requested to speak to Robbins to break the news about her new status. She didn't know he'd already heard through the grapevine, which was one of the reasons he denied her permission to see him. Another reason was that he was busy getting drunk at the NCO club. Another one of his buddies bit the trainee dust.

At 1700 hours, immediately following the command of Fall Out, Audi and Cassidy walked into the bay, both very serious. They ordered the women to sit on the floor in front of them. When the subject of the impromptu meeting turned out to be fraternization, Dale's stomach knotted. How ironic that Cassidy was standing before her, looking guilty because she was about to assist Audi in lecturing Dale and the others regarding cadre consorting with trainees.

Cassidy seemed unusually tongue-tied as she danced around the issue. Audi finally took pity on her and came right out and told the women that an NCO was caught the night before with a trainee in his office, in a compromising situation. Dale stopped staring at Cassidy, who never once glanced Dale's way, and sought out Shannon. When she spotted her, a row over, Shannon was looking back at her.

What had they missed? Who was involved? Did Bishaye already know? Was it one of their drill sergeants? Dale looked around at the other women in the room and could not see anyone absent from the group, except for Wachsman and she was on CQ duty. If one of the Alpha women had been directly involved she would not have been present during the discussion. By the time Cassidy and Audi left, everyone knew someone had fraternized with someone and got caught but the drill sergeants didn't say who the culprits were.

The bay was abuzz with speculation. The odds were on Bradbury and a Bravo female. Dale hoped it wasn't Holmquist and was grateful it wasn't Cassidy. When Wachsman returned upstairs after her shift, she put an end to the conjecture.

"The trainee was a girl named Resington from Charlie. The NCO was Casey, the Supply Sergeant. I guess he wanted her to dirty up his sheets." Wachsman had been in the Orderly Room when the details came in.

The piece of news made the lieutenants breathe a little easier but they agreed they needed to start splitting up again. Even though it hadn't been Alpha Company this time, it was still Tenth Battalion that was involved and Dale wondered how Bishaye was handling it.

Shannon took off to the Pizza Place with Wachsman and a few others and then they went to Averill to stay at the Journey Inn.

Dale spent the night in the barracks with a majority of the others but not before closing down the EC.

"Good morning, ladies," Cassidy's voice drifted down from the ceiling of the bay. She had just begun her shift as the Staff Duty NCO. "I need volunteers for police call, to empty trash and to sweep the patios. Those of you with hangovers, like Walker and Oakes, try not to fall down the stairs this morning and please come fully clothed. Oh...and watch that sun...it's awfully bright." The bitch box clicked off and Cassidy's sweetly sarcastic voice was gone.

Dale looked at her watch and chuckled. It was 0830 hours. Too bad Shannon wasn't there to hear Cassidy. She would have enjoyed that. Dale proceeded to haul trash with Beltran and sweep the north patio. Had it not been Cassidy, she never would have volunteered.

She knew she needed to get information about the latest

fraternization incident and her best bet would be Tierni, who was the CQ. She could have cornered Cassidy at some point but she didn't know how to ask specific questions without it coming back around to the two of them and Dale knew she was still vulnerable, still very much drawn to Cassidy's magnetism. One of the ways Dale could keep herself away from pursuing Cassidy was to find excuses not to be alone with her. Had she known Cassidy was going to be Staff Duty NCO, she would have volunteered to go to town and asked Shannon to stay on post.

She waited until she saw Cassidy leave the Orderly Room to, hopefully, make the rounds to all the Battalion CQ offices, which should keep her away a while. Dale carried her broom back to the Orderly Room, put it away and sat down near Tierni. "Heard anything else about Casey and the Charlie female?" Dale said, after they got all niceties out of the way.

"Nope. They've been very careful about what they've said in this office."

Dale nodded in comprehension. "I like Sergeant Casey."

"Me, too, but bringing her up to his office? That was just stupid," Tierni said.

"You know how it is when you get caught up in that passionate urge," Dale said, almost wistfully.

"Yeah, *I* know how it is but he's supposed to know better."

"Has anyone said what might happen to him?"

"Not that I've heard," Tierni said. Dale was about to ask Tierni if she knew what happened to the female caught with Casey when the door opened and Cassidy walked back in.

Both Dale and Cassidy showed surprise to see one another. It was Cassidy who recovered first. She picked up some papers from Sergeant Fuscha's desk. "Private Oakes. You're up and about. I am impressed."

"Thank you, Drill Sergeant. I'm not even hungover," Dale announced and then wondered why.

"That's not normal for you, is it?" It was clear Cassidy was joking with her.

"Not on a weekend, Drill Sergeant." Tierni said.

"Where's your partner in crime?" Cassidy asked. She appeared to be checking items on one paper against a list on another, so she was not looking at Dale.

PERMISSION TO RECOVER

"In town somewhere, probably where everyone else is."

"The Journey Inn, I presume. Ah, the memories. How come you didn't go?" Cassidy looked up at Dale.

"I didn't feel up to it, Drill Sergeant."

"Are you okay now?" Cassidy asked, with just the right amount of concern in her voice.

"Yes, I'm fine, thank you."

One of the male trainees opened the Orderly Room door and stepped halfway inside. "Drill Sergeant, the Dayroom is locked."

"Yes and it will stay that way," Cassidy said. "Whoever was in there last night made a mess of it and left it that way. Until it gets cleaned up, it will be off-limits." She turned to the trainee. "You want to volunteer to get it squared away?"

He shook his head. "Not really…"

Cassidy went back to inspecting her documents. "Then it stays locked. Use of the Dayroom is a privilege, not a right. If you abuse it, you lose it."

"Yes, Drill Sergeant." He closed the door.

Cassidy looked up at Dale. "Private Oakes, have you volunteered for any details today?"

"Yes, Drill Sergeant, I did trash and swept the north patio."

"Very good. Now you can do more service to your country and clean up the Dayroom."

"But I wasn't in the Dayroom last night," Dale protested.

Cassidy squinted at her. "I never said you were. What does that have to do with me giving you an order?"

"Nothing, Drill Sergeant, I'll get to it right away."

"Thank you, Private Oakes." Cassidy held her clipboard at her side. "Private Tierni, I will be in the XO's office, working on some pressing paperwork. I don't want to be disturbed unless it is an emergency. If I get a phone call, take a message."

"Yes, Drill Sergeant," Tierni answered. After Cassidy had closed the door behind her to the other office, Tierni said. "What'd you do to piss her off?"

"Nothing that I know of," Dale said.

Except reject her desire for further trysts.

Tierni handed her the key to the Dayroom. "Have fun. Run that back to me when you're done."

Dale took the key and headed out the door. "Will do."

Dale entered the Dayroom, shut the door, and locked it behind her. When she flipped the light switch on, she looked around to find that the room was spotless and everything was in its place. She put her hands on her hips. "What the —"

Just as the realization hit her, she heard the door open from the back hallway that led to the A-10 offices. The lights were switched off and Dale was backed up against the wall.

As much as Dale wanted to resist Cassidy, she could not. The drill sergeant put her hand to the side of Dale's face before she leaned in and pressed her lips against Dale's.

The kiss was as fiery as the others they had shared. It was long and heated and elicited immediate excitement in both women. When their lips parted, Dale was panting. "I thought we decided that —"

"No, *you* decided," Cassidy whispered and kissed down the side of Dale's neck. "I never agreed to anything."

"This is crazy, J.J.," Dale said, her voice hushed. "This is too risky."

"You and I have the only two keys." She swooped in for another kiss. Dale tightened her hold as she dissolved into Cassidy's touch.

Dale knew she should have balked, should have pushed her away but her body responded to the beautiful woman kissing her, not to the sensible thoughts rapidly fleeing her brain. She was surprised at how out of control she became in Cassidy's embrace.

"God, I want you." Cassidy's declaration and hot breath on Dale's ear made Dale even more defenseless. Cassidy lifted Dale's sweater over her head and dropped it on the floor by their feet. She caressed Dale's breasts through her bra.

Dale's body thrummed with arousal. It felt as though every inch of her body tingled. "We can't…do this…again."

That's just heavenly, I hope she never stops.

"Yeah, we can." Cassidy reached around and unhooked Dale's bra. It joined her sweater. Cassidy appeared to enjoy Dale's reaction to her attention almost as much as Dale did.

"We're going to smell like sex," Dale panted as Cassidy kissed one nipple, then the other.

PERMISSION TO RECOVER

"I have a clean uniform, underwear and a shower in the First Platoon office upstairs." She unbuttoned and unzipped Dale's jeans. "And you can go upstairs and shower." She pushed Dale's trousers and panties down so that they pooled around her ankles. Cassidy dropped to her knees, parted Dale, and began to feast.

In no time at all, Dale's legs quivered and the trembling shot through her entire body. The potency of her climax took her by surprise, as did the weakness in her lower body afterward. She was able to keep her orgasm quiet, which shocked her as much as pleased her.

Cassidy kissed her way back up Dale's body and lavished attention once more on Dale's breasts as her fingers entered Dale. Three different times during this blissful act, the door handle rattled but neither woman could stop the momentum of the thrusting or the suckling. Cassidy was certainly good at what she did and knew exactly where to touch Dale to get her to react.

Dale felt herself ready to peak as fire surged to every nerve ending and back to her center. She knew she was going to cry out this time so she muffled her vocalization into the curve of Cassidy's neck. For several minutes, she held onto Cassidy just to let her breathing regulate.

Cassidy stroked Dale's back and held her. She kissed Dale's shoulder and looked into her eyes. "I'm sorry."

"For?"

"I know you didn't want to be with me again and saying I couldn't help myself may be the truth but it wasn't right that I did what I just did."

"J.J., it's not that I didn't want to be with you again. I wanted it very much. As you can see, I really didn't resist you. I just don't want us to get caught. In fact, with what just happened to Sergeant Casey, I'm just really surprised you took this chance."

"You are worth it."

Dale wanted to cry. "No. I'm not. I'm not worth what would happen to you."

"If we keep quiet about this, nothing will happen. Look, Dale, this is just as shocking to me as it is to you. I have never fraternized with a trainee before. I would never, not in my wildest dreams, risk my career for a score. I've never been anything less than professional and I didn't get to where I am by doing stupid things but I can't stay away from you," Cassidy said.

"I'm not here for much longer, why would you risk your career for something as temporary as this?"

Cassidy bent to pick up Dale's bra and sweater and handed them to her. "I was hoping you wouldn't consider this temporary, that maybe when they cut your orders, I could put in for where they send you."

Dale pulled up her pants and refastened her jeans. The shock of Cassidy's confession was evident on Dale's face. "You what? Are you saying you want to be with me?"

"I told you this wasn't just sex to me, Dale. I don't do that, I thought I made that clear. I would never take a chance like I'm taking on someone who I'm just casually drawn to." Cassidy was defensive and Dale's eyes filled up. Cassidy reached out a finger and wiped a tear away. "What's wrong?"

Dale was so strong except when it came to matters of the heart. "I've never had anyone say something like that to me and actually mean it. I'm the girl who's always the other woman, the other girlfriend, the one in the background. I'm always the one who is good enough to fuck but never to openly date or marry. So…I just thought—"

"That I was just like the men in your life? Well, I'm not. And why *anyone* would not cherish you is beyond me."

Dale finished dressing. "You don't really know me."

"I want to." Cassidy pulled Dale close again and kissed her passionately. "I want to know everything about you because I'm attracted to everything about you." She kissed Dale's forehead. "I got the feeling that it was the same for you. If you tell me I'm wrong and I was just a conquest then I'll leave you alone."

"You're the first woman I've ever been with, J.J. Doesn't that scare you?"

"No."

"It scares me."

Cassidy took a step back. "Oh. Okay." She looked devastated. "I thought, well, I…I guess I was a fool." She reached for the back door handle. "I'm sorry, Private Oakes, it won't happen again."

"J.J., wait," Dale said. Cassidy didn't turn around. "I...I do want to be with you. I *like* being with you...but this is moving way too fast." She moved up behind Cassidy. "I am not scared that I'll want to see what else is out there, I'm scared that I won't be who or what you want me to be and we'll both get hurt."

And that's the least of my worries.

Dale was still stunned that Cassidy admitted to being that *into* her. Either there was something wrong with Cassidy or Dale's luck had really changed.

"You'll never know until you try, will you?" Cassidy turned around. She pulled Dale into a hug. "I know this is unusually fast but there is just something about you that I can't let go of. I have never been like this with anyone. My last relationship was more of a convenience than anything else. The one I had before her was really good...until it wasn't anymore. There are never any guarantees, especially in the military, especially for a couple like you and me. I'm not asking you for any promises or commitments, all I'm asking is that you give us a chance to see where we can go."

Dale squeezed her tighter. "You are so jaw-dropping gorgeous, it's just hard to believe you want me." She broke the embrace and held Cassidy's face in her hands. "You can have anybody you want."

"I have what I want right here in front of me." Cassidy kissed her, this time with more admiration than passion.

Dale's knees weakened again. "God, J.J., you make it very hard to fight you."

"Make love not war," Cassidy said and winked.

"Maybe we should hold off on our, um, meetings until I at least graduate LE School and get my orders."

"I can probably abstain until you get out of school but when your orders get cut, you'll be gone as soon as the next day. So, how about if we make a date for the night after your final class."

"I think I can do that." Dale smiled. "What if I get stationed in Greenland or something? You'd still want to pull a tour there?"

"Somebody would have to keep you warm," Cassidy said.

"We wouldn't be able to openly live together, though."

"It's a lot easier than you think. You just have to be

discreet."

Dale chuckled and gestured around the Dayroom. "Oh, you mean like this?"

Cassidy laughed. "Maybe a little more careful than this." She opened the door to the hallway. "Don't hang around with anyone too long before showering. You do have that distinct scent about you."

"Thanks to you. Don't you want —?"

"Raincheck," Cassidy said. She closed the door behind her and locked it.

Dale leaned against the door. Her mind went back to Cassidy devouring her and her heart fluttered. She sighed and walked to the door that led outside and stopped, burying her head in her hands.

What the fuck do I do now?

Dale returned the Dayroom key to Tierni and made it upstairs to shower before she got close enough to anyone that they could guess what she had been doing. The encounter with Cassidy left her adrenalized yet empty. What she really wanted to do was get back together with Cassidy and spend the rest of the weekend in bed with her. She knew that was impossible.

She did accept an offer to attend a company party at a motel other than the Journey Inn on the outskirts of Averill. The Red Clay Motor Rest was a small establishment made up of twelve individual cabins and certain members of Alpha-10 had reserved them all a week earlier. Most of the troops from the company planned to be there at some point during the night.

Dale shelled out a percentage of the money to share a cabin with Tierni, Tramonte and one of the marine females, Endres. Endres learned during Bivouac that she would not get any special treatment or privileges because she was in a different branch of service so she jumped right in and pulled her own weight. She tried to make friends her second or third day in Alpha-10 but she was slow to be accepted because of the initial actions of her sister marines. Endres persevered and it didn't take her long to finally be accepted. Neither Dale nor Shannon had spent much time with her but Tierni and Tramonte had spoken very highly of her.

PERMISSION TO RECOVER

The party at the Red Clay grew rapidly and the darker the sky got, the wilder the party became. Even though Shannon had to report to CQ at 0600 the next morning, she intended to show up and hang out for a few hours before heading back to the barracks. She had planned to accompany a group of five to the party as soon as they finished their pizza at the Pizza Place. She never got the chance.

The six Alpha trainees had just finished their pizza and a pitcher of beer and were about to walk back to the company area to call a taxi to take them to the Red Clay when Steele and Mark Morse, the A-10 male she was dating, burst through the door of the Pizza Place.

"Any fucking Bravo-10 douche bags in here?" Morse yelled. He was livid. If there were any members of Bravo there, no one spoke up and admitted it.

Shannon stopped Steele. "What's the matter with him?"

"Wotek's in the hospital. He got the shit kicked out of him by four Bravo guys."

Questions resounded from everyone in the group.

"The only information we have is that three Bravo guys held him and one beat him up on the track behind the EC, supposedly because he danced with a Bravo girl who was dating one of the four guys," Steele said.

"They beat him up for a dance?" Bigfoot was incensed.

"That's what we heard," Morse said.

Shannon then saw the meaning of all hell breaking loose. With few exceptions, the twenty-seven Alpha men and six Alpha women left in the Pizza Place, were out for blood. They wanted to find these guys and do to them what they had done to one of their own.

Bigfoot appointed himself the leader and sounded as though he were heading up a posse. If the rumored story was true, Shannon did think the Bravo boys deserved what they got but if Bigfoot got the current, willing group as stirred up as he was, the results might be *more* than they deserved. Revenge was certainly sweet but not if the payback resulted in the dismissal from LE School of everyone involved. When they got back to the company area, Shannon was happy to see Cassidy walk out of the CQ office at that precise moment, as the attitude of the Alpha trainees had taken on a mob mentality.

"*At ease!*" Cassidy commanded and, even as furious as they all were, they all quickly assumed the position and stopped talking. "Okay, listen up. I just got back from the hospital. Private Wotek is in good condition. He has some bruising, minor lacerations but no internal injuries. He does have a mild concussion so he will be at the hospital for another day or two for observation. There is no reason Private Wotek can't continue training, uninterrupted, once he is released."

"What about the guys who put him in the hospital, Drill Sergeant?" Morse asked.

"When we find them, we will deal with them. I want you all to stay out of it. I know you're pissed off. *I'm* pissed off. If one guy is going to go this nuts just because a girl he has the hots for danced with someone else and three other boneheads don't have the presence of mind to make individual choices to either stop it or walk away and report it then these are not the people we want as military police officers. The knowledge that they're going to be kicked out of LE School when they are so close to graduating should be rewarding enough. Especially since Wotek will be fine."

"What if you don't find out who they are, Drill Sergeant? People can stick together and get pretty tight-lipped about stuff like this," Bigfoot said.

"Sergeant Jessup and Sergeant Bradbury are with the female who was allegedly the cause of all this stupidity now. If she doesn't tell them anything then when Wotek gets out of the hospital, we will walk him through the Bravo ranks and he can personally pick them out."

"Wotek is our friend, Drill Sergeant," Steele said. "Somehow I feel like we're letting these Bravo guys off the hook."

"You're not, believe me, Private Steele." Amid the grumbling, Cassidy raised her voice until it was quiet again. "Let me put it to you this way. Anyone who goes out looking for these guys and lays one hand on them will also be immediately cycled out of LE School. You people are almost at the end of training! When are you going to start thinking like cops?"

PERMISSION TO RECOVER

Bradbury jogged up the stairs to the north patio, followed by Jessup. Jessup said something to Cassidy that only she could hear. She nodded and then addressed the trainees. "Why don't all of you go on to your party now and leave this mess to us. Be careful in town. Don't get too rowdy and do not get arrested! We don't need our entire company eliminated from LE School. Dismissed!"

The trainees dispersed as Cassidy, Bradbury and Jessup walked into the Orderly Room. When Shannon followed them in, Cassidy looked at her, curiously.

"What do you need in here, Walker?"

"I was designated to call the taxi, Drill Sergeant," Shannon responded.

Cassidy motioned for her to use the phone on the CQ's desk. As Shannon dialed the number by heart, she also turned her head so that the ear that wasn't on the phone faced the drill sergeants. Their voices were hushed but Shannon could still hear them.

"She didn't tell us anything," Bradbury said. "She said the males involved threatened to shut her up permanently if she opened her mouth."

"Nice bunch you got there, Jane." Cassidy sighed. She looked back and forth from Bradbury to Walter Jessup. "Any idea at all who these goons might be? Or are all your guys aggressive, uncivilized thugs?"

"And what were you doing with the crowd you just dismissed? Organizing a church choir?" Bradbury's eyebrow arched in sarcasm.

"I have a good idea who it might be," Jessup said. "There's a male in my platoon who can't stop the wisecracks and has an entourage of three parrots. He needs a serious Kiwi injection."

"Why haven't you given him one?" Cassidy asked.

"If he can drop the attitude, he'll make a fine MP."

"Not with his obvious issues. He'll be nothing but a bully with a badge. The MP Corps doesn't need any more of those," Cassidy said.

"Who is it, Walt?" Bradbury asked.

"Paulsen."

"That redneck? I should have known," Bradbury said.

"I'm not positive it's him, I'm just making a calculated guess," Jessup said.

Shannon knew who Paulsen was and would have bet her next paycheck it was him. She was in the vicinity when he'd made an absolute nuisance of himself with Pam Ryan, who told him to fuck off in no uncertain terms. She recalled him telling Ryan that he would get her before he PCS'd out of there and show her what she was turning down. Paulsen was flanked by three jerks who agreed with everything he said. A week later, he apparently forgot all about Ryan and set his sights on Navarrette, who just pretended he didn't exist. That did not set well, either and after punching out a pinball machine, he left the Pizza Place.

"Did she give you any indication at all as to where those guys might have gone?" Cassidy asked. She crossed her arms. "Private Walker, how much longer are you going to be on that phone?"

"Just until I get confirmation that the taxi is on its way, Drill Sergeant," Shannon answered. The dispatcher had already confirmed and hung up but Shannon didn't want to miss the drill sergeants' conversation.

"She was scared to death, J.J., she didn't say much of anything. She was close to being hysterical," Bradbury explained.

"Then I guess we just wait until they sign in Sunday night and nail them then."

"Don't count on it, J.J.," Jessup said. "If the only witnesses are our girl and your boy, I doubt it will go anywhere. These guys will be thick as thieves and the more time that kid spends in the hospital to think about it, the more I bet you'll find him with sudden amnesia."

"No. I have no doubt Wotek will identify them." Cassidy sighed. "I don't believe this. Less than a month to go before my company is all graduated from LE School and not only do we have an incident like this but I'm probably going to have to go into Averill tomorrow and bail half my company out of jail."

Seeing the quizzical stares of Bradbury and Jessup, she explained. "They are on the outskirts of town where they rented an entire motel to throw a company party. It's to celebrate making it so far. Why'd this crap have to happen on my duty night?"

When Shannon saw Bradbury and Jessup leave, she walked back into the Orderly Room. The taxi had arrived and picked up the last of the Alpha trainees who were going to the party and the CQ and CQ runner were out of the office. She spotted Cassidy seated at Sergeant Fuscha's desk.

"I thought you were going to the party," Cassidy said.

"I was but I had a better idea. I took a quick run to the EC and Paulsen and his friends have gone back there."

"You were listening to our conversation earlier?"

"I spent a lot of time on hold. I couldn't help it."

Cassidy nodded. "Well, thanks for the tip, Private Walker. I'll send the MPs around to pick them up."

"Drill Sergeant, if I may...won't it be Wotek's word against their word?"

"Probably."

"What if I can get them to admit they did it?"

"Absolutely not, Walker. I don't want you or the others involved in this."

"We're already involved, Drill Sergeant. If the MPs pick them up and Wotek IDs them, they'll get kicked out of LE School but can't they come back and say there's no real proof and really make this a headache for everyone?"

"It's a possibility." Cassidy scrutinized her. "What are you suggesting?"

"Let me try to get them on tape, outright admitting that they did it."

"No. You are not trained for undercover work and I will not allow one of my trainees to be put in that kind of danger."

"Drill Sergeant, you asked when we were going to start thinking like cops. Well, I'm thinking like a cop. If you won't let me do it, does McCullough have any female MPs?"

"Not at the moment, at least that I'm aware of."

"Then why can't we call the MPs, have them put a tape recorder on me and agree on a signal for when I have the confession and they can then come in and do all the dirty work? At least then, there is no question of whether or not Paulsen did it."

Cassidy was silent as she thought about Walker's offer. "What if Paulsen recognizes you?"

"All the better. He knows I'm just another trainee."

"But he wouldn't admit to thumping an Alpha trainee to another Alpha trainee."

"He would if the circumstances were right."

Cassidy was quiet again as she contemplated Shannon's suggestion. She picked up the phone and placed a call to Captain Colton.

Colton's abrupt authorization surprised both Cassidy and Shannon but then Shannon realized Colton probably hoped she got the hell beat out of her, as well. He made sure Cassidy had Shannon sign a waiver that she would not hold the company responsible for any injuries she might sustain while performing this task. Cassidy assured Shannon that they would do everything possible not to let her get hurt but if she was, the Army would still take care of her. Shannon signed it and hoped that Colton wouldn't contact Bishaye who would definitely put a halt to what Shannon was about to do.

Military Police Investigators came to the company area and affixed a small tape recorder and microphone to Shannon's midsection. They explained to her what she needed to get the perpetrators to say before she gave them the approved gesture to let them know it was time to move in.

They sent Shannon back to the EC and a plain clothed member of MPI followed her in a minute later and stayed at the bar.

The four men had just ordered another large pitcher of beer that indicated they planned to stay a little while longer. Shannon decided to test the water first. The main disco room was packed so Shannon nonchalantly strolled up to their table and stood, appearing as though she were watching the dance floor. Not even a minute passed before Paulsen took notice of her.

Paulsen and his cohorts shouted out raunchy come-ons to Shannon and she responded with flirty smiles and winks. By the next song, Paulsen had motioned for her to sit down and join them. The conversation was mostly overpowered by music so everyone had to shout to one another to be heard. Shannon hoped the recorder was able to pick up their dialogue. The adhesive that held the instrument to her body had already started to irritate the area it was affixed to and her skin itched. Hopefully, she could hold out until she got them to say something incriminating. She had already politely refused the invitation to dance twice. The last thing she needed to do was jar the tape recorder loose. She told them she'd much rather sit there and drink with them.

"Just how drunk do you want to get?" one of the men at the table asked her.

She parted her lips and licked them slowly. "Drunk enough."

Paulsen and his group hooted and hollered like a tribe of pleased barbarians at Shannon's words and gesture. Then the young man seated next to Paulsen spoke. "Hey…aren't you in Alpha-10?"

The table discussion stopped abruptly and all eyes were on her. Shannon looked at the male who asked the question and ever-so-innocently replied, "Yes, I am. Why?"

The boys eyed one another apprehensively. Paulsen spoke up, his tone cautious. "Didn't you hear?"

"Hear what?" Shannon batted her long eyelashes, trying her best to look demure. "Did something happen?" She leaned in closer to Paulsen.

"When was the last time you were back in your company area?" Paulsen asked.

"1700 on Friday. What is this? Twenty questions? What's with you guys? Lighten up, would you? If you guys don't want to have fun then I'll find some guys who do." She flashed a dazzling, promising smile. It was enough to make four horny men drop their collective guard.

"What kind of fun did you have in mind, babe?" Paulsen asked. He reached out and started to stroke her arm.

Shannon acted her most seductive. "You seem like a smart boy. What kind of fun do you think?" She noticed the scrapes and swelling on Paulsen's knuckles.

More cheers came from the men at the table. Paulsen's hands suddenly went to his lap to cover his reaction. "This must be my night."

The young man on the other side of Shannon looked at Paulsen. "Yeah! Why don't we teach those smart-ass Alphas a real lesson and fuck with one of *their* girls?"

Paulsen shut him up with a sharp stare and Shannon looked at Paulsen and smiled, sweetly. "And just how do you boys plan on fucking with me?"

"In a good way, believe me. I guarantee you'll love it," Paulsen said. He put his face close to Shannon's. "What do you say we have ourselves a little party here and now?"

"My favorite two ways to have a party," Shannon said, wantonly. "Right here at this table?"

"No, the green grass motel, outside, behind the club. You know where the track is?" Paulsen asked.

Shannon nodded. The scene of the crime. Perfect.

"Behind the bleachers, near the woods. It's a real quiet time of night. No one to bother us." He looked up at his friends at the table. "I go first. Anyone have a problem with that?" No one admitted to having a problem with it. He returned his attention to Shannon. "You don't have a problem with that, do you?"

Shannon knew she had to be careful of how she chose her words, to avoid entrapment. "A problem with what?" She smiled at him, knowingly.

"A problem with this." He took her hand and placed it on the hard bulge in his pants. He moved her hand over as much of his erection as the material holding it in would allow.

Shannon removed her hand. "Oh my." She reluctantly admitted to herself that she was rather impressed with what she had felt. Too bad he was such a narcissistic asshole. "You'll excuse me while I hit the ladies' room first? I should get rid of all this beer." She stood up and he grabbed her wrist.

"I'll wait for you outside the bathroom," Paulsen said. "I want to make sure you don't get distracted along the way."

"Suit yourself," Shannon said, pleasantly. She wanted to give the MPI investigator a head's up that they were taking the situation outside. She hoped, when he would see them all walk out the door together, he would be astute enough to get the word to his co-workers in the parking lot and not just assume she had botched the case. She wished she were partnered off with Dale. They could read one another's mind. Dale would know what to do and if things quickly went to shit, Dale could fight like a ninja.

Shannon saw the agent watch her walk to the bathroom. She prayed he was still there when she came back out. Regardless, she knew Cassidy, who was in the unmarked MPI van in the parking lot, would not allow anyone to leave until either Shannon gave the investigators the sign or was back in the van. Cassidy was responsible for Shannon's welfare on this little endeavor, there was no way Shannon would be off anyone's radar for too long.

Once in the stall, Shannon checked the tape recorder to make sure it was still working and that it still had enough room to catch another half hour or so of conversation. When everything checked out okay, Shannon buttoned back up and put her jacket back on, leaving it open in the front.

Paulsen put his arm around Shannon when she exited the ladies' room and escorted her out the door. She was relieved to see the investigator still at the bar. Shannon and Paulsen walked around to the back of the building, across an access road, a dirt track, and a field to a set of bleachers with a thicket of trees behind it, which led to dense woods. Shannon asked if anyone had sneaked out anything to drink and two of the four men produced bottles of beer. Shannon asked to take a drink out of one so that she could 'wet her whistle.' She really wanted to waste a little time so that everyone could, fingers crossed, sneak in place for when she gave the signal.

"Drink that up, babe, we got us some partying to do," Paulsen told her as he finished a cigarette.

"You know, I overheard some girls talking in the ladies' room about a fight earlier that took place right here, right where we're standing. Damn, wish I'd been here earlier. I hate to miss a good fight," Shannon said.

"You are my kind of chick," Paulsen said, as the other guys laughed. "You would have loved this fight. It involved one of your Alpha boys."

"Oh, really? Were you there?" Shannon looked around at the group.

"Yeah, you could say that," one of the males said.

"Wait a minute...you didn't beat up one of the guys in *my*

company, did you?" Shannon asked.

Paulsen stepped forward threateningly. "What if we did? That's not going to change our plans."

"Maybe it will," Shannon challenged. "Tell me who it was. I might be glad, even more grateful, that you did."

"I think his name was Wotek," Paulsen said. "A Polack. No great loss."

"Wotek, huh?" Shannon repeated. "I know him. He's a nice guy. What'd he do to you?"

"How about we stop the questions and start the action." The others agreed with Paulsen as he closed in on Shannon. Fortunately, as he moved in, he couldn't resist the urge to brag. "You see, your buddy, Wotek, couldn't keep his hands off my girl. I told him to stay away from her but he just had to dance with her one more time. So we taught him a lesson, didn't we guys?" The other three men agreed with him.

"What did you do to him?" Shannon asked.

"We beat him to a pulp and put him in the hospital," Paulsen said. His patience with waiting had clearly run out. He grabbed Shannon and pushed her to her knees as he unzipped his fly. "Now, shut up and blow me."

Shannon knuckle punched him in the crotch. "Consider yourself blown," she said. She watched him bend double and fall to the ground. She stood up as the three others started toward her. Before Shannon had a chance to give the investigators a sign to close in, the area was lit up by three vehicle spotlights and MPs surrounded the group. Two MPs helped Paulsen up as Shannon walked away toward Cassidy who walked in with one of the investigators.

"Hey! This is entrapment!" Paulsen protested. "She enticed us here with promises of sex!"

"No, you just assumed you were going to get sex," Shannon said.

"Fucking bitch!" Paulsen spat out.

"Shut up, Paulsen, that's not what you're getting busted for and you know it," Cassidy said. She looked at Shannon. "How are you doing?"

"I'm okay, Drill Sergeant. I need to get this tape recorder off me, though. It's itching like crazy."

"I think that can be arranged." Cassidy called one of the investigators over. They walked Shannon back to the van and removed the recording device. The tape was played back to see if they got what they needed. The first part was partially muffled and distorted because of the noise in the EC but they still could decipher the pertinent dialogue. Everything said outside was clear and damning.

Cassidy drove Shannon back to the company area. "You took quite a chance, Walker. What made you think those guys would talk?"

"I've been around Paulsen a couple times before. He hasn't noticed me but I've noticed him because he's an obnoxious braggart. I knew he wouldn't resist a chance to boast. He'd had enough to drink and I was pretty sure he'd gloat. I really like Wotek, Drill Sergeant. He's a sweetheart and he didn't deserve what those assholes did to him. I wanted to do something because I know Wotek and he would have let it go."

"You really think he would have?"

"Yes. I think he will just want to forget it and get out of here ASAP and that would leave them free to do it again."

"You did a fine job, Walker. I will definitely write up a recommendation based on your performance and initiative tonight to accompany your records to your first permanent duty station."

"Thank you, Drill Sergeant." Shannon thought that was nice of Cassidy, even though she didn't need it.

Cassidy parked in the Alpha Company lot. "What were you before you enlisted? A vigilante? All I can say is that I'm glad you're on our side." They got out of the car. "If I wouldn't be accused of fraternizing, I'd buy you a drink."

"And I'd take it," Shannon said. They walked toward the Battalion Orderly Room. "Now what?"

"Oh, now you get to see the real heart of police work, the endless filing of reports. We have some paper to write, Private Walker."

By the time Shannon finished writing her statements, it was two-thirty in the morning. Cassidy bet Shannon she'd never make it up in time for CQ duty at 0600. Cassidy didn't know Shannon well enough to never bet her on anything.

Bleary-eyed and cranky as all get out, Shannon was up, awake, in uniform and in the A-10 Orderly Room at six o'clock A.M. At 0601, Cassidy walked into the office, holding two huge Styrofoam cups of hot coffee. She set one down on the desk Shannon was sitting behind.

"I brought an extra cup of mud, just in case, but I still can't believe you're here," Cassidy said.

"If I could have gotten out of this, I would have," Shannon told her.

Cassidy pulled up a chair and sat across from the undercover lieutenant. "You remind me a lot of me when I was going through basic and LE School."

"Oh, no," Shannon said. She let the steam from the coffee drift up into her face. "I hope that doesn't mean I'm going to grow up to be a drill sergeant."

Cassidy smiled. "I like you, Walker. I don't know why. There's something about you. You've got balls, my friend. Big ones. Both you and your buddy, Oakes, and that's good because you're going to need them."

Shannon just nodded. "I don't know about that. I just have never been able to tolerate bullies very well."

Cassidy looked around and lowered her voice. "That's probably why you and Sergeant Ritchie used to lock horns."

Shannon glanced at Cassidy, mid-sip. She swallowed. "How did you know about that?"

"Sergeant Ritchie. There aren't a lot of secrets in a basic training company. Anyway, Sergeant Ritchie said he thought you were going to make one hell of an MP. He said you really seemed to know right from wrong and that you didn't take shit off anybody."

"Ritchie said that?" Shannon was astounded.

"Drill Sergeant Ritchie," Cassidy corrected her. "Yes, he did."

Shannon went back to sipping her coffee. She looked at her watch. "Where the hell is Beltran? She supposed to be my runner."

"Private Beltran is at the MP Station, trying to explain to them why she finds it necessary to carry a concealed switchblade on her person," Cassidy said.

"No kidding. When did this happen?"

"About ten minutes ago. I was on my way to get the coffee and she was on her way downstairs, I guess to report for duty, when she tripped and the knife came flying out of her shirt. She wouldn't give me a reason for possessing the weapon so I hauled her back to battalion and called a patrol in to question her."

Shannon shook her head. "That girl..." She placed the brassard on her upper arm that alerted everyone to the fact that she was the A-10 CQ. "I didn't think she was going to make it here this morning because she was sick. Last night at the Pizza Place, she kept complaining of throat trouble."

"Knowing Beltran, several people are probably trying to cut it."

Shannon laughed. She genuinely liked Cassidy and wished the E-6 had been there from the beginning. It may not have bode well for Dale, seeing the relationship that developed between them but Cassidy would have definitely been better for the morale of the women. She sat back and studied the drill sergeant. Yes, she could see Dale and Cassidy together and then she was amazed at how smoothly she had transitioned Dale from men to women. Maybe that really was where Dale was meant to be.

Dale was back in the company area just in time to meet her partner coming off CQ. She followed Shannon upstairs to the nearly empty bay, where Shannon collapsed, face down, on her bunk. Dale sat on the floor and leaned her back against Shannon's locker.

"I hear congratulations are in order. Good job."

"Thank you," came the muffled response.

"So much for keeping a low profile."

Shannon raised her head off the pillow and looked at Dale. "It's too late for that. You blew it with Kirk, we both blew it at the EC that night and you aggravated it further in self-defense class. So I figured, what the hell? I knew I could get those little bastards and I did." She rested her head back down on the pillow, still facing Dale. "No one suspects a thing and I would have done the same thing six years ago if the same thing happened. No big deal."

"Hey, I didn't have a choice in self-defense class, you know that," Dale said defensively. "You should have called me at the motel. I would have come back and helped you out."

"I didn't have time. Besides, it was good for me. It relieved the boredom and got those old survival juices flowing. Now if you'll excuse me, I'm exhausted. I'll probably sleep right through until tomorrow morning."

"The party was great, thanks for asking," Dale said, sarcastically.

"Great."

"How was Cassidy through it all?"

"I'm her new hero," Shannon mumbled.

"Oh, man, don't say that, you'll make me jealous."

"Dale, please, can we save this oh so interesting conversation for another time?"

Dale stood up and lightly slapped Shannon's foot. "Of course. Sleep well, hero. I'll talk to you in the morning."

CHAPTER FIFTY-THREE

Dale mastered Ticket Writing class the next day and Shannon easily got through Processing A Drunk Driver class.

After 1700 formation, they spent most of their evening at the EC, playing pool. They held the table most of the night as partners. There was not a lot of competition and that left them easily bored, so they left the club early and started walking back toward the barracks.

"Did you ever notice that going to the EC is always a lot straighter than coming back?" Dale observed.

"Some nights are more crooked than others," Shannon agreed. "Weekend nights are definitely the worst." She stopped to light a cigarette. They strolled in silence for a short while. "Have you thought about calling her when we're finished here?"

Dale looked at the ground. "Who?"

"You know damned well who. Cassidy."

"I don't know."

"Yeah, I guess it might be awkward. You know how flings in training environments are."

"She told me Saturday again that she didn't feel that way about us. You know, casual."

"Saturday? When did you two get time alone to talk on Saturday?" Shannon asked.

Dale smiled and blushed. "Um…we, uh…"

Shannon grabbed Dale's arm and stopped. "You didn't!" Dale wore a silly smirk. "You did! Where? When?"

"She sent me to the Dayroom to clean it and met me there."

"Something tells me no cleaning got done."

"No cleaning was needed. It was already squared away."

"I can't believe she took the chance with you after Casey just getting caught."

"It's probably the best time. The cadre would think no one

would be stupid enough to do the same thing the day after that."

"How was it? The sex?"

"Unbelievable. Jesus, Shannon, I honestly never knew it could feel as good as she makes it."

"Dale, you should see how you light up when you talk about her."

"She wants more, Shannon. She wants to put in for wherever I'm assigned after I leave here."

"Wow. Dale, she's really serious about you," Shannon said, shocked.

Dale chuckled at Shannon's incredulity. "Yeah, I'm stunned, too."

Shannon swatted her partner. "Stop that. I'm not surprised because she likes you that much, it's just that, well, it's awfully soon."

"I know. It scares me."

"Scares you how?"

"In a good way. I could get used to waking up next to her. That scares the hell out of me." They began walking again. "She's so sincere and I hate lying to her. When she finds out everything, I know she's not going to want anything to do with me."

"How do you know? You're getting out. You'll be a civilian so you don't have to worry about fraternizing with an enlisted soldier."

"Even if she does forgive me for not being able to be up front with her, we're both women. Uncle Sam doesn't exactly welcome that with open arms."

"True but obviously she has managed so far. Something tells me she could manage with you, too."

"Why are you so hot to get us together?" Dale asked, curiously.

"I don't know," Shannon said and shrugged. "I like you, I like her, I think you two would be good together and you should give it a chance. You have nothing to lose at this point. Let something nice come from your Army experience for a change."

"Some nice things have come from it, like my friendship with you and my friendship with Anne."

"Yes," Shannon said, slyly. "But our interaction with you

doesn't satisfy you the way Cassidy's does."

"It does…just in a different way."

Shannon was suddenly alert and she touched Dale's arm and pointed. Dale looked in the direction Shannon indicated. "There's Casey. Let's go see if he'll talk to us."

The two agents jogged up to Sergeant Casey, who seemed to be headed toward the Tenth Battalion barracks. Casey turned around when he heard the running and his name being called. Casey stopped to let them catch up.

"Hi, girls," Casey said. "On your way in for bed check?"

They both nodded as they caught their breath.

"You two amuse me," Casey said and smiled. "You're the subject of many entertaining stories while having a beer at the NCO club."

"Really?" Dale asked.

"Yeah. You two know how to have a good time and, at the same time, not lose sight of what you're at McCullough for. Must be because you're older than the average trainee."

"Yeah, we're ancient," Dale said and smirked.

"Sorry to hear about your problems," Shannon said finally, as the three of them started to walk again.

Casey shook his head. "What can I say? It happens all the time around here. I was just stupid enough to be caught."

"Did she come on to you?" Dale asked.

"We sort of came on to each other," Casey answered. "She always flirted with me and I flirted back. Normal stuff. Then we had an opportunity to take it farther. I should have taken her off-post, that's all."

"Do you think you were set up?" Shannon wanted to get right to the point. She lit up another cigarette.

Casey lit his cigarette off Shannon's. "No. Not by her, anyway. Hell, no. This has been going on between us almost every night for three weeks. If it was a set up, it wouldn't have gone on that long. Every cycle, members of the cadre of each company usually nails a trainee or two. It's kind of like a game of Russian Roulette. Take Bradbury for instance. She normally never nails less than two trainees per cycle. She's very lucky, almost a Goddamned legend around here. They're always female but, hey, she gets what she wants."

"What about our company?" Dale asked. She prayed he wouldn't say anything bubble-bursting about Cassidy. "I heard we were jinxed but no one's been doing shit this cycle."

Casey snickered. "That's what you think. I know for a fact one of your girls is doing someone and it's not a drill sergeant, either. This chick is going for the brass. I saw them together two Wednesday nights in a row at Clancy's, a bar in town that's usually not populated by military personnel."

Dale and Shannon looked at one another, then back at Casey. "Who's the girl?" Shannon asked.

Casey seemed hesitant. "You've got to promise me you won't say anything."

"Why?" Dale asked. "What difference does it make now? You're already burned, why not bring an officer down with you, especially for the same infraction?"

"I suppose you have a point," Casey said. "Except I don't know if I want to ruin this girl's chance to get wherever it is she wants to go."

"Come on, Sergeant Casey, who's the girl?" Shannon persisted. "If it ever comes out that we know, we'll never say it came from you. Promise."

Casey studied them both. It was evident the anticipation was eating them both alive. "Well...all right. I don't remember her name but you sure can't forget a face and a body like hers. You know, that knock-out blonde who's always so quiet."

Dale and Shannon looked at him, speechless, then looked at one another other with the same stunned, indignant, and disappointed expressions. "Michaelson?" Dale asked, weakly.

"That's it!" He snapped his fingers. "Yes. Michaelson."

"Are you sure?" Shannon pressed him.

"Oh, I'm sure. She's the one who was in the barracks with that black chick who killed herself early in the cycle, right? The one CID took all the statements from?"

"That's definitely her," Dale said. "Who was she with?"

"Now that's a little more difficult to pinpoint. This guy is a full-bird colonel, I know that. At the angle I saw them from, I only saw his rank, not his nametag."

"He was in there in uniform?" Dale asked.

"In his fatigues. You know the high brass. They're so cocky, they think they're untouchable. I've seen him before but I don't remember where. He's an old dude, though, too old for a stone fox like her. Funny what attracts some people, huh?"

"Hilarious," Dale muttered. She looked at Shannon. Clearly, they both felt they'd been had.

"Hey, Walker, by the way, great job on that thing you did two nights ago," Casey said.

"Thanks."

"Okay, girls, this is my stop," Casey said. They hadn't realized they had arrived at Tenth Battalion already. "Thanks for the escort."

"Anytime," Shannon told him. Dale was too preoccupied to speak. Shannon grabbed Dale's upper arm and pulled her up the street, toward the LE School. "Let's walk a bit."

"Michaelson? It's fucking Michaelson?" Dale kept her voice controlled even though she felt anything but in control. "I can't believe it. Of all people. And who was the officer?"

"Maybe someone who doesn't know she's a GI. At first, even a second glance, if she's out of uniform, it wouldn't be my guess."

"But she lied to us, Shan. She told us she went to the rec center on Wednesdays."

"Maybe she does. Maybe she goes to Clancy's afterward."

Dale went on as though she hadn't heard her. "You know, it actually crossed my mind to follow her but I really believed her, especially after she asked me to help her and spar with her." Dale shook her head in doubt. "I knew she was ambitious but I didn't think it would be this way. The power of rank can be so seductive, though."

"I still wonder who the colonel is."

"He could be anybody. Do you know how many full-birds are running around on this post?"

"Then I guess we'll have to go see for ourselves Wednesday night. I have fireguard from five to seven. I'll have to meet you there," Shannon said.

Dale glared at her. "No, no, Shan, this could be our big break. Fuck fireguard!"

"If I can get someone to switch with me, I will, but you know how difficult that is."

"Shannon —"

"Dale, what if neither shows up? Then I have been AWOL from duty and nothing to show for it. We don't know what's going on here and we'd better play it safe. You get there, plant yourself and wait. On the chance they may have already gotten there, I'll just wait outside for you. I agree, this could be it but let's not get ahead of ourselves. I'll leave it to you to get in touch with Bishaye."

"Something isn't adding up here, Shannon. She doesn't tie in to Henning."

"Not yet, anyway, but this is the first solid bite we've had. Michaelson has obviously been deceiving everyone."

"Maybe this old colonel came on to Henning once and she turned him down."

"I suppose that could be it, Dale, but an officer as high-ranking as a colonel wouldn't have to go through all this to get even. He would just use his rank against her to retaliate and not have to bother with this garbage. It's got to be something more." Shannon noticed the troubled expression on Dale's face. "It could be anything. Hell, we know nothing about Deborah Michaelson and the bottom line is, it could be nothing more than what Casey thought it was: a girl who is gold digging."

"It could be but I don't buy it." She looked directly into Shannon's eyes. "I've got a real bad feeling about this and I can't shake it."

Dale tried to call Anne Bishaye five different times the next day but could not reach her. Bishaye was out of the office all day on business and no one answered her home number later on.

The two CID agents attended their classes but it was nothing more than going through the motions. They kept as close an eye on Michaelson as they could without raising suspicion. If anyone noticed anything out of place, it had to be that both Dale and Shannon were peculiarly serious the entire day.

For some unexplainable reason, both undercover lieutenants felt they had hit onto something big, something that may have been totally unrelated to the reasons they were brought there. It only complicated matters when they ran into Casey later in the mess hall and he supplied them with the name he couldn't come up with the night before.

"Sedakis," Dale repeated later, her tone one of disbelief. It was fifteen minutes before bed check and they stood on the far end of the north patio, keeping this conversation quiet. "I can't believe this. This is becoming a can of worms right before our eyes. What has she been doing meeting with him once a week? Shannon, what is really going on here?"

"I don't know. Didn't you say you had a run-in with Sedakis in the past?"

"Not really. He was some kind of adjutant to the Provost Marshal's office at Fort Ord when I was there. He sped through the main gate, nearly running me down one night and I ticketed him for it. It was just that one incident. I did bust the headlight out of his new car, accidentally on purpose, though. He and Anne got into it but she stuck by me. She was a little annoyed with my procedure but my citation was solid so he dropped it. Honestly, I'd never even seen the guy before that night and the only thing I had against him was that he thought he was exempt from following post rules. Soon after, I heard he had PCS'd to Fort Sam in Texas. The next time I heard his name was when Anne brought it up to me in October."

"And now he's the regional CID commander here. This doesn't make sense. Isn't he the one who's ultimately responsible for bringing us here?"

"Yes."

Shannon watched Dale, curiously. The look on Dale's face was deeply contemplative. The gears were turning and Shannon knew Dale was working hard to connect the dots. Suddenly, a look of sheer panic hit Dale's thoughtful features and she started to hyperventilate. "Dale, what's the matter?"

Dale dropped to her knees as though punched in the gut and the wind had been knocked out of her.

"Dale! What —?"

"Oh my God. Shannon. It can't be."

"What?" Shannon was trying hard not to panic at the behavior of her partner.

"It's a set up." Dale's voice was flat.

"No shit," Shannon said.

Dale grabbed Shannon and pulled her down to the cold, hard, concrete patio floor with her. "No! *We* have been set up! *I'm* the lieutenant Carolyn Stuart was talking about! Not Henning, me. Revenge against *me* and Sedakis is behind it."

Shannon had never seen Dale look so terrified. "Why? Because of a citation? Dale, don't be ridiculous."

"No, listen to me," Dale said as they stood up and moved to the picnic table. "It just hit me what's happening here. I was on an assignment in Texas called the Eagle Project. It had a longer name than that but we called it the Eagle Project. I was sent to Fort Sam Houston to infiltrate this group the military suspected of dealing arms to South America. There's been this little civil war going on there for years and -"

"Yes, I know all that. Get to the point."

"This particular South American group was paying our particular group beaucoup bucks to transport all kinds of new, state-of-the-art combat weapons down to them by helicopter. There were a few minor sales before I hooked up with the group but this deal was major. Big time. We were going to supply them with unmarked M16s, grenade launchers, claymores, and ammunition up the ying-yang. Well...I was the one who led that bust. I was the one responsible for that expensive deal falling through but because we had to move in on it so fast or risk losing it, we went in before the mastermind and the guy flying the helicopter showed up. They were both reputed to be military and they got away."

"And you think Sedakis was the mastermind?"

"I *know* he was, Shannon, I know it in here," Dale said and tapped her chest. "That was a major, *major* setback for him, that deal going belly up like that. I'm sure he not only lost a lot of credibility with the bad guys but he lost a lot of money, too."

"Aren't you overlooking something important here? Bishaye brought us down here and supposedly, it was all Henning's idea."

"No, that was all very clever on Sedakis' part. He knew Henning would have to eventually think about suggesting using trained people to find out what was going on. It was all too suspicious not to. He knew she would have to run it by Bishaye and he had to know Bishaye would look for people she could trust. He makes sure I'm on the available list, maybe even suggests my name to her, I don't know. Let's say, for argument's sake, he goes ahead and sets up this elaborate scheme, establishing a pattern, enough times for the higher ups to think it's not a coincidence and -"

"You actually think he would just randomly ruin innocent men's lives like that?" Shannon asked.

"An officer as ruthless as Sedakis? Yes, I do. Enlisted men — and women — are expendable when you are working toward a specific goal. In this case, *I'm* the goal. Okay, so then, he gets Anne involved by putting my name out there as still available. He knows she's going to request me. She does, he protests slightly, she insists and he relents. Here, we are."

"Dale, that's just too out there. I need more. The investigation will need more."

"No, Shannon, I remember now. I remember everything about the night I was attacked and the car ride where they thought they killed me. It just surged into my memory at the realization that Sedakis was behind it. Those men were hitting me with the butt of the pistol and, before I lost consciousness, one of them said, 'Sedakis wants her body dumped out here.' It's as if it was yesterday."

"You just remembered that tonight?"

"Just now. When I was sharing my theory about Sedakis, it was like the block cracked and it all came flooding back. You know, the past two weeks, the word 'dock' kept coming into my head and I didn't know why. And now it makes sense. My brain was trying to work through the trauma. SeDOCKis. It had to be that."

"Okay. So, we're here. What do you think he planned to do?"

"Kill us," Dale stated, blankly. "Maybe not you but definitely me. I bet if we did a little research, we would find that Ms. Michaelson is probably a professional he hired to get rid of me."

"Dale, do you realize how paranoid this all sounds?"

"I do. But I've earned the right to be paranoid."

"Why go to all this trouble? Why not just pick you off in Vermont?"

"I don't know. Maybe so he could have the pleasure of watching it done. Or knowing for a fact it *had* been done. I haven't figured that part out yet."

Shannon was reeling from Dale's revelation but her watch said they had to get upstairs for bed check. "We've got to get to the bay."

"Or what?" Dale smiled, wryly. "They'll write us up?"

"Look, as it stands right now, this is all great in supposition but we need proof. I think we should keep our mouths shut, go to Clancy's tomorrow night, catch them in act and take it from there. Try to get Bishaye to sit there with you to witness it."

"If I can get in touch with her. She's been in a ton of meetings lately."

"It's getting to the end of Tenth Battalion's cycle. She's got to be swamped in paperwork. If she can't get free, we'll have to do it ourselves."

Dale sat in a dark corner of Clancy's bar. She tried to stay calm but found herself fidgeting, nonetheless. She felt emotionally prepared for what the evening would bring yet, at the same time, still had not come to grips with the revelation of the night before. She wanted this episode of her life finished and wanted Sedakis to pay, if he really was behind it all. And she knew, beyond a reasonable doubt, that he was. Shannon was right, though.

Dale's gut instinct wouldn't cut it, they needed irrefutable proof.

She had arrived early enough to get herself well hidden. She ordered a beer but didn't drink it and hoped, by some miracle, that Shannon had found someone to trade fireguard duty with and she would show up soon.

Dale thought about taking a drink from her beer, torn between needing something to calm her down and feeling the need to keep a clear head. She still hadn't decided which need was greater when Deborah Michaelson walked into the bar. The part of her that hoped Sergeant Casey was mistaken died a very quick death.

The development of Michaelson and Sedakis was strange enough but within minutes of Michaelson's arrival, Anne

Bishaye walked through the door, clearly searching for someone. Dale wondered if Shannon had been able to contact Bishaye at the last minute. She thought about trying to signal Bishaye to let her know where she was but the possibility of revealing her presence to Michaelson was too risky, so Dale stayed concealed.

She nearly fell out of her chair when she saw Bishaye recognize and then approach Michaelson. They stood close to one another and conversed as though they were well acquainted. Whatever was going on, Bishaye had an urgent expression as she spoke and Michaelson kept nodding.

Dale continued to watch, speechless, consciously making sure she remained out of plain view. Just before the meeting ended, Dale's gaze followed Bishaye's hands as they took an envelope out of her jacket and handed it to Michaelson. Michaelson looked at it, skeptically, but then put it away, inside her own coat. The entire exchange took less than five minutes, then Bishaye left Clancy's.

Taken completely by surprise, Dale had no idea what to do. Her first instinct was to run after Bishaye but she decided, instead, to wait and see what Michaelson did.

Deborah Michaelson finished her drink and walked out the door. Dale left her corner table with the speed of a bullet. She pushed through the crowd until she was outside and spotted Michaelson in the parking lot. Dale sprinted up behind her. "Michaelson!"

Michaelson whirled to see Dale approaching her and looked as shocked to see the undercover lieutenant as Dale had been to see Bishaye. Her reflective expression then turned suspicious. "Oakes...where did you come from?"

"Inside. What's going on with you and the Battalion Commander, Deb?" Dale sounded angry and confused.

"What do you mean?"

"Don't fuck with me, Michaelson, this is too important. What did she give you? What's in that envelope?"

Michaelson's voice and attitude became chilling. "That's none of your business." The look on her face stopped Dale cold. Dale had seen that look before. It was worn by the men who tried to kill her at Fort Jackson.

Dale swallowed the instantaneous fear she suddenly felt. "It is my business, it is *very much* my business." Her mind raced wildly as she scanned the parking lot.

"What's the matter, Oakes," Michaelson taunted. "Can't stand the thought of Bishaye being involved with anyone but you?"

Dale took a step back and scrutinized her. "Who are you?" There were traces of genuine panic in her voice.

A slight smile crossed Michaelson's lips. "Rest assured, you will never know, Lieutenant Oakes." She started to walk away from Dale but Dale grabbed her arm and turned her around. Michaelson shook her off.

"How do you know I'm working with Bishaye? How do you know my rank? What the hell is going on here?"

Michaelson shoved Dale away from her. "Don't you ever put your hands on me again. I know all about you, Oakes, I know all about you and why you're here. I know your reputation and I know you're nobody to mess with. Guess what? You've met your match in me."

Dale didn't back down. "Who the hell are you working for?"

Michaelson smile came back. "Don't like this, do you, Oakes? You don't like not knowing anything, do you?" Michaelson leaned in close. "I love it."

Dale had lived seventeen weeks with this woman and now she was looking at a total stranger. Michaelson's hostility was strong and Dale wasn't quite sure how to handle her. She tried being rational. "Deborah, all I want to know is why you're meeting with Anne Bishaye. You obviously accepted something of importance from her. I have a professional interest in anything the colonel does since you know I work for her."

"Fuck off, Oakes. This is between Bishaye and me. If she'd wanted you in on this, she would have told you."

"Then where does Sedakis come in?"

"Oh, so you know about him, too. Gee, Oakes, you *are* good."

"Since I apparently wasn't clear the first time, I'll ask you again: Who the hell are you working for?"

Michaelson tapped the space in her jacket where she had placed the envelope. "Right now I'm working for the highest bidder." She turned from Dale and started to walk away but Dale wasn't about to let her get too far.

She took a step closer to Michaelson who, with split-second timing, whirled and grabbed Dale's hand, pulled Dale to her and spun Dale around, trapping Dale up against a car. A little taken aback by Michaelson's undeniable strength, Dale did her best to maintain a calm exterior. Dale's voice was coolly even. "Now it's your turn to take your hands off me." Michaelson tightened her grip. "Come on, Deborah, I don't want to fight with you. I'm not that foolish. I just want to find out what's going on here."

"Why do you think I'd tell you?"

"What do you have to hide from me?"

"Nothing. I just don't feel what I do, no matter who it's with, is any of your business." They had started to attract minor attention. Michaelson relaxed her grip and positioned herself against Dale so that it appeared they were getting intimately physical as opposed to dangerously physical. Michaelson chuckled. "You like this, don't you, Oakes? My hands on your body. Too bad I don't like women like that, although, if asked to do you, I might have been able to manage it very well."

How does she know about my sexuality? Just what, exactly, has Anne told her and why? "If you say this is none of my business, then why are you baiting me?"

"Because I like it, Oakes. It's what I do." Michaelson ran her hands across Dale's body in a cursory search for weapons.

"I'm clean," Dale assured her, then mustered up all her strength and pushed herself away from the car, in turn, knocking Michaelson off-balance. In an admirable display of speed and fluidity, Dale had Michaelson trapped on the ground. They were now hidden behind several parked cars and out of view of the entrance to Clancy's, so the incoming and outgoing patrons could no longer see them. Dale's voice was now tight. "Cut the bullshit, Michaelson. I want answers and I want them now!"

Michaelson struggled and tried to get up but Dale smashed her back down to the pavement. "Get off me, Oakes."

"Not until you give me some answers."

"Go ask Bishaye," Michaelson spit out. "I've got nothing to say to you."

"Then I guess I'll just hold you here until my partner arrives and go ask her myself."

"The hell you will," Michaelson said. She brought her knee up, caught Dale off-guard and flipped Dale over her head. Dale fell forward and the wrestling match had begun. As they tussled away from their unnoticeable spot, they started to draw a small crowd again. Michaelson had connected a few solid punches to Dale's face and body but not as many as Dale had deflected.

Dale and Michaelson were still fighting for control when a civilian black and white squad car drove into the parking lot. A taxi with Shannon in it pulled in right behind the patrol unit. The police officer and Shannon exited their respective vehicles at the same time.

Shannon was automatically curious as to what was going on to involve the police and a crowd. She circled the perimeter and wondered at the same time what was happening inside with Dale, Michaelson and Sedakis.

"All right, girls, break it up," the cop announced, as he moved through the crowd.

He reached the inner line of spectators just in time to see Dale take down Michaelson. Michaelson then kicked Dale away from her, which caused Dale to fall backward. When Michaelson got to her feet, she spied the officer.

"I said break it up!" He stepped between two people, out into the open, visibly taking this scene much too lightly. He stood there with his hands on his hips. "I enjoy a good catfight as much as anyone else but you girls really need to break it up."

"Goddamn it!" Michaelson cried, desperately. She reached down, underneath her bell-bottomed blue jeans where her boots were and came back up with a .22 Beretta in her hand.

Shannon broke through the crowd and into the clearing just as all this was taking place. She recognized the woman holding the gun and the woman it was aimed at. "Michaelson, no!" Shannon shouted.

Several cries of *'she's got a gun'* went through the now scattering crowd. If the people had not panicked, pushing Shannon backward with their movement, she might have been able to lunge for Michaelson.

Michaelson fired off a round at Dale, who rolled out of the

way as soon as she saw a glint of metal. By this time, the patrol officer had unholstered his .38 and had drawn down on Michaelson. "Drop it!" He ordered her.

She turned toward him and brought the weapon with her. She did not look ready to surrender and the officer took no chances. He fired off a round and the impact of the bullet sent Michaelson sprawling backward onto the hood of a car.

Dale and Shannon ran to her from different directions. Shannon kicked the Beretta toward the officer and they helped her to the ground.

"Back away from her! Now!" The officer's gun was still pointed at Michaelson.

"We're CID agents!" Shannon shouted to him "This is military business. Call an ambulance. Now!"

The cop kept his .38 trained on the group of three but he used his portable, hand-held radio to request back-up and medical help. When that was done, he returned his radio to its pouch and stepped closer to them. "Let me see some badges and ID."

"We don't have any on us," Dale said. She cradled Michaelson with her own body. She held direct pressure to Michaelson's wound. It didn't seem to help. "We're undercover."

"Call Colonel Anne Bishaye at –"

"No!" Dale cut Shannon off. She told the officer to have his station contact Karen Henning at home and have her meet them at the scene with their badges and IDs.

His weapon was still out but he did as Dale asked.

"What the fuck happened here, Dale?" Shannon asked. Her partner had a wild look on her face.

Dale ignored her. "Deborah, you've got to tell me what's going on here. What is worth getting shot over?"

"Oh, God. God, it hurts," Michaelson said. Her breaths were coming in short gasps and she had already lost what looked to be an unhealthy amount of blood. "Maybe…you didn't…meet your…match."

"Deb, what were you doing in there with Bishaye?" Dale's tone had gentled.

Shannon stared at Dale. "Bishaye?"

"She...she was giving...giving me money." Michaelson was now clammy and very pale.

"For what?"

"I...was...hired to kill...you before...the...cycle ended." The pain was obviously becoming unbearable. "Kirk...was a...a...part of it...too...but she...was too young...couldn't handle it...not after...she...got...to know...you. So I...I had...to carry it...out...by myself."

"Kirk?" This appeared to be too much for Dale, Tears welled up in her eyes. "Bishaye hired you? Why would Bishaye want me dead?"

"I don't...know. Sedakis...was running...the...show but...Bishaye was always....the one I got...the money...from. Sedakis would say...do it this week...and Bishaye...would...show up with more...money...and...say 'Wait'." Michaelson was getting weaker.

"Where is that fucking ambulance?" Shannon yelled to the cop.

"It's on its way." He still had his weapon aimed at them.

"Hang in there, Deb, you're going to be okay," Dale told her, unrealistically, as she smoothed her hair. "Did they tell you why they wanted me dead?"

"I..didn't...ask. All...my...concerns...were...monetary. It...wasn't...personal, Oakes. I...kind...of...liked...you."

They heard the sounds of sirens in the distance.

One of the paramedics walked up behind Dale and put his hand on her shoulder. Dale and Shannon had given their statement to a group who included two MPs, two CID agents, four civilian police officers and Karen Henning who was still in a state of shock.

"Lieutenant Oakes?"

"Yes?" Dale turned to the paramedic.

"I'm sorry, Lieutenant. She just died."

Dale nodded and bowed her head. She turned back to the group and Henning touched Dale's arm, supportively.

Shannon looked at Dale. "I need to speak to Lieutenant Oakes, privately."

"Sure," one of the CID agents said.

PERMISSION TO RECOVER

Shannon escorted her partner far enough away from cluster of authorities so that they could talk without being overheard. They stopped by an unmarked police car and Dale leaned against it. "I can't believe it, Shan. I can't believe she would do this to me…"

"Don't you find it curious that in Michaelson's official statement, she never mentioned Bishaye's name once? Just Sedakis."

"Curious, yes, but it sure as fuck doesn't leave her off the hook. I know what I saw and what I heard Michaelson say," Dale said.

"Look, why don't I tell them and go with them to pick up Bishaye?" Shannon offered.

"No," Dale said and shook her head, adamantly. "No. I will get her. It's only right. I want to know why she was involved in all this. I still need answers. You go with them to get Sedakis and I'll get Anne."

"By yourself? I'm not so sure that's a good idea."

"Then bring the MPs with you after you pick up Sedakis. I have to face her alone."

Anne Bishaye unlocked her door and walked inside her house. She removed the keys from the lock, shut the door and turned on the kitchen light. She put the carton of milk she was carrying in the refrigerator and headed toward the hallway. Her eyes suddenly focused on a figure standing in her living room. Her hand instinctively went to the light switch but the intruder's voice broke the silence before she could turn the light on.

"It's me."

Bishaye recognized Dale's voice and proceeded to turn on the light. Her hand was still in the vicinity of her chest and throat. "You frightened me, Dale. How did you get in…" She stopped at the sight of Dale's bruised face and then her gaze fell to the large stain of blood on Dale's shirt and jacket. "Oh, my God, Dale, what happened?" The alarm in her voice sounded genuine. She moved to Dale quickly with the clear intent of embracing her.

"Stop!" Dale barked. Dale shut her eyes and with tears welling up caused by anger and betrayal, she stepped back, out of Bishaye's reach. "Don't do this to me. Don't lie to me anymore and don't pretend you care about me." Dale's voice was shaking.

"What are you talking about? Of course I care about you. For Christ's sake, Dale, what happened?"

"Stop it! I know everything! I was there at Clancy's tonight. I saw you with Michaelson. I know what's going on."

Bishaye looked dazed. "Michaelson?"

"Yes. Michaelson. She's dead, by the way...but she was expendable. Isn't that right, Eagle?"

The words 'she's dead' came out and hit Bishaye like a slap but when Dale used the code word that was the root of the revenge set up, she turned sheet white.

"Yeah." Dale watched her closely. "You should get nervous because Michaelson made a deathbed confession. You're in some pretty deep shit, Eagle."

"Stop calling me that."

"Why?" Dale winced in pain. Grappling with Michaelson had re-injured her bad foot and it was now throbbing. She shifted her weight, babying the ankle. "I'm so sick of having my body battered because of you," Dale said. Bishaye took a step closer to her and Dale put her hand up to halt her. "Don't push your luck, Colonel Bishaye."

"You won't hurt me," Anne said, in a whisper.

"Just like you weren't going to hurt me? Of course, you already tried once and failed."

"What does that mean?"

"Memory loss? That was my problem, not yours. Silver Lincoln Continental? Florida plates? The officer's club at Jackson?"

"Have you lost your mind? I wasn't anywhere near there. Why would you think I would try to kill you?"

"*Stop lying to me!*" Dale yelled, agonizingly. "Why, Anne? Why you and why me? I idolized you, *I loved you!* You could have made me into anything, I was a fucking puppet for you!"

"A puppet with a mind of her own, I'm afraid." Anne appeared to have resigned herself to no longer deny her involvement. "What happened to Michaelson?"

"She tried to shoot me. She missed. Of course, I didn't have my gun, so an Averill cop did the honor." Dale shifted her weight again. "I think, at the very least, you owe me an explanation."

Bishaye reluctantly stood still. "This Eagle thing was so much bigger than you can imagine…"

"I know. Sedakis should be filling his pants right about now. Jesus. Why didn't you just leave me in Vermont? I never would have figured it out. It would have been so much better, especially for you and me."

"No. No, they were going to kill you, regardless. They wanted you dead. You knew too much. They knew at some point your mental block would crack and you'd remember all the numbers on that Florida license plate or conversations in the car that could come back to them. You never would have left it alone then, Dale. You would have investigated everything and exposed them. They were going to arrange an accident in Vermont and because you weren't on active duty, no one would have suspected them. So I talked them into this situation. I was so insane about it that I think they did it just to humor me. We had to set it up first but, in developing the plan and having you brought here, I figured I was buying you time."

"Why was Michaelson here?"

"She wasn't my idea. She was hired by Sedakis, originally, just to keep an eye on you."

"Like she kept an eye on Carolyn Stuart?"

"Stuart panicked. She was going to blow the whole thing but her death was a surprise to me, too."

"Just like Kirk?" Dale shook her head, disgusted. "You recruited a child to do your dirty work! You ruined the careers of six innocent men!"

"Some of them weren't that innocent. It was just a matter of time before they got caught for fraternizing."

"That's no excuse! And, believe me, knowing their lives were wrecked so you could allegedly save mine? You've put that on my shoulders. Sorry, but I'm not grateful to you for that."

Bishaye's jaw shifted. "I never expected any of this to get so out of hand."

"That's it? That's all you have to say?" Dale wiped a tear away from her eye as Bishaye stayed silent. "Why? Why did it have to be you?"

"Why did it have to be *you* who made the bust on that damned Eagle project?" Anne countered.

"I'm a cop, Anne! That's what I do, remember? I work for Uncle Sam on the side of the good guys. You used to work for the good guys, too. What happened? What were you doing mixed up in that shit, anyway? You were always one of the most, law-abiding, honest people I've ever known."

"It was Jack! It was *his* involvement! I didn't know he was going to pilot that helicopter until after everything came down. I was still at Ord when all this happened and he was supposed to have been in Texas on business for his school. He's my husband, for God's sake! I love him. I felt I had to protect him."

"And who's supposed to protect you?" Dale asked her, sincerely curious.

"Dale...I grew up in the fifties. I grew up in an Ozzie and Harriet/Father Knows Best home. I was raised in a traditional setting. You marry for life and you stand by the man you marry."

"Oh, bullshit, Anne! Nothing about your marriage is traditional. You're in the military, you run a battalion. You make more money than Jack does. You have no kids. Stop me when the tradition peeks in here..."

"You're right, I didn't do everything just like my parents did but I do love my husband and I am loyal to him."

"He betrayed his country and because you didn't turn him in, knowing what he and Sedakis were involved in, you're just as guilty." Dale took a deep breath, let it out and studied Bishaye. "You should have left him, Anne."

"There were circumstances –"

"Circumstances that override being a traitor and treason?" Dale was incredulous.

"Not everything is black and white, Dale. If anyone should know that by now, it's you." Bishaye's tone held a hint of bitterness.

"Anne - nothing, *nothing* is this gray! It's one thing when you sell out on yourself but you had no right to sell out on me!"

Bishaye saw that even if Dale did intend to take her into custody, she'd have to be able to focus on her first, an impossibility considering the amount of tears in Dale's eyes. She seemed surprised to see Dale this hurt or this vulnerable. She reached over, took Dale's hand, and pulled her into an embrace. Bishaye held her tightly. "I am so sorry. I wanted to warn you so many times. I know you will never understand and I'm not trying to excuse myself but I *was* doing everything I possibly could to get you to come out of this alive. I just didn't know what to do. I was caught in the middle of an impossible situation." She kissed the top of Dale's head. "You know how I feel about you."

"I thought I did."

Bishaye leaned back, lifted Dale's chin to kiss her. Dale fiercely pushed her away.

"No! You can't do this to me anymore! I won't allow you to manipulate me any longer. Two weeks ago, you might have gotten away with it but not now. Things have changed. I changed! Everything's changed." Dale was surprised at how relieved she was that she no longer felt the intense craving for the colonel. She still loved Bishaye, despite everything and that was confusing for her, because at this moment, love was the furthest thing from her mind. How could a successful outcome of catching the criminals feel like such a failure? Instead of the festive debriefing she had pictured and counted on with a woman she cherished and idolized, Dale now had to deal with mourning that loss and coping with the duplicitously flawed human being Bishaye really was.

Dale watched through tear-blurred vision as her one time mentor turned and walked back to the bedroom. Curious, Dale followed her and was struck temporarily mute as she watched Bishaye grab a travel bag and start to pack.

Dale wiped her eyes on her sleeve. "What are you doing?"

"What does it look like I'm doing?" Bishaye rifled through her bureau, grabbing essentials and throwing them into the tote. "How long do I have until the cavalry gets here?"

Dale stepped into the room and held up her hand. "Wait, wait. Do you think I'm just going to let you go?"

Anne stopped packing and straightened up. "I just told you that I'm as much of a victim here as you are. Surely you aren't going to allow me to be apprehended…"

Dale's mouth dropped open. Suddenly her anger overrode her hurt and shock. "Are you fucking kidding me? You had choices, Anne, and you made the wrong ones and then continued to make the wrong ones until people ended up dead. You can say you did it for me all you want, the ends don't justify the means. You weren't stopping Sedakis from having me killed, you were prolonging the prelude. If this was about saving me, that's exactly what you would have done, the minute you found out what Sedakis was planning, regardless of Jack's involvement."

"So I'm supposed to choose you instead of my husband?" Her tone seemed to be a cross between incredulous and cautious.

"You're supposed to do what's right and lawful, not become a criminal yourself!"

Anne sighed loudly, shook her head, and resumed loading her grip. "Then I guess you're just going to have to physically stop me. I will not go to Leavenworth for this."

"That's for a court martial to decide," Dale told her, stiffly.

Anne picked up her bag, moved to her bathroom, removed a pre-packed sundries container from a cabinet and pushed that into the tote before zipping it up. She then walked toward Dale with a determined stride and a hard expression. "Get out of my way, Dale," Anne said, her voice a low growl.

Dale shook her head. "No."

Anne took another step and moved her arm to shove Dale away from her. Dale grabbed Anne's hand and twisted it in a direction it was not designed to go. When Anne swung her tote toward Dale's head, the younger woman ducked and turned, stuck her leg out and swept the colonel's feet out from under her. She still had possession of Anne's hand which now rendered the colonel immobile. "You're going to break my arm," Anne said, pain clearly showing in her voice and expression.

"I could. And I will if I have to."

"Dale, please! Think of what you're doing. At the very least, you'll be sentencing me to life in prison and, at the most, death by a firing squad." Dale didn't ease up. "Let's go away together, you and me. We'll start that life you want," Anne puffed out, sounding desperate.

Dale's rage rose at Bishaye's flagrant attempt to manipulate Dale's feelings and she almost did break Anne's arm. Dale yelled at the top of her lungs to rid herself of the angry energy. "Don't. You. Dare. Jesus Christ! I don't even know who you are. How could I have let you play me all these years? I don't want you anywhere near me ever again!" Dale was crying again, feeling like one big open emotional wound.

The sound of sirens grew, signaling that back up was on its way. "Dale, I am begging you," Anne whimpered.

"What else are you mixed up in, Anne? What else have you and Jack done? This set up was not the work of an amateur. What else will I find out when the investigation really starts?" Dale's voice was controlled but still shaking.

"You don't...you'll never understand," Anne said, defeated.

"You're right."

"Well, that was a scene I never thought I'd see," Shannon said to Dale. She sat on the front stoop with Dale while Bishaye's house was searched. "Neither one of you looked too happy with the other."

Dale's head was resting on her wrists as her knees supported her forearms. "There wasn't anything to celebrate." Her voice was muffled as she never raised her head.

"How could she do this to you?"

Dale shook her head. "She told me I'd never understand. I don't think I ever will."

"Did she even try to explain why?"

Dale raised her head and stared out at the patrol car that held Lieutenant-Colonel Anne Bishaye in the back, handcuffed, behind a division screen. She then looked at Shannon. "She said Sedakis was the mastermind and Jack was the pilot who was to transport the weapons. She tried to tell me we were both victims of circumstance."

"And you're not buying it," Shannon said as a statement.

Dale looked back in the direction of the patrol car. "Obviously not."

Shannon put her arm around Dale. "Maybe it's true."

Dale put her head back down. "Doesn't matter at this point. I've done my job, I've arrested the suspect. Let JAG sort out her guilt or innocence."

"I honestly can't imagine how betrayed you must feel."

"No, I assure you, you can't," Dale said, softly.

Three hours later, after all the statements were taken, all the stories were told and retold and all the secrets were out in the open, Dale and Shannon were transported back to the barracks where they were to collect their belongings and leave Tenth Battalion.

Henning was waiting for them at the Orderly Room when they walked in. Since there was such confusion at the crime scene, she waited to return their badges, IDs, and personal weapons until they got back to A-10. This action happened in front of a dumbfounded Minty, the CQ.

Henning hadn't had the chance to converse with either Shannon or Dale privately yet. She had shown up after the incident to confirm the identity of the two agents to the police. She had tried to phone the Battalion Commander before she left her apartment. Now she was relieved there had been no answer.

"I'm really sorry," Henning said to Dale regarding Bishaye. "I had no idea."

"I know." Dale was subdued. She glanced at Minty, who was clearly mystified by the activity in the CQ office. Minty also focused quite intently on the blood on Dale's clothes.

"What happens now?" Henning asked.

"We pack, we leave, we're out of here, goodbye, and good luck. Or is that good riddance." Shannon put her ID and badge away. She opened the cylinder of her .38 and checked to make sure it was loaded. After she closed it, she looked at Dale.

"I have to take a shower and get out of these clothes before I go anywhere," Dale said, despondently.

It had just hit midnight by the time they got upstairs to the bay. When they walked in, they saw that everyone was wide awake and waiting for them. The lights were on and there were gasps at the sight of Dale's shirt.

"What in Christ's name is going on with you two?" Belinda Ryder asked them. "An hour ago, Holmquist comes up here with the MPs, cuts Michaelson's lock off, clears her locker out, strips her bunk and doesn't say a word. Then you two don't show up for lights out or bed check...what's happening? And where the hell is Michaelson?"

Both agents were quiet as they opened their lockers and started packing their belongings. As Dale went to take a shower, Shannon stripped her bunk and separated her civilian and military items.

"This isn't fair, Walker. Talk to us. No one can believe you'd wait until the end of LE School to mess up like this." It was Chrissie Wachsman who asked this time. "Why'd they come and take all of Michaelson's stuff?"

"Michaelson won't be coming back," Shannon said, finally.

"Did she go AWOL?" Sherlock asked.

"Sort of," Shannon answered.

"What does that mean? Either she did or she didn't."

Shannon hesitated. She wondered just how much she was supposed to say. "Okay, listen up. Michaelson wasn't who she pretended to be and she got caught. She went a little nuts and was shot by an Averill cop." There were several gasps.

"Shot? How is she?" Wachsman asked.

"She's, uh...she died."

While the women reacted to this, Dale emerged from the shower in her robe and toweling off her hair. Her bruised face was quite colorful now.

"Are you going to pack up your military shit?" Shannon hollered to her.

"Not only no but hell, no," Dale said. She reopened her locker. "I hope I never see any of it again and I don't care what they do with it."

Shannon finished packing and brought her suitcase down to Dale's bunk. Kotski walked next to Shannon. "Where are you two going? Are they kicking you out? What did you two do?"

Dale had just finished tucking her shirt into a clean pair of jeans when they heard the barracks door slam open and a cry of "Man on the floor!" went out. The trainees assumed the position of Parade Rest and Holmquist walked to Dale and Shannon. He stood face to face with them, stood at Attention and saluted them.

"At ease," Dale said.

He relaxed and looked at the trainees. "You heard her. At ease." Everyone took up their previous informal stance but now they were even more curious about what was going on. Why had Holmquist just saluted two soon-to-be ex-trainees? "Do they know yet?"

"About Michaelson, yes. About us? No," Shannon answered.

"What about you?" Snow asked, just as Dale put her badge and ID in her breast pocket. She then donned a shoulder holster and put her .38 in it. More gasps. "Why do you have a weapon?"

Shannon displayed her ID and Badge to everyone. "We're lieutenants and CID agents. We've been here on assignment since 22 November. The assignment was officially terminated at 2224 this evening. We are leaving here now and moving on to our next assignment." She looked at Dale. "For one of us that means freedom as a civilian."

The look of astonishment on the faces of their barracks mates was expected. Some looked at the two lieutenants with a new admiration while a few others, Kotski, Mroz Wachsman, Tierni and Tramonte, looked as though they had been deceived.

"We couldn't say anything," Dale explained. "We didn't know who to trust. This situation was very delicate. Even the drill sergeants didn't know." Dale slipped on her jacket, which hid her revolver nicely. They both picked up their suitcases.

"You've all been wonderful to go through training with," Shannon said. "You've all got great potential. I'd be proud to be partnered off with any one of you." She turned and looked directly at Snow. "Even you. I'll tell you up front: I don't like you but despite that, I think you're going to make a hell of a cop."

Snow smiled. She extended her hand to Shannon, who shook it. "Thank you."

Shannon looked at Chillemi. "You need to work on your people skills."

When the hugging was over, Holmquist escorted them downstairs to the Orderly Room. Henning was still there and so was Minty who grinned and shook her head.

"Y'all are lieutenants working undercover? I should have known with all that you two got away with," Minty said.

Shannon smiled back at her and both lieutenants wished her good luck. They shook hands with Henning. "I'm going to have one more smoke and then we'll come back in and call a cab."

"Certainly," Henning said.

They left their suitcases in the CQ office and walked outside for Shannon to have one last cigarette, one last time in the Alpha-10 company area. "Where are you going now?" Shannon asked.

"Back to Vermont. If CID needs anything else, they can call me there. What about you?"

"I have to go back to Texas and wait to see if I'm going to stay there or be cut new orders."

"What about tonight? We could get a room in town and wind down."

"Maybe another time," Shannon said.

"Why? Do you have plans?"

"No," Shannon answered. "But you do." She nodded her head toward the A-10 parking lot where they saw a vehicle race up to the front. It was Cassidy's car.

Dale looked at Shannon. "What? How did she find out -?"

"I called her."

"When?"

"About ninety minutes ago, when you were giving a statement. She knows everything now. And look, Dale...she came anyway."

Dale's attention was drawn to Cassidy who ran up to her. Cassidy looked like she wanted to wrap Dale up in her arms but she obviously wasn't sure of the protocol in the new situation. She stopped just short of invading Dale's space. Shannon nodded to Cassidy and left them alone.

"Hi," Dale said.

"Hi," Cassidy said back. "You look like hell."

"I feel worse." Dale smirked.

"Were you just going to leave?" Cassidy looked frantic but she kept her voice calm.

"I wasn't sure you'd want anything to do with me."

"Dale, I mean, Ma'am –"

"You can still call me Dale. If the rank didn't make a difference one way, it shouldn't make a difference the other."

Cassidy stared at the patio floor. "How much trouble am I in?" She looked up at Dale.

"You aren't. Did you think you were part of my assignment?"

"I'd hoped not. Walker...I mean, Lieutenant Walker indicated I was not."

Dale shook her head. "Everything that happened between us on my part was genuine."

Cassidy looked relieved and managed to smile. "Mine, too. Dale, I understand why you had to stay quiet, why you had to keep your cover. I don't blame you."

"Really?"

"Really. Would you...could we go get a cup of coffee and talk?"

Dale hesitated, not because she had doubts about the woman standing before her, but because she wasn't sure if she wanted coffee or a good, stiff, drink. "I would love to go have coffee with you. We have a lot to discuss. But I'm not sure right now is the time." Dale paused. "My whole life was turned upside down tonight. Everything I thought to be true and just, turned out to be a lie and...I'm not sure about anything anymore."

"Just coffee, Dale. I think we both need to figure some things out," Cassidy said.

Dale was pensive, then she nodded. She had let go of so much tonight and she was not willing to let go of the only real thing she may have had left. "Where would we go?"

"How about my place? You said you liked my coffee."

Dale smiled for the first time since the night's events began. "That would be lovely. But I think I am going to stay with Walker for the night, whether she wants me to or not. We still have business to discuss. Maybe tomorrow we could meet for coffee. There are some things you need to know...about my relationship with Colonel Bishaye."

Cassidy seemed to search Dale's face for a clue as to what that may have meant.

Dale looked around the company area, then back at Cassidy. "It's complicated but I think you should know. And now she's unpredictable. She knows about us and I wouldn't be shocked, at this point, if she uses that information against you to hurt me. I wouldn't want any backlash coming back on you."

"I'm not worried," Cassidy said, not looking away from

Dale's eyes.

"You should be. This is still the military and it's a violation of the UCMJ. Don't risk your career."

Just as Shannon exited the Orderly Room and began crossing to Dale, they heard the unwelcome bellow of Senior Drill Sergeant Ritchie.

"Sergeant Cassidy, it is 0145 hours! What are these two trainees doing here and not upstairs in their bunks? Don't we have bed check anymore?" Ritchie stomped up to Cassidy. "I got a call from the Staff Duty NCO that there's been an emergency with our company. Let me guess – it has to do with these two bitches," he said, glaring back and forth from Shannon to Dale.

"Uh, Sergeant Ritchie -" Cassidy started but Dale put her hand up to stop her.

By the time Shannon reached their little group, she had her badge out. She stopped in front of the senior drill sergeant and thrust her shield and ID in his face. "That's *Lieutenant* Bitch to you, Sergeant," Shannon said, with a certain amount of glee in her voice.

"Get that..." Ritchie began then it obviously registered what Shannon had said to him and he looked closely at the information in Shannon's grasp. "What —?" He visibly swallowed.

"At ease, Sergeant Ritchie!" Dale commanded.

He looked at Dale. "You, too?"

"You mean, you, too, *Ma'am*, don't you?" Dale asked.

Ritchie then focused on Cassidy, bewildered. "What's going on here?"

Cassidy shrugged as Shannon put her ID away. "You were given a command, Sergeant Ritchie."

Reluctantly, Ritchie moved into the required position. "My apologies, Ma'am," he said, tightly.

Dale turned her back to the senior drill sergeant and looked at Cassidy. She nodded her head toward the parking lot and then mouthed the words. "I'll see you tomorrow."

Cassidy inclined her head and then came to the position of Attention and saluted both Dale and Shannon. It wasn't necessary, as neither lieutenant was in uniform but it was a respectful gesture, nonetheless. "Goodnight, Ma'ams. It has been an honor."

"As you were, Sergeant Cassidy," Dale said out loud, then winked at her. Cassidy couldn't stifle her grin as she backed away and walked toward her car.

Both Dale and Shannon then focused entirely on Ritchie. "Let's go to the laundry room, shall we?" Shannon suggested. They flanked the senior drill sergeant to the aforementioned destination and closed the door behind them.

"You have to understand, I didn't know," Ritchie said, sounding defensive.

"And that's an excuse for your abuse?" Dale asked him.

"Abuse? I was turning undisciplined civilians into soldiers. I did nothing wrong." He couldn't keep the sneer out of his voice.

"The IG said differently and we saw with our own eyes and heard with our own ears your sexist bullshit. Despite that, a majority of the women still completed training and will graduate from LE School. I personally experienced your blatant misogyny and discrimination…or did you forget that?" Shannon asked him.

"I don't recall any such incident," Ritchie said.

"Of course not," Shannon said and smirked.

"You know what else you don't seem to recall, Sergeant? Protocol. You have disrespected our rank by not addressing us accordingly," Dale said. She looked at Shannon with a gleam in her eye.

"Get your ass in the position of Attention, Sergeant!" Shannon barked at him.

Ritchie glowered at her and slowly assumed the proper stance.

"Not good enough," Shannon told him, her mouth right next to his ear. "Get your ass down and knock me out twenty-five!"

The senior drill sergeant looked at her shocked. "Are you…?"

"Now, Sergeant Ritchie!" Dale yelled into his other ear.

He stood there, defiantly. "I guess it will be your word against mine if I don't."

Dale looked at Shannon. "I'll go into the CQ office and activate the bitch box and call an emergency muster. We can play it that way."

Shannon nodded and then said to Ritchie. "Your choice.

Either you do it for us or you do it in front of the entire company."

Veins started showing in Ritchie's neck. His nostrils flared and his skin turned beet red. He grimaced, snorting out breath through his nose and then lowered himself into the front leaning rest position. He started pushing up and got to the silent count of six before Dale stopped him.

"You aren't counting off, Sergeant," Dale said. "Start over."

He readjusted his position and started again. "One, two, three."

"Ah, ah, ah. That's one, ma'am, two, ma'am. Start again."

Ritchie moved back into the starting position, took a deep breath, got his bearings, and began again. "One, Ma'am. Two, Ma'am."

Shannon smiled at Dale then looked down at the senior drill sergeant. "Isn't this exciting, Sergeant Ritchie? Now you won't have to embarrass yourself in front of your subordinates. We're being a lot more respectful to you than you've ever been to anyone else while we've been here."

He continued to count.

"Maybe from now on you'll be more professional and not inject your personal opinions and feelings into training. You'll just do your job, without prejudice, because you never know who's watching."

By this time he had reached his twenty-fifth push-up. He made the mistake of starting to stand. "Whoa, Sergeant! You have not been given permission to stand. For that, you need to knock out another twenty-five."

They both observed Ritchie struggle though the required additional amount.

"Now, do you know what to say?"

Ritchie had worked up a sweat. He looked up at the two lieutenants, who watched him, expectantly. "Seriously?" Then he saw Dale's eyes turn into slits. "Ma'am?"

Shannon made a slight hand gesture to get Dale's attention. "Why don't you get going so you can get to that appointment?"

Dale cocked her head. "That meeting will take place tomorrow. You're stuck with me tonight."

Shannon grinned. "I think I can handle that." She looked

down at Ritchie. "Now, what are you supposed to say, Sergeant Ritchie?"

As Dale left the laundry room en route back to the CQ office to call a cab, she heard the senior drill sergeant say, "Thank you for conditioning my mind and body, Ma'am. Sergeant Ritchie requests permission to recover."

EPILOGUE

May, 1985
Ventura, California

JJ Cassidy, at the end of her morning jog, sprinted away from Ventura beach. She jogged up California Street by Carrow's Restaurant, the 101 overpass, the convenience store and the Bombay Bar and Grill. Dump trucks full of sand were being unloaded onto the Bombay dance floor, setting up for their annual Malibu Beach Party. She turned right onto Santa Clara Avenue, passed the Midwick Residential Hotel, the bank and then dashed across the parking lot down three more blocks to her bungalow.

Dale was progressing through the slow, deliberate exercises of Tai Chi when her lover of nearly eight years walked through the door. JJ silently approached Dale, who was facing away from her, stood impossibly close and precisely mimicked all of Dale's movements of the ancient Chinese martial art. They looked like they were performing an erotic pas-de-deux.

"You're finally up," JJ stated, trying to break Dale's concentration.

"Shhh."

"How much longer? I'm hungry."

"You had breakfast an hour ago," Dale reminded her.

A sensuous chuckle bubbled up from behind Dale that made her smile.

"Oh. That kind of hungry. You're all sweaty."

"That's never bothered you before."

"Let me finish."

"I always let you finish," JJ whispered seductively into

Dale's ear.

Dale shivered and stopped her ritual motions. "I see a very long shower in our immediate future." She turned and was captured by JJ's arms. A passionate kiss followed, accompanied by sexually arousing caresses. "You're exhausting me. No more Monica Nolan for you before lights out."

"Jogging makes me horny," JJ said and smiled.

"Breathing makes you horny," Dale countered. "I'm shocked either of us ever gets out of bed."

"Well, if you don't want to…" JJ's voice trailed off in mock disappointment and she tried to step back.

Dale grabbed JJ's arms and refit them back around her waist. "I never said that." They both laughed and began kissing in earnest as JJ waltzed them back toward the bedroom.

Former lieutenant Shannon Walker, now a captain in the US Army, with the married last name of Morales, stopped in front of the cozy little cottage that seemed to her to be the epitome of California beach living. It was a one-story dwelling with a broad porch, a sloping roof, wrought-iron gates and window bars. Brightly colored tiles and pavers covered the small courtyard and stoop. The exterior walls were white stucco, half-barrel tiles in warm earth tones were layered over the roof and three exposed wooden beams showed above the door. She ascended the three steps to the veranda and knocked on the teal-painted, wooden door frame.

Dale and JJ were just about to get down to intimate business when they both heard an insistent knock on the door. They both yelled 'go away!' at the same time and then laughed. The knock persisted.

Frustrated at the intrusion, JJ who was still completely dressed, got up and answered the door, surprised to see Shannon standing there in her full-dress uniform.

"I interrupted something again, didn't I?" Shannon said to her. "I'm always interrupting you two. Do you ever stop humping like bunnies?"

JJ smirked. "Dale, it's for you," she shouted. She held the

door open and stepped aside for Shannon to enter.

"Now, why do you say that? I could be here for you."

"Not unless you want to enlist. But since that ship has sailed –"

"Shannon?" Dale threw a t-shirt on, quickly moving to the living room. "What's with the pickle suit?"

"What's with the shorter, spikey hair?" Shannon parried.

"Billy Idol, baby," Dale responded, imitating Idol's trademark snarl.

"More like that singer on Knot's Landing but, yeah, it took me a while to get used to it. I really like it now." JJ closed the door. "Can I get you anything to drink? Coffee?"

"Got any alcohol?" Shannon asked.

"Rubbing or liquor?" JJ asked. Shannon raised an eyebrow. "Just thought I'd clarify."

"It's nine on a Friday morning. Why would you need alcohol?" Dale asked, cautiously.

"It's not for me, it's for you," Shannon told Dale.

Dale and JJ exchanged grim expressions. "Oh, fuck. What's happened?" Dale asked.

Seated on the couch, Dale held tightly onto a tumbler half-filled with bourbon. JJ stood behind Dale, resting her hands on the back of the sofa, waiting for Shannon to drop the bombshell they knew was the reason for her unannounced visit.

The last time this happened, it was to give Dale the news that Anne Bishaye had tried to fake her own suicide in an effort to escape. She was found out and placed safely back in solitary confinement for thirty days. Since that attempt, three years ago, it was assumed she was being a good little prisoner.

Dale blinked, awaiting the news Shannon seemed to be anxious yet hesitant about telling her. Telling *them*, as Dale and JJ were a committed team and anything that affected one, had an impact on the other. Both Dale and JJ knew that Shannon's husband, Lorenzo, was privy to any and all information concerning Anne Bishaye, Jack Bishaye and Raymond Sedakis, the last three traitors apprehended who were participants in the Eagle Project, a plot to sell modern US military weapons to South American guerillas.

JJ took the bull by the horns. "Why didn't you let us know you were coming?"

"Because I didn't know myself until last night."

"What's going on?" JJ asked.

"Just don't kill the messenger," Shannon said.

"It's Bishaye again, isn't it?" Dale asked quietly.

Shannon took a deep breath, then proceeded. "Renzo received the news that Bishaye escaped custody last Friday. We haven't been able to find her."

"How does someone escape Leavenworth?" JJ asked.

Dale just sat there, staring at Shannon. She finally tossed back the drink in her hand, swallowing the contents of the glass in two gulps. "She's been running around for a week and you're just telling me now?" Dale asked, incredulously.

"How did she escape?" JJ asked again, more loudly.

"As best we can tell, it was between morning and afternoon headcount. We believe she hid in the laundry facility, kicked out one of the small ventilator windows and climbed out, escaping through the reconstruction mess of towers two and three." She looked at Dale. "As I said, I just found out myself, last night, after my husband and everyone else involved was briefed and came up with a plan. And don't think I'm not livid he didn't tell me sooner."

"What's the plan?" Dale asked, hoarsely.

"To use you as bait to lure her out into the open."

"Absolutely not!" JJ roared.

"That's what I told him," Shannon countered, soothingly. "But they really believe Dale's the ticket."

"I wouldn't even know where to start looking," Dale said.

"Don't even think about it," JJ warned.

"A CIA operative thinks he spotted her in Nicaragua."

"He *thinks?*" Dale repeated. "And what am I supposed to do? Wander around Managua until she allows me to find her?"

"Dale..." JJ started to pace behind her, barely containing her anger and growing panic.

Dale put her hand up in a halting motion. "She's too smart. Because of the circumstances of her apprehension, the first thing she would think of is you using me to track her down. Does Renzo not think I'd be walking into a trap?"

PERMISSION TO RECOVER

"He said you were clever enough to outsmart her," Shannon said.

"No," JJ said sharply. "This should not even be a discussion."

"Look," Shannon said, directly to Dale, "if it makes any difference, I told them you wouldn't agree to this."

"Then why are you here?" JJ snapped.

"Because I was ordered to," Shannon told her calmly. She acknowledged the thick tension that now blanketed the room. "Okay. How about I pop over to the San Buenaventura Café, grab some breakfast and let you have some privacy to discuss the offer and settle the stress between you? I'll come back in a couple of hours and, if you decide to do this mission, I'll give you the details."

"Why don't you give them to me now?" Dale asked

"Because they are need-to-know. If you aren't going to accept this assignment, you don't need to know."

"There have been two attempts on your life because of that woman. And don't you dare defend her again by blaming those actions on her husband. She knew what she was doing. She had choices. She made the wrong ones. You had choices and made the right ones."

"And it's haunted me ever since," Dale stared at Shannon's retreating form through the screened door, watching her get into her car.

"I know that. What I've never understood is why. She used you. She used your love for her against you." JJ sat on the couch next to Dale and took both of her hands in hers. "Baby, you know you don't owe Uncle Sam or Bishaye anything." Her voice had a pleading quality to it. "You earned your separation papers and this peaceful life you've chosen that we have built together. Bishaye is the military's problem, not yours."

After Shannon's car drove away, Dale turned and looked at her cherished lover. "What if she's not in Nicaragua? What if she's here? I'm sure she still has enough loyal military ties to track us down. How will I ever be able to live with it if she resurfaces and does something to you just to get back at me?"

"And how am I supposed to live with it if she kills you this time? You may be ready to make that sacrifice but I am nowhere near ready for you to make that sacrifice. And I never will be."

"I don't think she'd kill me. She had plenty of chances before and she didn't do it."

JJ looked at Dale in disbelief. "You are so intelligent and so naïve sometimes. You don't think, since you would be the only thing standing between her freedom and being brought back to Leavenworth, she wouldn't do whatever she had to do to stay free?"

"You know, maybe it's not about her getting revenge on me. Maybe it's about me getting revenge on her."

"Dale, I am begging you to leave it be. We'll watch each other's six and keep our heads on swivel. Let Shannon, Lorenzo and that whole group keep track of her whereabouts. That's their job."

"Do you want to live the rest of your life that way? Always looking over your shoulder?"

"No. I want to live the rest of my life with you, alive and well, by my side. I refuse to let the last six years with you be for nothing." Distraught, she pulled Dale into an urgent embrace.

Dale returned the hug, deep in thought. Finally Dale said, "What if you could come with me?"

The End (?)

ABOUT THE AUTHOR

Author Cheyne Curry

Cheyne Curry is a combat-trained, US Army Military Police veteran. She is also a former entertainment security manager employed by Time-Warner (Warner Bros. Studios in Burbank and HBO Corporate HQ in New York City) and a Central Station Supervisor for an armed patrol company that serviced high profile clients and celebrities in Hollywood, CA. She was born in Vermont, raised in New York and spent a majority of her life being bicoastal. She recently relocated from the Midwest to the Southwest with her wife, Brenda, and their two fur babies, Mesa Antiope (a Black Mouth Cur) and Riley Villanelle (a black Bombay). Permission To Recover is her sixth published book.

OTHER TITLES BY CHEYNE CURRY PREVIOUSLY PUBLISHED

<u>Clandestine</u>   Tia Ramone is a gritty, self-destructive, ex-CIA operative who seeks absolution in a bottle. Jody Montgomery is a naïve heiress to a vast fortune, married to a man she discovers she really doesn't know. Tia's and Jody's paths cross in a sinister plot they are forced to take part in. With both their lives at stake, can the clandestine meeting that brought them together ultimately be the bond that saves them?

<u>The Tropic of Hunter</u>   Hunter Roberge left Otter Falls, Vermont when she was 18, to get away from a life of scandal and judgment. Sixteen years later she returns to her 'hometown' for the funeral of the one person who condemned her the most: her mother. Being bequeathed the family house is just the first in a long line of mysteries that unravel the fabric of everything Hunter believes to be true. Can the support of a childhood acquaintance keep her on the right track or will she once again fall victim to her mother's hatred?

<u>Renegade</u>   What would you do if one minute you were in the 21st century and the next you were in the 19th? One day you're driving a Mustang and the next day you're riding one? Dirty cop Trace Sheridan faces this dilemma as she moves from a present day mob war to a range war over a hundred years in the past. The year is 1879, when cattle barons, crooked lawmen, saloons, painted ladies, cowboys and Indians ruled the Wild West, and laws were only as strong as the gunman who upheld them. In Sagebrush, the town and the sheriff belong to the Cranes, who take what they want or bad things happen. Trace finds this out firsthand when she ends up on the land of Rachel Young, a struggling ranch woman who won't give in to the merciless cattle baron and his obsessed son. For some unexplainable reason, Rachel trusts the enigmatic Trace who uses 21st century sensibilities to battle 19th century turmoil, while Trace is forced to keep the secret of her origin from the attractive and vulnerable Rachel. Renegade is a story of redemption in its purest form as Trace discovers what truly matters in life and how past really is prologue.

PERMISSION TO RECOVER

TITLES CO-AUTHORED BY ROSELLE GRASKEY AND CHEYNE CURRY:

The End – Book One of the Sanctuary Series

The power flickers, the ground shudders and when the backup system kicks in, Lieutenant Jessica Baumer and Staff Sergeant Branna Maguire do a routine equipment analysis and check the exterior surveillance cameras, where they see an unspeakable horror. Somehow, someway, without warning, the world as they know it has ended.

Originally used as 'lab rats' for an isolation experiment they are in their subterranean data center at Fort Hood, expecting another "normal" exercise. It becomes anything but.

There is little information available, which leads to many questions and no satisfactory answers. What happened and, more importantly, are there other survivors?

When the air begins to clear, Baumer and Maguire set off on a journey to one of the few areas of the country rumored to be less devastated by the attack. Their mission is to reach sanctuary, a safe compound and shelter run by a woman who has been preparing for a catastrophic event for years.

The law and order of modern day civilization has ceased to exist and the situation has become where only the strong and clever will make it to another day. Others join the two soldiers along the way; some are skilled, some are in need of rescue and others, who will stop at nothing to be masters of this new world order.

Do Baumer and Maguire, even with their military training, have the capability and fortitude to make it to sanctuary? And, if so, what will they find when they get there? Will it be the hope for civilization and America's future?

Or is this really The End?

CHEYNE CURRY

The Resistance – Book Two of the Sanctuary Series

The Resistance (Book Two of the Sanctuary Series) picks up where Book One, The End, left off...

With the release of her repressed memories, Branna Maguire returns to Sanctuary with warnings of deeper, more widespread dangers than that which they were already aware. Now she must answer the summons to rescue the mysterious Dr. Elaine Madras.

Along with a reluctant Devon Prescott, they trek to just outside Yakima, Washington, connecting with Dr. Madras to discover her reported situation isn't what it seems. Maguire then learns of their next mission: mount a resistance to combat the forces that orchestrated Day Zero.

Made in the USA
Middletown, DE
13 September 2024